In *Darwinism's Struggle for Survival* Jean Gayon offers a philosophical interpretation of the history of theoretical Darwinism. He begins by examining the different forms taken by the hypothesis of natural selection in the nineteenth century (Darwin, Wallace, Galton) and the major difficulties which it encountered, particularly with regard to its compatibility with the theory of heredity. He then shows how these difficulties were overcome during the seventy years which followed the publication of Darwin's *Origin of Species*, and he concludes by analysing the major features of the genetic theory of natural selection as it developed from 1920 to 1960. This rich and wide-ranging study will appeal to philosophers and historians of science and to evolutionary biologists.

Darwinism's struggle for survival

Cambridge Studies in Philosophy and Biology

This major new series will publish the very best work in the philosophy of biology. Nonsectarian in character, the series will extend across the broadest range of topics: evolutionary theory, population genetics, molecular biology, ecology, human biology, systematics, and more. A special welcome will be given to contributions treating significant advances in biological theory and practice, such as those emerging from the Human Genome Project. At the same time, due emphasis will be given to the historical context of the subject, and there will be an important place for projects that support philosophical claims with historical case studies.

Books in the series will be genuinely interdisciplinary, aimed at a broad cross-section of philosophers and biologists, as well as interested scholars in related disciplines. They will include specialist monographs, collaborative volumes, and – in a few instances – selected papers by a single author.

Forthcoming
Richard Creath and Jane Maienschein *Epistemology and Biology*
Steven Orzack and Elliott Sober *Adaptationism and Optimality*

Darwinism's struggle for survival

Heredity and the hypothesis of natural selection

Jean Gayon

Université Paris 7 – Denis Diderot

Translated by Matthew Cobb

CAMBRIDGE UNIVERSITY PRESS
Cambridge, New York, Melbourne, Madrid, Cape Town, Singapore, São Paulo

Cambridge University Press
The Edinburgh Building, Cambridge CB2 8RU, UK

Published in the United States of America by Cambridge University Press, New York

www.cambridge.org
Information on this title: www.cambridge.org/9780521562508

Originally published in French as *Darwin et l'après-Darwin: une histoire de l'hypothèse de sélection naturelle* by Editions Kimé, Paris, 1992 and © Editions Kimé 1992

First published in English by Cambridge University Press 1998 as *Darwinism's struggle for survival*

This digitally printed version 2007

A catalogue record for this publication is available from the British Library

Library of Congress Cataloguing in Publication data

Gayon, Jean.
[Darwin et l'après Darwin. English]
Darwinism's struggle for survival: heredity and the hypothesis of natural selection / Jean Gayon; translated by Matthew Cobb.
p. cm. – (Cambridge studies in philosophy and biology)
Includes bibliographical references (p.) and index.
ISBN 0 521 56250 3 hardback
1. Natural selection – History. 2. Evolution (Biology) I. Title. II. Series.
QH375.G3813 1998
576.8′2 – dc21 97-20559

ISBN 978-0-521-56250-8 hardback
ISBN 978-0-521-03967-3 paperback

This book is supported by the French Ministry for Foreign Affairs, as part of the Burgess Programme headed for the French Embassy in London by the Institut Français du Royaume-Uni.

For

Georges Canguilhem†

and

François Dagognet

Contents

Illustrations

Preface to the English edition

The most direct and simple way of presenting the aim of this book is to tell the story of how a young Frenchman became fascinated by Darwinian science. In following the path that led me to 'Darwinism', it will become clear that this book is not a cultural, social or institutional history of science – all of which might be implied by a title that contains an 'ism'.

The parents of this book were failure and disappointment. Having begun studying philosophy, I followed the advice that Georges Canguilhem – the only teacher who really made an impression on me – used to give to his students in philosophy, and enrolled in a science faculty with the aim of receiving the ordinary thorough education of a French biology student. In four years' study at what at the time was the best Parisian science university, I never once heard talk of evolutionary theory. I had hundreds of hours of lectures on zoology, botany, paleontology and genetics. I learnt all sorts of things about the succession of organisms in the history of life, about the evolution of the eye, of the kidney, of algae and of vertebrates, but I never once learnt about the theory of evolution. Quite simply, evolution was a fact. It was so obvious that it was not even discussed. As a result I barely suspected that a theory even existed. During an oral exam, an eminent paleontologist reproached me for accidentally using the term 'adaptation' – I ought to realise once and for all that this word was 'unscientific and metaphysical'.

At the end of this period, I wanted to know more, so I enrolled in the equivalent of a master's degree in evolutionary genetics, in a very serious laboratory which was nevertheless somewhat exotic by local standards, or at least the kind of place to be avoided if one wanted a brilliant career as a biologist. It was there that I discovered population genetics, paleobiology, Dobzhansky, Mayr and Simpson – in other words, the 'modern synthesis', or what the French call – somewhat more grandly – the 'synthetic theory'. (They then go on to cast doubt on the idea that it is indeed a scientific theory.) I was amazed by this encounter and decided to abandon my studies in biology and to do a doctoral thesis in the history of science, taking as my subject the 'synthetic theory'. My dream was to clearly identify the structure of this theory, and to trace its genesis. In 1984, thanks to a Fulbright scholarship, I spent some time in the USA, where I delved into the archives of Dobzhansky, Mayr, Huxley, Simpson and a number of others. After three months' work, a scenario began to emerge, somewhat different from what I had imagined. It became clear that the 'synthetic theory' – or at least what had been called such since the 1950s – was less the offspring of Julian Huxley (who had coined the term 'modern synthesis' in 1942) than that of the determined work of a group of American biologists. During the Second World War they had decided to make their disciplines coexist and their doctrines congruent, dividing up the work in a typically American way (setting up committees and making decisions like 'The Eastern states will deal with botany, the Western states with zoology . . .').

As I became aware of the local historical dimension of the 'synthetic theory', the genuinely theoretical aspects of the synthesis appeared increasingly confused. The great intellectual adventure seemed more and more to be an episode of sociology. A genuine historian would have been happy. This philosopher was deeply disappointed. One day, I told Richard Lewontin of my fears, hoping that he would contradict me. But he only made things worse: 'No,' he said, 'I think your hypothesis is correct. There is no synthetic *theory*. It's a question that needs to be dealt with by the social history of science.'

On my return to France, I found myself faced with a major problem. To write a social history of the synthesis, it would be better to live in the USA, and to have enough time and money to travel widely. Unfortunately, my professional situation did not enable me to do this. But I couldn't shake off the philosophical dream of a heroic history of science, in which the good ideas triumph over the limits of individuals and of society. Beginning to despair, and because I still had to finish my doctoral thesis, I decided to write a kind of introduction to the project that I had neither the time nor the means to realise. Hence this book. I gathered together my ideas on what seemed to me to be essential from the point of view of the conceptual and methodological history of Darwinism, isolating one narrow question within which a genuine rational debate had taken place. Rather than dealing with Darwinism in general, I decided to concentrate on the principle of selection. Rather than dealing with the *theory* of natural selection in its totality, that is as a vast reconstruction of natural history, I restricted myself to what Darwin had called the 'mere hypothesis' of selection. Finally, I only dealt with this hypothesis in terms of its relation to heredity, that is in the light of what in Darwin's time was not a science, but which during the 20th century was to become the most fundamental of the life sciences. Compared with my initial project (a history of the modern synthesis), the object had shrunk in the space of scientific knowledge, but had spread out in its temporal axis.

Having paid this price, it was possible to make a rational reconstruction of the history of Darwinism. By 'rational reconstruction of history', I mean a history of the theoretical or experimental difficulties that produced a rational debate and also its solutions. The history set out here is not a history of events as they *had* to happen, to paraphrase Lakatos. It is a history of events as they happened, along the axis of concept and method (I know that many contemporary historians will find this old-fashioned). I accept in advance the criticism that I have described an internalist history and, worse perhaps, a logicist version of an internalist history. I wanted to reconstruct the rational history of the hypothesis of natural selection for two very different reasons. The first was contextual: having grown up in a country where, for many people, the Darwinian theory of evolution is not a science but at best metaphysics and at worst mere ideology, I wanted to refute this opinion. Secondly, as a historian of science, my argument in favour of an internal and conceptual history is limited, but essential. Conceptual history is not opposed to the social history of science. But the social history of science attains its goals all the more effectively if it has a pre-existing reconstruction of the intellectual debate involved in that history. The internal history of concepts and methods is no doubt a simplification which sooner or later will be corrected and filled out by the social history of sciences. But it sets out the boundaries, and that is not so bad. This book is thus a kind of outline drawing. I have paid particular attention to the more or less explicit and sometimes contradictory conceptions of science that the Darwinians adopted in their study of the hypothesis of selection.

This work would not have existed without the example and advice of Georges Canguilhem, who convinced me I should go and learn some science; without Jean-Claude Beaune, who one day convinced me I could write the first word; and finally and above all without the unflagging support of François Dagognet. The idea of a philosophical study of Darwinism grew out of my contact with Claudine Petit and Jean-Michel Goux at Université de Paris 7. They initiated me into population genetics and the modern theory of evolution. In the history of sciences insider accounts generally do not radically modify the lessons that can be drawn from documents, but they do have the enormous advantage of directing one's attention. In this respect I gained a great deal from Jean Genermont, Pierre-Henri Gouyon, Max Hecht, Maxime Lamotte, Richard Lewontin, Ernst Mayr, Paul Siegel, Henri Tintant and Bruce Wallace. I am particularly indebted to Michel Delsol, who more than anyone else helped me to see that the real questions are not a matter of words.

Throughout this work, the subject forged strong and rich friendships: I am extremely grateful to Richard Burian, Claude Debru, Jonathan Hodge, Hervé Leguyader, Everett Mendelsohn, Paul Mengal, Françoise Parot, Michael Ruse and Michel Veuille. Their ideas and suggestions can be found in the pages of this book, and their support carried it to fruition. The finalisation of the French manuscript owes a great deal to Elisabeth Gayon-Valsecchi, who read it from beginning to end. The English version owes its existence to Jonathan Hodge and Michael Ruse. The English text represents a complete revision of the French edition. Matthew Cobb, who translated it, Marjorie Grene, who reread the translation, and Jane Van Tassel, the copy-editor, cannot be thanked enough for their work. On a whole series of points the translation has improved the original idea. Finally, I would like to thank the Université de Bourgogne (Dijon) and in particular its helpful and able librarians Françoise de Boissieu and Eve Justrabo, who gave me access to documents that are hard to find in France.

Introduction

The history of Darwinian theories of evolution is highly complex, but it is possible to discern a common conceptual position that unites them all. 'Darwinian' refers to any interpretation of evolution that involves the gradual modification of species, predominantly guided by a process of natural selection operating on a field of intra-populational variation.

Since the early 1980s, it has often been argued that this gradualist and panselectionist view is in decline.[1] Although the aim of this book is not to analyse this current debate,[2] it is worth referring to because it echoes a far graver crisis that affected Darwinian evolutionism at its very beginnings and which, curiously enough, has generally been ignored by modern critics. As Peter Bowler has rightly noted in *The Eclipse of Darwinism*,[3] too many modern opponents of Darwinism present themselves as struggling against a dogma that the scientific community has supposedly never seriously questioned, thus giving credit to the idea that the theory of natural selection has developed in a linear and peaceful fashion since Darwin's time. This gives rise to an ahistorical and flat version of 'Darwinism' which often leads on to criticisms of its supposedly 'non-scientific' nature. Such a misreading lies at the root of much of the epistemological literature which, following Karl Popper, accuses natural selection of being 'unfalsifiable'. Leaving aside the fact that Popper's arguments and their development have not always been understood, they have nevertheless done much to reinforce the image of Darwinism as a kind of modern myth; like psychoanalysis or Marxism, Darwinism is allegedly outside the realm of scientific rationality, a theory that has been supported by an unending series of confirmations but has never been seriously subjected to the test of falsification.[4]

This epistemological fable does not withstand historical analysis. Twentieth-century theoretical Darwinism was the outcome of a long and intense crisis that is difficult to conceive of today. In the early years of the century, the 'death' of Darwinism as a model for explaining biological evolution was announced with regularity.[5] In 1924 Erik Nordenskiöld devoted a whole section of his monumental *History of Biology* to 'the decline of Darwinism'[6] and concluded the entire volume with the following words:

The whole of this problem of evolution is of course highly involved and its discussion must, as far as our own times are concerned, terminate in a number of unanswered questions. First of all, selection; that it does not operate in the form imagined by Darwin must certainly be taken as proved, but does it exist at all? ... The competition between individuals, in which Haeckel thought he saw true selection – does it exist at all, or is it only imaginary, as O. Hertwig affirmed? ... These and many other questions it is for the future to answer.'[7]

This position is far from unique. It reflects a widespread feeling expressed in the evolutionary literature from 1890 to at least 1930. But it is important to note exactly what people meant by the 'decline of Darwinism'. There was no question of criticising the prestige of Darwin the naturalist, nor his evolutionary vision of biology, nor even the usefulness of the hypothesis of natural selection. The decline of Darwinism was virtually always attributed to the experimental refutation of the hypothesis of 'natural selection' in the highly restrictive sense that Darwin had intended, that is as a factor of progressive evolution,[8] acting gradually[9] upon infinitesimally small inherited variations,[10] and constituting the dominant force (or 'paramount power') in the modification of species.[11] In *La genèse des espèces animales* (1932), Lucien Cuénot summed up what it was about Darwinism that most biologists at the beginning of the 20th century considered to be dead: 'We must completely abandon that part of the Darwinian hypothesis that concerns the slow, invisible and certain effect of selection on floating variation, which gradually determines the evolution of the species towards perfection.'[12]

The supposed death of Darwinism was linked to the idea that gradual, slow selection was incompatible with the findings of the new experimental science of heredity. For several decades, the central core of Darwinian evolutionism was widely doubted. But it was precisely this central core that was to be rehabilitated and finally vindicated, first by the theoretical conjectures of population genetics in the 1920s, and then, from the 1940s onwards, by what is known as the 'modern synthesis'.[13] Julian Huxley used the evocative term 'the eclipse of Darwinism' to express the temporary disappearance of the schema of gradual selection.[14] A key participant in the restoration of this schema, Huxley considered that, at the beginning of the 20th century, the rumours of Darwinism's death were greatly exaggerated.[15] However spirited this defence of Darwinism might be, it is too biased to be historically valid. With hindsight, we can see that the crisis of Darwinism that took place nearly a hundred years ago was indeed extremely profound. So much so that it can legitimately be asked whether there was any relation at all between the theoretical universe of Darwin's doctrine of natural selection and that of the 'renovated' Darwinism that was constructed from the 1920s onwards.

The aim of this book is to understand why the 'eclipse' of Darwinism took place in the 1900s. This will require a rational reconstruction of the history of natural selection, even if this will be far from the final word on the history of 'Darwinism'. In order for this aim to become a reality, the precedents and the consequences of the 'eclipse' will have to be studied. The result is a century-long panorama of the theory of natural selection. Thus this study begins many decades before the period of decline, starting with the work of Darwin himself. The long initial crisis of Darwinism was not only the result of external factors such as 'resistance' or the existence of rival evolutionary paradigms; it was also a consequence of a range of problems that were intrinsic to Darwin's central hypothesis. These difficulties, most of which were linked to the concept of heredity, could not be resolved by the biology of the time.

Nevertheless, Darwin's contemporaries were obliged to tackle this new problematic and to make it a priority of experimental biology. By a curious twist of fate, the new science of heredity seemed at first to disprove natural selection, thus opening a long period of doubt as to the validity of Darwin's hypothesis. When, in the 1920s, the whole question appeared in a different light – that of a *genetic* theory of selection – natural selection was restored to its former position, but at a price. This subtle change is dealt with in the last two chapters, which examine the structure and internal tensions of the modern – or genetic – theory of selection, once Darwinism's founding crisis had been overcome. The relationship between heredity and selection is thus at the heart of this history of more than a century of Darwinian theory. This choice has doubtless led to an important number of conceptual and historical aspects of 'Darwinism' being passed over. On the other hand, however, it permits the clear identification of the main theoretical and experimental track along which theoretical Darwinism developed. By taking this approach, we can study the central theoretical core of the Darwinian theory of the modification of species, the long crisis that undermined it as the new science of heredity took shape, and the renewed Darwinism that was the end-product and that was to dominate the theory of evolution throughout the 20th century.

Before outlining the scenario of rational reconstruction that forms the framework of this book, it is worth dealing with some potential objections to this approach. Some of these criticisms are linked to the current state of Darwinian historiography; others are related to methodological debates in the history of science (for example, it is arguable whether the history of science is, or should aim to be, a 'rational reconstruction').

A first potential objection would be to attack the criteria used here to identify 'Darwinian' theories of evolution. As explained above, 'Darwinian' is taken to mean any interpretation of evolution as being the product of the gradual modification of species, predominantly guided by a process of natural selection functioning on a field of intra-populational variation. Apart from the fact that this definition employs certain terms ('evolution' and 'population')[16] that are not generally to be found in Darwin's writings, it could be argued that it is over-restrictive, as it would tend to exclude much that is considered as 'Darwinian' in both science and, more generally, in 19th- and 20th-century culture. Darwinism has had a massive influence on the whole of contemporary history, with numerous consequences for a wide range of fields of intellectual enquiry and of social practice. Any historian who intends to deal with all the historical manifestations of 'Darwinism' has first to tackle their extreme heterogeneity. This point, which is obvious as far as 'Darwinian'[17] social theories are concerned, applies equally to the biological theory of evolution. In recent years a substantial literature has developed on the 'reception' of Darwinism in various national scientific communities at the end of the 19th century.[18] These studies have generally come to the same conclusion: although Darwin's writings were rapidly, immediately and widely disseminated, a series of heterogeneous, and sometimes incompatible, interpretations of the 'Darwinian' theory of evolution rapidly sprang up. From this point of view, the narrow definition of Darwinism given above is inappropriate. For example, from the very earliest days there existed saltationist versions of Darwinism, as well as interpretations of the 'struggle for existence' that put the emphasis on the struggle *between* species rather than within them. Other fundamental distortions of Darwin's ideas could equally well be pointed out. They all lead to the inevitable conclusion that while the *Origin of Species* was rapidly accepted as an 'exemplar' (to borrow Thomas Kuhn's term) in the field of organic evolution, 'Darwinism' as such nevertheless remained a thoroughly heterogeneous tradition for many years.[19]

There is a simple answer to this objection. There is, of course, much that can be gained from a comparative study of science that claimed to be 'Darwinian'. The various studies of the 'reception' of Darwinian ideas have opened one of the richest chapters in the contemporary history of biology. But they are no more historically legitimate than an approach that is centred on the experimental destiny of Darwin's hypotheses. From this point of view, history shows that the theory of selection was continually enriched through the study of intra-populational hereditary variation and of the gradual modification of species. This theoretical framework, as important for Darwin as it was experimentally problematic, had a number of phases – biometry (at the end of the 19th century), then population genetics and finally the 'synthetic theory of evolution'. This is the approach to evolution that, for over a century, has become increasingly identified with the 'Darwinian' science of biological evolution. If this were not the case, it would be hard to understand the attacks by certain contemporary biologists on the 'Darwinian paradigm'. When, for example, the advocates of 'punctuated equilibrium' or the supporters of an 'epigenetic' theory of evolution criticise 'Darwinian' orthodoxy, their explicit target is an approach to biological evolution that is dominated by the genetic study of intra-populational variation and by the conviction that selection operating at this level is the prime mover of the gradual modification of species. The reconstruction of the Darwinian theory of selection put forward in these pages has real historical depth, tracing the development of the key concepts as the history of science demands. In this respect it is no less genuinely 'historical' than an approach centred on the reconstruction of the sociological contexts (scientific, institutional and cultural) in which Darwinism developed.

A survey of the current historiography of Darwinism could give rise to another criticism. One of the fundamental arguments of this book is that the modern theory of natural selection arose out of a long initial crisis, the first signs of which can be discerned in Darwin's writings, and which had barely been overcome by the 1920s. It could be argued that while this was indeed a crisis of evolutionary explanation, it would be wrong to concentrate on the internal history of the Darwinian theory of selection and to neglect the general climate of conflict and uncertainty that characterised the whole of evolutionary biology from the middle of the 19th century up until 1930. Peter Bowler[20] has interpreted 'the eclipse of Darwinism' using terms borrowed from the epistemology of Thomas Kuhn. According to Kuhn, rather than studying the *logical structure* of scientific theories, historians of science and epistemologists should consider scientific theories as *paradigms*, as exemplary models for the 'normal' scientific activity of problem-solving.[21] Kuhn has argued that one of the preconditions for the establishment of a new paradigm is a period of crisis. In such a period, research is characterised by the failure of the 'normal' activity of problem-solving, and a wide range of different theoretical explanations are put forward to account for the same data.[22] In other words, the period of crisis that precedes the emergence of a new paradigm is marked by the absence of a paradigm and by a state of theoretical uncertainty. This situation is classically characterised by a sharp conflict between competing theories, none of which is sufficiently consolidated to serve as the starting point for the ordinary activity of research. Applying these epistemological categories to the history of Darwinism, Bowler has interpreted the 'eclipse of Darwinism' that took place around the turn of the century as being linked to a period of deep theoretical uncertainty amongst evolutionists. In the span of a few decades, evolutionary thought was marked by major theoretical alternatives: orthogenesis, neo-Lamarckism and mutationism. From this point of view, the early crisis of Darwinism appears to be the result of the division of the field of evolutionary theory between competing positions, none of which was able to win out. In other words, this crisis can be attributed to a general climate of eclecticism and scepticism with regard to the theory of evolution.

This analysis is indeed pertinent from a sociological point of view, but it cannot explain the crisis of the Darwinian interpretation of evolution that took place around 1900. The emergence of powerful anti-Darwinian schools of thought no doubt provides a way of evaluating the disappearance and temporary decline of the theory of natural selection in this period.[23] But the crisis of Darwinian theory cannot be explained merely by reference to competition from rival theories such as neo-Lamarckism or orthogenesis. The theory of evolution by natural selection was not simply eclipsed by evolutionary doctrines that corresponded as well – if not better – to the ideological mood of the times. For reasons that can be determined, Darwinism underwent a profound internal theoretical crisis, and was subsequently able to overcome this crisis. It is for this reason, and on this basis, that a rational reconstruction of the history of natural selection is both possible and desirable.

What of a more fundamental objection, touching the methodology of the history of science? The history of science might not necessarily have as its objective a 'rational reconstruction'. From the outset, the method employed here needs to be made clear. Particular attention is paid to the conceptual and empirical difficulties encountered by the Darwinian theory of selection, in terms of an epistemological history. The unique aspect of such an approach to the history of science is that it takes into consideration the difficult question of the validity of different theories. The methodological starting point is that science is, among other things but above all else, a problem-solving activity. This implies that whatever the weight of sociological and ideological determinants of a given field of research, it must be possible to describe the history of a scientific theory as an enterprise within which *some* problems are posed in such a way that they can be solved.[24] Of course, this approach is by definition retrospective, because, as has often been pointed out, the historian of science cannot do without a 'frame of reference'.[25] But for the same reason, this approach considers the history of science as an empirical discipline, because it cannot be guaranteed in advance that a given aspect of the history of science can be rationally reconstructed.

The importance of such methodological remarks should not be exaggerated, however. Although the term 'rational reconstruction' has been borrowed from the philosopher Imre Lakatos,[26] I do not agree with him that such a reconstruction can do without historical evidence and that events *had* to happen in the way that they did. Thus, in dealing with the history of the theory of natural selection, a large place is given to the doubts expressed by those who constructed the rationality of the principle of natural selection. But a special emphasis is placed on the theoretical and experimental difficulties that could be rationally resolved, that is on those problems that were not specifically linked to the historical conditions in which they were encountered. It is one thing to say that a given understanding developed owing to given historical conditions; it is quite another to argue that the validity of this knowledge is limited to the epoch in which it was formed.[27]

Thus a rational reconstruction in no way excludes the social reconstruction of the history of science.[28] Historians of science have recently tended to set these two approaches in opposition. In the present case, there can be no doubt that concepts such as heredity, population and selection were deeply rooted in a history that was extremely 'social' and often highly 'ideological'. Throughout this work, the reader will find a series of references to the social, political and historical context that both enlighten and guide the rational reconstruction of the history of the hypothesis of natural selection. For example, it would have been impossible to analyse the contributions made by the English school of biometry and statistics without taking into account the fact that some of its founders were involved in the eugenics movement. But it is precisely in this literature, so unpleasantly framed in ideological terms, that we can find indicators of a rationality that might go beyond ideology, for the simple reason that 20th-century evolutionary biology and genetics both absorbed the heritage of biometry.

The methodology used in the pages that follow is that the history of sciences, like all history, is 'social'. Scientific activity develops in given social and historical conditions, but this does not at all imply that the validity of knowledge is rigidly dependent upon the conditions in which that knowledge was constructed. Or, to put it another way: in order for the history of science not to lose itself in a morass of relativist descriptions of customs and beliefs, in order not to abstract the history of science from everyday human history, we have to allow for a certain retrospectivity and avoid thinking that ignorance of the current state of science in some way helps the historian.[29] Relatively few episodes of the history of what we call science can resist resolution by a strong dose of hindsight. These are the episodes that merit a rational reconstruction of their origins and development. The theory of selection and its evolution from 1859 onwards is such a case. It really is that simple.

The hypothesis that runs throughout this work is based on the observation that despite the intuitive correctness of the principle of natural selection, between seventy and eighty years were to pass before the principle was established as both an empirical reality and a theoretical possibility.

In 1937, at a meeting of the Zoology section of the Royal Society of London, D. M. S. Watson opened a session devoted to the 'present state of the theory of natural selection' with the following words:

> The theory of natural selection is the only explanation of the production of adaptations which is consonant with modern work on heredity, but it suffers from the drawback that by the introduction of related subsidiary hypotheses it becomes capable of giving a theoretical explanation of any conceivable occurrence, and that the scope and indeed the validity of its basal assumption of a selective death-rate determined by a favourable variation has not been yet established.[30]

Despite the fact that, even at the time they were uttered, these assertions were somewhat overstated, they nevertheless give a good summary of the three difficulties that had dogged eight decades of selectionist theory: the suspicion that this was an *ad hoc* principle, the difficulty of reconciling the theory of selection with that of heredity, and the fundamental problem of providing an experimental proof of natural selection. Here are three fruitful lines of attack for a reevaluation of the history of natural selection.

Furthermore, a link existed between these three problems which was present from the very beginnings of the theory of selection, a link that can be found in the work of Darwin himself.

How Darwin conceived of and named the theory set out in the *Origin of Species* is particularly revealing. Darwin considered that his magnum opus was 'one long argument' in favour of a single theory.[31] If we consider the name that Darwin gave to this theory and the method he explicitly outlined, a number of problems become apparent. The initial thesis advanced in the *Origin of Species* is that of the genealogical significance of classification – the diversity of life is the product of gradual, branching evolution. Darwin formulated this argument under the title 'descent with modification'. The other thesis put forward in the *Origin of Species*, it is classically argued, is an explanatory hypothesis of the mechanism that produces the modification of species – natural selection.

But such a presentation, however pedagogic, fails to emphasise the link between the theory of 'descent' and that of 'natural selection'. In the conclusion of the *Origin of Species*, Darwin clearly describes the theory put forward in his book as a 'theory of descent with modification by variation and by natural selection'.[32] The *Origin of Species* was 'one long argument' in favour of this theory. To fully appreciate Darwin's designation of his theory as a 'theory of descent with modification', it has to be realised that he did not have a clear and empirically founded theory of heredity. Despite the fact that he regularly invoked the 'powerful principle of heredity', without which the concept of natural selection could not have been elaborated,[33] Darwin frequently admitted that this principle was, in fact, totally mysterious[34] and that, for the time being, he would have to be satisfied with the idea of heredity as consisting of the resemblance between the members of a given line, in other words with a genealogical concept of heredity. It is thus hardly surprising that Darwin sometimes wrote of his theory as being one of '*inheritance* with modification'.[35] The phrase 'theory of descent with modification by variation and by natural selection' thus covers a theoretical position that is much more complex than traditional textbook presentations would tend to suggest. Instead of stating that Darwinism consists of a fact (common descent) and a causal hypothesis (natural selection), it is probably better to say that Darwin's summary description of his theory marks the convergence of two theses, neither of which was intuitively obvious either to Darwin or to his successors. The summary in fact expresses two different ideas that are inseparable in Darwinian language. The first, and more literal, is that natural selection is the cause of the modification and diversification of species; in other words, the theory of selection is a meta-theory of classification[36] which interprets taxonomy by a hypothesis of the transformation *and* of the division of species. The second idea is revealed if 'descent' is replaced by 'heredity', following Darwin's occasional usage. The formulation thus becomes a theory of *heredity* modified by *variation* and by *natural selection*. Considered from this angle, Darwin's theory is a causal theory of the transformation of characters.

In Darwin's time, all three terms of this theory – variation, heredity, natural selection – presented difficulties for an empirical science of the modification of species. Heredity and variation posed problems because Darwin could only advance conjectures as to their nature and functioning, which placed him in an uncomfortable position with regard to his ambition of satisfying Newton's imperative of invoking only '*verae causae*'.[37] Even more important, in the years that followed the publication of the *Origin of Species* it rapidly became apparent that not all hypotheses on the nature of heredity were equally compatible with the modification of species by selection. Very early on, the principle of natural selection thus came up against a fundamental objection: while heredity was a prerequisite for the Darwinian concept of selection, there were hypotheses of heredity that would render ineffective the process of selection.

Nevertheless, it could be argued that the concept of natural selection was at least plausible, as long as there was some kind of hereditary variation. This is probably true; however, the theory of natural selection presented another methodological difficulty which has not been sufficiently studied, perhaps because of Darwin's genius for argument. Throughout Darwin's entire work, there is not one *direct* empirical proof of the efficacy of natural selection, or even of its existence. Not once, neither in the *Origin of Species* nor in any subsequent work, did Darwin present a single empirical example of natural selection.

This leads naturally to the question of Darwin's scientific method. This method has often been described as a subtle mixture of induction and deduction which constitutes a puzzle for epistemologists because it is marked both by a purely inductive scientific practice and by the experimental method. It is no accident that Darwin placed an aphorism from Bacon at the beginning of the first edition of the *Origin of Species*. In his *Autobiography*, Darwin recalls this choice and explains its significance:

> My first note-book was opened in July 1837. I worked on true Baconian principles, and without any theory, collected facts on a whole-sale scale, more especially with respect to domesticated productions, by printed enquiries, by conversation with skilful breeders and gardeners, and by extensive reading. When I see the list of books of all kinds which I read and abstracted, including whole series of Journals and Transactions, I am surprised at my industry.[38]

> As in several of my books facts observed by others have been extensively used, and as I have always had several quite distinct subjects in hand at the same time, I may mention that I keep from 30 to 40 large portfolios, in cabinets with labelled shelves, into which I can at once put a detached reference or memorandum. I have bought many books and at their ends I make an index of all the facts that concern my work ... Before beginning on any subject I look to all the short indexes and make a general and classified index, and by taking the one or more proper portfolios I have all the information collected during my life ready for use ... My mind seems to have become a kind of machine for grinding general laws out of large collections of facts.[39]

These autobiographical confessions might appear surprising, coming as they do from a naturalist who is traditionally presented as a fieldworker and an intelligent observer. In fact, they reveal Darwin's research strategy. Darwin's major methodological principle – presented explicitly in 1868 at the beginning of an account of the hypothesis of pangenesis – was to 'enumerate the groups of facts' and 'bring them under a single point of view'.[40] This formulation is reminiscent of one of Buffon's arguments[41] which, in Darwin's hands, takes on a particular significance. The

Darwinian strategy of establishing hypotheses involved arguing along two complementary lines of attack. The first step was to demonstrate that the established facts were incompatible with the alternative hypotheses, and thus to render his hypothesis plausible. But he also examined *all* the facts that could be thought to contradict his hypothesis and showed that, in fact, the hypothesis explained them. This is the method that produced the hypothesis of natural selection put forward in the *Origin of Species*, and the now-discredited hypothesis of pangenesis outlined in *The Variation of Animals and Plants under Domestication*. The fact that the same method elicits admiration from the contemporary reader in the first case, and astonishment in the second, is another question. The key point is that, in both cases, Darwin used the same method, advancing no direct empirical proofs either of natural selection or of pangenesis.

These observations about the *title* and method of Darwin's transformationist theory outline the *problems* that Darwin left to his successors. The first task was to find empirical proofs of natural selection. For many years, the search for these proofs continued to employ the method of indirect corroboration used by Darwin, which involved looking for facts that could only be explained by natural selection. Thus first Bates (1862) then Wallace (1867)[42] put forward the phenomenon of mimicry as the first 'proof' of natural selection, because, unlike other adaptations, it was *a priori* impossible to explain it by the effects of use and disuse. But it is obvious that this was in fact merely an 'illustration' or at best an indirect proof of the process of natural selection.[43] Despite the fact that from 1859 onwards the vast majority of Darwinian biologists adhered to the hypothesis of natural selection because of its explanatory power, this kind of proof by consequences was unsatisfactory. In physics 'principles' are genuine axioms. The 'principle of natural selection', however, is the result of a deduction made on the basis of a number of empirical premises, some of which, such as the disequilibrium between the rate of reproduction and the availability of subsistence, were perfectly plausible and could be documented by 19th-century science. But this was not the case for the very special notions of heredity and variation that were required by Darwin's hypothesis. To be firmly established, natural selection required the development of an experimental science of variation and of heredity which quite simply did not yet exist. Hence the importance of finding a direct empirical proof of the process of natural selection. This in turn required the clarification of the different elements of the hypothesis, which tended to become confounded with the proof of the possibility of the hypothesis. The major problem which faced those post-Darwinians who tried to prove or disprove the principle of selection was to measure to what extent this or that hypothesis of hereditary variation was compatible with Darwin's principle. Moreover, proving the possibility of natural selection initially required it be formalised in mathematical terms. Without such a step, it would be impossible to master the complex components of the concept of selection (for example, the rates of mortality and of reproduction, the distinction between the effects of selective and stochastic biases, the relationship between selection and crossing system, etc.). For this reason, the first real attempts at proof only appeared with the biometricians, at the end of the 19th century. But experimental biology also had to clarify the mechanisms of hereditary variation and transmission. This only took place with the development of the Mendelian paradigm in the first two decades of the 20th century.

In fact, there were two crucial moments in the long quest for a *demonstration* of the hypothesis of natural selection. The English school of statistics and biometry can claim to have made the first systematic exploration of the twin problems of a quantitative formulation of the theory of natural selection and its articulation with the theory of heredity. The genesis of 20th-century Darwinism cannot be understood without this crucial phase of post-Darwinism. For around thirty years (1870–1900), the theory of natural selection found its operational expression in this tradition. In studying the complex history of statistics and biometrics at the end of the 19th century, historians (and evolutionists) have been confronted with two contradictory temptations. On the one hand it has often been considered a dead-end in the history of Darwinism, which could be summarised by the fact that it was outmoded, the closed world of statistical theory having been, at best, a sterile phenomenalist science and at worst a façade for a naked ideology (eugenics). On the other hand, researchers have been struck by the methodological rigour of scientists like Weldon and Pearson, and by the determination with which they set out to prove both the theoretical possibility and the empirical reality of natural selection, at a time when most Darwinian evolutionists were satisfied with both the 'utilitarian' obviousness of Darwinism and its universal plausibility. The history of biometry shows that the tradition that did most to support and maintain the Darwinian model, and in fact elaborated the mathematical tools that established the theoretical basis of natural selection,[44] was also the very same school that had to be transcended in order for 20th-century Darwinism to be established.

The second moment in the search for a proof of the hypothesis of natural selection took the form of the synthesis of Darwinism and the experimental science of heredity in the first decades of the 20th century. This synthesis was neither simple nor easy, and it did not take place without difficulty. Mendelian genetics was initially widely considered to be incompatible with the theory of evolution by natural selection. More or less absorbed into the sphere of the 'mutationist' theory of evolution, Mendelism appeared as an experimental refutation of the Darwinian hypothesis because it showed that the hereditary material had a discontinuous nature. This view of variation went against that of Darwin, whose hypothesis was based on the selection of 'infinitesimal' variations, or what Bateson called 'continuous variation'.[45] It thus tended to credit the idea that selection could do nothing more than sift out, *a posteriori*, forms which it had played no part in creating. Nevertheless, it was Mendelian genetics that was to finally enable the hypothesis of natural selection to overcome its long initial crisis. However, this rehabilitated and 'Mendelised' Darwinism also turned out to be a Darwinism that had been profoundly reorganised in its internal theoretical structure.

This overview provides the general framework of the present work, which deals not so much with the *theory* of natural selection as with the *hypothesis*. This philosophical distinction between natural selection as a 'mere hypothesis' and as a 'theory' was made by Darwin himself. Selection as a 'mere hypothesis' is the conclusion of an argument, the premises of which deal with the rate of reproduction of organisms, and the existence of inherited variations that are more or less advantageous. Natural selection as a 'well-grounded theory' includes all the verifiable consequences of the theory in the whole of biology.[46] The theory of natural selection owed its success to its remarkable explanatory power and to its ability to unify natural history. Indeed, this is why Darwin spoke of a *principle* of natural selection. The *hypothesis* of natural selection, however, required that both its possibility and its reality be subject to long and detailed investigation. The vulnerability of this hypothesis, and in particular its dependence on notions such as variation and heredity which at the time did not have a clear experimental status, explains the profound crisis that affected theoretical Darwinism from the very beginning, despite the fact that so many naturalists had adopted it as a meta-theory of natural history. This book reconstructs the specific history of the hypothesis of natural selection. In other words, emphasis is placed on the historic analysis of the content of Darwin's central *hypothesis*, rather than on the massive reorganisation of natural history that formed the *theory*. Heredity thus occupies pride of place in this study, precisely because it is one of the key preconditions of the hypothesis, and not one of its consequences.

The first two parts reconstitute the conceptual configuration that enabled the relation between selection and heredity to be studied first by Darwin, then from 1865 to 1910 (the years respectively of the first publication by Galton on heredity and the beginning of population genetics) by others during the long years of crisis and doubt. This conceptual configuration will doubtless appear strange to contemporary readers. To varying degrees, turn-of-the-century biologists were all haunted by the idea that the 'principle of heredity' might turn out to be a 'force' that acted against the main principle of Darwinian evolution in such a way that heredity would either nullify the effects of selection or favour the return of ancestral characters.

The first part of the book explains how this concern was already present in the works of Darwin himself. To the extent that he linked the fate of the principle of natural selection to that of the 'great principle of heredity', without having an operative representation of the latter, Darwin exposed his theory of modification to attack from contemporary concepts of heredity, all of which were more or less marked by the idea of a 'reversion'. This aspect of Darwinian thought is extremely important for an epistemologically oriented history of Darwinian science. Far from being an *a priori* truth (as was the case with Spencer's reinterpretation of natural selection as 'survival of the fittest'), Darwin's hypothesis was continually threatened with refutation.

In the second part, we will see how, under various names and theoretical forms, be they anatomical or experimental, the theme of 'reversion' or 'regression' constituted the conceptual matrix of the post-Darwinian crisis that threatened the principle of selection. This archaic, polymorphous and confused concept was the focus of all the problems experienced by the post-Darwinians in trying to understand the relation between the 'principle of heredity' and the 'principle of selection'. The idea that these two principles might be in competition was the main obstacle that the concept of selection had to overcome in order to become truly intelligible. It was in this atmosphere that the formal (in fact statistical) theory of natural selection was initially elaborated. Particular attention will be paid to Galton's criticisms of the Darwinian concept of selection, and to the rise of biometry, both of which deserve to be reevaluated. The problem of reversion made a final reappearance during the initial period of Mendelian genetics, in the years following 1900. Before becoming the citadel of reconstructed Darwinism, Mendelism initially appeared to have dealt Darwinism its death-blow. For this reason the subtle reciprocal theoretical reelaboration undertaken by Darwinism and Mendelism will be examined in some detail.

The third part of the book deals with the theoretical universe of the new Darwinism that was developed from the 1920s onwards, that of the genetical theory of natural selection. In this last section, close attention will be paid to a new theoretical field, that of the specific theory of the hypothesis of selection. Despite being firmly anchored in the experimental biology of heredity, and despite its close association with 20th-century neo-

Darwinism, this new theoretical field, classically termed 'population genetics', took the form of a science of the conditions of the possibility of evolution.

A final word on the limited ambitions of this study, which is not *the* history, but rather *a* history, of the concept of selection in evolutionary theory. There are several reasons for this precision: on the one hand, as has been repeatedly pointed out in this introduction, this history is fundamentally a history of the *hypothesis* of natural selection rather than of the *theory* of natural selection in general. Furthermore, by focusing on the conceptual framework of heredity rather than, for example, on ecological or biogeographical aspects, those elements of natural selection which, initially the most obscure, eventually turned out to be the simplest and clearest are placed centre stage. Had an ecological approach been chosen, the opposite would probably have been the case.

Part 1

The Darwinian hypothesis

This first part characterises the Darwinian theory of natural selection by comparing it with the analogous theory of the modification of species put forward by Alfred Russel Wallace. Although the two men put forward similar *theories*, their central *hypotheses* were not the same. Historians have tended to underestimate certain important differences which separated Darwin and Wallace. Two major differences are traditionally underlined, differences that were openly and clearly stated at the time: (1) Darwin used the term 'selection' whilst Wallace did not; (2) an essential element of Darwin's argument was a comparison between domesticated and wild species.

These differences have a direct relationship to Darwin's emphasis on the inheritance of individual differences in his concept of selection. The disagreement between Wallace and Darwin was also expressed in a complex discussion of 'reversion': Wallace invoked this argument as part of his opposition to the analogy of domestication, while Darwin invoked the analogy of domestication to reply to the objection of reversion. He also used domestication to underline the importance of individual inheritance, which he considered to be more important than racial or species-based inheritance (Chapter 1).

The key question of inheritance also led Darwin to a rigorous consideration of the nature and the causal status of the units of selection. Darwinian selection is a selection that operates on *individual heritable variations*, due to the chances of survival and reproduction that these confer on *individual organisms*, thus modifying *races* or *varieties* (and consequently the species, since races are 'incipient species', as Darwin put it). Through this strict ontological schema, Darwin's principle became a conjecture that was open to the test of experience, unlike the *a priori* conceptions of selection that can be seen in most of Darwin's contemporary readers, in particular in Spencer. For this reason, particular emphasis is given to the ontology of the Darwinian concept of natural selection (Chapter 2).

However, by linking the fate of the hypothesis of selection to individual heredity, Darwin exposed it to obstacles that could not be overcome within his own theoretical apparatus. These problems can be seen with particular clarity in the objections to Darwin's work made by Fleeming Jenkin in 1867. In a famous article that has rarely been analysed with the care it deserves, Jenkin showed that, considered in quantitative terms and with reference to a population, the Darwinian concept of heredity implied the existence of forces of reversion that would make natural selection ineffective. Jenkin's argument, which Darwin immediately recognised as being the most important criticism he had encountered, will be analysed in some detail. This objection, and Darwin's attempts to reply, contain the first clear signs of the crisis of the hypothesis of the modification of species by selection (Chapter 3).

1 Wallace and Darwin: a disagreement and its meaning

1.1 Similar theories, different hypotheses

The story of how Charles Darwin (1809–1882) and Alfred Russel Wallace (1823–1882) independently developed the theory of natural selection – give or take a name – is well known and has been widely studied.[1] This chapter concentrates on those methodological and conceptual elements that are crucial for the theory of natural selection.

First of all, the facts. While disputes over priority are generally not particularly important for the history of science, chronology can be decisive for the epistemological history of concepts. In 1858, while on an expedition in the Malay archipelago, Wallace sent Darwin a short essay entitled 'On the Tendency of Varieties to Depart Indefinitely from the Original Type', and explicitly asked him to show it to Charles Lyell, if he thought it justified. Wallace's article contained a conception of the formation of species that had astonishing similarities with the theory Darwin had long developed and had called 'natural selection', but of which he had not yet published the slightest outline. As soon as Darwin received Wallace's letter (18 June 1858), he wrote to Lyell and then to Hooker, both of whom had been long aware of Darwin's work. In these letters, Darwin expressed his surprise at the similarity of Wallace's ideas to his own, as well as his dismay at being beaten to publication. He also explained that he wished it to be acknowledged that he had been the first to develop the idea, as shown by his 1842 'Sketch' and by an 1844 'Essay', in which he had used the term 'natural selection' (which Wallace had not used) and had outlined the fundamental elements of the theory. Darwin had sent a copy of his 1844 'Essay' to Hooker and then to Lyell, and left instructions for its publication in the case of his death. Within a few days, Hooker and Lyell had worked out a way of satisfying both Darwin's demand for priority and his moral scruples with regard to Wallace (he had immediately asked Lyell to publish Wallace's essay). On 1 July 1858, a mere thirteen days after his receipt of Wallace's letter, Lyell and Hooker made an oral presentation to the Linnean Society of London, consisting of three papers under the joint signatures of 'Charles Darwin, Esq., FRS, FLS & FGS and Alfred Wallace, Esq.' These were an 'extract' of Darwin's 1844 'Essay', described as 'an unpublished work by C. Darwin on species', a summary of an 1857 letter from Darwin to the American naturalist Asa Gray which gave a schematic presentation of six aspects of the Darwinian theory of the transformation of species, and Wallace's essay. The three documents were given a common title: 'On the Tendency of Species to Form Varieties; and on the Perpetuation of Varieties and Species by Natural Means of Selection'.[2]

It has often been said that history has been unjust to Wallace. We use the word 'Darwinism', goes the argument, to describe a theory that was in reality equally and simultaneously the work of Wallace. Some commentators have gone so far as to argue that Darwin, not content with stealing Wallace's justified priority in publication, copied the principle of divergence, which plays such a prominent role in the *Origin of Species*, in particular in the chapter on natural selection.[3] Such remarks are grossly exaggerated. In terms of claims to priority, the matter can be settled by a cursory examination of the 1842 'Sketch' and the 1844 'Essay', despite the fact that neither was published. Both these texts contain ideas that had already been clearly expressed in Darwin's 1838–9 notebooks, showing that Darwin apparently developed his ideas some twenty years before Wallace, there being no trace in the latter's writings of any elements of Malthusianism before his 1858 essay.[4] While it is true that Wallace had prior claim to the publication of the concept of divergence, using the image of an ancestral tree in an 1855 article,[5] Camille Limoges has pointed out that Darwin had a clear priority in terms of the development of the idea, again as shown by his notebooks.[6]

Apart from the 1842 'Sketch' and the 1844 'Essay', there is another document which settles the question of priority: a critical edition of a manuscript that reveals why Darwin put off publishing his ideas on the modification of species for so long, and how he was taken by surprise by Wallace in 1858. When Darwin received Wallace's essay (June 1858), he had already written eleven chapters of a 'big book', begun in 1856, which was intended to prove a 'theory of descent with modification through natural selection'. By 1858, the manuscript was 225,000 words long, and would have been around 375,000 words long had the projected fourteen chapters been written. In a detailed edition of that part of the manuscript which corresponds to the *Origin of Species*, R. C. Stauffer has shown that Darwin's projected book would have been over twice as long as the *Origin of Species* (150,000 words), of which around one-half (80,000 words) is a condensed version of the 1856–8 manuscript. Furthermore, the first two chapters of the projected book corresponded to an outline (140,000 words) of his 1868 work on the *Variation of Animals and Plants under Domestication* (315,000 words), which was itself a development of the first chapter of the *Origin*, 'Variation under Domestication' (11,000 words).[7] In comparison, Wallace's article is only eight pages long, and there is no trace in any of his previous writings of any hypothesis equivalent to the Darwinian principle of natural selection. These facts show plainly who had priority.

The unexpected arrival of Wallace's article clearly altered Darwin's plans for publication. But even more importantly it changed his strategy of exposition. The major work that Darwin called 'my big species book'[8] and which he had still not renounced hope of completing in 1868,[9] was never published. It would have included all the elements of the *Origin of Species* (1859), of *The Variation of Animals and Plants under Domestication* (1868) and a part of *The Descent of Man, and Selection in Relation to Sex* (1871). In other words, it would have dealt with all the interrelated aspects of the evolution of domesticated species, of the evolution of animals and plants in their natural state, and of human evolution. It would have had a substantial bibliography, and lists of observations from the field and from collections, as was the case for *all* Darwin's books *except* for the *Origin of Species*. The *Origin* contains no bibliography and no figures (except for the famous tree diagram illustrating the principle of divergence) and does not contain the pages and pages of lists that fill up Darwin's other books; furthermore, it does not deal with the problem of the evolution of man. In the Introduction to the first edition, Darwin explicitly described the *Origin* as an 'Abstract' (490 pages long!) of a book which he had been obliged to postpone publishing: 'This Abstract, which I now publish, must necessarily be imperfect ... No one can feel more sensible than I do of the necessity of hereafter publishing in detail all the facts, with references, on which my conclusions have been grounded; and I hope in a future work to do this.'[10]

In fact, Darwin was grateful to Wallace for obliging him to write the *Origin* in such a way that argument takes pride of place over academic respectability. As he wrote in the conclusion: 'This book is one long argument.'[11] The hasty writing of the *Origin* enabled Darwin to separate the *hypothesis* from the *theory* of natural selection. No doubt the hypothesis was not as firmly rooted in empirical data as he might have wished. It cannot be emphasised enough that there are no proven examples of natural selection in the *Origin*, and that the hypothesis is based upon variation and heredity, both of which remained mysterious in terms of their physiological mechanisms. On the other hand, the theory, of which natural selection was the organising hypothesis, was ready for publication. The *Origin of Species* is above all a massive reconstruction of natural history, based on the unifying capacity and exploratory power of the hypothesis of natural selection. The argument that led to this reconstruction, involving such diverse facts as the geographical distribution of species, their anatomical 'affinities', their embryology, their paleontological history, their adaptation to the environment, etc., had been ready for some time when Darwin decided to 'summarise' it.

Wallace knew all this quite well, or rather he understood it very quickly once he read Darwin's book. He rapidly described himself as a 'Darwinian' and finally, despite reservations, adopted the term 'natural selection'.[12]

If there is one problem in the debate over priority, it is conceptual. From the moment he received Wallace's essay, Darwin continually claimed that the article contained 'exactly the same theory [as his own]'.[13] While this is fundamentally true in one sense, in another it is fundamentally false. Darwin and Wallace agreed on the general implications of the principle of selection, in other words on the *theory* of natural selection, but they did not agree on the meaning of the central *hypothesis*. Although they agreed that natural selection implied a fundamental reorganisation of natural history, they did not have the same understanding of the process of selection, or rather of what Darwin called the process of selection.

Was there a significant difference between the theory of the modification of species presented by Wallace in 1858 and that which Darwin had described since 1842 as 'selection'? Although Darwin's previous writings (from the 1842 'Sketch', through the 1844 'Essay', to the 1856–8 manuscript) could be cited, a simple comparison of the two extracts of Darwin's work that were published in 1858 alongside Wallace's essay provides sufficient evidence.

Historians have generally repeated the two facts noted by Wallace and Darwin: there are a remarkable series of terminological and conceptual similarities between their two sets of ideas, but Wallace did not use the terms 'selection' or 'artificial selection', and even less so the specifically Darwinian term 'natural selection' (or 'natural means of selection'). The second point is linked to the fact that Darwin considered that artificial selection was fundamental for proving his causal hypothesis of the modification of natural species, while Wallace formally rejected any similarity between domesticated and natural species.

This raises two linked questions which will be examined separately:

- To what extent did Darwin and Wallace really develop analogous theories of the modification of natural species? This can best be answered by abstracting from the question of the relevance of domesticated species.

- Why was the analogy of domesticated species rejected by Wallace while it was essential for Darwin?

Darwin and Wallace put forward independently formulated theories of the origin of *natural* species. The convergences are indeed remarkable. Both authors presented their hypothesis as a 'principle'[14] with two epistemological aspects. First, a series of empirical premises that rendered the hypothesis of 'natural selection' (or Wallace's equivalent) probable. Second, phenomena that were explained by the hypothesis and thus provided independent confirmation. This second aspect is similar to the way the principles of Newtonian mechanics were proven *a posteriori* by the number of empirical laws that they led to.[15] Thus, for both Darwin and Wallace, the theory was divided into arguments related to plausibility and arguments related to the explanatory power of the central hypothesis.[16]

Many of the points related to 'plausibility' are virtually identical. Both repeat De Candolle's famous aphorism that 'nature is in a state of war', and both give the metaphor a precise sense that they describe as 'the struggle for life' (Darwin)[17] or 'the struggle for existence' (Wallace).[18] Both these concepts imply that the quantity of food available for any given species is strictly limited,[19] while population size can be expected to expand exponentially.[20] However, natural populations are generally constant in size,[21] indicating that a substantial number of individuals must die.[22] Furthermore, Darwin and Wallace both considered that adaptations occurred in an environmental context that was itself constantly changing.[23] Finally, *both* men considered that organs and instincts show very small interindividual variations that affect both the vigour of individual organisms and their 'chance of survival'.[24] On the basis of these premises, Darwin and Wallace deduced that there existed an indefinite tendency for species to form new varieties and to perpetuate them. This common hypothesis appears in the title of the three papers jointly presented by Darwin and Wallace. The title also included the word 'selection' as the means by which new varieties are perpetuated, but this term was linked to ideas that were solely Darwin's.

There were also striking convergences concerning what the hypotheses were supposed to explain (their 'explanatory responsibility', to borrow Jonathan Hodge's formula).[25] Both Darwin and Wallace argued that two major consequences flowed from their principle. The first was the necessary adaptive 'progress' of surviving races or species. Both authors use exactly the same terms: 'progress', 'progression', 'improvement', all specifically applied to 'adaptation'.[26] The second consequence was 'divergence': Darwin and Wallace both argued that species tended to branch out indefinitely, producing new species with characters that were increasingly different from each other. In the *Origin of Species*, the principle of divergence is dealt with at great length as Darwin tries to explain the asymmetrical tree-like diagram of classification on the basis of the biogeographical consequences of natural selection: a successful species will tend to occupy the largest possible number of 'places' in nature, thus diversifying. Thus natural selection not only explains the 'modification' of species, but also their diversification.[27] This idea was already present in Darwin's 1857 letter to Asa Gray, an 'abstract' of which was presented in the joint communication of 1858:

Every organic being, by propagating so rapidly, may be said to be striving its utmost to increase in numbers. So it will be with the offspring of any species after it has become diversified into varieties, or sub-species, or true species. And it follows, I think, from the foregoing facts, that the varying offspring of each species will try (only few will succeed) to seize on as many and diverse places in the economy of nature as possible. Each new variety or species, when formed, will generally take the place of, and thus exterminate its less well-fitted parent. This I believe to be the origin of the classification and affinities of organic beings at all times; for organic beings always *seem* to branch and sub-branch like the limbs of a tree from a common trunk, the flourishing and diverging twigs destroying the less vigorous – the dead and lost branches rudely representing extinct genera and families. [Original emphasis][28]

It has been argued[29] that Darwin took the idea of the principle of divergence from an 1855 article by Wallace.[30] While this is quite possible, it will not be discussed here, except to point out that Wallace's contribution to the joint presentation of 1858 includes a discussion of 'divergence':

This new, improved, and populous race might, itself, in course of time, give rise to new varieties, exhibiting several diverging modifications of form, any of which, tending to increase the facilities for preserving existence, must, by the same general law, in their turn become predominant. Here, then, we have *progression and continued divergence* deduced from the general laws which regulate the existence of animals in a state of nature, and from the undisputed fact that varieties do frequently occur. [Original emphasis][31]

To summarise, Wallace and Darwin put forward a hypothesis that had two aims: first, to explain the origin of adaptations and second to provide a temporal representation of life as a series of random branchings. It should also be noted that Wallace and Darwin had both developed a theory that revealed the error of the classic taxonomic distinction between 'varieties' (or races) and species. Both of them considered that there was a process that tends 'to convert varieties into species',[32] the two classes of groups of organisms having only a genealogical reality.

The 'theories' put forward by Darwin and Wallace in 1858 were thus marked by major convergences. Neither the two authors, nor readers at the time, nor historians since have had the slightest doubt that the two men's writings contained the same *theory*. The ideas developed by Darwin and Wallace were equivalent in that they each contained a theory with an identical *structure*. On the basis of two fundamental hypotheses that, at first glance, seemed similar, Darwin and Wallace put forward two sets of arguments aimed at establishing both the plausibility of the central hypothesis and its explanatory power. In both cases, the ideas put forward led to the same unified vision of life, in which notions of descent, of modification and of divergence were interrelated.

However, there were two significant differences between the ideas of Darwin and Wallace. These did not initially relate to their disagreement over the usefulness of domestication as a model for understanding the modification of species in the wild, but were focused on the role played by Malthusian population pressure. They also related to the concept of heredity, which was fundamental for Darwin from the outset but was absent from Wallace's ideas.

Although Wallace did not cite Malthus (even though he had read him), there can be no doubt that the idea of an imbalance between demography and the means of subsistence plays as important a role in his writings as in those of Darwin. Both authors considered that population pressure was a force that could change the characters of a species owing to differences in the utility of these characters. But, as Peter Bowler has pointed out, the effect of this pressure was not the same for Darwin as for Wallace.[33] Darwin, citing Malthus, considered that 'the amount of food for each species must, *on an average*, be constant, whereas the increase of all organisms tends to be geometrical, and in a vast majority of cases at an enormous ratio'.[34] This suggests that competition operates mainly when the size of a race or variety is constant. Darwin did not argue that competition (or 'the struggle for life') does not function in periods where the abundance of food enables population size to expand, but rather that competition is at its strongest under the normal condition of limited resources. Wallace approached the question of limited food resources from another point of view. He concentrated on the growth (positive or negative) of population size due to the availability of food and the capacity of the species to exploit it. Darwin and Wallace thus invoked two different phenomena in order to define the struggle for existence. Darwin considered groups of organisms that were approximately stable in size and conceived of competition as taking place *between individuals*. Wallace, on the other hand, thought that the 'struggle for existence' involved competition *between populations*. According to the Darwinian schema, competition led to the modification of the race, while for Wallace it led to changes in the population size of several races. These two opposed views need to be teased out further.

According to Darwin, the struggle for life takes place principally and fundamentally between individuals. From 1844 onwards, he made this point quite clearly:

Can it be doubted, from the struggle each individual has to obtain subsistence, that any minute variation in structure, habits, or instincts, adapting that individual better to new conditions, would tell upon its vigour and health? In the struggle it would have a better *chance* of surviving; and those of its offspring which inherited the variation, be it ever so slight, would also have a better *chance*. [Original emphasis][35]

Darwin did not reject the idea of 'competition' between varieties or species. Quite the opposite: he explicitly proposed it.[36] But, for Darwin, competition (or 'the struggle for existence') never took place between taxonomic groups as such, but rather between *individuals* of a given group, or of different groups. Thus, in the *Origin of Species*, Darwin argued that 'the most important' aspect of the struggle for existence was 'the life of the individual and its success in leaving progeny'.[37] The highly restrictive 'Malthusian' schema set out by Darwin flowed from this starting point. The intensity of competition depends on two factors. First, the nature of the entities involved: competition is more intense where the forms of organisms are most similar and where they are struggling for strictly identical 'places' in nature, or, in other words, similar natural resources.[38] The struggle for existence is therefore at its strongest between members of the same species, or between varieties of the same species.[39] Similarly, the struggle is sharper between species of the same genus than between more distantly related species.[40] For Darwin, competition does not occur between 'forms', or types, but between more or less complex measurable groups of individuals.[41] The second decisive factor affecting the intensity of competition is the availability of subsistence resources, or more generally all the conditions that determine the size of a group. Competition is more severe when the group is obliged by circumstances to be stable in size than when it is expanding.[42] This way of dealing with the problem of competition also leads to the conclusion that competition is at its most severe between individual members of the same species.

However, Darwin's 'individualistic' starting point shows itself most clearly in the concept of natural selection. The Darwinian definition of natural selection involves only three factors: 'variations', 'individuals' and 'races' (or 'varieties'). In this three-level schema, only 'variations' are selected. By 'variations' Darwin meant 'individual differences',[43] particular structures or behaviours that could be hereditarily transmitted ('any variation that is not hereditary is of no interest to us').[44] These variations, more or less independent of each other, constitute the 'raw material' of natural selection, which gradually 'accumulates' those that are advantageous within a line of related individuals.[45] The Darwinian hypothesis of natural selection is thus intimately linked to the idea that transmissible characters are separate, atomised. From this it follows that, strictly speaking, Darwinian selection involves the selection *of* variations, *within* the race, *because of* the advantage that these variations give to individuals. In other words, Darwinian selection requires us to distinguish what is selected (individual heritable variations), what is the target of the causal action of competition (individual organisms) and what is modified (the race).[46]

The Darwinian hypothesis of natural selection thus stands or falls with the concept of heredity, understood as the ability to transmit individualised and individual characters. Darwin clearly characterised his concept of selection as early as his 1844 'Essay', in the passage cited in part above:

> Can it be doubted, from the struggle each individual has to obtain subsistence, that any minute variation in structure, habits, or instincts, adapting that individual better to the new conditions, would tell upon its vigour and health? In the struggle it would have a better *chance* of surviving; and those of its offspring which inherited the variation, be it ever so slight, would also have a better *chance*. Yearly more are bred than can survive; the smallest grain in the balance, in the long run, must tell on which death shall fall, and which shall survive. Let this work of selection on the one hand, and death on the other, go on for a thousand generations, who will pretend to affirm that it would produce no effect, when we remember what, in a few years, Bakewell effected in cattle, and Western in sheep, by this identical principle of selection? [Original emphasis][47]

Darwin's individualistic concept of natural selection coincides with another aspect of the 1858 texts that has often been noted. It has rightly been remarked that one of the differences between Darwin and Wallace was the fact that 'sexual selection' was present only in Darwin's writings. 'Sexual' selection must necessarily take place between individuals belonging to the same reproductive community. It would not make sense, for example, to speak of sexual selection occurring between races or species. The general hypothesis of 'selection' as the means of transforming species thus has far-reaching implications: 'sexual' selection, like both 'natural' selection and 'artificial' selection, only makes sense in terms of heritable variations between individuals.

The concepts put forward in Wallace's essay are substantially different, and can easily be characterised with reference to Darwin's work. At first glance, it might appear that there is a strict equivalence between Wallace's ideas and the Darwinian conception of 'natural selection'. And indeed, certain passages of Wallace's 1858 essay do suggest the idea of the modification of species based upon individual variation, in particular those sections where he defines the 'struggle for existence':

The life of wild animals is a struggle for existence. The full exertion of all their faculties and all their energies is required to preserve their own existence and provide for that of their infant offspring ... The numbers that die annually must be immense; and as the individual existence of each animal depends upon itself, those that die must be the weakest – the very young, the aged, and the diseased, – while those that prolong their existence can only be the most perfect in health and vigour – those who are best able to obtain food regularly, and avoid their numerous enemies. It is, as we commenced by remarking, 'a struggle for existence', in which the weakest and least perfectly organized must always succumb.[48]

However, this individualistic definition of the struggle for existence is not clearly linked to the 'principle' that runs through the article. This is evident from the first description of that 'principle': 'There is a general principle in nature which will cause many *varieties* to survive the parent species, and to give rise to successive variations departing further and further from the original type.'[49] This formulation has two parts. The first is the principle of the survival of 'varieties', or in Darwinian terms, a principle of interracial, rather than interindividual, selection. The second part is more ambiguous. In order to understand exactly what is meant, it is necessary to refer to the quantitative aspect of the process, the description of which takes up most of the essay. Wallace is exclusively interested in the growth (positive or negative) of the 'population' *of* a species (i.e. the population size of the species). He focuses on the 'abundance' or 'rarity' of groups of organisms. This explains why he often used the term 'population', which Darwin did not in 1858, and which is extremely rare in his later writings, in particular in the *Origin of Species*.

This terminological nuance is more important than might at first appear. Nineteenth-century English usage of 'population' had two senses: a descriptive sense (the number of individuals composing a group) and a dynamic sense (to people or to grow in number). In his *Essay on the Principle of Population* (1798), Malthus used the word in both senses, playing on the ambiguity. Concepts such as 'the power of population' or the idea that resource limits could be considered as 'a powerful brake acting constantly on the population' can only be understood in the context of 'population' as an intrinsic tendency of 'peoples' to increase their number.[50] Although Darwin and Wallace both based their arguments on the Malthusian 'population principle', that is, on the natural law that maintains a balance between the 'unequal powers' of the 'population' ('multiplying' power) and the means of subsistence ('additive' power), they used this principle in different ways. Darwin focused on races or species of a constant size, subject to environmental conditions that were approximately constant. That, no doubt, is why he did not use the Malthusian term 'population'. From this point of view, changes in the size of groups are a consequence of the process of natural selection: races or species that have been 'improved' by natural selection will have a greater chance of reproducing and spreading. But, logically speaking, 'improvement' (adaptive modification) precedes demographic expansion.

To a certain extent, Wallace approached the problem from the opposite end. He was primarily interested in demographic competition between varieties and between species as such. 'It is an uncontested fact,' he wrote, 'that varieties appear frequently.'[51] Any new 'variety' composed of 'individuals' with certain advantages will have a tendency to grow in size, to the detriment of other varieties. Wallace's schema, repeated over and over again in his essay, is well summarised in the following paragraph, in which competitive superiority is clearly considered as residing in 'varieties as such':

> If ... any species would produce a variety having slightly increased powers of preserving existence, that variety must inevitably in time acquire a superiority in numbers ... This new, improved, and populous race might itself, in course of time, give rise to new varieties, exhibiting several diverging modifications of form, any of which, tending to increase the facilities for preserving existence, must, by the same general law, in their turn become predominant.[52]

This formulation and others like it show that Wallace considered that the fundamental process involved in the modification of species is the 'power of the growth of populations',[53] a power that is proportional to the ability of these populations to have the greatest growth in a universal struggle to occupy ecological space. This schema does not clearly involve the Darwinian notion of the modifying power of competition between individuals. Although Wallace occasionally evoked the existence of an individual struggle for existence, he did not make it an operative element in his principle of modification. Rather, it occupied the place of what modern quantitative ecology would call synecological relations (e.g. Lotka's and Volterra's equations for predator–prey relationships). Thus, when Wallace referred to 'the doctrine of probabilities and means',[54] unlike Darwin he did not see the theoretical implications of the spread of individual variations in a natural race (a 'population' in the modern sense of the term). Instead he was suggesting that, on average, the modified varieties that survive tend to replace the species they came from.[55] In contemporary terms, Wallace constructed a model of the transformation of species by intergroup selection, in which evolutionary success was determined by 'the comparative abundance or scarcity of the individuals of the several species',[56] not by the slow internal modification of varieties or races. Wallace thus employed the Malthusian idea of demographic limits imposed by the subsistence necessary for the total living community occupying a given ecological space: 'the animal population of a country is generally stationary, being kept down by a periodical deficiency of food, and other checks'.[57] According to Wallace, it is in this general demographic context that the relative abundance of species and varieties, in other words their 'power of population', changes.[58]

In one sense, this conception of 'natural selection', if it can be called such, seems familiar to the modern evolutionary biologist, in particular because of its emphasis on the dynamics of ecological equilibrium, rather than on intra-populational and intra-specific competition. It is also reminiscent of modern debates about the 'units of selection'. However, the historian has to take care not to mistake what may in fact be an archaism for an anticipation of future problems and concepts.

And it appears that, on this fundamental point, Wallace's 1858 essay was indeed an archaism rather than a brilliant prefiguration. Wallace failed to distinguish clearly between 'varieties' and 'variations'. There is no trace in his explanation of the modification of species of anything similar to the idea of the gradual transformation of 'varieties' or 'races' by the selective accumulation of what Darwin called 'individual variations'. 'Varieties depart indefinitely from the original type [of species]', as the title of Wallace's essay puts it. But that simply means that, for Wallace, species give rise to varieties which themselves give rise to new varieties, and so on. This is the background to Wallace's definition, cited above: 'There is a general principle in nature which will cause many *varieties* to survive the parent species, and to give rise to successive variations departing further and further from the original type' (original emphasis).[59] It is particularly striking that Wallace's idea of extremely small variations, despite being expressed in identical terms to those employed by Darwin ('variations . . . however slight . . .'),[60] does not involve the same entities or units. For Wallace, it is not the individual but the 'variety' (in the sense of 'race') that presents the 'slightly increased powers of preserving existence'.[61] Nevertheless, Wallace did not think there was any fundamental difference between 'variety' and 'variation'. He considered that 'variety' and 'race' were identical terms, and insisted on their demographic and genealogical significance. 'Variation', meanwhile, was an anatomical concept of use to systematists (variations are defined for 'parts' of an organism). But Wallace argued as though these were merely two different methodological approaches to a single phenomenon: varieties produce variations,[62] variations define varieties.

This suggests that Wallace made only a partial break with what Ernst Mayr has called the 'typological' species concept. To be fair, however, Wallace's essay is in one sense a sustained attack on the idea of the 'original and permanent distinctness of species' and on their 'permanent invariability'.[63] All species give rise to varieties that differ from the original type and which, sooner or later, will replace it; there is no real distinction between species and varieties. On this point, the views of Darwin and Wallace were identical. They both considered that it was pointless to try and find a criterion for differentiating species from varieties. Species are not more stable than varieties and are therefore not 'types'. However, it can be argued that Wallace used 'typological' concepts when he dealt with variation. Because for Wallace variation could only be considered in terms of 'varieties', not individuals, it was impossible for him to take the decisive step towards a genuinely 'populational' representation of evolution and life. Although Wallace did use the term 'population' in 1858, he only saw it in terms of numbers, not in terms of variability.

To sum up this comparison of Darwin's and Wallace's theories of the modification of natural species as presented in the seminal texts of 1858, we can say that Wallace's essay does not contain the strict equivalent of the Darwinian hypothesis of natural selection. In particular he did not put forward the idea of population pressure acting on the distribution of individual variations. In terms of the history of science, the differences between Wallace and Darwin can be considered from two perspectives.

From the point of view of contemporary debates about the 'units' and 'levels' of selection, the different ideas of Darwin and Wallace are reminiscent of current discussions as to the possibility of a 'group selection' different and separate from the 'individual selection' that Darwin defended so vigorously. This is not an example of retrospective projection: on several occasions, Wallace and Darwin disagreed over the possibility of selection operating on groups of organisms, in particular in terms of the evolution of altruistic behaviours.[64] Right from the outset, therefore, there were two competing interpretations of the hypothesis of natural selection. This sheds an important light on the degree of convergence between the two men. Basing their arguments on the same empirical premises, Darwin and Wallace constructed two versions of a hypothesis of such explanatory power that it completely altered the whole of natural history. If this overall vision can be described as 'the theory of natural selection', then in 1858 Darwin and Wallace had indeed arrived independently at the same 'theory of natural selection', of which Darwin was to summarise the epistemological structure ten years later, in the Introduction to *The Variation of Animals and Plants under Domestication* (1868):

In scientific investigations it is permitted to invent any hypothesis, and if it explains various large and independent classes of facts it rises to the rank of a well-grounded theory. The undulations of the ether and even its existence are hypothetical, yet every one now admits the undulatory theory of light. The principle of natural selection may be looked at as a mere hypothesis, but rendered in some degree probable by what we positively know of the variability of organic beings in a state of nature, – by what we positively know of the struggle for existence, and the consequent almost inevitable preservation of favourable variations, – and from the analogical formation of domestic races. Now this hypothesis may be tested, – and this seems to me the only fair and legitimate manner of considering the whole question, – by trying whether it explains several large and independent classes of facts; such as the geological succession of organic beings, their distribution in past and present times, and their mutual affinities and homologies. If the principle of natural selection does explain these and other large bodies of facts, it ought to be received.[65]

If we ignore for a moment the reference to domesticated races, this passage expresses precisely the common conception that Darwin and Wallace had from the outset, and the ideas that attracted all those biologists who subsequently accepted the Darwinian vision of evolution.

But there is more. Darwin's hypothesis – unlike that of Wallace – was based on the confused but clearly stated principle of heredity, or more precisely of the inheritance of individual differences. On the basis of this principle, Darwin was able to construct the highly restrictive concept of selection operating by and upon individuals. Furthermore, because Darwin linked selection with heredity, natural selection was rapidly exposed to theoretical and experimental criticisms that would not have been raised had it been merely a vague metatheory of natural history. Darwin thus placed the problem of heredity at the centre of biological theory. The only serious way he had of understanding the question of heredity was through the example of the variation of domesticated species. For Darwin, examples of human (or 'artificial') selection constituted the most direct experimental argument available that could support the hypothesis of natural selection, for which there was no experimental evidence. From this second point of view, Darwin and Wallace constructed a theory with the same overall structure and explanatory ambition, but on the basis of substantially different hypotheses. An essential aspect of Darwin's hypothesis was that it was based on the inheritance of individual differences.

Wallace's and Darwin's competing conceptions can thus be seen in a different light. In his 1858 essay, Wallace also considered that variation in domesticated species was of fundamental importance, but in order to reject any analogy between the natural and domesticated states. This clearly stated disagreement between the co-discoverers of the principle of selection reveals why the very term 'selection' was so important for Darwin.

1.2 Wallace's rejection of the model of domesticated species

Wallace's essay was not a direct criticism of the model of domesticated species employed by Darwin in setting out his hypothesis of natural selection. First and foremost, of course, in 1858 Wallace was not aware of Darwin's speculations about the modification of species. Second, although Wallace's 1858 essay does contain a criticism of the analogy of domesticated species, it is related to the way in which domesticated species were considered at the time by most naturalists, notably as a direct experimental argument in favour of the stability of species. However, Wallace's critique was sufficiently general to include in its scope Darwin's methodological inclinations, and Darwin considered it necessary to respond on several occasions. It is worth reconstituting this initial debate.

What was Wallace's argument against the validity of the analogy of domesticated species? His 1858 essay opens with the following statement:

> One of the strongest arguments which have been adduced to prove the original and permanent distinctness of species, is that *varieties* produced in a state of domesticity are more or less unstable, and often have a tendency, if left to themselves, to return to the normal form of the parent species; and this instability is considered to be a distinctive peculiarity of all varieties, even of those occurring among wild animals in a state of nature, and to constitute a provision for preserving unchanged the originally created distinct species.
>
> In the absence or scarcity of facts and observations as to *varieties* occurring among wild animals, this argument has had great weight with naturalists, and has led to a very general and somewhat prejudiced belief in the stability of species. Equally general, however, is the belief in what are called 'permanent or true varieties', – races of animals which continually propagate their like, but which differ so slightly (although constantly) from some other race, that the one is considered to be a *variety* of the other; which is the *variety* and which the original *species*, there is generally no means of determining, except in those rare cases in which the one race has been known to produce an offspring unlike itself and resembling the other. This, however, would seem quite incompatible with the 'permanent invariability of species', but the difficulty is overcome by assuming that such varieties have strict limits, and can never again vary further from the original type, although they may return to it, which, from the analogy of the domesticated animals, is considered to be highly probable, if not certainly proved.
>
> It will be observed that this argument rests entirely on the assumption that *varieties* occurring in a state of nature are in all respects analogous to even identical with those of domestic animals, and are governed by the same laws as regards their permanence or further variation. But it is the object of the present paper to show that this assumption is altogether false, that here is a general principle in nature which will cause many *varieties* to survive the parent species, and to give rise to successive variations departing further and further from the original type, and which also produces, in domesticated animals, the tendency of varieties to return to the parent form.[66]

This passage poses the question of the relevance for the naturalist of data from domesticated species. From the outset, Wallace made clear that the aim of his article was to attack the assumption of the 'stability of species'. The 'fixist' naturalists whom Wallace was addressing were steeped in a secular tradition of study of 'varieties', based on the work of Linnaeus and the experience of hybridisers. According to this tradition, species were not considered in terms of absolutely rigid characters, but as spheres of variation with limits that could not be transcended. Lyell summed up this doctrine perfectly in his book *The Principles of Geology* (1830–3): 'There are fixed limits beyond which the descendants from common parents can never deviate from a certain type.'[67] Wallace made a clear reference to this doctrine in the passage quoted above (see the end of the second paragraph).

During the 19th century, this view of species was founded primarily upon the experience of animal breeders and horticulturists. Wallace examines the two apparently contradictory conclusions drawn by 'fixist' naturalists on the basis of the observation of domestic varieties. The first is that varieties are unstable and tend to 'return to the parent form'. The obvious implication of this assertion is that truly domestic *varieties* do not exist, only a superficial and artificial variation from the only form that clearly exists in nature: the species' 'wild type'. Wallace, a biogeographer who had studied the reports of many expeditions, must have known of the anecdotal reports of sailors and travellers on the fate of animals abandoned in far-off countries. Livingstone, Blyth and indeed Darwin himself had described in their notebooks what happened to rabbits or pigs that had been abandoned or had escaped: wherever they found themselves, these animals had developed similar characters and had apparently returned to a wild state. Pigs were the most often cited example. Pigs that had gone wild were small and furry – they had 'reverted' to the state of wild boars.

The second classic argument dealt with by Wallace is somewhat different, and at first glance appears to be opposed to the first. 'Equally general, however, is the belief in what are called "permanent or true varieties", – races of animals which continually propagate their like, but which differ so slightly (although constantly) from some other race, that the one is considered to be a *variety* of the other.'[68] Faced with such an example, explains Wallace, the classic conception of the species argues that variation, even though it might be spontaneous, always happens in the same way (for example black lambs in flocks of white sheep).

To summarise, for the 'fixist' naturalists the function of the domestic model was to show that 'varieties', whether they are stable or not, tend to revert to the original type or 'normal form' of the species. Faced with this kind of argument, Wallace's strategy was quite simple. He accepted the validity of the doctrine of the reversion of domestic varieties: 'Domestic varieties, when turned wild, *must* return to something near the type of the original stock, *or become altogether extinct*' (original emphasis).[69] But he rejected the application of this principle to species in the natural state. Natural varieties tend to vary indefinitely from the original type and cannot revert, he argued. In other words, Wallace completely rejected the idea that variation in domestic races could be a model for understanding natural variation: 'No inference as to varieties in a state of nature can be deduced from the observation of those occurring among domestic animals. The two are so much opposed to each other in every circumstance, that what applies to the one is almost sure not to apply to the other.'[70] However, the aim of Wallace's article was precisely to underpin this rejection of the domestic model by setting out a single principle that could explain the contradictory behaviour of domestic and natural varieties. In other words, Wallace's 'principle', traditionally considered as being identical to Darwin's 'natural selection', was intended to explain why the 'reversion' of natural varieties is *impossible*, while that of domestic varieties is *necessary*.

The reasoning is disarmingly simple, and has often been taken (mistakenly) as proof that Wallace's ideas were indistinguishable from those of Darwin. Domestic animals, explains Wallace, are fed, and are protected against predators and the vicissitudes of the environment. They are thus sheltered from the effects of the struggle for existence. Population pressure cannot advantageously modify their characters. For their own ends, humans encourage characters that have no significance for the natural economy of the species, but different variations develop spontaneously, making a domestic animal in the wild unable to resist the competition of wild-type animals. Wallace thus explains how domestication develops characters that would be useless in the natural state. In the absence of the regulatory control of the domestic environment,[71] animals tend to produce unusual and unstable variations, the stability of their forms being entirely due to their artificial rearing conditions.

In terms of what it reveals about his attitude to fundamental Darwinian concepts, Wallace's argument is extremely problematic. He attacks the model of domestic varieties in the name of what he later called 'the principle of utility',[72] according to which no organ or attribute can exist in a natural species unless it is or has been useful to the organisms that possess it. This was probably one of the most obvious points of convergence between Darwin and Wallace, to the extent that in his later writings Wallace presented the principle of utility as an immediate implication of the Darwinian concept of natural selection:

Perhaps no principle has ever been announced so fertile in results as that which Mr Darwin so earnestly impresses upon us, and which is indeed a necessary deduction from the theory of Natural Selection, namely – that none of the definite facts of organic nature, no special organ, no characteristic form or Marking, no peculiarities of instinct or habit, no relations between species or between group of species – can exist, but which must now be or once have been *useful* to the individuals or the races which possess them. [Original emphasis][73]

This declaration alludes to the following passage in the *Origin of Species*:

Every detail of structure in every living creature (making some little allowance for the direct action of physical conditions) may be viewed, either as having been of special use to some ancestral form, or as being now of special use to the descendants of this form – either directly, or indirectly through the complex laws of growth.[74]

Why is the same principle associated with radically different concepts in Darwin's and Wallace's discussions of domesticated species? For Wallace, one of the main consequences of the principle of utility was that species do not really evolve under domestication. Once the artificial effect of human intervention has been removed, they either revert to their 'original type' or they become extinct. No similar statement can be found in Darwin's writings. On the contrary, the Darwinian hypothesis of natural selection is intimately linked to the idea that domestic races are modified in the same irreversible way as natural races. This aspect of Darwin's hypothesis will be dealt with in more detail.

1.3 Darwin: the relevance of the model of domestication

In a remarkable study of the genesis of the concept of natural selection, Camille Limoges has sharply criticised 'an over-long tradition of the history of Darwinism that has taken for granted the heuristic step from the artificial to the natural'.[75] He puts forward three arguments to underline the fact that the example of domestic species was not a necessary part of the development of the theory of natural selection. In the first place, Limoges notes that in the 1838–9 *Notebooks on Transmutation of Species*, Darwin did not use the term 'natural selection'. He did however develop the concept, for example when biogeographical data led him to argue that adaptation is always relative. Species are not 'by nature' adapted to this or that situation; an absolute adaptation would be predefined, and thus non-modifiable. The analogy with artificial selection came later, in 1842,[76] in what Limoges describes as a rhetorical explanation of the theory.[77] Limoges's second point is essentially psychological. The problem, he argues, is to understand why Darwin maintained the expression 'natural selection' despite criticisms of its naive anthropomorphism. For Limoges, academic considerations played a fundamental role. By maintaining this term, Darwin emphasised his priority, underlining the originality of his idea.[78] Finally, Limoges argues that Darwin behaved like 'a teacher defeated by his own pedagogy, . . . unable to understand that . . . the way in which he presented his ideas had an effect on his explanations of natural selection'.[79] Forgetting the real origin of the theory, Darwin ended up reconstructing it on the basis of the metaphor of a 'natural selector'.[80] This explains why Darwin characteristically talked of natural selection as an 'agent' or a 'power', which he even went so far as to personalise.[81]

Limoges's arguments are telling, in particular the first of them. It is indeed significant that Darwin and Wallace both discovered the concept of natural selection in the context of a study of biogeography. However, this point needs to be qualified by another, which is at least as important. While it is true that Darwin did not use the term 'selection' before 1843, the 1838–9 *Notebooks* reveal throughout an interest in 'heredity' and in the animal-breeding, horticultural, veterinary and medical literature that dealt with this problem at the time. The Darwinian concept of selection was thus not only developed in an ecological framework, it was *also* developed, and in this it was quite specific, in the context of a detailed study of heredity in domestic races.[82] As for the other two points put forward by Limoges, it seems unlikely that 'strategic' and 'pedagogic' considerations carry sufficient epistemological weight to explain the usefulness of the analogy of domestic races in the theory of selection. It seems more probable that domesticated species were *methodologically* necessary for the development of Darwin's theory. For this reason, I have focused on this element of Darwin's ideas with reference to Darwin's final theory, rather than on the genesis of the doctrines involved in that theory.

Darwin considered that Wallace's opposition to the domestic model was of fundamental importance. This is relatively difficult to detect, because both authors did their best to cover up this initial difference, and not to make it public. However, Darwin made three replies to Wallace and more generally to all those who doubted the validity of the domestic model. He first sought to respond to Wallace's use of facts, in particular the reality of the 'tendency to revert' allegedly shown by domestic varieties. More important, Darwin gave two reasons why he emphasised the analogy between natural modification and the modification of domestic races. The first was that domestic species demonstrate that a non-reversible 'hereditary modification' is possible. The second was that artificial selection also clearly demonstrates how small variations can accumulate in species and change their 'type'. Domestic races can thus be considered an empirical test of the heredity of variations and the power of selection, both of which were fundamental for Darwin's concept of natural selection. The following sections deal with these arguments, and more particularly with the diverse facets of Darwin's thoughts on domestic varieties and their evolution.

1.3.1 Do domestic varieties show a tendency to revert?

From the Introduction to the *Origin of Species* onwards, Darwin solemnly outlined his conviction that 'the attentive study of domestic animals and cultivated plants gives the best chance of resolving the obscure problem' of 'the means of modification and of coadaptation'. 'I dare to express my conviction in the great value of such studies, even though they have generally been neglected by naturalists.'[83] A few pages later, in the chapter 'Variation under Domestication', Darwin makes a transparent but indirect allusion to Wallace's article. The terms used are so close to those of Wallace that there can be no doubt as to their target: 'Having alluded to the subject of reversion, I may here refer to a statement often made by naturalists – namely, that our domestic varieties, when run wild, gradually but certainly revert in character to their aboriginal stock.'[84] The 'statement' that Darwin attacks is in fact virtually a word-for-word quotation from Wallace, who wrote in 1858: 'Domestic varieties, when turned wild, *must* return to something near the type of the original stock, *or become altogether extinct*.'[85]

Darwin's response betrays his embarrassment. First of all he casts doubt on the empirical importance of the 'statement': 'I have in vain endeavoured to discover on what decisive facts the above statement has so often and so boldly been made.'[86] But he leaves the question of fact there. On the other hand, he puts forward a strange argument in which he claims that, even if it were true that domestic species that had gone wild reverted to the original wild type[87] or became extinct[88] (this alternative is again a direct reference to Wallace), this would not affect the validity of the inference made from domestic varieties to species in general:

> If it could be shown that our domestic varieties manifested a strong tendency to reversion, – that is, to lose their acquired characters, while kept under changed conditions, and while kept in a considerable body, so that free intercrossing might check, by blending together, any slight deviations of such case, in such case, I grant that we could deduce nothing from domestic varieties in regard to species. But there is not a shadow of evidence in favour of this view: to assert that we could not breed our cart- and race-horses, long- and short-horned cattle, and poultry of various breeds, and esculent vegetables, for an almost infinite number of generations, would be opposed to all experience.[89]

The conclusion is clear. Darwin admits that, in practice, little was known about domestic species that had 'reverted to the natural state'. On the other hand, everyday experience showed that domestic races conserved the characters that interested the breeder or the cultivator. Domestic races were thus the best proof of a dual principle of the modifiability and the inertia of 'varieties'. Varieties change, but the change tends to be conserved, whatever the force that produced it. This is what Darwin meant by his repeated use of the term 'heredity', a term that breeders understood so well, but which naturalists grasped so poorly.[90]

This was the theoretical context in which Darwin replied to Wallace and to other naturalists who rejected the model of domestic species. The study of domestic races was crucial, because it directly refuted the notion of a 'type', replacing it by the twin concepts of variation and heredity, which in turn led to the idea that the modifiability of species is a fundamental and regular phenomenon. For Darwin, domestic races constituted a direct proof of the modifiability of species. This is the first point he makes in the *Origin of Species*: 'I shall devote the first chapter of this Abstract to Variation under Domestication. We shall thus see that a large amount of hereditary modification is at least possible.'[91]

In *The Variation of Animals and Plants under Domestication* (1868), Darwin felt it necessary to respond directly to the argument about the 'reversion to type' of domesticated species, using empirical data. The first volume of this work, entirely devoted to a species-by-species review of studies on the modification of domestic races, records a relatively large number of observations of domestic races that had reverted to a wild state. Darwin admits the reliability of some of these accounts, in particular about pigs, which he considered the most well-documented example. But he also provides many examples of animals and plants which either did not change or which were transformed in various ways according to the varying wild conditions in which they found themselves. He devotes particular attention to the rabbit, for which many examples had been known for many years. In many regions of the world, European colonisation was accompanied by the proliferation of escaped domestic rabbits in places where they had not previously existed. During the voyage of the *Beagle*, the young Darwin became interested in the fate of certain species of rabbit that had gone wild. In his account of his visit to the Falkland Islands, he had noted with amusement that 'French naturalists' like the illustrious Cuvier had previously identified as a 'new species' one of the varieties (black) that, in the Falkland Islands, coexisted and interbred with another variety (grey).[92] In other words, on the same island domestic varieties gone wild had given rise to one variety that had 'reverted' and to another that was sufficiently unusual for naturalists to consider it a 'new species'. Thirty years later, in his book *Variation*, Darwin felt confident enough of the facts to reject the argument that domestic varieties would necessarily revert to the original type. Reversion to 'the natural state' did not mean reversion to a 'type'. There is no intrinsic tendency to reversion.

In this veiled polemic against Wallace, the most intriguing element is the fact that Darwin not only never publicly criticised Wallace, but even went so far as to give him credit for having refuted the idea that domestic races would necessarily revert to type:

It has often been argued that no light is thrown on the changes which natural species are believed to undergo from the admitted changes of domestic races, as the latter are said to be mere temporary productions, always reverting, as soon as they become feral, to their pristine form. This argument has been well combated by Mr Wallace ('Journ. Proc. Linn. Soc.', 1858, vol. iii, p. 60); and full details were given in the thirteenth chapter, showing that the tendency to reversion in feral animals and plants has been greatly exaggerated, though no doubt it exists to a certain extent. It would be opposed to all the principles inculcated in this work, if domestic animals, when exposed to new conditions and compelled to struggle for their own wants against a host of foreign competitors, were not modified in the course of time.[93]

This passage from the *Variation*, together with another similar one,[94] is the only point at which Darwin makes an explicit reference to the article that Wallace sent him in 1858. Note that he gives a page reference; this page is precisely where Wallace states that 'domestic varieties, when they turn wild, *must* return to something near the type of the original stock, *or become altogether extinct*' (original emphasis).[95] Darwin thus makes Wallace say the exact opposite of what he did in fact say! The reasons for all this are not clear. It is probable that Darwin wanted to cover up the difference and to reinforce the cohesion of the natural-selectionist 'clan'.

1.3.2 Theoretical advantages of the domestic model

According to Darwin, there were two main advantages inherent in the model of domestic species. First, the proof that a 'hereditary modification' on a grand scale was 'at least possible'. Second, the demonstration that selection – any kind of selection – in the form of 'the accumulation of slight variations' could be successful.[96] Historians have previously concentrated on the second aspect, no doubt because it corresponds to one of the elements of Darwinism that has best resisted the rigours of time. On the other hand, Darwin's speculations on heredity are generally discreetly ignored; not only did they lead him to put forward a hypothesis that today is wholly discredited (i.e. 'pangenesis'), they are also often obscure, archaic and disjointed. But it is precisely these passages that explain Darwin's insistence on the validity of the domestic model.

At the end of the 19th century, it was widely considered that Darwin had made 'heredity' a central problem of biological theory. This was absolutely true. By reexamining and reinterpreting ideas that had generally been the domain of the techniques of domestication (breeding, horticulture, medicine and veterinary science), Darwin carried out a genuine revolution in the naturalist's conception of variation. His attempts to place heredity at the heart of naturalist theory involved reinterpreting categories that had traditionally been used to conceptualise biological diversity: 'type', 'species', 'variety', 'race'. Darwin's concept of heredity, although it had been formulated well before 1859, appears in the *Origin of Species* only in the form of brief but precise allusions. The *Variation*, however, contains a far more developed version of his ideas on heredity. This doctrine was coherent, and was fundamentally linked to the concept of selection. It can be schematically summarised in two ways: the relation of heredity and individual variation, and the interconnection of heredity and modification.

1.3.2.1 Heredity and individual variation. To the modern reader, Darwin's conception of heredity appears archaic, not only because of the kind of data on which it is based, but also because of the specific theory of heredity he eventually put forward ('pangenesis'). Despite these problems, Darwin's work is remarkable because of the way in which it isolated 'heredity' as a unique and fundamental problem of biology. From a catalogue of curiosities and whims of nature assembled by breeders, veterinary surgeons, horticulturists and physicians, Darwin was able to draw out the fundamental problem of heredity.

From these observations and empirical generalisations, Darwin retained first and foremost the idea of the 'force of inheritance':

> It is hardly possible, within a moderate compass, to impress on the mind of those who have not attended to the subject, the full conviction of the force of inheritance which is slowly acquired by rearing animals, by studying the many treatises which have been published on the various domestic animals, and by conversing with the breeders.[97]

This presentation of heredity as a 'force' or as a 'powerful tendency' returns over and over again in Darwin's writings. But what did this force consist of? Simply that 'like begets like'.[98] This might seem trivial; after all, the fact that 'generation' produces organisms that are similar to their parents was hardly a 19th-century discovery. However, Darwin underlines the fact that this saying came from breeders, not naturalists:

> If animals and plants had never been domesticated, and wild ones alone had been preserved, we should probably never have heard the saying, that 'like begets like'. The proposition would have been as self-evident as that all the buds on the same

tree are alike, though neither proposition is strictly true. For, as has often been remarked, probably no two individuals are identically the same. All wild animals recognise each other, which shows that there is some difference between them; and when the eye is well practised, the shepherd knows each sheep, and a man can distinguish a fellow-man out of millions on millions of other men.[99]

What exactly did Darwin take from the experience of breeders? The reference to 'like breeds like' appears provocative. Darwin was well aware that it was a favourite saying of fixist naturalists. Linnaeus quoted it at the beginning of his *Systemae Naturae* (1735), in the very section where he insisted that 'there are no new species' and that in every species unity kept order.[100] A few lines later, Darwin explains the consequences for breeders of 'like breeds like'. By focusing on heredity rather than on the traditional concept of 'type', Darwin argues that breeders in fact subverted the traditional meaning of the saying: 'The saying that "like begets like" has, in fact, arisen from the perfect confidence felt by breeders, that a superior or inferior animal will generally reproduce its kind; but this very superiority or inferiority shows that the individual in question has departed slightly from its type.'[101] The use of the word 'type' in this passage is particularly noteworthy. Darwin only rarely used this term, and always with some care. Certain passages of the *Origin of Species* explain on what basis one could use this term, which was so dear to morphologists. For Darwin, the term 'type' could only be used on condition that the overall attributes of organisms or groups of organisms were understood as the consequence of common descent. In other words, classification has an objective significance only in so far as it constitutes a reconstruction of the genealogy of living organisms. Ahistoric typologies and the logic of 'forms' have no place in natural classification.[102] Nothing could be more opposed to Darwin's thinking than the 'transcendental morphology' put forward at the beginning of the 19th century by Goethe, Geoffroy Saint-Hilaire or Serres.

The vocabulary used by Darwin in the passage quoted above is particularly revealing. In the way breeders understood it, the saying 'like breeds like' is fundamentally hereditarian. It is based on the conviction that a superior or inferior animal will reproduce 'its own kind'. Darwin did not say 'its own type', even though breeders and naturalists spontaneously used this kind of vocabulary. Instead, he deliberately used the term 'kind', which suggests the idea of a collection of objects having a distinct character, a common *difference*. And it is precisely the transmission of a difference that is implied by 'like breeds like'. The experience of breeders shows that an animal does not transmit its 'race', 'species' or 'type', but rather those characters that make it 'superior' or 'inferior', in other words, those by which it differs from the 'common type' of the race or species.

Darwin shows that the terms used by breeders concealed something that went against their spontaneous tendency to speak of 'types'. Animals may 'breed true to type' or 'breed true', but, as Darwin noted, breeders have always been particularly interested in obtaining the maximum number of animals that correspond to individuals considered 'superior'. Whatever such 'superiority' may consist of, whatever 'model' breeders might have in their heads, they will come up against the law of nature which acts such that the transmission of qualities is individual. That is what breeders called 'heredity'.

This was the most important lesson Darwin drew from the experience of breeders, and more generally of domestication, be it of animals, plants or humans. 'Heredity' grouped all the examples of 'resemblance' that could not be explained in terms of 'types'.[103] Heredity, like variation, is individual; an individual inherits a particular character, not a type.

Nevertheless, heredity does not consist simply of the resemblance between parents and offspring. As Darwin notes, the 'force' of heredity, although incredibly strong, is also 'capricious': 'When a new peculiarity first appears, we can never predict whether it will be inherited.'[104] Furthermore, two identical parents do not necessarily produce an offspring that is like them for this or that character: 'If both parents from their birth present the same particularity, the probability is strong that it will be transmitted to at least some of their offspring.'[105] These examples show that the phenomena of heredity cannot be grasped simply through the idea of similarity between individuals. Examples of heredity are in fact closely linked to individual variation. In his book *Variation* Darwin insists on two major categories that could reveal the nature of this link: the transmission of monstrosities (or 'extraordinary particularities' or 'sports') and reversion. Neither of these categories directly reveals the nature of heredity, but they both lead to an understanding of a fundamental element of the problem of heredity.

In his discussions of heredity, Darwin always begins by drawing attention to exceptional structural anomalies such as polydactyly or albinism. This was not because he considered them as being of decisive importance in the modification of species; far from it. For Darwin, of course, small variations, not 'sports'[106] (major variations), were the raw material of evolutionary innovation. But these examples provided a crucial example of the originality of heredity. From the 18th century onwards, it was well known that certain rare characters (such as polydactyly) appeared repeatedly, but irregularly, in some families, but were non-existent in practically all others. Reexamining this problem in the *Variation*, Darwin developed a quantitative and explicitly populational analysis, which in itself was relatively unusual:

Let the population consist of sixty millions, composed, we will assume, of ten millions families, each containing six members. On these data, Professor Stokes has calculated for me that the odds will be no less than 8333 millions to 1 that in the ten million families there will not be even a single family in which one parent and two children will be affected by the peculiarity in question. But numerous instances could be given, in which several children have been affected by the same rare peculiarity with one of their parents; and in this case, more especially if the grandchildren be included in the calculation, the odds against mere coincidence become something prodigious, almost beyond enumeration.[107]

This argument had already been put forward, albeit in a truncated form, in the *Origin*,[108] as well as in the 1856–8 manuscript, where a reference to Stokes can also be found.[109] Darwin's conclusion was that the distribution of major anomalies, well known to physicians and veterinary surgeons, was due neither to the direct action of external conditions (because the same anomaly could be found under different external conditions), nor to a random conjunction of heterogeneous conditions. These anomalies were due 'to unknown laws acting on the organisation or the constitution of the individual', and were based upon 'that which the members of the same family inherit from some common point in their constitution'.[110] But the fact that the character is not systematically transmitted to all the offspring of a given family means that the phenomenon cannot be considered as the simple reproduction of a type. Having established the nature of the phenomenon, on the basis of a list of classic examples taken from veterinary, horticultural and medical sources, Darwin merely had to generalise: if major anatomical deviations are 'hereditary', then so too must be the infinitely small 'slight variations'.[111] This conclusion did not necessarily follow, but Darwin had a series of other facts to support it: the successes of artificial selection showed that individual animals and plants 'tend' to transmit their 'superiority' or their 'inferiority'.

Darwin dealt at some length with a second category of unusual facts – examples of reversion.[112] The critical importance of reversion in Darwinian thought has already been mentioned; not only did Darwin consider that reversion was a 'great fact of heredity', he also felt that it probably gave the best insight into the nature of heredity. Reversion, or 'atavism', is the name given to the phenomenon whereby, for a given character, a child resembles not its parents, but one of its ancestors. Reversion thus describes the fact that the character 'jumps' one or more generations. In certain cases, the character that reappears has not been observed in any relative, and apparently refers to some ancient state of the race or species.

Using examples of reversion, Darwin made a series of striking observations about heredity. First, heredity cannot be defined simply by the resemblance of parents and offspring, because certain unusual cases resemble distant ancestors or collaterals. Second, Darwin did not interpret reversion as a force that would periodically revert individuals to the type of the race or species. Far from considering that reversion was a proof of a racial heredity that transcended individuals, Darwin used it to argue that individual heredity was a mosaic of characters, some of which are visible, others, extremely numerous, being invisible.[113] Darwin described these invisible characters as 'latent' or 'dormant'.[114] Thus each sex 'latently' possesses the characters of the other sex. Normally these characters do not develop, but they are nevertheless transmitted to the next generation. Any character can be 'transmitted' without 'developing', and may be revealed only at a later generation. Reversion thus constitutes a striking example that 'proves to us that the transmission of a character and its development, which ordinarily go together and thus escape discrimination, are distinct powers; and these powers in some cases are even antagonistic, for each acts alternately in successive generations'.[115] For Darwin 'the principle of reversion ... is one of the most wonderful of the attributes of Inheritance',[116] precisely because it reveals that heredity must not be confounded with development.

This was the conceptual framework built by Darwin on the basis of the experience of breeders, gardeners, doctors and veterinary surgeons. In these lists of nature's whims and curiosities Darwin found evidence for an intrinsic link between heredity and individual variation. He thus focused on *individual variability* rather than on *typological diversity*, and the former could be seen to be the source of the latter. Darwin did not have an operative model of the link between variation and heredity, but his work shows the outline of a heuristic conception that at the end of the 19th century was to have an immense influence on experimental biology. This consisted of two propositions:

- Heredity, although expressed as a measure of resemblance, is in fact the *probability* of resemblance between parents and offspring. An experimental study of heredity would thus have to begin by allowing for differences *from* parents and *between* offspring.
- By concentrating on the reality of reversion, which he interpreted as the effect of latent characters, Darwin advocated a particulate view of heredity.

These two points show that Darwin's view of heredity was not as archaic as might be suggested by his use of the term 'force'. In fact, they summarise the two main concepts used at the end of the 19th century to unravel the enigma of heredity. They were, no doubt, implicit, if not explicit, in the science of the time: Galton's idea of heredity as a probability of resemblance came thirty years after Quételet's first hesitant steps in statistical biology, while the idea of latent characters was put forward at the same time as Mendel published his essay on hybridisation (1865) and the first elements of cell theory were discovered. But it is to Darwin's credit that he insisted that heredity – a notion that was primarily used by breeders, horticulturists and physicians – was an essential and unavoidable conceptual problem for experimental biology and for natural history.

The relation between this study of heredity and the Darwinian theory of selection is plain. Darwin was able to develop the hypothesis of the gradual and cumulative modification of races and species by the selection of individual variations as early as 1844 because he had already broken with the idea that biological identity is defined by the *type* of the variety or species. This conceptual framework was reinforced and strengthened by his subsequent work, and enabled him finally to understand that natural selection and artificial selection are linked not only by a metaphor, but also by a common theory.

1.3.2.2 Heredity and modification. Are all these ideas really present in the development of Darwin's work, and if so, were they both unique and necessary? Darwin not only considered that heredity was a strictly individual power of transmission of characters, he also often wrote as if 'the great principle of heredity' applied equally to races, to species and to all taxonomic groups superior to individuals. After all, the hypothesis of selection (natural or artificial) could be summarised thus: characters that appear as individual variations are gradually 'fixed' in a reproductive community (a line) under the effect of forces (human methods of improvement, or the struggle for existence) that affect the chances of survival and reproduction. The characters that are thus fixed are no less 'hereditary' than variable characters. It is just as legitimate to speak of 'hereditary differences' between races, species, genera etc., as between individuals. Darwin was quite explicit on this point,[117] which agrees with his tendency to identify 'heredity' and 'descent' (see Introduction above). It is thus perhaps false to say that Darwin argued in favour of a strictly 'individualistic' concept of heredity, as opposed to 'racial' or 'specific' heredity.

Yet again, the answer to this problem can be found in Darwin's study of breeding. Darwin did not simply copy down what he read and heard from breeders; he subjected it to a critical analysis. In the 19th century, two concepts were particularly influential among breeders. One was called 'the doctrine of the constancy of the race'. It stated that the degree of 'certainty' of heredity of the characters of a given animal depended on the purity of its descent, so that heredity was considered as a 'force' or a 'power' of varying intensity. Heredity, it was argued, is as 'strong' as a line is 'pure'.[118] This theory was at its apogee in the years when Darwin was reading works on breeding, around 1840–50. Another doctrine, often found in 19th-century works on breeding, stated that the 'force of heredity' was proportional to the age of the character. In other words, the longer a character has been transmitted, the more it will be perpetuated. Heredity was thus considered in the same way that physicists conceived of momentum: the longer the force acts, the stronger it is, just as the kinetic energy of a falling body increases over time. These two doctrines, which were in fact very similar, tended to present heredity as a kind of inertia to be found in the types of races or species. Darwin explicitly criticised both these conceptions, rejecting the idea that there existed a specifically racial heredity that transcended individual heredity.

This is shown by Darwin's attitude to the doctrine of the constancy of pure races. In the first chapter of the *Origin*, he says that he had been 'struck' by the fact that all the breeders and the cultivators of plants he had met or read were 'firmly convinced that the several breeds to which each [had] attended, [were] descended from so many aboriginally distinct species'.[119] This belief, he pointed out, often became a doctrine: 'The doctrine of the origin of our several domestic breeds from several aboriginal stocks, has been carried to an absurd extreme by some authors.'[120]

The belief may have been 'absurd', but Darwin considered it to be extremely important. Breeders and horticulturists were convinced that their races were stable, despite the fact that these races were largely human creations. Even a cursory study of the history and geography of domestic races reveals that the 'constancy' of races, far from being an effect of the 'purity' of a given type, in fact reveals the intrinsic 'force' of 'heredity'. For Darwin the 'force' of heredity is not a function of the purity and constancy of a racial type; it is compatible with the modifiability of races. The fact that breeders and horticulturists spontaneously believed in the stability of domestic races was thus a powerful argument against those naturalists who thought that domestic varieties tended to 'revert' to the species type.

Darwin's remark casts an interesting light on the first half (chaps. 1 to 12) of the *Variation*, in which he reviews a wide range of data from a long list of domesticated species, both animal (dogs, cats, horses, donkeys...) and vegetable (wheat, corn, cabbages, potatoes...). For the most part, these examples support the theoretical speculations in the second part of the book, which deals with the causes of variation, the nature of heredity, hybridisation and selection. What is particularly remarkable is the way in which they are presented. Darwin paints a typically 'naturalist' portrait of the history of each species or group of domesticated species. In each case, he lingers over the genealogy of the races, their geographical origin and dispersion, giving the first volume of the *Variation* the form of a biogeographical study of domesticated plants and animals. He regularly reminds the reader of his objective: to show that the forms considered by breeders and horticulturists to be 'distinct species' are in fact modified geographical races.

The *Variation* thus contains an approach that is strictly complementary to that of the *Origin*, where the hypothesis that 'varieties' are 'incipient species' is based on a range of indirect arguments, all of which are aimed at making up for the impossibility of directly studying genealogies. In the *Variation*, however, such a study was often possible, because the relations between the different domesticated organisms were either known or could be easily and reliably reconstructed. Darwin is thus able to show that the 'distinct species' of the breeders can most often be interpreted as 'geographical races'. If it is possible, *at least in the case of domesticated species*, to arrive at a 'genealogy' which corresponds to real lines of individual 'parents' and 'offspring' (and not merely to a succession of hypothetical 'forms'), then this constitutes a powerful empirical argument showing that the general theory of the modification of species is not only a theory of 'descent with modification' but also, as Darwin sometimes put it, a theory of 'heredity with modification'.[121]

This is why Darwin used the widespread belief in the 'constancy' and naturalness of domestic races. The belief was fundamentally mistaken, but, in its own way, it showed the relationship between heredity and modification.

This is further shown by Darwin's sharp criticism of the other widely held view of heredity: that the 'force' of heredity was proportional to the antiquity of the character. In the *Variation*, this argument is put forward in a paragraph with the revealing title 'The fixity of characters', where Darwin examines 'a general belief amongst breeders that the longer any character has been transmitted by a breed, the more fully it will continue to be transmitted'.[122]

This was only one aspect of the 'doctrine of constancy', which also implied that the efficacy of heredity was due to the purity of descent or the age of the line. Darwin attacked this belief, using the breeders' own examples. The everyday practices of breeding and horticulture did not at all confirm that 'inheritance gains strength simply through long continuance'.[123] Darwin accepted that if a domestic race was improved by eliminating individuals that deviated from the desired form, then the race would tend, over time, to become fixed. But this artificial selection, even if it presupposes the heredity of the characters that are to be fixed, does not imply that heredity becomes stronger over time. In order to measure the hereditary 'force' of a given character, what counts is not its age, but rather what happens when a new character develops, either spontaneously or through a cross. On this point, the experience of breeders and horticulturists showed that a new character can become rapidly fixed, or show major variations, or not be transmitted at all. Thus '[there does not appear] to be any relation between the force with which a character is transmitted and the length of time during which it has been transmitted'.[124] In fact, the whole history of domestication shows that the idea of heredity as a power that becomes stronger with time is utterly false. Domestication shows that 'scarcely any degree of antiquity ensures a character being transmitted perfectly true'.[125] This is the remarkable lesson that Darwin drew from the experience of the breeders: heredity is not a force of fixity, but a force of fixation. To use a metaphor from physics, which is not be found in Darwin's writings, heredity can be considered as being analogous to inertia, but not at all to momentum. Even though he used archaic terms like 'force', Darwin's conception of heredity was not energetic ('heredity as force'), but 'material' and 'particulate', as shown by his theory of pangenesis.[126]

The advantages of the example of domestication were not only metaphorical. For Darwin, domesticated animals and plants constituted the most direct proof available of the modifiability of species. He had a reply to those like Wallace, who considered that domestic varieties were unstable, and that they would inevitably revert to the primitive wild state if the constraints of domestication were abolished. Domestication, Darwin argued, can be considered as a gigantic experiment that proves the possibility of the non-reversible modification of species. Perhaps most domestic varieties would become extinct in the wild.[127] But if this were the case, it would merely constitute a supplementary proof of the ability of animal and plant races to *fix* characters.

Darwin analysed this capacity of permanent fixation in terms of 'heredity', which he imposed as a way of grappling with the modification of species. By making heredity a primary subject for scientific research, Darwin sought to undermine typological representations of the future of species. It was not sufficient to point out that there was no sharp boundary between 'varieties' and 'species'; ever since Lamarck many naturalists had admitted as much. It was also necessary to identify the level of organisation at which characters were transmitted. Agricultural practice showed Darwin that the old adage 'like begets like' was only really true with regard to the individual and differential transmission of characters. Neither species nor races inherit characters; only individuals do. It is at this level that modification needs to be considered. Darwin thus conceived of the selective process – 'artificial', 'natural' or 'sexual' – precisely in terms of 'heredity'. For Darwin, selection in general was nothing other than the gradual accumulation of 'individual' variations. He repeated this point in the *Origin of Species*: the 'principle of heredity' is a prerequisite for selection. For Darwin, natural selection is not the selection of varieties (Wallace), nor, to be accurate, the selection of individuals (Spencer), but rather a bias in favour of (small) individual hereditary variations.

Darwin was not the only person to have thought of the idea of the 'survival of the fittest', yet he was the only person to link this hypothesis to a consideration of the nature of the natural entities that survived. He found the answer in the treatises of breeders, gardeners, veterinary surgeons and physicians: the entities involved were differential, individually transmitted, always 'latent' to some degree, but not really 'fixed'. Of course, Darwin's conception of heredity was unclear, confused and influenced by archaic ideas. But it was precisely these unclear points that were to be so crucial for the validation of the theory of natural selection in the six or seven decades following the publication of the *Origin.*

We have seen the fundamental interrelationship of the concepts of variation, heredity and modification that was presupposed in the Darwinian hypothesis of selection. It remains to be demonstrated that the analogy of artificial selection, and thus the very term 'selection' itself, was indeed of such fundamental importance for Darwin's theory of the transformation of species.

1.3.2.3 Natural selection and artificial selection. In order to complete this discussion of the role of domestication in Darwin's thought, the relationship between 'artificial' and 'natural' selection needs to be clarified. As soon as the *Origin of Species* was published, Darwin's use of 'selection' was criticised as being naively anthropomorphic. Wallace, for example, used this argument in 1866 in an attempt to persuade Darwin to abandon the term. He claimed that the term 'selection' inevitably implied an element of choice, and thus of thought and direction. The constant comparison between natural selection and selection carried out by man, added Wallace, inevitably leads to the natural process being considered a personified agent who 'prefers', 'seeks the good of the species', etc. The term 'natural selection' was thus an 'indirect and incorrect' 'metaphorical expression' that it would be prudent to abandon in favour of another, more neutral, term, such as Spencer's 'survival of the fittest'.[128]

Darwin had replied to such criticisms in advance, as early as the third edition of the *Origin of Species* (1861):

In the literal sense of the word, no doubt, natural selection is a misnomer; but who ever objected to chemists speaking of the elective affinities of the various elements? – and yet an acid cannot strictly be said to elect the base with which it will in preference combine. It has been said that I speak of natural selection as an active power or deity; but who objects to an author speaking of the attraction of gravity as ruling the movements of the planets? Every one knows what is meant and is implied by such metaphorical expressions; and they are almost necessary for brevity.[129]

This passage requires no comment. All scientific terms have their origin in some metaphor or other, traces of which often remain. From this point of view, all scientific theories contain an element of anthropomorphism. The real problem is not so much whether or not a metaphor is used, but rather whether the metaphor is justified, in other words, whether the analogy has a real basis. A comparison is epistemologically justified if one can show that two objects have a genuine identity or common nature. If this is not the case, the comparison will be mere rhetoric.

Darwin devoted some serious thought to the relation between artificial and natural selection, though not in his occasionally polemical responses to the charge of anthropomorphism. His examination of this problem is to be found in his key writings, in two different forms. On the one hand, Darwin developed the idea of a general theory of selection. On the other, in the *Variation*, he examined the problem in terms of the objective relation between the two kinds of selection, when they find themselves side by side, or indeed interacting, within domesticated species. Darwin therefore considered that the common use of the term 'selection' in 'artificial selection' and 'natural selection' was conceptually justified, that is, it was based on a legitimate metaphor.

1.3.2.4 The basis of the metaphor: a general theory of selection. There is a perfectly clear reason why Darwin never abandoned the term selection. On the basis of his practice as a naturalist and as a theoretician of domesticated species, he was convinced that a general theory of selection could be established. This means that *from a certain point of view*, 'artificial selection', 'natural selection' and even 'sexual selection' all describe a single process. The pages of the *Variation* that deal with selection open with these words:

The power of Selection, whether exercised by man, or brought into play under nature through the struggle for existence and the consequent survival of the fittest, absolutely depends on the variability of organic beings. Without variability nothing can be effected; slight individual differences, however, suffice for the work, and are probably the chief or sole means in the production of new species.[130]

Darwin here defines a necessary condition for selection: there must be variability, and that variability must be individual and hereditary ('Any variation which is not inherited is unimportant for us').[131] Selection, be it natural or human, does not create variation, but does require it.

The general process of selection consists of the 'accumulation' of certain variations provided by nature, through the conservation and preferential (or even exclusive) reproduction of individuals possessing these variations. This description of selection obviously applies as much to artificial selection as it does to natural selection. It is directly linked to the manner in which Darwin – unlike Wallace – conceived of his central hypothesis of the modification of species. In nature as in domestic breeding, selection does not merely consist of the sorting out of varieties or races, but of their modification through the accumulation of infinitesimally small differences: 'If selection consisted merely in separating some very distinct variety, and breeding from it, the principle would be so obvious as hardly to be worth notice; but its importance consists in the great effect produced by the accumulation in one direction, during successive generations, of differences absolutely inappreciable by an educated eye, differences which I for one have vainly attempted to appreciate.'[132]

The unity of the concept of selection in general can be summed up as follows: selection is a power that can modify (or conserve) species by the gradual accumulation (or the elimination) of hereditary variations that it does not produce. This unity appears even more clearly in certain more radical passages where Darwin presents selection in general as the main power controlling the modification of species. In the *Variation* there are two chapters that deal with selection. The first is entitled 'Selection by Man', the second, simply 'Selection'. This second chapter closes with the following words:

Throughout this chapter and elsewhere I have spoken of selection as the paramount power, yet its action absolutely depends on what we in our ignorance call spontaneous or accidental variability. Let an architect be compelled to build an edifice with uncut stones, fallen from a precipice. The shape of each fragment may be called accidental; yet the shape of each has been determined by the force of gravity, the nature of the rock, and the slope of the precipice, – events and circumstances, all of which depend on natural laws; but there is no relation between these laws and the purpose for which each fragment is used by the builder. In the same manner the variations of each creature are determined by fixed and immutable laws; but these bear no relation to the living structure which is slowly built up through the power of selection, whether this be natural or artificial selection. If our architect succeeded in rearing a noble edifice, using the rough wedge-shaped fragments for the arches, the longer stones for the lintels, and so forth, we should admire his skill in a higher degree than if he had used stones shaped for the purpose. So it is with selection, whether applied by man or by nature; for although variability is indispensably necessary, yet, when we look at some highly complex and excellently adapted organism, variability sinks to a quite subordinate position in importance in comparison with selection, in the same manner as the shape of each fragment used by our supposed architect is unimportant in comparison with his skill.[133]

These lines are instructive in several respects. They contain the obvious and explicit use of a metaphor, and what is more, of a technical metaphor. This metaphor in fact confirms everything that is implied by the vocabulary used in the *Origin of Species* to describe natural selection, in particular the presentation of selection as an 'agent' or a 'power'. Far from backing down, Darwin repeated and emphasised his point. This might be considered surprising, given that the *Variation* was published after Darwin had been attacked for anthropomorphism and had decided to take account of these criticisms in the *Origin of Species* (third edition, beginning of chap. 4: see note 128 above) and in the Introduction to the *Variation*, thus minimising their impact. Upon closer inspection, the metaphor Darwin uses in this text has two notable aspects. First, selection is compared to 'construction'; it is from selection that the 'structure' emerges. For Darwin, this was a particularly forceful way of distancing himself from Spencer's metaphor of 'sifting', with its implied imagery of a sieve (see Chapter 2). If a metaphor is required to explain selection, it should be that of the architect making use of what is to hand, not that of a 'sifting'. Second, this metaphor is a metaphor that applies to *two* kinds of selection, natural *and* artificial. Its aim is to bring out something that is common to both types of selection, and thus to discredit the idea that in 'natural selection', 'selection' was nothing but an image. In other words, if a metaphor was necessary, better to choose one that made the concept of selection clearer. The metaphor of construction shows the difference between variation and the modification of species. Modification presupposes the existence of individual, hereditary variations that it does not create. In this sense, selection is not creative. 'Selection' (human or natural) describes the complex of forces that determine the fate of variation, in other words its elimination or its diffusion in a population (or a 'race' in Darwinian terms). In this second sense, selection *is* creative, just as an artisan creates a form even though he or she does not create the raw material. Thus Darwin did not hesitate to say that even though variability is a necessary precondition for selection, it plays only a 'subordinate' role in the modification of species. This conceptual scheme is independent of the fact that artificial selection is intentional, whereas natural selection is not. It is also independent of the utility of the variations that are retained by selection. The forms and capacities constructed by artificial selection are not fundamentally chosen for their utility for the animal concerned.

This also applies to sexual selection: even though this form of selection is 'natural', it often conflicts with the adaptation of the species, which, in the last analysis, is controlled by 'natural selection' in the strict sense of the term.

Once this framework has been established, it is an open question as to whether the power of selection is greater than that of variation. Suppose, for example, that variation were to be directed and abundant. In this case, selection would not necessarily operate. This was a fundamental problem for Darwin (see Chapter 3) which continued (and continues) to haunt evolutionary theory, taking a variety of forms (for example the theories of 'orthogenesis' at the end of the 19th century, and more recently Kimura's theory of neutral molecular evolution).

It seems clear, however, that if Darwin maintained the term 'selection' despite all the pressure and criticism, it was first and foremost because he had in mind a general theory of selection. In other words, if he used a metaphor, it was because that metaphor was conceptually justified.

1.3.2.5 The objective relation between natural and artificial selection. Some aspects of Darwin's study of natural and artificial selection do not deal with questions of comparison or metaphor, but rather with the relation between the two forms of selection. As has already been suggested, Darwin was less interested in domesticity as a possible metaphor for nature than as an object that could be integrated into a scientific theory. This had important philosophical implications, which need to be understood in terms of evolutionary theory. On the one hand, Darwin thought there was *no absolute distinction* between artificial and natural selection. On the other, he suggested that there might be an *interaction* between the two forces. These two positions raise substantially different problems.

The argument that there is no sharp distinction between natural and artificial selection is by far the clearer of the two theses, but it was probably not the more profound. Rather, it was part of a rhetorical strategy that was aimed at making natural selection intuitively more plausible to the reader. Darwin distinguished two methods of selection carried out by human beings: 'methodical' selection and 'unconscious' selection. The former corresponds to what is generally called 'selection' in breeding and horticulture, where 'a man ... systematically endeavours to modify a breed according to some predetermined standard'.[134] This is what happens, for example, in breeding ornamental species, and what from the 18th century up to the present day has been called the 'improvement' of animals and plants. 'Unconscious selection', by contrast, involves 'preserving the most valued and destroying the less valued individuals, without any thought of altering the breed'.[135] Unconscious selection is practised spontaneously by breeders who simply try to maintain the quality of the breed, by removing either malformed or weak individuals or simply those that do not conform to the breeder's idea of a vigorous and lucrative animal.

Using the kind of gradualist schema to be found throughout his work, Darwin easily showed that there is a wide variety of possible intermediate forms between the two types of artificial selection. All domestication inevitably involves selection. 'Unconscious' selection eventually produces domestic races which have been substantially modified by the intervention of breeders and horticulturists. This is shown by historical descriptions of animals: domesticated species have changed enormously over the last few thousand years. On the basis of this slow and gradual unconscious selection, Darwin put forward a suggestive argument that supported the concept of natural selection: without meaning to, human beings have modified a large number of species.

In the *Variation*, Darwin carried this argument to its logical conclusion: not only does unconscious selection imitate natural selection, but the boundary between the two is often difficult to detect. This is the case, for example, when peasants favour races that are highly adapted to local conditions:

> In Great Britain, in former times, almost every district had its own breed of cattle and sheep; 'they were indigenous to the soil, climate, and pasturage of the locality on which they grazed: they seemed to have been formed for it and by it' (Youatt on Sheep, p. 312). But in this case we are quite unable to disentangle the effects of the direct action of the conditions of life, – of use or habit – of natural selection – and of that kind of selection which we have seen is occasionally and unconsciously followed by man even during the rudest periods of history.[136]

Strictly speaking, there is no clear difference between artificial and natural races, and it has to be admitted that the most primitive of peoples, and indeed certain animals, practise a form of 'unconscious' selection. A good example is given by a subject that Darwin studied passionately: slave-making ants.[137] In order to understand how 'the wonderful instinct of making slaves' could have evolved, Darwin compared two slave-making species, *Formica sanguinea* and *Formica rufescens*, which show different degrees of dependence on their slaves. *F. sanguinea* participates with the slaves in nest-building and finds its own food, while only the slaves look after the larvae. *F. rufescens* shows a much greater dependence; not only are these slave-making ants incapable of feeding their offspring, they are also incapable of building their own nest and even of feeding themselves. Darwin deduced that the evolution of the 'slave-making instinct' in ants must have been gradual, and he put forward a theory to explain the origin of the habit. It is often the case, he argued, that non-slave-making ants will steal pupae from other species in order to feed themselves. On this basis, he drew the following conclusion:

It is possible that pupae originally stored as food might become developed; and the ants thus intentionally reared would then follow their proper instincts, and do what work they could. If their presence proved useful to the species which had seized them – if it were more advantageous to this species to capture workers than to procreate them – the habit of collecting pupae originally for food might by natural selection be strengthened and rendered permanent for the very different purpose of raising slaves ... I can see no difficulty in natural selection increasing and modifying the instinct – always supposing each modification to be of use to the species – until an ant was formed as abjectly dependent on its slaves as is the *Formica rufescens*.[138]

It could be argued that this passage contains a thoroughly 'naturalist' model of natural selection. Darwin is not far from saying that artificial selection is an effect of co-evolution which can be explained in the last analysis by natural selection. There exist a large number of similar examples which show that Darwin tended to consider artificial selection as a special case of natural selection.

Much more interesting, however, is Darwin's outline of the interaction between artificial and natural selection. Although his remarks on this question generally go no further than mere allusions, they pose a problem which, in the light of the later development of the theory of selection, was to become a major question in agronomy. The idea is particularly simple. Darwin envisaged the possibility of 'Natural Selection, or Survival of the Fittest, as affecting domestic productions'. In the paragraph that bears this title,[139] Darwin warns that 'we know little on this head'. But he immediately goes on to state that 'natural selection must act on our domestic races'.[140] Even in countries that have been civilised for many centuries, even where man tries as much as possible to control the conditions of rearing and cultivation, natural selection must act, sometimes favouring, sometimes opposing the action of artificial selection. Of course, the interaction between the two processes appears most clearly when they act in opposite directions:

When man attempts to make a breed with some serious defect in structure, or in the mutual relation of the several parts, he will partly or completely fail, or encounter much difficulty; he is in fact resisted by a form of natural selection. We have seen that an attempt was once made in Yorkshire to breed cattle with enormous buttocks [this would today be called a Charollais], but the cows perished so often in bringing forth their calves, that the attempt has been given up.[141]

Natural selection thus poses limits on the action of artificial selection; not everything is possible. This makes sense, but left on its own it leads to obvious problems. In arguing that natural selection limits, channels and controls artificial selection, one is rapidly drawn to the conclusion that artificial selection can do nothing that is not tolerated by natural selection. From this point of view, artificial selection would only be able to produce superficial changes, minor variations on a theme that was fundamentally set by nature. To accept this point is to accept Wallace's principled objection to the domestic analogy. Wallace's argument was essentially this: domestic races are unstable and tend to revert to type, but because the species 'type' is continually changing because of the effects of universal competition, the 'tendency to revert' simply means that the evolutionary line is determined by natural selection (or at least by what Darwin termed natural selection), and that artificial selection only produces noise, interference, minor and unimportant variations.

Darwin avoided this trap. Of course, he admitted, 'natural selection often determines man's power of selection'.[142] That is, 'natural selection often checks man's comparatively feeble and capricious attempts at improvement'.[143] Man does as much as he can, but there is not much he *can* do when faced with natural selection. But to get to the heart of the matter, this argument has to be recast in terms of 'utility'. Natural selection works for the good of each organism, while artificial selection favours characters that are useful for man. That does not imply that natural utility and usefulness for man are necessarily incompatible. There is nothing to stop natural selection from favouring the action of artificial selection. It is well known, for example, that in nature many animals and plants are subject to a strong selection pressure in favour of early reproduction. Breeders and horticulturists have obtained some of their most stunning results precisely by selecting for early reproduction. Pigeons, rabbits, chickens and annual or biennial plants such as asparagus, artichokes, Jerusalem artichokes or potatoes[144] are all excellent examples of the way in which artificial selection follows the path of natural selection. And yet, in these examples, the two forms of selection are clearly distinct. In reality, explains Darwin, the debate about the 'utility' of characters is biased by an obsolete doctrine. Naturalists believed that the visible characters of animals – colour, ornaments, external form – did not have a vital significance. This leads to the conclusion (which Darwin admitted to having toyed with) that natural selection determines only the conformation of the essential – internal – organs and does not affect the external or 'special' characters.[145] This doctrine is false on two accounts. First, it is difficult to be certain that a particular character is of no selective value and thus escapes the action of natural selection. More often than not, such judgements are based on our ignorance of the physiology and ecology of the organism, and the character only appears to be 'neutral'.

On the other hand, in the *Variation* Darwin cites a large number of examples which show that artificial selection can lead to substantial changes in the internal organs. For example, when producing a race such as the Polish Cock, breeders not only selected a range of brightly coloured feathers, but also a totally deformed head, and produced a race of chickens that had entirely lost the instinct of incubating its eggs.[146] Thus 'we ought to be extremely cautious in judging what characters are of importance in a state of nature to animals and plants, which have to struggle for existence from the hour of their birth to that of their death, – their existence depending on conditions, about which we are profoundly ignorant'.[147]

Darwin did not go much further in his consideration of the interaction between artificial and natural selection. However, his preliminary studies did lead to an important series of questions. At the end of the 19th century, it became vital to measure the importance of natural selection in animal husbandry. Those who supported Darwin's position, that is, the integration of the concept of natural selection into the theory of artificial selection, denounced the positions of the breeders, which they described as 'formalism'.[148] 'Formalism' stressed the significance of the animal's most visible characters, in particular fur or feather colour and external morphology, in other words all those generally qualitative characters that are listed in the rules of agricultural breeding competitions. In general, formalism went hand in hand with an obsession over pedigrees and myths of 'racial purity'. Against 'formalism', those breeders who agreed with Darwin put forward the notion of 'utility value', that is, the totality of characters that contribute to the vigour, reproductive power and, ultimately, the economic value of an animal. The obsession with pure types, apart from often producing severe physiological abnormalities, is extremely costly. In the 20th century, the idea of 'utility value' or 'intrinsic value'[149] has been given theoretical expression in 'overall fitness', a parameter that summarises the viability and the fertility of the animal. From a strictly theoretical point of view, this concept has proven somewhat problematic. In practice, however, the use of the concept by breeders has been accompanied by an increasing attention to the many factors that affect the animal's physiology (immunology, digestive physiology, stress resistance, etc.), the sum of which determine its 'breeding value'. In this light, the truly secondary characters are the ornamental ones. The key factors have become those which naturalists consider essential: the overall physiology and ecology that tend to increase an animal's chances of survival and reproduction. It goes without saying that in following this method, the breeder pays little attention to 'type', and considers domestic organisms in a methodological and conceptual framework that is identical to that of the Darwinian biologist. 'Artificial' selection appears as a sort of imitation of 'natural' selection with the aim of increasingly 'adapting' domestic animals to their domestic environment.

We have seen that in 1858 Darwin and Wallace did not in fact put forward equivalent hypotheses to explain the modification of species. Wallace interpreted the transformation of species as the consequence of competition between varieties ('races'). The most advantaged varieties would grow in number, to the detriment of others, and would finally eliminate them. Darwin considered that this view contained an important element of his theory of the modification of species, but that his central hypothesis was different. The hypothesis of 'natural selection' explains the modification of 'varieties' themselves, by the accumulation of inherited individual differences. However, apart from this difference over the central *hypothesis*, Darwin and Wallace agreed on the general structure of the *theory* of natural selection, which they both thought would lead to a total reorganisation of natural history.

In 1859, Wallace raised a series of objections to any theory of the modification of species based on observations of domesticated species. Imagining that he was merely attacking a minor argument of 'fixist' theorists, Wallace unwittingly opposed Darwin's conception in terms of both its method and its conceptual content. Darwin considered this criticism sufficiently important to warrant a series of responses in the *Origin* and in the *Variation*. He initially used a series of facts to oppose the idea that domesticated races, released in the wild, would revert to their original wild type. But in particular he showed how contemporary data on the domestication of species constituted crucial tests of the hypothesis he sought to defend.

The use of the domestic analogy was not a pedagogic device. It was methodologically essential; without it, the subtle interrelationship between variation, heredity and modification, so characteristic of the Darwinian hypothesis of selection, would have been nothing more than empty speculation without any empirical content. In fact, it is extremely unlikely that the hypothesis could have been developed without this supporting evidence. Once the conceptual framework was in place, experimental biology had to clarify the exact nature of the individual 'heredity' that was required by Darwin's hypothesis.

2 The ontology of selection

In the preceding chapter it was argued that Darwin's concept of selection was thoroughly individualistic, that is, that Darwin considered natural selection to be a process that acted within groups rather than between groups. This theoretical choice led to a subtle but real distinction between Darwin's concept and that of Wallace, as shown by Darwin's interest in 'heredity'. However, it remains to be demonstrated that this 'individualistic' approach is in fact present in Darwin's writings or his ideas. Indeed, it is extremely tempting for the modern reader to project contemporary disputes about the 'units' and 'levels' of selection onto the founding documents of Darwinian evolution. It is in this modern context that the term 'Darwinian selection' has become a synonym for 'individual (or intra-group) selection', as against various hypotheses of 'group selection'. However, such antinomies can only be taken as an historical starting point if they are at least partially justified by a close examination of the texts. This chapter shows how the question of the levels of selection was fundamental to Darwin's ideas and outlines his explicit and thorough study of the significance and implications of the 'individualistic' conception of selection.

2.1 Two aspects of the problem

To what extent, and in what way, does the concept of natural selection refer to the *individual*? The concept of selection, and even more so that of natural selection, is richer than may at first appear: although some might argue that it is self-evident and has no need of explanation and proof, this is only because of a series of verbal oversimplifications. Spencer's reinterpretation of natural selection as 'survival of the fittest' is undoubtedly the best historical example – indeed, Spencer considered that it was an *a priori* truth that did not require any empirical proof. In 1864, in the same book in which he made the proposal to rebaptise natural selection 'survival of the fittest', he wrote:

The survival of the fittest, which I have sought to express in mechanical terms, is that which Mr Darwin has called 'natural selection, or the preservation of favoured races in the struggle for life'. That there is going on a process of this kind throughout the organic world, Mr Darwin's great work on the *Origin of Species* has shown to the satisfaction of nearly all naturalists. Indeed, when enunciated, the truth of his hypothesis is so obvious as scarcely to need proof. Though evidence may be required to show that natural selection accounts for everything ascribed to it, yet no evidence is required to show that natural selection has always been going on, and must ever continue to go on. Recognising this as an *a priori* certainty, let us contemplate it under its two distinct aspects.[1]

Shortly after this passage, Spencer points out that the Darwinian principle of natural selection should be recognised as 'a certain and *a priori* truth', and argues that the only serious question is that of which facts this principle can explain. The criticism of 'tautology' is thus scarcely new. For Spencer, however, this was not a weakness but a strength. He considered that his term – 'survival of the fittest' – removed all the anthropomorphic connotations raised by Darwin's choice of 'selection'. Wallace, as has already been seen, had used this argument in order to try and persuade Darwin to abandon the term 'natural selection'. He only half succeeded: from 1869 on Darwin occasionally used Spencer's formulation together with his own, but merely as a verbal concession.[2]

Why did Spencer think that the term 'survival of the fittest' was so 'obvious'? The principle of the 'survival of the fittest' is based on a remarkable philosophical pirouette which altogether abstracts itself from any reference to biology whatsoever. Spencer's formulation does not specify whether the 'fittest' are individual organisms, races, species or anything else. It does not indicate in any way, shape or form to what exactly it might apply. This is because Spencer considered that Darwin's concept was merely one illustration among many of a more general principle – which he claimed to have originated – that of 'segregation', or 'sorting'. This applied *a priori* to any entity that might be preserved or eliminated owing to the action of a certain 'field of forces'.[3] This applied, for example, to atoms, to electrical particles, to bodies subject to gravity, to stars, to nervous impressions, to members of social communities, to peoples and even to the works of civilisation. It is easy to understand why, for Spencer, the Darwinian concept of 'natural selection' was merely one manifestation of this great general *a priori* truth, and that it was unnecessary to consider whether the principle applied to individual organisms, to races, to species or to other units: it necessarily and *a priori* applied to all of them. Or at least this is how Spencer's formulation was understood by the followers of Darwin who were able to unite around it.

Darwin himself, however, did not consider that natural selection was a truism. As he frequently pointed out, 'natural selection' is a handy term that enables the rapid expression of a complex hypothesis, based on a series of empirical generalisations (the tendency of all species to show a geometric increase in population size, the absolute limits to subsistence, the existence of hereditary variation for the vast majority of biological characters, the affirmation that these variations affect the probability of survival and reproduction of organisms, the analogy with domestic selection). What are the ontological implications of this hypothesis?

A hasty reading of Darwin's writings can lead to the impression that the ontology of selection is unclear. Consider, for example, the following formulations:

- 'This preservation of favourable *variations* and the rejection of injurious *variations*, I call Natural Selection' (added emphasis).[4]
- '. . . natural selection acts by life and death, – by the preservation of *individuals* with any favourable variation, and by the destruction of those with any unfavourable deviation of structure' (added emphasis).[5]
- 'This preservation, during the battle for life, of *varieties* which possess any advantage in structure, constitution, or instinct, I have called Natural Selection' (added emphasis).[6]
- 'The theory of natural selection is grounded on the belief that each new variety, and ultimately each new *species*, is produced and maintained by having some advantage over those with which it comes into competition' (added emphasis).[7]

These quotations seem to suggest that Darwin had at least four answers to the question of *what* is selected: 'variations' (the context clearly indicates that these are 'individual differences'), individuals, varieties (which Darwin does not seem to distinguish from 'races')[8] and species. It should be noted that this imprecision, be it real or apparent, is also present in the very title of the 1859 magnum opus: *On the Origin of Species by Means of Natural Selection: or, The Preservation of Favoured Races in the Struggle for Life*. This title can be understood at least two ways. The first suggests that natural selection involves 'the preservation of favoured races in the struggle for life', and thus consists of the selection of races. The second (which was clearly Darwin's) suggests that species, which cannot be distinguished by any absolute criterion from 'races' (or 'varieties'), are the result of a process of modification by 'natural selection'.

Another way of grasping the potential ontological ambiguity of natural selection is not to ask, 'What is selected?', but rather, 'What does natural selection act in the interest of?' This way of posing the question underlines the problem of 'utility'. In this respect, the numerous passages that Darwin begins with 'natural selection acts for the good of. . .' only raise more problems. These texts can be classified into four groups:

1. Darwin often prudently says that natural selection can only act 'for the good of each or for the good of each being'. Sometimes, however, a

more unusual formulation appears. For example, chapter 4 of the *Origin* begins as follows: 'Natural selection can act only through and for the good of each being.'[9] The context reveals that behind this ambiguous formula Darwin really means '*for* the good of the species' and '*through* the good of individuals'. In other words, something takes place that leads to the improvement of the species (or race), but the means by which this happens is individual advantage. From a causal point of view, the important question for Darwin is the second one: at what level must the advantage exist for a selective process to begin? In other words, the key question is not '*for* the good of whom?' but '*through* the good of whom?' This subtlety has frequently been lost on commentators, who have tended to focus on the apparently finalist argument.[10] Furthermore, in terms of the units that express the advantageous or disadvantageous variations that determine the 'chances of survival and reproduction', Darwin envisaged several different possibilities. These will be discussed below.

2. In virtually all of Darwin's writings, 'advantage' is discussed at the level of the individual organism. In other words, selection acts causally at the level of the individual. The following passage is typical of Darwin's most frequently expressed attitude: 'It is the steady accumulation, through natural selection, of such differences, *when beneficial to the individual*, that gives rise to all the more important modifications of structure, by which the innumerable beings on the face of this earth are enabled to struggle with each other, and the best adapted to survive' (added emphasis).[11]

3. Nevertheless, Darwin did occasionally consider that characters may be selected because they are advantageous to 'the community'. In particular, this is the case for instincts and for sociality. In the *Origin of Species*, Darwin evokes this possibility with regard to social insects, affirming, for example, that certain instincts that lead to the death of the individual (for example, the instinct that drives the bee to sting and thus to die) could have been constructed by selection because they were 'useful to the community'.[12] In *The Descent of Man* a similar hypothesis is used to explain the development of certain moral virtues, such as courage, obedience and faithfulness.[13]

4. Finally, Darwin sometimes considers – but always rejects – the hypothesis that natural selection could operate on characters that are advantageous to the species as such. He mused at some length as to whether interspecific hybrid sterility could have been constructed by natural selection because it is 'advantageous to the species'.[14] Obviously, sterility cannot be considered to be advantageous to the individual.

Thus the question of the entities involved in natural selection is complex. Darwin's thinking initially appears eclectic, if not ambiguous, whether from the point of view of the units that are selected or from the causal perspective of the entities that, having a given differential advantage, are the target of natural selection.[15]

Things are in fact very different. The apparent ambiguity exists because Darwin's different positions have been taken out of context. The whole point of Darwin's hypothesis is to be found in its narrow and absolutely explicit limits, which both give it an empirical meaning and forbid us from considering it a bland truism that can be applied to everything.

2.2 The units of selection

In its canonical form, the Darwinian definition of natural selection requires only three units: 'variations', 'individuals' and 'varieties' (or 'races'). Although there is a semantic minefield of ambiguity lurking beneath the terms 'variations' and 'variety',[16] to Darwin's credit he imposed a clear three-level schema which can be reformulated as follows: heritable character, individual, population. These three levels of description of the process will be examined, retaining Darwin's original terms.

Strictly speaking, only variations are selected: they constitute the 'raw material' of natural selection, which 'accumulates' them within a line of related individuals when they are advantageous.[17] Unlike naturalists of the time – and especially botanists – Darwin did not use the term 'variation' as an expression of a temporary change due to external physical circumstances (for example the reduced height of plants that grow at high altitude).[18] On the contrary, he focused on hereditary variation: 'any variation which is not inherited is unimportant for us'.[19] It would, however, be overstating the case to suggest that Darwin had a particulate representation of heredity. But the hypothesis of natural selection clearly points in the direction of an atomisation of transmissible characters. From the standpoint of heredity, Darwin apparently considered individuals as mosaics of partially independent characters. This interpretation is confirmed *a posteriori* by the speculation about 'pangenesis' to be found in the *Variation*.[20]

It follows that, strictly speaking, natural selection does not produce its effects on individuals but on lines of related individuals which accumulate the variations transmitted from individual to individual. Darwin's definition of natural selection, to be found in chapter 4 of the *Origin of Species,* is particularly important:

Can we doubt (remembering that many more individuals are born than can possibly survive) that *individuals* having any advantage, however slight, over others, would have the best chance of surviving and of procreating their kind? On the other hand, we may feel sure that any variation in the least degree injurious would be rigidly destroyed. This *preservation of favourable variations and the rejection of injurious variations*, I call Natural Selection. [Added emphasis][21]

There are a large number of similar passages in which, starting from competition between individuals, Darwin ends up with the idea of the selection of variations.

To be absolutely precise, *individual variants* are not selected. Heritable variation intervenes in the concept of natural selection in two ways: first as something that is propagated in lines of individuals, second as a property that influences the probability of the survival and reproduction of individuals: 'Natural selection can act only by the preservation and accumulation of infinitesimally small inherited modifications, each profitable to the preserved being.'[22] From a philosophical point of view, this schema contains two references to individuality – first, to the hereditary 'atom' that is selected, and which, over time, accumulates in the race. This 'atom' is the slight variation that occasionally appears in organisms, or, in Darwin's words, an 'individual difference'. ('Under the term of "variations", it must never be forgotten that mere individual differences are always included.')[23] For natural selection, therefore, the organism is a mosaic. But it is also clear that natural selection also involves an unavoidable reference to the individuated totality of 'organisms': heritable particularities are selected in the form of factors affecting their chances of survival and reproduction.

Natural selection is thus 'individualistic' in two ways. First, selection operates on an *n*-dimensional field of characters and in this way atomises individual organisms; second, the success of a given 'atom' of variation is expressed above all through the longevity and reproductive success of these individual organisms. As Darwin put it: 'natural selection acts by life and death, – by the preservation of individuals with any favourable variation, and by the destruction of those with any unfavourable deviation of structure' (added emphasis).[24]

The status of 'varieties' in this schema is not clear. In strictly semantic terms, the notion of variety is typological: it classically refers to a group of individuals which share a certain property that distinguishes them from other members of the same species. Darwin is often uncomfortable with this timeless notion of variety (variety as a logical class). He thus subverts it by explaining the origin of varieties. At the beginning of the chapter of the *Origin of Species* devoted to 'Variation under Nature' (chapter 2), Darwin subtly insinuates that if variations are heritable, then the term 'variety' should be applied to any line of individuals that presents a 'variation'.[25] But a large number of variations can be observed within the individuals of a given 'race'. Thus 'races' or 'varieties' are not only capable of growing in size, but may also gradually 'form' and change.[26] The 'variety' or 'race' is not what is selected, but is rather the product of natural selection.

If this analysis is correct, the canonical concept of Darwinian natural selection can be expressed in two 'strong' formulations:

- Natural selection is not the selection of individuals but of 'variations', that is, of hereditarily transmissible peculiarities of structure or behaviour.
- The unique criterion for the retention of variations is that of the advantage conferred upon individuals in their struggle for survival and reproduction. In other words, natural selection deals only with individual advantage: it acts only through the 'preservation' (i.e. the survival and reproduction) of individuals, in proportion to the advantageous or disadvantageous variations they possess.

To be sure, these two formulations reduce the flexibility and ambiguity of Darwin's original position. In fact, the differentiation of 'variation' (as in character) and 'variety' (as in population) is not consummated in Darwin's writings, owing to the lack of both a theory of heredity and a statistical approach to populations (a word which is extremely rare in Darwin's writings). However, this objection is, in fact, superficial. It in no way invalidates the assertion that Darwinian selection is a selection *of* variations, *in* a race, *due to* the advantage conferred on individuals. Furthermore, we should not overlook the fact that although such a schema might appear biologically confused because of the absence of a solid theory of heredity, it is clearly applicable to the modification of domestic races. In 1859, Darwin could not present a single direct example of natural selection, but he had thoroughly imbibed a mass of agricultural, zoological and horticultural literature which led him to realise that hereditary variations tend to reappear periodically, that they can affect all organs, and that the work of selection always applies simultaneously to several variations in order to improve the race. He thus had good reason to imagine that if there was a 'natural means of selection' (to use the term that preceded 'natural selection') this means should also act gradually and cumulatively on 'traits' or 'characters'. In this context, variation could be considered the 'raw material' upon which natural selection would operate. The (numerous) archaisms that can be detected in Darwin's vocabulary of variation and heredity are therefore not sufficiently damning to undermine the structural coherence of the hypothesis that he constructed.

Darwin's contemporaries were fully aware of these contradictions. In 1893, Spencer produced a remarkable document which clearly shows that the question of the units of selection was at the heart of Darwinism from its very origin. In this article, published ten years after Darwin's death, the ageing Spencer tried to express as clearly as possible the difference between the Darwinian principle of natural selection and his own principle of 'the survival of the fittest'. The main interest of this document lies in the fact that Spencer uses the Darwinian term 'selection' and raises the problem of *what* exactly is selected. It is hardly surprising that, once again, the relation of natural and artificial selection is at the heart of the matter. Here is an extract from this astonishing and little-known paper:

> Artificial selection can pick out a particular trait, and, regardless of other traits of the individuals displaying it, can increase it by selective breeding in successive generations. For, to the breeder or fancier, it matters little whether such individuals are otherwise well constituted. They may be in this or that way so unfit for carrying on the struggle for life, that, were they without human care, they would disappear forthwith. On the other hand, if we regard Nature as that which it is, an assemblage of various forces, inorganic and organic, some favourable to the maintenance of life and many at variance with its maintenance – forces which operate blindly – we see that there is no selection of this or that trait, but that there is a selection only of individuals which are, by the aggregates of their traits, best fitted for living. And here I may note an advantage possessed by the expression 'survival of the fittest'; since this does not tend to raise the thought of any one character which, more than others, is to be maintained or increased; but tends rather to raise the thought of general adaptation for all purposes ... Survival of the fittest can increase any serviceable trait only if that trait conduces to prosperity of the individual, or of posterity, or of both, *in an important degree* ... That which survival of the fittest does ... is to keep all faculties up to the mark, by destroying such as have faculties in some respect below the mark; and it can produce development of some one faculty only if that faculty is predominantly important. It seems to me that many naturalists have practically lost sight of this, and assume that natural selection will increase *any* advantageous trait. Certainly a view now widely accepted assumes as much. [Original emphasis][27]

Thus Spencer considers natural selection (or rather the survival of the fittest) merely to be a negative regulation that eliminates the unfit and maintains the norm. If the norm changes, this must take place clearly and massively; in other words, selection will only positively favour those characters that already appear as adaptations. The Darwinian schema is very different: it is based on the idea of a gradual and opportunistic construction of adaptations on the basis of virtually undetectable advantages. From this point of view, adaptation is not the starting point but rather the end-product of a temporal process. Similarly, selection does not involve the sorting of individuals, but rather a greater or lesser bias in the probability of survival and reproduction of individuals, in proportion to the more or less advantageous heritable traits that are transmitted from parents to offspring.

2.3 Who benefits from natural selection?

It has not yet been shown to what extent, and in what way, Darwin thought that individual organisms – and *only* individual organisms – profit from natural selection. This is an extremely delicate question, despite (or perhaps because of) the apparent simplicity of the language of utility. The virulence of recent controversies over 'group selection' shows that this is a fundamental theoretical question, as open today as it was in 1859. From beginning to end, Darwin adopted an extremely firm tone on this question, arguing that individual organisms, although they are not strictly speaking the *things* that are selected, are the only conceivable targets for the causal action of natural selection.[28] Nevertheless, no one was more aware than Darwin of the terminological, conceptual and empirical difficulties raised by his intransigent position.

2.3.1 What does 'utility' mean?

> 'Man selects only for his own good: Nature only for that of the being which she tends.'[29]

The most important difference between natural and artificial selection lies in the principle of utility. What exactly did Darwin mean by 'Nature [selects] only for the good of the being which she tends'? Is this some kind of teleological exaggeration? Does natural selection, like human selection, see into the future?

And when Darwin wrote that 'natural selection can act only *through* and *for* the good of each *being*',[30] was he referring to the individual that presents a given advantageous adaptation? This would often be a justifiable interpretation. For example, in the conclusion to the *Origin of Species*, the operation of natural selection is characterised as 'the accumulation of innumerable variations, each good for the individual possessor'.[31] However, there is a logic to the use of a term as general as 'being'. *A priori* there is no reason not to think that natural selection can develop adaptations that benefit the community (as, for example, in the case of social insects)[32] or, more generally, the species as such. Darwin only rarely employed such formulations, but certain statements show what he was prepared to accept on the subject, and what he considered to be out of bounds: 'What natural selection cannot do, is to modify the structure of one species, without giving it any advantage, for the good of another species; and though statements to this effect may be found in works of natural history, I cannot find one case which bears investigation.'[33]

Darwin is unclear about what natural selection acts for the good of. In reality, this ambiguity covers up two difficulties. The first is related to the historicity of the process of natural selection. It makes no sense to speak of natural selection acting 'for the good' of the individuals upon which it operates: if a given individual (not an individual in general) lives longer and reproduces more, this is not because of natural selection. Quite the opposite; it is because there is variation in the ability to survive and reproduce that there is a process of natural selection. Another way of highlighting the absurdity of the formulation is to consider the mass of individuals that do not survive for long, or do not reproduce, or both. Natural selection, which operates just as much on them as on other organisms, clearly does not act 'for their good'. That is why Darwin uses the unusual construction of natural selection only acting 'through and for' the good of each. Natural selection acts 'through' individual advantage. This is a causal assertion: a given peculiarity has a greater or lesser probability of being present in subsequent generations because it affects viability and reproductive success. In an extreme case, it will spread throughout the race, and ultimately the whole species. 'Advantage' has thus been transformed into 'adaptation', and it is not incorrect to speak of natural selection having worked 'for' the good of the race or species.

The significance of this conceptual construction needs to be thoroughly understood. According to the Darwinian view of modification, the 'good' of the individual is simply measured by the volume of his or her offspring. This involves a subtle subversion of the idea of utility: the *utility* of a character is expressed in its *selective* value, that is in the effective transmission of the character. This conception of utility evokes an economic model: individual advantage can be considered an investment, and its representation in offspring (through the effect of cumulative selection) a profit.

Because of this, a parallel has often been made between Darwin's ideas and one of the fundamental dogmas of liberalism: each person working for his or her own good contributes to the enrichment of all.[34] This interpretation is sufficiently well known not to require further explanation. However, the doctrine that sheds most light on the concept of natural selection is in fact that of compound interest. This constitutes a strictly analogous process: in the struggle for life, and in the process of selection which results, everything takes place as though, in varying, organisms make investments that can produce positive or negative benefits. All these investments and benefits are expressed in a common currency, which in each generation is the contribution to the creation of the following generation. From this point of view, Darwinian utilitarianism looks suspiciously like 'an econometric treatment of demography'.[35]

The concept of utility in the theory of natural selection raises another problem. Adaptations are often presented as being either specifically advantageous for individuals, or as being comprehensible only in the light of the 'community' (e.g. social instincts) or even the 'species' (e.g. interspecific hybrid sterility). This poses a major problem for the theory of natural selection: is it conceivable that a variation could spread in the offspring of an individual because it is advantageous to the community (or the species) without necessarily being advantageous to the individual? To take the argument to its extreme: is it possible that natural selection could lead to the consolidation of traits that are *advantageous* for the community but *disadvantageous* for the individual?

This question is essential for the Darwinian theory of the modification of species. Right from the first edition of the *Origin of Species*, Darwin declared that this kind of question was 'by far the most serious special difficulty which [his] theory has encountered'.[36] He used three different examples to explain his views: neuter (or sterile) insects in certain orders of social insects, hybrid sterility and social instincts in man. It is in these passages where Darwin discusses potential alternatives to individual selection that he in fact shows himself to be most intransigent on the question.

2.3.2 *Natural selection and 'neuter' insects*

In Darwin's 1856–8 manuscript, the question of the origin of neuter or sterile insects takes up twenty-one pages, within a chapter devoted to the 'difficulties of the theory of natural selection'.[37] At the end of a long explanation, Darwin concludes: 'I have discussed this case of neuter social insects at great length, for it is by far the gravest difficulty, which I have encountered.'[38] In the *Origin of Species*, the same discussion, considerably abridged, is inserted not in the chapter on the difficulties of the theory, but in the chapter on instinct.[39] Although the same arguments are present, the new location raises doubts as to Darwin's theoretical intentions. Rather than being devoted to social insects in general, this section discusses the origin of neuter insects, that is of sterile individuals that are regularly produced in large numbers in each generation, and are often morphologically very different from the fertile individuals that have produced them (e.g. 'worker' bees).

Darwin's analysis has two parts: first he wonders how natural selection could have rendered sterile a significant proportion of the female ants present in each generation. He then considers how neuter (or sterile) insects may be different in other respects from fertile individuals. Curiously, the first question did not preoccupy Darwin much. So little, in fact, that in the *Origin* it was reduced to a brief allusion:

How the workers have been rendered sterile is a difficulty; but not much greater than that of any other striking modification of structure; for it can be shown that some insects and other articulate animals in a state of nature occasionally become sterile; and if such insects had been social, and it had been *profitable to the community* that a number should have been annually born capable of work, but incapable of procreation, I can see no very great difficulty in this being effected by natural selection. [Added emphasis][40]

As it originally appeared in the 1856–8 manuscript, the reasoning is as follows. In a large number of species, it is extremely frequent for sterile females to be born as a result of changes in external conditions. For example, in the common wasp, females become larger and more fertile as the season progresses.[41] The food received by larvae can equally affect subsequent adult fertility.[42] The important point here is that Darwin did not consider that individuals would be sterile because of their hereditary constitution, but because of a 'treatment' which could quantitatively and qualitatively alter fertility in a community of individuals. In other words, Darwin did not argue that there has been a selection of hereditary variations that render their possessors sterile – to accept such a possibility would be purely and simply to negate the very concept of natural selection. However, he was careful not to mention this possibility and merely dealt with the possible advantage to the community of a large number of females being sterile. In such a community, most of the individuals will not waste time or energy reproducing, and will work for the prosperity of the colony. In this way 'natural selection or the struggle for life, would ensure the continuance or the increase of the same treatment, so that the degree of sterility or the number of sterile individuals might be increased'.[43] Natural selection will tend to increase this sterility, both qualitatively and quantitatively, for as long as it is to the advantage of the community.

The second aspect of the discussion of the origin of neuter insects gives a very different impression. Worker ants and worker bees are not only sterile, they also present a large number of morphological and behavioural characters that distinguish them from their fertile parents. Could these characters have been selected? At first sight, this appears impossible. How could natural selection have accumulated small variations that are specific to sterile individuals, if these individuals are precisely unable to transmit those variations? How can we explain the fact that worker ants are wingless? The traditional Lamarckian explanation invokes disuse. But, Darwin objects, this is absurd, because 'it is just the wingless individuals which can never leave offspring!'[44] This therefore appears to be an example that cannot be explained either by a Lamarckian process[45] or by natural selection.

At this point, Darwin introduces an explanation which, under the cover of broadening the individualistic concept of natural selection, in fact restates it without any concession whatsoever: 'This difficulty, though appearing insuperable, is lessened, or, as I believe, disappears, when it is remembered that selection may be applied to the family, as well as to the individual, and may thus gain the desired end.'[46]

But what is meant by such a family selection? As is often the case, Darwin resorts to a comparison with artificial selection. In cattle-rearing, farmers select for a quality of meat that can only be fully appreciated in animals that have been castrated. Such animals, of course, do not reproduce. But the breeder has confidence in the 'great principle of heredity' and isolates the parents of the bullocks that correspond to his desired type. The farmer thus selects variations advantageous to man by selecting related animals, not the individual expressing the desired character, which is killed and prevented from reproducing.

The same reasoning can be applied to colonies of social insects, with the caveat that selection operates not to the advantage of man but to that of the 'animal community'. If natural selection has gradually produced the differentiation of social insects into castes, it has been through its action on the family, and for the benefit of the community:

> As with the varieties of the stock [in animal breeding], so with social insects, selection has been applied to the family, and not to the individual, for the sake of gaining a serviceable end. Hence we may conclude that slight modifications of structure or of instinct, correlated with the sterile condition of certain members of the community, have proved advantageous . . .[47]

In this passage the interest of the 'community' is not considered as independent of the reproductive strategies of individuals. In the case of caste insects, the 'good of the community' coincides precisely with a unit that has a clear genealogical sense: the 'family'. The advantage can thus be definitively and unambiguously expressed in the language of individual reproductive success: the prosperity of 'colonies' is nothing other than that of the individual fertile males and females that produce the differentiated colony. This is exactly what Darwin expresses at the end of the sentence which was in part cited above: '. . . consequently the fertile males and females have flourished, and transmitted to their fertile offspring a tendency to produce sterile members with the same modifications'.[48] The origin of social insects can thus be interpreted as being a 'long-continued selection of the fertile parents which produced most neuters with the profitable modification'.[49]

This is an orthodox restatement of the individualistic concept of natural selection: natural selection acts only by the accumulation of variations that are advantageous to the individual, but this advantage is ultimately expressed by the success of the individual in leaving offspring. This requires a blurring of the concept of an individual organism, because the colony appears to function as a kind of gigantic and complex corporal appendage of the reproducing individuals. However – and this is extremely significant – Darwin does not resort to the classic interpretation of insect colonies as 'super-organisms' and thus some kind of superior order of 'individuals', precisely because this would not help in the slightest in understanding the principle of natural selection. Darwin is here principally interested in the fact that natural selection, whatever the form that 'advantage' takes, can nevertheless be expressed in the language of 'heredity':

> I have discussed this case of neuter social insects at great length, for it is by far the gravest difficulty, which I have encountered; so grave, that to anyone less fully convinced than I am of the strength of the principle of inheritance, & of the slowly accumulating action of natural selection, I do not doubt that the difficulty will appear insuperable.[50]

2.3.3 *Natural selection and hybrid sterility*

Darwin used the problem of interspecific hybrid sterility to restate forcefully his individualistic interpretation of natural selection. A study of the successive versions of the theory of natural selection reveals that, although the final rigorous outline of the theory is present in the 1856–8 manuscript, it is passed over relatively rapidly, and only becomes fully explicit in the fourth edition of the *Origin of Species* (1866).

In its final form, Darwin's thesis is particularly audacious, especially in the light of modern evolutionary theory: 'First crosses between forms sufficiently distinct to be ranked as species, and their hybrids, are very generally, but not universally sterile . . . The sterility of first crosses and of their hybrid progeny has not, as far as we can judge, been acquired through natural selection.'[51] The first of these two affirmations is present in the first edition, and, for a naturalist, goes without saying. The second is an addition to the fifth edition (1869), although the idea was already present in previous versions. For example, it can be found in a minor detour in the 1856–8 manuscript,[52] and in the 1859 edition it is outlined in the opening lines of chapter 9:

> The importance of the fact that hybrids are very generally sterile, has, I think, been very much underrated by some late writers. In the theory of natural selection the case is especially important, inasmuch as the sterility of hybrids could not possibly be of any advantage to them, and therefore could not have been acquired by the continuous preservation of successive profitable degrees of fertility.[53]

Curiously, however, in this first edition, Darwin does not develop the affirmation that natural selection is not at the *origin* of hybrid sterility, but rather concentrates on the *causes* of hybrid sterility, that is on its diverse contemporary expressions (anatomical, physical and behavioural barriers, etc.).

Darwin's prudence is understandable. The whole argumentation of the *Origin of Species* is criss-crossed by the repeated affirmation that there is no qualitative difference between 'species' and 'varieties'. Species are fixed varieties; varieties are 'incipient species'. Once this point has been established, the causal argument of natural selection can function: as a hypothesis, natural selection explains the formation and modification of varieties (or races) by the accumulation of individual variations; as a theory, it explains the formation and modification of species, and, ultimately, of the infinite diversity of forms thrown up in the history of life. In other words, the *theory* of natural selection is an amplification of the *hypothesis* of natural selection, founded precisely on the assertion that there is no fundamental difference between a 'variety' and a 'species'.

It is at this point that hybridisation constitutes a potentially embarrassing problem. In general, varieties are fertile when crossed, and give rise to fertile offspring. Interspecific crosses, however, are generally sterile, and when they are not, it is the offspring that are sterile. Traditionally, hybrid sterility defined the boundaries of the species, and was thus interpreted as a 'special quality' of the species as such. This led to an objection that threatened Darwin's theory: either Darwin rejected the idea that hybrid sterility was a 'special quality' of species, in which case he had to deal with the reality of species intersterility, or he accepted that hybrid sterility distinguished species from varieties, in which case he had to explain whether (and how) natural selection could be at the origin of this sterility.

In most of the passages in which he deals with hybridisation, Darwin prudently limits himself to replying to the first problem: sterility of first crosses and of offspring is not peculiar to species, which are only quantitatively different from varieties. In this way the question of the 'origin' of hybrid sterility is not raised and there is therefore no reason to speculate on its current 'causes'. That is why, in the first edition of the *Origin of Species*, the problem of the origin of hybrid sterility is largely avoided, an omission which could appear surprising in a book on 'the origin of species'.

From the fourth edition (1866) onwards, the relation of hybrid sterility to natural selection becomes central. Significantly, the paragraph initially entitled 'Causes of the Sterility of First Crosses and of Hybrids' becomes in the fourth edition 'Origin and Causes of the Sterility of First Crosses and of Hybrids'. In this completely revised paragraph, Darwin explicitly discusses the question whether natural selection can, strictly speaking, be considered the *origin of species,* with species taken to be groups of organisms isolated by sterility barriers. The study of this possibility leads him to affirm without the slightest hesitation that the hypothesis is incompatible with the concept of natural selection.

This remarkable argument deserves to be examined in detail. Better than any other, it reveals how the concept of natural selection was conceived in terms of individual advantage:

> At one time it appeared to me probable, as it has to others, that this sterility might have been acquired through natural selection slowly acting on a slightly lessened degree of fertility, which at first spontaneously appeared, like any other variation, in certain individuals of one variety when crossed with another variety. For *it would clearly be advantageous to two varieties or incipient species*, if they could be kept from blending, on the same principle that, when a man is selecting at the same time two varieties, it is necessary that he should keep them separate ... It may be admitted, on the principle above explained, that *it would profit an incipient species* if it were rendered in some slight degree sterile when crossed with its parent-form or with some other variety; for thus fewer bastardised and deteriorated offspring would be produced to commingle their blood with the newly-forming variety. [Added emphasis][54]

In other words, hybrid sterility, although disadvantageous in immediate demographic terms, could in the long run favour the species by preventing a general loss of vigour and fertility. This implies that natural selection operates on species as such, considered as units that prosper relative to others. But Darwin categorically rejects this hypothesis:

> But he who will take the trouble to reflect on the steps by which the first degree of sterility could be increased through natural selection to that high degree which is common with so many species ... will find the subject extraordinarily complex. After mature reflection it seems to me that this could not have been effected through natural selection; for it could not have been of any direct advantage to an individual animal to breed poorly with another individual of a different variety, and thus to leave few offspring; consequently such individuals could not have been preserved or selected.[55]

To summarise: in a discussion where it would appear that natural selection operating at the level of the species would be an inevitable conclusion, Darwin explicitly rejects this possibility. The species would certainly gain an advantage from hybrid sterility, but this advantage has no causal significance, and is not sufficient to define the process as being one of natural selection. If there were natural selection in favour of interspecific hybrid sterility, it would imply that individuals had an advantage in being sterile, in other words in leaving less-numerous offspring, and, ultimately, in leaving none at all. It is impossible to imagine what advantage there might be to *individuals* in being less fecund than their relatives.

But perhaps sterility has been selected by a process similar to that which produced neuter castes in social insects? In other words, is the species not a 'community' which, in order to prosper, requires that certain individuals be sterile? The argument does not hold. It could only work for those species where the members live in society. But hybrid sterility affects all species: '... an individual animal not belonging to a social community, if rendered slightly sterile when crossed with some other variety, would not thus itself gain any advantage or indirectly give any advantage to the other individuals of the same variety, thus leading to their preservation'.[56] The problem of hybrid sterility thus provided Darwin with the opportunity to clarify and strengthen his individualistic interpretation of natural selection. It enabled him to emphasise an aspect that was often poorly understood by contemporary readers: from the point of view of natural selection, the ability to survive was not only defined by the vigour and longevity of the organism, but also by its fecundity, in other words by its possibility of being 'represented' in future generations. This debate also shows that Darwin avoided a hasty use of utilitarian arguments. Intersterility, no doubt, is advantageous to the species, but this does not necessarily imply a process of natural selection.

Darwin's firm formulations in the fourth edition of the *Origin of Species* did not fail to provoke a critical reaction from Wallace. We have already seen that the hypothesis put forward by Wallace in 1858 involved a conception of selection operating on groups (see section 1.1). It is thus scarcely surprising that a polite but vigorous polemic broke out between the two men over the problem of interspecific hybrid sterility. This controversy has been well analysed by Michael Ruse.[57] In response to Darwin's affirmation that natural selection could not explain hybrid sterility, Wallace replied that it was perfectly conceivable, on condition that one understood that hybrid sterility, despite not being advantageous to those individuals that cross, could nevertheless favour the species as such, if its exploitation of ecological space was taken into account:

I do not see your objection to *sterility* between allied species having been produced by Natural Selection. It appears to me that, given a differentiation of a species into two forms each of which was adapted to a special sphere of existence, every slight degree of sterility would be a positive advantage, not to the *individuals* who were sterile, but to each form. If you work it out, and suppose the two incipient species, A, B, to be divided into two groups, one of which contains those which are fertile when the two are crossed, the other being slightly sterile, you will find that the latter will certainly supplant the former in the struggle for existence, remembering that you have shown that in such a cross the offspring would be *more vigorous* than the pure breed, and would therefore certainly supplant them, and as these would not be so well adapted to any special sphere of existence as the pure species A and B, they would certainly in their turn give way to A and B. [Original emphasis][58]

Although the details of Wallace's speculation were far from clear, his theoretical intention can be summarised as follows: hybrids, being intermediate between the two parental forms, will not be adapted to the ecological niche of either parent. It is thus advantageous to two recently differentiated forms for their ecological identity to be protected by a restriction on hybrid fecundity. Darwin repeatedly replied that, if crosses took place (which was the initial hypothesis), then it was impossible to conceive of how crossed individuals could gain an advantage from producing sterile offspring.

Darwin's discussions of sterility will not be dealt with in any more detail. Suffice it to say that, dealing head-on with the question of *the origin of species* as such, Darwin did not hesitate to argue that, in their most distinctive character (intersterility), species are not the product of natural selection. Natural selection forms and modifies varieties which, accidentally, become intersterile, and thus deserve to be called 'species'. But it does not construct the intersterility barriers, because this would imply that a species is formed because of the reproductive success of the least fertile individuals. In Darwin's naturalistic ontology, there is, strictly speaking, no 'good of the species'. Although he may have used the expression (relatively frequently, in fact), it was usually in a transitive sense: that which is good for individuals (that which serves their propagation) leads also to the prosperity of the race, and of the species.

2.3.4 Natural selection and the moral faculties of man

There is at least one context in which Darwin openly envisaged that natural selection might, indeed must, act not at the individual level but at that of the group: the evolution of man. Thanks to the manuscripts which have survived, it is known that, very early on, Darwin thought there was no reason not to apply natural selection to the origin of humans, and in particular to our intellectual and moral qualities.[59] Nevertheless, the *Origin of Species* prudently leaves the question open, and it is only in *The Descent of Man* (1871) that it is dealt with. It is highly significant that it is in this particular context, and in no other, that Darwin clearly conceded that natural selection could act to the advantage of the community and against that of the individual. However, Darwin's caveats and restrictions are such that it can legitimately be asked whether the hypothesis of 'tribal' selection advanced in *The Descent of Man* does not ultimately reinforce the individualistic concept of natural selection.

The only example for which Darwin enlarged the principle of natural selection to the preservation of what is advantageous to the community as such, was the development of 'social and moral qualities' in man. Among these qualities, there are some, such as sympathy, faithfulness, obedience and courage, for which it is difficult to conceive how they could have developed gradually as a result of the individual advantage they might confer. Such qualities clearly operate for 'the general good', which Darwin proposes to interpret not – to paraphrase the philosophers – as 'the general good of the species' but, in agreement with his usage in the *Origin of Species*, as 'the welfare of the community': 'The term, general good, may be defined as the means by which the greatest possible number of individuals can be reared in full vigour and health, with all their faculties perfect, under the conditions to which they are exposed.'[60] Against the utilitarian philosophers, in particular John Stuart Mill, whom he explicitly attacked, Darwin refused to consider that morality and sociability rested upon 'the base principle of selfishness'.[61] The anti-utilitarianism of Darwinian ethics can be discerned in two theses:

1. Social and moral actions should not be exclusively interpreted in terms of pleasure or pain. To think otherwise is to accept the view that moral faculties have no hereditary basis, and are acquired during life.[62] To this Darwin replied that although the force of habit, sanctioned (and not motivated) by pleasure and pain, has considerable importance for human behaviour, the existence of social instincts should not be ignored. He remained evasive about which moral qualities are genuinely instinctive and which an indirect consequence of other instincts and of experience,[63] but he clearly rejected the idea that, from a moral point of view, each individual can be considered at birth a *tabula rasa*.

2. Unlike the utilitarian philosophers, who considered the 'general good' as a more or less universal and abstract extension of individual interest, Darwin limited the concept to the 'community', and tried to explain the origin of moral faculties in the language of natural selection.

The Descent of Man is a vast fresco describing the conditions that presided over the development of the 'moral sense' in man. The fundamental thesis, clearly expressed at the beginning of the third chapter,[64] and repeated in the fifth chapter[65] and in the book's overall conclusion,[66] is that the acquisition of the moral sense was fundamentally based on two preconditions: (a) the pre-existing possession of effective 'social instincts', and (b) the complete development of man's intellectual faculties (in particular of language) and of the ability to represent the past and calculate the future.[67] To these two fundamental conditions, Darwin added a third – habit – which in the individual reinforces the products of the joint action of the two other conditions.

This simple and even obvious thesis does not reduce the 'moral sense' to social instincts, but requires that it be constructed on the basis of them. It does not derive moral principles from the faculties of representation, abstraction and communication, but neither can it do without them. Finally, it does not reduce moral behaviour to the reinforcement of 'habits', but also invokes the role of instincts. On close examination, it becomes apparent that Darwin was somewhat embarrassed about deciding whether a given moral behaviour should be interpreted according to one or another of these different emphases. These categories have a tendency to merge into one another, and it is sometimes extremely difficult to distinguish the causes in a given case: 'In many circumstances it is impossible to decide whether certain social instincts have been acquired through natural selection, or are the indirect result of other instincts and faculties; or again, whether they are simply the result of a long-continued habit.'[68] Darwin's view is highly flexible and tends to agree with all possible interpretations: instinct, reason and thought all play a role and interact.

But how does natural selection intervene? In particular, did Darwin call upon group selection to explain the origin of any of the components of the 'moral sense'? Take the case of the intellectual faculties: on this point, Darwin is clear – they are formed and improved under the pressure of a classic purely individual selection:

> These faculties are variable; and we have many reasons to believe that the variations will be inherited. Therefore, if they were formerly of high importance to primeval man and to his ape-like progenitors, they would have been perfected or advanced through natural selection ... We can see that, in the rudest state of society, the individuals who are the most sagacious, who invented and used the best weapons or traps, and who were best able to defend themselves, would rear the greatest number of offspring. The tribes which included the largest number of men thus endowed would increase in number and supplant other tribes ... It is, therefore, probable that with mankind the intellectual faculties have been perfected through natural selection.[69]

Thus Darwin had no hesitation in applying the orthodox concept of natural selection to intellectual faculties. This is also shown by his supreme indifference when faced with the argument developed by Galton in *Hereditary Genius* (1869), according to which the most intelligent individuals had a tendency not to marry, and not to have children. Darwin's reply to his cousin was straightforward: if geniuses have few children, the other members of their family nevertheless continue to produce offspring, and thus transmit the hereditary qualities of their exceptional relative. And he coldly adds: '... it has been ascertained by agriculturists that by preserving and breeding from the family of an animal, the desired character has been obtained'.[70]

What then of the other component of the moral sense, the 'social instincts' which are a necessary but not sufficient precondition of all morality? For Darwin, these instincts are nothing other than the hereditary dispositions that drive certain animals to live together in society. Among the higher animals, he fundamentally recognised two kinds of instincts – love and sympathy. The other social dispositions, either acquired or instinctive, are constructed on the basis of these two feelings.[71] Love is a 'feeling of pleasure from society' and 'is probably an extension of the parental or filial affections', which themselves 'apparently lie at the basis of the social affections'. 'We may infer that they have been to a large extent gained through natural selection.'[72] Here Darwin does not invoke a 'tribal' selection, distinct from individual selection (which may be extended to familial selection). The other fundamental social instinct – 'sympathy' – drives individuals of the same community to help each other. At this point in the argument it might be expected that group selection would be invoked to explain the genesis of instincts. Not at all. Instead Darwin invokes a kind of reciprocal altruism, where the individual's advantage merges with that of the community:

> In however complex a manner this feeling may have originated, as it is one of high importance to all those animals which aid and defend each other, it will have been increased through natural selection; for those communities, which included the greatest number of the most sympathetic members, would flourish and rear the greatest number of offspring.[73]

These two passages reveal the schema that Darwin had in mind to explain the development of intelligence. It would be the result of a classic individual selection (as shown by the beginning of the first quotation (note 69) or familial selection, which is in itself nothing more than a development of individual selection. In the second place (end of the first quotation), the 'tribes' which are improved by this process prosper and supplant others. This is not an example of group selection, but of a *consequence*, at the group level, of a selection process that remains thoroughly individual. Darwin thus replies, in passing, to Wallace, who in 1864 had been the first to explain the improvement of human intellectual capacities by natural selection – but, in the case of Wallace, this involved selection *between* races.[74]

Darwin appears to be on the verge of adopting a concept of group (or 'tribal') selection. Strictly speaking, this may indeed have been the case. But Darwin did not explicitly take this step. Darwin did not need to use the concept of group selection to explain the origin of feelings linked to relatives or the sympathy that solidarises the community, because he thought that the prosperity of individuals and that of the community coincide. At no point in *The Descent of Man*, or in the *Origin of Species*, is it envisaged as a general hypothesis that social instincts have been produced by natural selection operating to the advantage of a group and against that of individuals.

All this leads to the following striking observation: it is precisely at the point at which Darwin concluded his discussion of the 'moral sense' that for the one and only time in all his works he advances a conception of 'natural selection' operating by and through group advantage and against that of the individual.

The argument was put forward in the context of a discussion of the origin of those 'moral qualities' that Darwin considered to be most uniquely human: courage, sacrifice, fidelity. The origin of these qualities is initially presented as an enigma for the naturalist; the reasoning is the same as that used in the discussion of hybrid sterility:

It may be asked, how within the limits of the same tribe did a large a number of members first become endowed with the social and moral qualities...? It is extremely doubtful whether the offspring of the more sympathetic and benevolent parent, or of those which were the most faithful to their comrades, would be reared in greater number than the children of selfish and treacherous parents of the same tribe. He who was ready to sacrifice his life, as many a savage has been, rather than betray his comrades, would often leave no offspring to inherit his noble nature. The bravest men, who were always willing to come to the front in war, and who freely risked their lives for others, would on average perish in larger number than other men. Therefore it seems scarcely possible (bearing in mind that we are not here speaking of one tribe being victorious over another) that the number of men gifted with such virtues, or that the standard of their excellence, could be increased through natural selection.[75]

The proposed paradox has a very simple structure: the virtues of sacrifice and fidelity confer a considerable advantage on societies in their struggle with rival societies, but certain of these virtues reduce the probability of survival and reproduction of those who possess them. Consequently, one cannot explain their development *within* a society on the basis of natural selection: interindividual competition would tend to inhibit the development of such qualities. The only possible solution would be to admit the existence of a process of natural selection acting for and through the prosperity of groups, and able to develop those characters that are not advantageous to individuals. This is exactly what Darwin proposed, two pages further on:

Although a high standard of morality gives but a light or no advantage to each individual man and his children over the other man of the same tribe, yet an advancement in the standard of morality and an increase in the number of well-endowed men will certainly give an immense advantage to one tribe over another. There can be no doubt that a tribe including many members who, from possessing in a high degree the spirit of patriotism, fidelity, obedience, courage, and sympathy, were always ready to give aid to each other and to sacrifice themselves for the common good, would be victorious on other tribes; and this would be natural selection. At all times throughout the world tribes have supplanted other tribes; and as morality is one element of their success, the standard of morality and the number of well-endowed men will thus everywhere tend to rise and increase.[76]

This argument is unique in Darwin's work. Its philosophical structure is extremely interesting; Darwin seems to rally to the support of those of his contemporaries who argued that the result of the relentless struggle between peoples depended upon their intrinsic moral qualities. Wallace undoubtedly went furthest along this road when he affirmed in 1863 before the Anthropological Society that natural selection had ceased to act on the body of man, and now acted solely upon his moral qualities, which alone continued to evolve, thus playing a crucial role in the struggle between human races.[77] In this view of human evolution, moral qualities are given an openly reductionist interpretation. Curiously enough, this is exactly the opposite of the interpretation that appears in the passage from *The Descent of Man* quoted above. In Darwin's text there is no question of 'social instincts' in general, but rather of 'morality'. This is not merely a sloppy formulation; the passage immediately follows a sentence that acts as a general conclusion to the discussion of 'the moral sense':

Ultimately, a highly complex sentiment, having its first origin in the social instincts, largely guided by the approbation of our fellow-men, ruled by reason, self-interest, and in later times by deep religious feelings, confirmed by instruction and habit, all combined, *constitute our moral sense or conscience.* [Added emphasis][78]

It is impossible not to discern in this passage the spectre of 'culture' or, to use the preferred 19th-century term, 'civilisation'. And it is precisely in this context that Darwin unequivocally admits that it is possible that natural selection operates on groups and not on individuals. To repeat the point: the only moment in all his writings where Darwin clearly admitted this hypothesis was when he was trying to explain the origin of the attribute that he believed most distinguished man ('I fully subscribe to the judgement of those writers who maintain that of all the differences between man and the lower animals, the moral sense or conscience is by far the most important').[79] Thus it was when Darwin was obliged to speculate about the uniqueness of man that he abandoned his individualistic representation of natural selection.

Darwin the naturalist-cum-philosopher did his best to confuse the matter. It is tempting to read *The Descent of Man* as a description of a linear ascent towards humanity, totally dominated by the principle of natural selection. And this is indeed the case: natural selection explains each point. But at the final step the principle collapses. Seduced by 'morality', Darwin abandoned his thoroughly individualistic conception of natural selection.

Thus the discussion presented in *The Descent of Man*, far from invalidating the individualistic concept of natural selection, in fact confirms it, but at the price of a philosophical somersault. Having finally embraced philosophy in the pages of *The Descent of Man*, Darwin divides the kingdom of 'natural selection' into two regions: for brute Nature, selection acts only for the good of individuals; for civilised Man, it acts beyond individuals, and for an end that cannot be defined by sole reference to individuals.

This ultimate division of the concept of natural selection confirms the analysis put forward throughout these opening chapters. It has been argued that the concept of natural selection as constructed by Darwin led him to enthrone 'heredity' as the new crucial concept of theoretical biology. With Darwin, 'heredity' emerged as the new philosopheme of biological identity (that is, of identity from a biological point of view). Because the material of natural selection is hereditary variation, Darwin conceived of selection as a causal process acting upon individual organisms. In the case of man, 'heredity', which at the time was beginning to be identified with 'nature', gives way to another form of permanence: teaching or tradition or civilisation – whatever name one wishes to give it. In this context, natural selection, if it still operates, does not function via heredity. Darwin's position when he steps over the threshold into the realm of anthropology can be interpreted as a retrospective confirmation of his starting point: as a genuinely 'natural' process, natural selection acts only through the preservation of hereditary variations that are advantageous to individuals.

To conclude this analysis of the ontological implications of the original concept of natural selection, as part of his campaign against 'fixist' typological schemas Darwin developed a theoretical account of the historicity of forms that required three kinds of hierarchically organised units: 'variations', 'individuals' and 'varieties' (or races). Darwin's genius was to wrench this vocabulary from its semantic implications and to oblige it to express something very different from what might at first be implied: a 'variation' is not a 'variety'. Of course, Darwin sometimes followed common usage and employed the two words as though they were synonymous. Thus in the *Variation*, one can read, a few lines apart, that natural selection is both 'the preservation ... of varieties that present a given advantage'[80] and 'the preservation of variations that are advantageous to the individual'.[81] But it is also true that Darwin subverted this common usage and imposed another, intrinsically linked to his view that natural selection modifies varieties by accumulating variations that are advantageous to individuals. Heredity acts between these three units and should not be defined either as the type of the variety (and even less so of the species) or as the transmissible essence of an individual. The general process of selection is also defined in terms of these three units.

In modern evolutionary theory this concept is known as 'Darwinian selection', implying an opposition to other conceivable forms of natural selection, such as group or species selection. These two opening chapters have examined both the implications of this concept and the theoretical context in which it was elaborated. Our next task is to analyse the crisis which rapidly erupted. A particularly knotty epistemological problem emerges at this point: according to the generally accepted idea of the history of science as a 'rational reconstruction',[82] major scientific theories are adopted not because they are initially 'true', but because they give rise to a fertile programme of research. Imre Lakatos has argued that it is essential for such a programme to contain a 'negative heuristic' – a range of potentially revisable auxiliary hypotheses that protect the central core of the theory from attempts to refute it. The central propositions of the theory thus have time to take a coherent form and to enlarge their field of application, without being threatened by minor refutations. The Darwinian theory of natural selection appears to have met an exactly opposite fate. The *theory* was immediately applied to a wide range of subjects, extending to the whole of natural history and beyond. It was adopted wholesale by naturalists and also by the general culture of the time. The central hypothesis, however, in the extremely constrained form outlined by Darwin, was quickly exposed to major internal problems and went into crisis almost as soon as it was born. By rigidly linking the hypothesis of selection to the question of heredity, Darwin did indeed open a long-term research programme that was rich in ideas and potential. But in the half-century that was to follow, the hypothesis was exposed to the permanent threat of refutation. The power of the process of natural selection as conceived of by Darwin did not hold under all conceivable models of heredity. Worse, Darwinian selection was perhaps incompatible with Darwin's own representation of hereditary variation.

Darwin was convinced that there was no place for 'reversion to type' in his hereditarist conception of selection. He was wrong. The most telling criticism of the hypothesis of natural selection was to come from someone who had perfectly understood not only the implications of the individualistic conception of natural selection and the crucial role played within it by heredity, but also the real ambiguity of the new concept of variation which was required by this framework. The next chapter is devoted to this attack, made in 1867, which constituted the first indication of the crisis of the Darwinian hypothesis of selection.

3 Jenkin's objections, Darwin's dilemma

In 1867, Henry Charles Fleeming Jenkin[1] (1833–85), Professor of Engineering at Edinburgh University, published a review of the *Origin of Species* in which he raised a criticism of the theory of natural selection that was to have a major influence on both Darwin and his successors.[2] In the 1869 and 1872 editions of the *Origin of Species*, Darwin mentioned Jenkin's criticism in the chapter on natural selection and made explicit reference to his 'able and valuable article' published by the *North British Review*.[3] In a letter to Hooker written in 1869, Darwin declared: 'Fleeming Jenkin has given me much trouble, but has been of more real use than any other essay or review.'[4] Francis Darwin, in his biography of his father, confirmed this reaction: 'It is not a little remarkable that the criticisms, which my father, as I believe, felt to be the most valuable ever made on his views should have come, not from a professed naturalist but from a Professor of Engineering.'[5]

Despite being well known to historians of Darwinism, Jenkin's criticisms have nevertheless led to some highly contradictory interpretations. Some, such as Loren Eiseley, have argued that Darwin was so impressed by Jenkin's criticisms that he virtually renounced the theory of natural selection by developing his theory of pangenesis in *The Variation of Animals and Plants under Domestication* the following year. Eiseley argues that in putting forward this hypothesis Darwin had openly embraced a form of Lamarckism.[6] At the other extreme, Peter Vorzimmer has claimed that Jenkin's article had no lasting effect on Darwin's thinking.[7]

Both these positions are exaggerated. It cannot be seriously argued that Darwin adopted the theory of pangenesis as a consequence of Jenkin's 1867 review: leaving aside the fact that Darwin had written a manuscript on pangenesis in 1865,[8] there are many indications that he had always had in mind this kind of concept to explain the origin of variation. His grandfather, Erasmus Darwin, had formulated a similar theory, as early as 1801.[9] But it is equally one-sided to suggest that he was virtually unaffected by a criticism that he publicly described as 'valuable'. The real problem is whether the theory of natural selection was substantially affected by Jenkin's attack. Following Peter Bowler, it seems reasonable to argue that Darwin realised that Jenkin's critique meant that the description of natural selection would have to be recast, without either abandoning it or even thoroughly altering it.[10]

Unlike previous accounts of the Jenkin affair, this chapter stresses Jenkin's criticisms, rather than focusing on Darwin's reactions. Whatever Jenkin's motivations (and these were clearly reactionary), his criticisms of the theory of natural selection were sufficiently rigorous and robust to open a second phase in the development of Darwinian rationality, within which the personality and role of Darwin began to be less important. Jenkin's critique was aimed at the blind spot in the theory of selection: he emphasised its dependence on a certain class of theories of variability and heredity and pointed out their quantitative consequences. The strength of Jenkin's argument was to show how, under certain hypotheses of the nature of variation (those advanced by Darwin), and on the basis of a strictly quantitative explanation, reversion to type was an absolute necessity.

In this chapter Jenkin's criticisms that Darwin found particularly impressive will be examined first, followed by their probable consequences for the conception of natural selection, and the perspectives they opened for the development of a theory of selection and its relation to heredity in the years following Darwin's death.

3.1 Jenkin's objections

Jenkin's article leaves no doubt as to his intentions, or as to the manner in which he conceived of the question of the origin of species. Jenkin was convinced that species are 'types' and that their variation is strictly contained within fixed limits. Thus he contested that natural selection (which he accepted as both possible and real) can modify species and that classification reflects genealogy. For Jenkin, there is no common descent; species are not modified beyond the potential contained in their 'sphere of variation'. Selection, be it natural or artificial, is unable to form or modify populations, even though it may displace them within the 'sphere of variation'. Thus it cannot carry out the task given to it by Darwin.

Ernst Mayr has expressed his surprise that Jenkin's review could have made such an impression on Darwin and on his contemporaries. Jenkin, he argues, is just one of many examples of *essentialist* thought.[11] Jenkin's position does indeed contain the kind of argument used by Lyell – Darwin's intellectual guide – against the idea of the 'conversion' of species: 'there are fixed limits beyond which the descendants from common parents can never deviate from a certain type'.[12]

Mayr goes on to list a number of facts that reveal that Jenkin was not a 'competent naturalist' and that he may simply not have understood Darwin's ideas. Jenkin's criticisms are thus neatly classified as characteristic of a physicalist representation of science, unsuitable to the specificity and complexity of biology.

Mayr is no doubt correct about Jenkin's intentions. He may also be right about Jenkin's lack of qualifications as a naturalist. But he misses the essential point, which is that Jenkin, despite his unoriginal fixist arguments, launched a series of major methodological criticisms against Darwin's theory simply by pointing out the predictable empirical consequences of Darwin's starting point. This fits well with his academic training: Jenkin was not only an engineer, he was also a brilliant mathematician and a close friend of Kelvin, who also made significant contributions to economic theory.[13]

In his review of the *Origin of Species*, Jenkin develops three main objections:

1. He rejects the idea that species can indefinitely vary in a given direction; this objection is based on the results of animal husbandry and artificial selection.

2. He shows that, under the (Darwinian) hypothesis of blending inheritance, natural selection is probably not able to modify the 'mean species type'.

3. Noting the immense length of time required by Darwin for natural selection to produce the variety of living forms, he points out that this is incompatible with physicists' estimates of the age of the earth.

The final point does not need to be dealt with in any detail. Inspired by two articles published in 1862 by the physicist William Thomson (better known as Lord Kelvin),[14] a close collaborator and friend in Edinburgh, Jenkin discusses the thermodynamic consequences of the Kant–Laplace hypothesis on the formation of planets. According to this hypothesis, the earth was formed by the condensation of gases on the fringe of the primitive solar nebula. Given the earth's mass, the quantity of energy it still receives from the sun and the action of factors that could act to slow down cooling (superficial solid crust, atmosphere), it is possible to estimate the earth's temperature in the past. Lord Kelvin argued that the sun was formed at most 400 million years ago, but he considered that a much younger age (between 20 and 40 million years) was more likely. He also estimated that (because of excessive heat) it was inconceivable that any form of life had developed on earth before 100 million years ago. Since this calculation was subject to the same margin of error as that for the sun, it was probable that the whole of geological and biological evolution would have to fit into only several million years. If this figure was correct, it would rule out the long stretches of time required by Darwin (at least several thousand million years), and would also invalidate the stratigraphic dating techniques that were widely used at the time.

Darwin did not have much to say in reply to this argument, except to point out that stratigraphic dating was one of the best available tools for testing physicists' hypotheses. The controversy was only resolved at the beginning of the 20th century, by the discovery of radioactivity, which made it conceivable that the earth had condensed several thousands of millions of years ago, but had nevertheless not been covered by ice as Buffon had suggested in *Les époques de la nature*. Radioactivity provided a sufficiently rich source of internal energy to explain the fact that the earth had been maintained in a quasi-stationary thermal state for a very long time.[15] Furthermore, the Kelvin–Jenkin cosmological criticism was not only a threat to Darwin's theory: it menaced all contemporary models of paleontology. Despite the fact that it was wrong, this argument nevertheless reveals the quantitative and experimental approach taken by Jenkin in his critique of Darwin's work.

What of the two 'special' objections, which take up most of the article? The first concerns the target of natural selection: variation.[16] According to Jenkin, 'Darwin's theory requires that there shall be no limit to the possible difference between descendants and their progenitors, or, at least, that if there be limits, they shall be at so great a distance as to comprehend the utmost differences between any known forms of life.'[17]

To be fair, Darwin did not argue that species show unlimited variation,[18] but rather that they show sufficient variation for selection to be able to modify them without limit. This is the subject of Jenkin's frontal attack. Jenkin does not disagree that many animals and plants show variability, nor does he contest that selection can modify races, especially under domestication. Rather, he attacks the postulate that, once selection has acted, variability continues to be expressed in all directions, as before. Jenkin states that, in fact, the action of selection is to increase variability in the direction of a reversion of the character under selection. The more one selects, the more one approaches the 'surface' of the 'sphere of variation' that constitutes the species, and the smaller is the probability that new individual variations will be more distant from the centre, while the probability of retrograde variations increases:

> A given animal or plant appears to be contained, as it were, within a sphere of variation; one individual lies near one portion of the surface, another individual, of the same species, near another part of the surface; the average animal at the centre. Any individual may produce descendants varying in any direction, but it is more likely to produce descendants varying towards the centre of the sphere, and

the variations in that direction will be greater in amount than the variations towards the surface.[19]

In other words, the offspring of an individual will generally tend towards the population mean rather than towards a more extreme value. Jenkin coined the term 'reversion' to describe this principle, and it can be considered to be the first, qualitative, description of what Galton was later to call the *reversion* (1877) and then the *regression* (1885) of inherited characters towards the mean (see Chapter 4). Jenkin adds that his imagery of the 'sphere' constitutes a 'generalisation' of the traditional notion of the 'tendency to revert' which Darwin himself accepted.

The sense of Jenkin's criticism becomes clear: in these circumstances how can selection – natural or artificial – overcome the tendency to reversion? Or, to be more precise: can selection at one and the same time overcome normalising variability (that which tends to return to the centre of the sphere of variation) and produce a new omnidirectional variability, exceeding the current species norm? In fact, Darwin had already encountered this problem. In chapter 5 of the *Origin of Species*, which deals with the 'laws of variation', Darwin had drawn attention to the fact that in animal husbandry and in nature, intense selection is often accompanied by an increased variability, and by more frequent and intense tendencies to reversion. However, he added, there was no reason to think that selection was not stronger than both the tendency to vary and the tendency to revert. Whatever the source of hereditary variation (the direct action of external conditions, the effect of use and disuse, reversion, correlation of growth), its continued existence in the species was always subordinate to *a posteriori* control by selection.[20] Further, and specifically in the case of reversion, Darwin responded to this old fixist argument by repeatedly referring to the nature of the 'principle of heredity': heredity is not a virtual property of species, but a current relation between parents and offspring, such that its 'force' is not a function of the antiquity of the character.[21] Animal husbandry gives the best example of the power of the principle, because the method that produces domestic races relies heavily on the natural law of heredity.

The main interest of Jenkin's criticism is that he perfectly understood this aspect of Darwin's argument:

There is indeed one view upon which it would seem natural to believe that the tendency to revert may diminish. If the peculiarities of an animal's structure are simply determined by inheritance, and not by any law of growth, and if the child is more likely to resemble its father than its grand-father, etc., then the chances that an animal will revert to the likeness of an ancestor a thousand generations back will be slender. This is perhaps Darwin's view. It depends on the assumption that there is no typical or average animal, no sphere of variation, with centre and limits, and cannot be made use of to prove that assumption. The opposing view is that of a race maintained by a continual force in an abnormal condition, and returning to that condition so soon as the force is removed.[22]

Thus Jenkin clearly understood the fundamental link between selection and heredity, both conceived in terms of populations of individuals showing 'variations'. It was in this context that he developed his second criticism, which was by far the most telling. This quantitative argument examines different hypotheses as to the nature of heredity, draws out their consequences for the statistical distribution of a character in a population under selection, and turns on the question of the efficacy of natural selection as a factor of modification.[23] It can be considered the first clear formulation of a fundamental problem that both characterised and undermined the theory of natural selection in the half-century following Darwin.

Jenkin introduces his argument as follows:

Those individuals of any species which are most adapted to the life they lead, live on an average longer than those which are less adapted to the circumstances in which the species is placed. The individuals which live the longest will have the most numerous offspring, and as the offspring on the whole resemble their parents, the descendants from any given generation will on the whole resemble the more favoured rather than the less favoured individuals of the species. So much of the theory of natural selection will hardly be denied; but it will be worth while to consider how far this process can tend to cause a variation in some one direction.[24]

Jenkin's argument brings into play three separate points which could have been presented more clearly (to be honest, the text is difficult for the modern reader). Jenkin outlines three factors that could affect the 'efficiency' of natural selection:

1. Which hypothesis is taken to govern the transmission of inherited characters? Are offspring intermediate between the parental values, or is the inherited variation transmitted as a unity? To use a term that became current after Darwin's time, this question asks whether we are dealing with blending inheritance or with particulate inheritance, with a given character being preponderant.[25]

2. Is variation slight, or is it strongly marked? Does it consist of major anomalies – 'sports of nature' – or of what Darwin called 'individual differences' – small variations that affect a character? Extra fingers (hexadactyly in humans) or the drastic reduction in leg size in Ancon sheep are classic examples of major variations; continuous variation in the size of organisms is a classic example of an 'individual difference'.

3. Is a given variation, be it large or small, present in a large number of individuals, in a small number or even in only a single individual?

Most if not all commentators have failed to distinguish these three points in Jenkin's article, tending to suggest that his criticism crassly opposed major inherited anomalies, discretely transmitted, to Darwin's view of infinitesimally small variations, constantly arising and transmitted by blending inheritance. While Jenkin did indeed often tend in this direction, to reduce his argument to this point means that the structure of the article is lost and, more important, that it is impossible to understand why Darwin was so impressed by it. For clarity's sake, the various scenarios put forward by Jenkin will be analysed in a different order from the one he chose, the logical structure of his argument being more important than its literary presentation.

1. What is the mode of transmission of characters that vary? Jenkin shows that there is at least one case that is particularly favourable to the efficiency of the process of natural selection, where a difference is transmitted unaltered and in all cases to the offspring.[26] Nineteenth-century naturalists generally called this type of transmission 'prepotency', 'preponderance in the transmission of character' or sometimes (and this term has remained in use in genetics) 'dominance'. Examples of 'prepotency' are opposed to those where the offspring is intermediate between the parents.[27] In 'prepotent' transmission the advantageous variation will be transmitted to all offspring, and 'it follows, from mere mathematics, that the descendants of our gifted beast will probably exterminate the descendants of his inferior brethren'.[28] The 'substitution' will probably be very rapid, and in this case there is no problem in imagining that natural selection would be effective. The only difficulty foreseen by Jenkin is that it would be 'a stiff mathematical problem to calculate the number of generations required in any given case' for total substitution to take place.[29]

However, as Jenkin points out, Darwin's theory is 'surely not' that the advantage given by a sport is retained by all descendants.[30] The only hypothesis of heredity under which natural selection would clearly be effective is not that advanced by Darwin. To admit that selection only consists in spreading prepotent anomalies would be to accept that nature proceeds by leaps. Ironically, but extremely perceptively, Jenkin notes that this would lead Darwin into a kind of non-religious creationism:

> The appearance of a new specimen capable of perpetuating its peculiarity is precisely what might be termed a creation, the word being used to express our ignorance of how the thing happened ... Perhaps this is the way in which new species are introduced, but it does not express the Darwinian theory of the gradual accumulation of infinitely minute differences of every-day occurrence, and apparently fortuitous in their character.[31]

It is worth noting that this interpretation of natural selection was precisely that of Thomas Huxley, one of Darwin's most fervent supporters. As soon as the *Origin of Species* was published in 1859, Huxley wrote to Darwin: 'you have loaded yourself with an unnecessary difficulty in adopting "Natura non facit saltum" so unreservedly'.[32]

Jenkin put forward two other objections, based on the theory of intermediate (or blending) inheritance. His argument is both rigorous and precise, and lays bare an astonishing ambiguity at the heart of Darwin's theory. These two criticisms indeed suggest that Darwin did not have a clear idea of what exactly he meant by slight variations or 'individual differences', because he did not separate the question of the *degree* of variation (the magnitude of the deviation from the mean) from that of the *number* of individuals showing that variation. Whatever Mayr's doubts about Jenkin's qualifications as a biologist, it is clear that he was arguing from the point of view of a scientist familiar with statistical mathematics as it existed in the 1860s. The constant reference made to the 'average animal', which he identifies as being the 'common type', shows that he had read Quételet, and that he interpreted Darwin's arguments in this light. And when Darwin speaks of imperceptible 'individual differences', Jenkin sees something that Darwin does not: the distribution curve of the character within the population. If natural selection operates on this kind of variation, it affects a degree of variation that marks all individuals, each character being merely a point on a continuous distribution. That is why Jenkin divides his analysis of the 'efficiency of natural selection' into two halves. He first envisages a situation in which natural selection operates on a variation that affects all individuals (a 'continuous' variation, in modern terms), then a situation in which a variation (be it large or small) only affects one individual or a small number of individuals (in other words, an 'occasional' variation). The reason why so many modern readers of Jenkin have missed this point – despite the fact that it is clearly stated in his article – is simply because of a tendency to read Jenkin according to Darwin's methodological framework. But Darwin – and this is the whole point – did not look at the world in this way.

2. What of the objection – or rather the observation – of the efficacy of natural selection in the case of continuous variation? By 'individual differences' Darwin meant the 'numerous small differences' which, in a given species, affect virtually all characters, and which are inherited by offspring from their parents.[33] This kind of variation, which Darwin considered to be the main and sufficient material of natural selection, is, in modern terms, quantitative: a given animal has longer legs, runs faster, sees better, than another, and so on. Jenkin merely took Darwin at his word and gave this view a statistical interpretation. 'Individual differences' become 'common variation', which is thus common to all

animals for all characters. In other words, for each observable character, all individuals diverge more or less from the mean or common type. Darwin would probably not have disagreed with this position, which concurs with his own conception of 'infinitesimal variation'. But he must have been surprised by Jenkin's conclusions with regard to the quantitative significance of natural selection:

> It must apparently be conceded that natural selection is a true cause or agency whereby in some cases variations of special organs may be perpetuated and accumulated, but the importance of this admission is much limited by a consideration of the cases to which it applies: first of all we have required that it should apply to variations which must occur in every individual so that enormous numbers of individuals will exist, all having a little improvement in the same direction; as, for instance, each generation of hares will include an enormous number which have longer legs than the average of their parents, although there may be an equally enormous number who have shorter legs.[34]

This passage, which appears clear to the modern reader, becomes strange if placed in the context of Darwin's writings. Darwin never said that natural selection acted on a range of variation that affected all individuals. Although he had a continuist conception of variation within a species, he always argued as though particular variations ('peculiarities') appeared recurrently and spread throughout the race and the species, like the diffusion of atoms. The vocabulary he used – 'preservation', 'elimination' and 'accumulation' – reflects this conception. Strictly speaking, Darwin did not have a clear populational and biometric conception of varieties and species; Jenkin's arguments, however, are clearly influenced by this approach, albeit in an underdeveloped form.

Thus Jenkin admits the idea of a graduated and individual variation, and then immediately shows the implication of such variation for natural selection. If, for example, it is advantageous for a hare to have longer legs, then virtually *all* individuals are 'advantaged', even if only to an extremely small extent, with regard to other individuals, for the simple reason that virtually every individual is larger than some others. Jenkin is close to interpreting natural selection as the alteration of a Gaussian distribution in which either the norm (mean) is displaced or, on the contrary, selection favours the norm by eliminating extreme variants ('it is clear that it [natural selection] will frequently, and indeed generally, tend to prevent any deviation from the common type').[35]

Jenkin thus draws attention to the fact that the hypothesis of natural selection depends on the nature of the variation on which it acts. It is thus important to know how this variation is inherited, how it is distributed, how it is measured and on what parameters it acts. If selection acts on 'variability' as such (i.e. in modern terms, on the measure of dispersion), it is normalising. If on the other hand it acts on the mean value, then it is modifying.

It can of course be argued that Darwin was aware of these problems, at least on a qualitative level, and that Jenkin was far from being a 'biometrician'. But the point is that Jenkin's criticism indicates that the Darwinian argument would have to shift its ground: Jenkin effectively invited Darwin and his followers to provide a more rigorous *description* of what they meant by variation. Such a description would in itself be extremely valuable, despite the absence of a physiological theory of variation. Although in 1860 no one could say with any confidence whether hereditary variation was continuous or discrete at the most elementary level, it would nonetheless be useful to understand the implications of these two possibilities.

3. Finally, Jenkin's most famous criticism is that relating to rare or 'occasional' variants. Despite the way in which this point has often been presented, Jenkin's argument does not deal solely with 'sports' (major anomalies or monstrosities). It does indeed start with this category of variants, but it is then applied to all variation (even infinitesimal) that affects single individuals.[36] The spirit of the criticism is that Darwin was basing his position on an imprecise use of the 'doctrine of chance' (the theory of probabilities), which in fact he did not understand.[37]

Given that a new form, perhaps extremely advantageous, arose in *one* individual, what *chance* would it have of spreading and establishing itself in the species? Jenkin considered that in such a case there would be two obstacles to the success of natural selection. The first is that this variant would run the risk of being one of the enormous mass of individuals which, in each generation, do not reproduce:

> The advantage, whatever it may be, is utterly out-balanced by numerical inferiority. A million creatures are born; ten thousands survive to produce offspring. One of the million has twice as good a chance as any other of surviving; but the chances are fifty to one against the gifted individuals being one of the hundred survivors. No doubt, the chances are twice as great against any other individual, but this does not prevent their being enormously in favour of *some* average individual. However slight the advantage may be, if it is shared by half the individuals produced, it will probably be present in at least fifty-one of the survivors, and in a larger proportion of their offspring; but the chances are against the preservation of any one 'sport' in a numerous tribe. The vague use of an imperfectly understood doctrine of chance has led Darwinian supporters, first, to confuse the two cases above distinguished; and, secondly, to imagine that a very slight balance in favour of some individual sport must lead to its perpetuation. [Original emphasis. Note that there is apparently an error here. Given that Jenkin had just referred to 10,000 survivors out of 1,000,000, the second sentence should probably read: '. . . the chances are fifty to one against the gifted

individuals being one of the 1per cent survivors', or 'one of the ten thousand survivors'.][38]

For Jenkin, the 'two cases' distinguished here were two sides of the same statistical problem. Further on in his review, Jenkin explains that the sources of variation depend on the joint effects of two factors: the chance of extinction of the 'favoured variety', which is inversely proportional to its number, and the intensity of the advantage. The rarer the favoured variety, the stronger has to be the advantage (in modern terms, the selection pressure) for the variety to be able to prosper.[39] Darwin took this point extremely seriously. And he was quite right to do so, because it constitutes a major problem for evolutionary theory that has made its presence felt ever since. De Candolle, Galton and Lotka all subsequently refined Jenkin's point by showing that, owing to random variation of family size, all characters transmitted from parent to offspring must eventually disappear. Initially elaborated in the context of the extinction of surnames ('All names must disappear' – De Candolle), the argument was subsequently extended to biological inheritance. The biometricians, and in particular Pearson, considered it to be an important step in the development of a mathematical theory of natural selection. In the 20th century, population geneticists have made it one of their main research topics, under the title 'random genetic drift'. All these developments will be dealt with in subsequent chapters. Here it is sufficient to note that from 1867 on the question of the consequences of the interaction of sampling effect and natural selection was clearly posed.

There is, however, another obstacle to the establishment of an isolated variation, at least if one accepts Darwin's hypothesis of the offspring having an intermediate character. What would happen if the variation were not eliminated by chance, that is, if the individual were to reproduce? Jenkin deals with this point by using the same numerical example as before, where, in each generation, one individual out of a hundred survives and itself produces a hundred offspring:

[The favoured sport] will breed and have a progeny of say 100; now this progeny will, on the whole, be intermediate between the average individual and the sport. The odds in favour of one of this generation of the new breed will be, say, 1½ to 1, as compared to the average individual; the odds in their favour will therefore be less than that of their parent; but owing to their greater number, the chances are that about 1½ of them would survive. Unless these breed together, a most improbable event, their progeny would again approach the average individual; there would be 150 of them, and their superiority would be say in the ratio 1¼ to 1; the probability would be now that nearly two of them would survive, and have 200 children, with an eighth superiority. Rather more than two of them would survive; but the superiority should again dwindle, until after a few generations it would no longer be observed, and would count for no more in the struggle for life, than any of the hundred trifling advantages which occur in the ordinary organs.[40]

The argument here is quite unambiguous: under the hypothesis of blending inheritance, it is the question of number that prevails. Any advantage, if it is not absolutely necessary for survival but rather constitutes a relative advantage, will be purely and simply 'swamped by number'.[41] Since Jenkin's time, the expression 'swamping effect' has been retained to describe this major criticism, which was to weigh upon natural selection until the end of the 19th century.[42] This criticism is not dependent on the amplitude of the variation; it applies equally to a 'slight' or to a 'strongly marked' variation: 'If it is impossible that any sport or accidental variation in a single individual, however favourable to life, should be preserved and transmitted by natural selection, still less can slight and imperceptible variations, occurring in single individuals, be garnered up and transmitted to continually increasing numbers.'[43]

Darwin explicitly acknowledged the relevance of Jenkin's objections in the case of single or rare variations. The importance of this question is revealed by the following admission, appended to the fifth edition of the *Origin of Species*: 'Until reading an able and valuable article in the "North British Review" (1867), I did not appreciate how rarely single variations, whether slight or strongly marked, could be perpetuated.'[44]

To sum up, Jenkin raised two fundamental methodological questions. First, how is variation distributed in the population and what are the consequences of this? If variation is continuously distributed, then it has to be admitted that, in each generation, all individuals vary. To understand what this means, a statistical interpretation of both heredity and advantage is needed. If on the other hand variation is expressed only in a small number of individuals, this also has to be measured: how many individuals, what fraction of the total population, with what precise advantage, and with what probability of being eliminated by chance?

The second question is: which hypothesis of hereditary transmission is assumed? If inheritance with 'preponderance' is assumed, the success of the selection process is assured. Producing an algebraic model of this process, however, would involve highly complex calculations. But this was not the hypothesis favoured either by Darwin or by most of his contemporaries, for whom preponderance was the exception, offspring generally being 'intermediate between their parents'.[45] Under this hypothesis, if the offspring of individuals with extreme characters interbreed, the difference they express can be maintained. It is, however, extremely unlikely that crosses will take place on such a basis. In every other case, in large populations with free mating, isolated variations – whatever their size and their relative advantage – will be gradually diluted and will not affect the mean value of the character.

True, Jenkin raised these two sets of problems in a somewhat perfunctory manner, and in an article that has the air of a reactionary tract. His arguments, however, are extremely telling. Using modern vocabulary, they can be summarised as follows: continuity or discontinuity of hereditary variation, blending or particulate inheritance, the effect of the mating system, the quantitative nature of 'advantage' and the interaction between selection and sampling effects. The mere engineer who allegedly understood nothing of biology left an impressive list of problems for the future theoreticians of natural selection.

It is to Darwin's credit that he recognised that this article contained the most 'valuable' challenge to his theory, the implications of which he was undoubtedly ill equipped to fully understand, because of his own culture, experience and lack of statistical training. Indeed, the problems raised by Jenkin were to form a large part of the research programme of the future, first in biometry, then in the later science of population genetics.

3.2 Darwin's dilemma

Faced with Jenkin's criticism, Darwin realised that his conception of the relation between selection and variation was deeply flawed and that he had unnecessarily weakened the principle of natural selection by simultaneously holding two different positions. Darwinian natural selection is primarily a principle of the progressive accumulation of variations that appear in an isolated manner in individuals (we can call this a model of *diffusion* by replacement). But Darwin also presented natural selection as acting on 'infinitesimally small' variation, in other words, on continuous variation. According to this point of view, variation is widespread, and selection should thus operate through the *displacement* of the mean character. In other words, Darwin realised that he had wanted to combine the advantages of both continuous and discontinuous variation, and that his term 'individual differences' in fact covered a confusion between the number and the intensity of variation. This was Darwin's dilemma.

This interpretation has solid historical bases, as shown by a letter from Darwin to Wallace (22 January 1869):

> I have been interrupted in my regular work in preparing a new edition of the 'Origin', which has cost me much labour, and which I hope I have considerably improved in two or three important points. I have always thought individual differences more important than single variations, but now I have come to the conclusion that they are of paramount importance, and in this I believe I agree with you. Fleeming Jenkin's arguments have convinced me. [Original emphasis][46]

Wallace was intrigued by the sentence dealing with 'individual differences' and 'single variation', which is not very clear. He took it to mean that Darwin now considered that 'sports' (large and rare anomalies) were of major importance. This would accord with Jenkin's point of view, but was very different from Wallace's own position, as his reply makes clear:

Dear Darwin,– Will you tell me *where* are Fleeming Jenkin's arguments on the importance of single variation? Because I at present hold most strongly the contrary opinion, that it is the individual differences or *general variability* of species that enables them to become modified and adapted to new conditions.[47]

Two days later (2 February 1869), Darwin replied:

My dear Wallace,– I must have expressed myself atrociously: I meant to say exactly the reverse of what you have understood. F. Jenkin argued in the *North American Review* [*sic*; in fact, *North British Review*] against single variations ever being perpetuated, and has convinced me ... I always thought individual differences to be more important ...[48]

This clarifies what Darwin had meant in his ambiguous sentence of 22 January. He had not intended to say that single variations are 'of paramount power' but the opposite. To provide a paraphrase: 'I have always thought individual differences more important than single differences, but now I have come to the conclusion that *the former* are *even more important that I had ever thought.*' What Darwin had taken Jenkin's criticism to mean is that, under the hypothesis of blending inheritance, single variations would inevitably disappear, swamped by the large number of non-variant individuals (sampling effect) and diluted over the generations (blending effect). On the other hand, if a favourable variation was present in a large number of individuals, it would have a chance of being preserved.

Where was the problem? Immediately after the passage quoted above, Darwin continued his answer to Wallace thus (2 February 1869): 'I always thought individual differences more important, but I was blind and thought that single variations might be preserved much oftener than I now see is possible or probable ... I believe I was mainly deceived by single variations offering such simple illustrations, as when man selects.'[49]

In what way had Darwin been 'blind'? Clearly, he had not argued that sports played a major role in evolution. Quite the opposite, as shown by the two letters cited above. Like Wallace, he had always argued that 'individual differences' were the most important form of variation, because they formed a 'general variation' that affects all characters in a species and thus permits the simultaneous evolution of many characters.[50] So what exactly had Jenkin revealed to Darwin? In the two letters to Wallace cited above he uses the term 'single variations'. This could mean 'unique variations', but Darwin sometimes expresses himself as if this term were a synonym for 'occasional' variations.[51] When he wanted to explain his point, he always referred to monstrosities and to major 'structural deviations'. But the letters to Wallace show that the idea of 'single variations' is not linked to the qualitative role of the variation in the organism's physiology. For Darwin, extremely small variations could also be 'single' variations. Hence the major difficulty expressed by Darwin in the later editions of the *Origin of Species*, with explicit reference to Jenkin's article:

> [In previous editions] I saw the great importance of individual differences, and this led me fully to discuss the results of unconscious selection by man, which depends on the preservation of all the more or less valuable individuals, and on the destruction of the worst . . . Nevertheless, until reading an able and valuable article in the 'North British Review' (1867), I did not appreciate how rarely single variations, whether slight or strongly marked, could be perpetuated.[52]

This passage contains a major admission which in modern terms can be expressed as follows: 'I have always accepted that natural selection operates preferentially, if not exclusively, on characters expressing continuous variation, which I have called *individual differences*. But I argued as though natural selection accumulated these *differences* as if they were particles, which explains my constant use of terms such as *perpetuation*, *preservation*, *elimination* and *extermination* with regard to variations. I admit that in doing so I wanted to take advantage of all the ambiguity contained in the expression "individual difference". This formulation in fact encapsulates several notions that should be distinguished: the notion of a hereditary atom or particle that can be diffused in a population; the notion of a spectrum of continuous variation within a population (in this sense, all individuals are characterised by a certain deviation from the mean of the character); finally, the term "individual difference" can even cover that which I have always opposed, the "sport", which my friend Alfred Wallace calls "individual variation", and which refers to a unique (or very rare) variation which brusquely affects an individual. In fact, the intensity and the number (the frequency) of a "variation" are different kinds of questions. A slight variation may be rare or frequent, just like a large variation (traditionally called a "sport"). It would be useful to adopt a vocabulary that would enable us to distinguish two things: variation in terms of a class of individuals showing a heritable particularity, and variation as a spectrum of all the differences existing within a population. All these distinctions are somewhat obscured by my formulation "individual differences", which you will notice I have virtually always employed in the plural, a rhetorical device that no doubt helped me to avoid this difficulty.' This hypothetical declaration implies that Darwin did not have a genuinely populational conception of the process he was describing. The hypothesis of natural selection no doubt required such a theoretical understanding, but the man who had discovered the hypothesis was unable to provide it.

However, nothing remotely resembling the imaginary admission above can be found in Darwin's writings. Instead, Darwin's statements confirm the ambiguity of his conception of variation and heredity, as pointed out by Jenkin. In particular, there is Darwin's speculation on what he thought to be the physiological cause of variation, known as 'pangenesis', which was published a year after Jenkin's article, in *The Variation of Animals and Plants under Domestication* (1868).

It has sometimes been argued that this hypothesis was Darwin's response to the 'swamping effect'. From a strictly historical point of view, this is not possible. As has already been indicated, in 1865 Darwin wrote a manuscript entitled 'Hypothesis of Pangenesis' which contained most of what was to appear as chapter 27 of the *Variation*. Furthermore, as R. Olby has pointed out, in 1867 Darwin wrote a letter to Lyell in which he claimed that the idea was '26 or 27 years old'.[53] Thus pangenesis was not developed in response the problem of the 'swamping effect', but can be considered a kind of ready-made speculative response that Darwin employed in reply to Jenkin's arguments.

Pangenesis is a unifying hypothesis which was intended to explain the various forms of heredity and variation, and the profound unity of sexual reproduction, asexual reproduction and regeneration.[54] As the name indicates, the hypothesis argued that 'it is not the reproductive organs or buds which generate new organisms, but the units of which each individual is composed'.[55] In its general form, the hypothesis of pangenesis goes back to antiquity and can be seen throughout the whole of natural history.[56]

Darwin's particular contribution was to reinterpret the hypothesis in the new language of cell theory: all 'cells or units'[57] at all stages of their existence were supposed to emit 'gemmules' that contained protoplasmic characters. The gemmules were thought to accumulate in reproductive elements (seminal cells or the cells of generative buds). Sometimes they would fuse by 'elective affinity', and sometimes they would conserve their identity; they were also able, in the processes of individual development, to reconstitute cells that were the equivalent to their 'parent' cells. This model explained the fact that offspring are generally intermediate between their parents (some gemmules fuse); the reality of reversion (some gemmules do not fuse and reproduce in a latent state over the generations); and the existence of variability (sometimes due to reversion, sometimes due to the conservation in the gemmule of a modification acquired by its 'parent' cell, in other words the inheritance of acquired characteristics).

It should be noted in passing that in casting his model in this way, Darwin did severe violence to cell theory, since pangenesis implies that a cell can come from something other than a cell (a gemmule).

This outline reveals the existence of a subtle but deliberate link between this hypothesis and that of natural selection – between the two most unifying principles that Darwin developed. Darwin, however, did not make this relationship explicit. The word 'pangenesis' does not occur in the last two editions of the *Origin of Species* (1869, 1872), and no reference is made to selection in the chapter on pangenesis in the *Variation*. It appears that there is no relation between the doctrine of the origin of variation and that of its eventual fate, apart from a relation of subordination (variation is to selection as bricks and mortar are to the construction of a building).

Some changes that were made in later editions of the *Origin of Species* (1869, 1872) show that Darwin did in fact try to make a link between his two main hypotheses. In the passage in chapter 4 where he explains Jenkin's criticism with regard to the swamping effect, Darwin accepts the absolute legitimacy of Jenkin's argument that if a favourable variation appears in only one individual, 'the chances would be strongly against its survival . . . The justice of these remarks cannot, I think, be disputed.'[58]

If the selective process is to have any chance of succeeding, the same variation would have to appear simultaneously and recurrently in many individuals. At this point Darwin makes a surprising declaration, which raises the question to what extent he was prepared to abandon his doctrine of the primordial importance of omnidirectional individual variations. This declaration is the only response Darwin made to Jenkin; the passage cited is from the fifth edition (1869), two years after Jenkin published his article:

> It should not, however, be overlooked that certain variations, which no one would rank as mere individual differences, frequently recur owing to a similar organisation being similarly acted on . . . In such cases, if a varying individual did not actually transmit to its offspring its newly acquired character, it would undoubtedly transmit, as long as the existing conditions remained the same, a still stronger tendency to vary in the same manner. The conditions might indeed act in so energetic and definite a manner as to lead to the same modification in all the individuals of the species without the aid of selection. But we may suppose that the conditions sufficed to affect only a third, or fourth, or tenth part of the individuals, and several such cases could be given . . . Now, in such cases, if the variation were of a beneficial nature, the original form would soon be supplanted by the modified form, through the survival of the fittest.[59]

Does this argument refer to a marginal case, or does it imply a renunciation of the doctrine of the preponderant importance of non-oriented 'individual differences'? Although Darwin limits his remark to a certain kind of variation ('certain rather strongly marked variations', as the sixth edition has it),[60] these lines were clearly written under the pressure of the 'swamping effect' argument, which shows that enormous quantities of similar and simultaneous variants are required in order to shift the population norm. However, Darwin did not completely abandon his previous positions. He never seriously studied the problem of the minimal frequency of variation required for the selective process to succeed. Had he been asked, he would probably have argued that the theory of pangenesis could explain how, in the same conditions, large numbers of individuals varied in the same fashion and transmitted to their offspring a tendency to vary in the same way. After all, according to Darwin's hypothesis, each cell, at each moment, produces gemmules which bear the trace of its reactions to the internal and external environment, which may or may not be beneficial. That is why selection maintains its role. Even if variation is oriented, it may or may not be beneficial, and therefore subject to the action of natural selection.

These reflections, occurring late in Darwin's life, should not be misinterpreted. In particular, they should not be taken as indicating that Darwin had abandoned the explanation of adaptation by natural selection, for the simple reason that pangenesis is not an explanation of adaptation. Pangenesis only explains the transmission of acquired characters, whatever they are. In this sense, it agrees with Darwin's belief in the power of heredity. In fact, in the passage reproduced here, Darwin recognises that the efficacy of the selective process may depend on the hypothesis concerning the nature of inherited variation. Natural selection thus experienced its first major crisis on the very terrain that Darwin had imposed in order for it to be understood.

However, Jenkin's criticism of the hypothesis of selection also left open two potential escape routes. The first would have been to abandon the hypothesis of blending inheritance and to accept the idea that natural selection only operates on discrete entities that appear repeatedly. In this scenario, any variation with a slight advantage has a high chance of winning out. This was a reasonable answer to Jenkin's twofold argument on the swamping effect (concerned with numbers) and blending inheritance (which implies the dilution of the character). The second option would be to consider selection as acting on a genuine spectrum of continuous variation, on something having the character of a Gaussian distribution. But Darwin did not favour either of these alternatives. However, in the following half-century these two theoretical possibilities were to structure the theory of natural selection.

Part 2

Selection faced with the challenge of heredity: sixty years of principled crisis

The theory of natural selection had to deal with a major challenge which threatened it with extinction: the theories of heredity and variability. In one sense, this was only right – Darwin required that for selection to exist, there had to be variability and heredity. Of course, he did not imagine that these concepts, which he did so much to put at the centre of the biological stage, would come to threaten the principle of selection. The problem that dominated the theory of selection for half a century after Darwin can be summarised as follows: to what extent was Darwin's hypothesis of natural selection compatible with the reality of heredity?

In 1877, Galton was the first to formulate this requirement of compatibility, in the passage in which he put forward the project of a statistical theory of inheritance and that of a statistical theory of selection.[1] Galton later claimed that natural selection could not act effectively on 'individual differences' (in other words, on continuous variation) and argued in favour of a saltationist view of evolution through the selection of discontinuous variations. This thesis was the epicentre of an internal crisis which shook the Darwinian principle of natural selection for more than a half a century.

The English biometrical school, centred on Karl Pearson and Walter Weldon, developed in the wake of Galtonian statistics. With unprecedented rigour, Pearson and Weldon tried to establish the possibility of natural selection as Darwin understood it – acting gradually on continuous variations. Pearson paid particular attention to finding a theoretical justification for the possibility of natural selection, using complex statistical arguments. At the same time, the naturalist Weldon, using statistical tools forged by Galton and Pearson, tried to find unequivocal proof of at least one example of natural selection. But the biometricians' phenomenalist approach, linked to their argument with Mendelism, did much to discredit natural selection around the turn of the century.

The final assault came from the physiological and experimental theories of heredity in the years 1890–1910. The mutationists (de Vries, Bateson, Johannsen, Cuénot etc.), zealous converts to Mendelism, rejected the plausibility of Darwinian selection in the name of the argument of 'pure lines'. In genuine pure lines, there is an apparent (or 'phenotypic') variability which closely resembles Darwin's 'fluctuating' variation, but which is not heritable. Hence the affirmation that the Darwinian selection of individual differences quite simply had *no* modifying effect whatsoever. This explains why Mendelian genetics first appeared to have delivered the *coup de grâce* to Darwinism.

It was, however, Mendelism that was to provoke the renewal of Darwinism. Darwinism emerged from its confrontation with Mendelism as a binary theory, able to express the modification of species in a dual vocabulary – as the diffusion of hereditary atoms (genes) in populations and as the gradual modification of heritable quantitative characters. The genetic theory of selection involved both a kinetic description of the accumulation of favourable mutations, and a rehabilitation of the Darwinian idea that the modification of species is based on 'fluctuating' or continuous variation.

This is how the crisis of the selectionist hypothesis was resolved in the sixty or seventy years after the publication of the *Origin of Species*. The characters who appear in the pages that follow are well known in the history of statistics, of genetics and, in particular, of population genetics. In this respect, the present study owes a considerable debt to the pioneer work of William Provine.[2] However, this study differs from previous approaches in its emphasis on the reconstruction of the specific history of the concept of selection.

Three dominant conceptual themes need to be noted from the outset, because they form a unity of discourse during a period in which rationalisations were generally complicated, polemical and often confused. The first of these themes is that of the subordination of the theme of selection to that of heredity. This poses a major problem for a historical study of selection: the post-Darwinians, irrespective of whether they were critics or acolytes of Darwin, were first and foremost interested in heredity. So much so that the problem of selection was dealt with often merely as a consequence of the theory of heredity. This inversion of theoretical priorities was characteristic of the post-Darwinian period. The question of the origin of variation was no longer in a 'subordinate position', as Darwin put it.[3] It is thus scarcely surprising that the question of heredity is dealt with in some detail in Part 2.

The second conceptual theme is that of a continual questioning of the reality and role of the 'forces of return' in evolution. This line of investigation was polymorphous, vague and confused, but the theme is clearly there: reversion, return, regression (statistical or anatomical), atavism, ancestral heredity, recessivity. The presence of this vocabulary throughout the work of *all* post-Darwinian naturalists cannot be ignored. Modern biologists are quite right to be wary of this old terminology: they know (or, to be honest, they no longer know) how difficult it was to escape the traps contained in such language. The philosopher-historian, however, has to confront the confusion of the past and to reconstitute its ambiguous unity.

Where appropriate, the social context will be mentioned in the chapters that follow, but only if it clarifies a theoretical point. Although it would be too easy to explain the confusion of the post-Darwinian period by simply pointing out its ideological expressions, they cannot simply be ignored. Indeed, it would be very surprising if the controversies that opposed the partisans of Social Darwinism and the partisans of eugenics, or in other words, those who supported natural selection and those who supported artificial selection in human society, left no trace whatsoever on the theory of selection. The aim of this book is not to analyse these implications systematically, but some key examples will be explained in detail, in particular where major theoretical advances have gone hand in hand with distinctly unpleasant ideologies.

Part 2 opens with an analysis of Galton's work on heredity and selection, which was to be the source of both the biometrical rationalisation of selection and its mutationist opponent. Furthermore, it is in Galton's work that all the polymorphous themes of regression converge (Chapters 4 and 5). The following chapters examine in what way and in what forms the post-Darwinians established the possibility and the reality of the Darwinian hypothesis (Chapters 6, 7 and 8).

4 Galton and the concept of heredity

The work of Francis Galton (1822–1911) is particularly difficult to judge in terms of the history of science, and even more so in terms of the history of biology. It is tempting to ridicule the unbelievable arrogance of this Victorian aristocrat, the clear ideological bias that runs through his studies of heredity, the confusion that characterised his attempts at finding a mathematical expression for populational data, and finally the doubtful biological and statistical approach to so many aspects of human existence that formed the focus for all facets of his work. Through developments in genetics and the tragic consequences of eugenic propaganda, 20th-century history has rendered Galton's hereditarian ideas both outdated and suspect.

However, we need to get away from the idea of putting Galton's work and ideas on trial. In the 20th century 'heredity' has become so important – theoretically and practically – that the genesis of the concept needs to be thoroughly understood. The history of theories of heredity can no longer begin with the rediscovery of Mendel's laws in 1900, with all other ideas being seen as simply archaic. Mendelism undoubtedly represented a fundamental methodological break, but in the context of the idea – already current in the 1870s – that discovering the nature of heredity constituted a major challenge for biology. And in these terms, Galton's role was fundamental, for reasons which are inextricably linked to both theory and ideology.

Galton's ideas exercised a determining influence over the two traditions which, around the turn of the century, did battle over whether species could change through Darwinian selection. The English biometrical school looked to Galton as the inventor of the statistical methods it used to justify the Darwinian hypothesis of natural selection. But his influence was equally great on those such as Bateson, de Vries and Johannsen, who, initially as mutationists and then as the first Mendelians, claimed Galton's conceptual heritage and methods as their own, in order to prove the inefficacy of Darwinian selection. This apparently contradictory situation obliges us to understand the unique character of this theoretician, who was neither a naturalist nor a statistician nor a biologist, but who was nevertheless the first to gain widespread acceptance for a 'hard' concept of heredity, and who explained the necessity of translating the concept of hereditary variation into statistical terms.

This chapter shows how, beginning from a Darwinian starting point, Galton slowly constructed a theoretical schema in which the problem of heredity gradually took pride of place over the problem of natural selection. Galton's main theoretical contribution was to pose the problem of the relation between heredity and population. As R. C. Olby has pointed out, whatever Galton's motivation he should be credited with having established that heredity, variability and the reappearance of ancestral characters ('reversion') are different aspects of the same phenomenon, and not, as was generally thought at the time, distinct forces.[1] Galton continually reworked this idea, first in terms of a physiological theory of heredity, then in terms of a statistical theory of heredity. The former led him to reject Darwinian pangenesis; the latter led him to put forward a quantitative interpretation of what Darwin called 'small individual differences'. The whole was eventually synthesised into an abstract generalisation known as 'the law of ancestral heredity'. This chapter examines these three successive aspects of Galton's work. His direct attack on the Darwinian concept of natural selection is dealt with in the next chapter.

4.1 A speculative physiology of heredity

Galton's theoretical work on heredity spanned more than thirty years. The first elements can be found in an article published in 1865, 'Hereditary Talent and Character',[2] while one of his last published works on the question was a 'diagram of heredity', published in 1898 as an illustration of his 'law of heredity'.[3] After this point, and especially after the rediscovery of Mendel's laws, Galton abandoned the theoretical debate and concentrated on the eugenic implications of his hereditarian ideas.[4]

It is important to note that Galton's work was carried out at the same time as many other attempts at developing a theory of heredity. All of these theories were physiological, and proposed different ways of explaining the phenomena of heredity. As R. G. Swinburne has noted,[5] Galton's originality lay in the fact that he developed a quantitative and predictive theory, more precisely a *statistical* theory, based on the simplest facts of heredity: children look like their parents, but not completely; the children of exceptional parents are generally more 'mediocre'; characters may often jump a generation. By proceeding in this fashion, Galton, and the biometrical school that followed him, sought to respond to the highly varied nature of the different physiological theories of heredity and to develop a theory of heredity which would be as free as possible of all unjustified causal assumptions. It should nevertheless be noted that Galton initially expressed his views on heredity in terms which were completely independent of statistics.

Despite the fact that he was the grandson of Erasmus Darwin and a cousin of Charles, Galton did not come to the problem of heredity via the practice of natural history. Before 1865 – the year he published his first article on heredity – his complete bibliography lists only articles on geography and meteorology, which were largely inspired by an expedition to South Africa (1850–2). Galton's 1865 article was published six years after Darwin's *On the Origin of Species*. This book played a decisive role in Galton's subsequent interest in 'heredity', although in a very peculiar way, as he confessed to Darwin some years later: 'the appearance of your Origin of Species formed a real crisis in my life; your book drove away the constraint of my old superstition as if it had been a nightmare and was the first to give me freedom of thought'.[6]

The 'crisis' Galton refers to was a crisis of faith. As a geographer and ethnologist, Galton took Darwin's discussion of the origins of organs and instincts and applied this method to the development of religious feelings in man. In particular, he saw the Darwinian principles of heredity and of natural selection as a natural interpretation of the Christian doctrine of original sin. This can be seen in the following passage from *Hereditary Genius*:

> The whole moral nature of man is tainted with sin, which prevents him from doing the things he knows to be right. The explanation I offer of this apparent anomaly, seems perfectly satisfactory from a scientific point of view. It is neither more nor less than that the development of our nature, whether under Darwin's law of natural selection, or through the effects of changed ancestral habits, has not kept pace with the development of our moral civilisation . . . The sense of original sin would show, according to my theory, not that man was fallen from a high estate, but that he was rising in moral culture with more rapidity than the nature of his race could follow.[7]

Liberated by this realisation, Galton hastened to extend it to all religious and moral feelings, and more generally to the mental qualities of man. This generalisation was the subject of his first article on heredity, published in 1865 under the title 'Hereditary Talent and Character'. In this article, Galton sought to show that the mental and moral characters of humans are the product of the combined action of natural selection and heredity. In the list of behavioural traits and abilities which Galton thought were hereditary we find the following: intellectual ability, a predisposition to impulsive passions, parental love, the sense of original sin and the ability to pray. Galton never hid the fact that both the origin and the main motivation of his work could be found in his belief that mental and moral attributes are inherited. Paradoxically, therefore, reading the *Origin of Species* led Galton to do something Darwin initially did not do: extend the principles of selection and heredity to the mental capacities of humans.[8]

Whatever his original motivations may have been, Galton's 1865 discussion of 'hereditary talent and character' contains the outline of a general concept of heredity in which can be discerned the initial shape of three elements, each of which was to grow ever stronger in his work.

The first element is very close to what Weismann, twenty years later, was to call the doctrine of independence and continuity of the germ plasm. According to Galton, it was necessary to draw out the consequences of the idea that 'everything we possess at our birth is a heritage from our ancestors',[9] to accept the hypothesis that we are all nothing more than 'passive transmitters of a nature which we have received and which we have no power to modify'.[10] Curiously, Galton attributed this hypothesis to Darwin, and formulated it in terms of common descent:

> We shall therefore take an approximately correct view of the origin of our life, if we consider our own embryos to have sprung immediately from those embryos whence our parents developed, and these from the embryos of *their* parents, and so on for ever. We should in this way look on the nature of mankind, and perhaps on that of the whole animated creation, as one continuous system, ever pushing out new branches in all directions that variously interlace, and that bud into separate lives at every point of interlacement. [Original emphasis] [11]

In this passage Galton deals with the Darwinian principle of common descent in a manner that prefigures a rejection of the inheritance of acquired characters (compare with Figure 1 below).

The second general element contained in the 1865 article deals with the relation between racial inheritance and familial inheritance. Galton refused to restrict the phenomenon of heredity to characters that were always transmitted from parents to children. He emphasised the existence of an inherited transmission of individual qualities, characterised both by the variability shown by children from a given family, and a tendency to revert to ancestral types (closer to the mean, or 'meaner' in Galton's terminology).[12] Although the text is extremely vague on this point, it is possible to detect an initial outline of Galton's main idea that familial variability, racial variability and 'reversion' are merely different aspects of the same phenomenon, which can be termed 'heredity'. Galton explained how, although the children of exceptional parents are 'meaner' than their parents, they nevertheless resemble their parents more closely than they do average members of the population. It was on the basis of this observation that Galton put forward the first formulation of what he was later to call the eugenic project: 'if talented men were mated with talented women, of the same mental and physical characters as themselves, generation after generation, one might produce a highly-bred human race, with no more tendency to revert to meaner ancestral types than is shown by our long-established breeds of race-horses and fox-hounds'.[13] As for the means of such an improvement, as early as 1865 Galton proposed encouraging marriage between young men and women who had successfully passed a competitive examination dealing with every mental quality.[14]

Finally, the 1865 article contains the first exposition of what at the end of the 1890s was to become the 'law of ancestral heredity':

> The share that a man retains in the constitution of his remote descendants is inconceivably small. The father transmits, on an average, one-half of his nature, the grandfather one-fourth, the great-grandfather one-eighth; the share decreasing step by step in a geometrical ratio, with great rapidity.[15]

Galton gives no justification for this position, and indeed does not deal with it in any detail apart from deriding those who boast of their famous ancestors: 'the man who claims descent from a Norman baron, who accompanied William the Conqueror twenty-six generations ago, has so minute a share of that Baron's influence in his constitution that, if he weighs fourteen stone, the part of him which may be ascribed to the baron ... is only one-fiftieth of a grain in weight'.[16] Despite being an aristocrat by birth, and despite his inegalitarian and hereditarian convictions, Galton felt that 'hereditary nobility' was the bane of English society. However, the passage just quoted would barely merit attention if it did not constitute the first expression of an idea that was to become extremely important in the debates over heredity and selection that took place at the turn of the century. At first glance, the geometrical progression invoked by Galton ($\frac{1}{2}$, $\frac{1}{4}$, $\frac{1}{8}$, ... $\frac{1}{2}^{n}$) resembles the old idea of 'fractions of blood' and the terms 'thoroughbred', 'half-breed' etc. And indeed, Galton uses the language of the breeder, but with a completely different objective. Breeders spoke of the fractions of 'blood' in the context of the cross-breeding of races: at the second generation, a given horse was said to contain no more than a quarter of the 'thoroughbred' that had been cross-bred. Galton was not interested in this effect, but rather in discovering a formula or a law of heredity that would describe the contribution of each individual ancestor to the 'heritage' of the offspring. He was thus dealing with intra-racial inheritance. This idea contains the suggestion that, in his or her hereditary constitution, an individual 'inherits' as much from ancestors as from immediate parents, and that this constitutes a kind of racial inertia that is the cause of the observed tendency of children to return to the population mean.

The conceptual outline of Galton's ideas on heredity is thus relatively clearly prefigured in the 1865 text 'Hereditary Talent and Character'. From this article onward Galton's work was marked by the three characteristic elements of the general theory of 'heredity' that he was to elaborate over the next thirty years: (1) a physiological hypothesis: continuity of the germ line; (2) the need for a theory able to articulate resemblance and variability at the level of both family and race; (3) the notion of ancestral inheritance.

After the appearance of this article, Galton concentrated increasingly on research related to heredity. In one sense, his initial anthropological inspiration never waned: the inheritance of mental abilities was to be the subject of many of his publications. At first, Galton merely drew interminable genealogies of famous men.[17] Then, as he became more and more involved in the world of statistics, he came to explain the inheritance of intelligence in terms of psychometry[18] at the same time as he developed an increasingly effective and organised programme of eugenicist propaganda. By the end of the century, what we would today recognise as the eugenicist aspect of his work was solidly in place, both in terms of a methodological framework (the triad heredity–statistics–psychometry), and in terms of its institutional means (laboratories, learned societies, publishing house and soon learned journals).

This well-known aspect of Galton's work will not be dealt with in any further detail;[19] the rest of this section focuses on a few epistemologically important moments in the development of the general concept of inheritance. In the ten years that followed the 1865 article, this development principally involved a kind of speculative physiology, which led Galton to explore the paradoxical implications of the rejection of the inheritance of acquired characters.

It is true that these early articles do not express the radical rejection of the inheritance of acquired characters advocated by Weismann in the 1880s.[20] However, a doctrine of the continuity of the germ cells is clearly present, in the context of an open opposition to the Darwinian hypothesis of pangenesis. The beginning of the story is rather confused: in 1865, Galton attributed to Darwin the idea that the habits of a given individual are barely – if at all – transmitted to its descendants. On this basis, he drew the conclusion that 'our own embryos ... have sprung immediately from those embryos whence our parents were developed, and these from the embryos of *their* parents, and so on for ever'.[21] This position contains a relatively clear prefiguration of the idea of the continuity of the germ line. In

1868, Darwin outlined his famous 'provisional' hypothesis of pangenesis, which claimed that 'it is not the reproductive organs or buds which generate new organisms but the units of which each individual is composed'.[22] The 'units' in question were cells, which Darwin supposed continually developed 'gemmules' that 'were dispersed throughout the whole system' and congregated in the sex organs[23] so as to 'constitute the sexual elements', or the buds.[24] One of the implications of the hypothesis of pangenesis was that any structural modification of the 'matter' of cells, induced by use or by non-use, by food or by any external factor, would be found in the fabric of the gemmules, which continually budded from the cells.[25]

It is easy to imagine Galton's discomfort when he learned of this hypothesis. Pangenesis was incompatible with the idea of a continuous chain of embryos from one generation to the next. In *Hereditary Genius* (1869), Galton relegates the idea of the continuity of embryonic elements to a concluding paragraph, written in an allusive, poetic and equivocal style that is nevertheless revealing in terms of the author's metaphysical beliefs:

> We may look upon each individual as something not wholly detached from its parent source – as a wave that has been lifted and shaped by the normal conditions of an unknown, illimitable ocean. There is decidedly a solidarity as well as a separateness in all humans, and probably in all lives whatsoever; and this consideration goes far, I think, to establish that the constitution of the living universe is a pure theism, and that its form of activity is what may be described as co-operative.[26]

In this book, Galton in fact adopted the hypothesis of pangenesis and actually put it forward when he argued that the gemmules 'circulate in the blood',[27] a detail that Darwin had carefully avoided presenting in the *Variation*, but which, in his first exchanges with Galton, he had also been careful not to reject.

In fact, Galton was not convinced by the hypothesis of pangenesis, and as soon as *Hereditary Genius* was published he set out to test it experimentally. This involved carrying out massive blood transfusions in rabbits with different coat colours. The transfused animals were systematically isolated from the donors and then allowed to reproduce with rabbits of the same type. If mixed-race offspring had appeared, the hypothesis of pangenesis would have been experimentally confirmed. The experiment was carried out between 1869 and 1871 and aroused Darwin's interest, as shown by his correspondence with Galton.[28] In 1871, Galton published his results:

> ... I have bred eighty-eight rabbits in thirteen litters, and in no single case has there been any evidence of alteration of breed ... The conclusion from this large series of experiments is not to be avoided, that the doctrine of pangenesis, pure and simple, as I have interpreted it, is incorrect.[29]

Darwin's reaction was immediate. In a letter to the scientific journal *Nature*, he explained that he had never argued that the gemmules circulated 'in the blood'. First, this would make little sense for protozoans or plants. Equally, it was possible that the gemmules diffused 'through the tissues, from cell to cell, independently of the presence of vessels'.[30] This reply bore more than a passing resemblance to a sleight of hand, as Galton pointed out in his ironic reply published a few days later in the pages of the same journal.[31] Had not Darwin declared in the pages of *Variation* that gemmules were 'minute granules ... which circulate freely throughout the system'[32] and that they 'must be thoroughly diffused',[33] adding that this process was not improbable, 'considering ... the steady circulation of fluids throughout the body'?[34] With heavy irony, Galton admitted that he had misunderstood the hypothesis and the interest shown by Darwin in the transfusion experiment (the correspondence shows that he helped Galton choose and buy the rabbits, and that he himself raised some of them). Galton concluded that he had been 'sent on a false quest by ambiguous language', but that his work had nevertheless 'not been in vain'.[35]

On the basis of his transfusion experiments, Galton drew the prudent conclusion that the reproductive elements (the gemmules) were not located solely in the blood. Two interpretations were possible: either the gemmules were to be found solely in the sex organs or they were also present in the blood but under these experimental conditions they died so quickly that they did not have time to get to the sex organs. The stronger hypothesis for a conception of heredity was obviously the former; it was this hypothesis that Galton studied and found more and more evidence for. In this he was soon to be followed by a growing number of cytologists, who put forward new arguments against pangenesis: for example, the problem of the placental barrier,[36] or that of the incompatibility of the Darwinian notion of the gemmule with the dogma of cell theory (the sex cells, like all other cells, are always the product of other cells, and never of 'gemmules').

A year after the controversy with Darwin, Galton published an article entirely devoted to heredity in general, without any particular concentration on mental abilities. The article was given the deliberately provocative title of 'On Blood-Relationship'.[37] As with all Galton's fundamental and innovative texts, it is rather difficult to read, but is nevertheless fascinating because of its paradoxical conclusions in terms of the link between 'kinship' and 'heredity'. The difficulty of the text comes from the fact that Galton uses a physiological language to put forward problems which in fact require a statistical theory of heredity. The article opens with these words:

> I propose in this memoir to deduce, by fair reasoning from acknowledged facts, a more definite than now exists understanding of the meaning of the word 'kinship'. It is my aim to analyse and describe the complicated connexions that bind an individual, hereditarily, to his parents and to his brothers and sisters, and, therefore, by an extension of similar links, to his more distant kinsfolk. I hope by these means to set forth the doctrines of heredity in a more orderly and explicit manner than otherwise practicable.[38]

Galton's fundamental hypothesis was that each individual is composed of two parts: one is latent and 'only known to us by its effect on his posterity'; the other is 'patent' and 'constitutes the person manifest to our senses'.[39] These two parts are formed by discrete elements, largely independent of each other. In this text, Galton remains evasive as to the anatomical nature of these elements, but insists that they all come from lineages that diverge from a common group of elements contained in the fertilised egg. Some of these original elements develop into 'organic units', or cells, which, organised together, constitute the differentiated body, while others remain in the 'latent' state in the reproductive organs. These latent elements are not 'organic units' (cells), but 'germs', which may develop into organic units.

The distinction between 'patent' and 'latent' elements was the inspired result of a deduction from the observation of 'reversion', that is, of characters that disappear for one or more generations, only to reappear subsequently. It should be remembered that for Darwin, reversion constituted a marginal example of 'heredity' that was nevertheless extremely significant, because it leads to the distinction between development and hereditary transmission. Galton adopted this idea and generalised it. For him, reversion was one of the most banal expressions of heredity. In any given case, the reversion effect becomes mixed up with the variability that can be observed between offspring from the same parents: each parent transmits to descendants qualities that he or she does not express in his or her 'manifest person', but which may be expressed in one or other of the children, or in a more distant descendant. In this light, reversion and variability are two aspects of the same phenomenon, 'heredity'. Reversion shows that in each individual there is a 'latent' variability that is substantially greater than that of the elements which form the manifest person.[40] This implies 'that large variation in individuals from their parents is not incompatible with the strict doctrine of heredity'.[41]

However, Galton should not be given more credit than is his due. To what extent does the theory of heredity outlined in his 1872 article 'On Blood-Relationship' really prefigure Weismann's distinction between the 'soma' and the 'germen'?[42] Weismann's 1883 doctrine affirmed that the

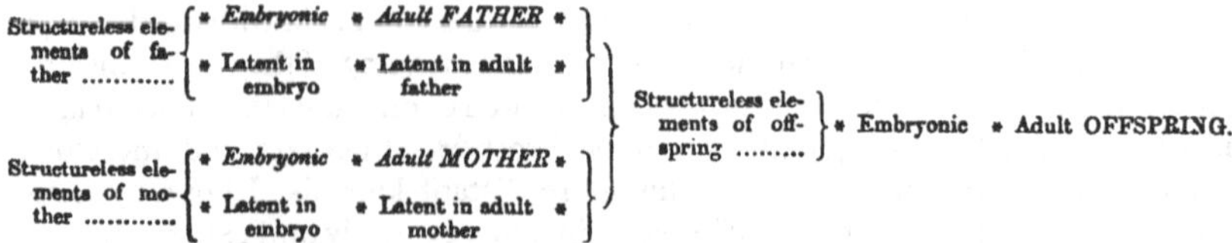

Figure 1 Schematic representation of a physiological hypothesis of heredity (F. Galton, 'On blood-relationship', 1872). The 'unstructured elements' are the elements contained in the fertilised egg before any embryonic differentiation has taken place. The elements in italics indicate 'weak lines of heredity'.

'germ line' is continuous, independent and virtually immortal. This doctrine was expressed in terms of cell theory, and implied a total rejection of any effect of the somatic cell lines on the germ line. This is the precise implication given by Weismann to his rejection of the inheritance of acquired characteristics. Galton did not go so far. In 1872 he did not exclude the possibility that the 'patent elements' had a weak effect on hereditary transmission, as is shown by Figure 1. In this diagram, the elements labelled in italics are those of the visible body and indicate 'weak lines of heredity'. This amounts to admitting that the somatic tissues may occasionally form 'germs' that participate in the formation of sex cells. However, three years later, in 'A Theory of Heredity' (1875), Galton's tone with regard to the theory of pangenesis became substantially harsher: 'developed' germs (those that gave rise to differentiated tissues) were now conceived of as being sterile, and fecundity was the sole property of the non-developed elements, which were based in the sex organs.[43]

Despite these reservations, Galton and Weismann clearly had an identical conception of the germinal elements. They both thought that during ontogeny differentiating cells retained only certain elements of the ovarian germ plasm: for Galton, the 'patent' elements were the result of a sampling of the 'latent' elements,[44] exactly like Weismann's germinal idioplasma. In this respect, the two authors shared the same archaic vision of the relation between development and heredity. Biological inheritance is represented as a kind of plant sucker which advances indefinitely in time, producing individuals in each 'generation' who, in their 'manifest person', only partially represent the original plasma.[45]

Galton's physiological speculations on heredity thus included an intuition that was extremely close to the doctrine of the continuity and independence of the germ line. However, the most remarkable aspect of Galton's work was the statistical method it led to. If heredity does indeed consist of 'latent' elements that are far more varied than the apparent characters of individuals, then it can no longer be considered to be the mere degree of resemblance between the *persons* of parents and children. This is why Galton suggested in 1872 that the formation of a child should be seen as a random sampling of the latent characters of the two parents. In making this suggestion, he used the image of an urn. From the point of view of heredity, the parents could be thought of as two 'urns' containing balls; the child can be thought of as a random sample of a common urn that contains all the parental balls. The most fundamental expression of heredity is thus the difference between the child and the two parents, or rather, between the child and the mean value of the two parents.[46] Heredity therefore expresses itself first and foremost in variability: because of their latent characters, each parent can be seen as a reservoir of characters which, being more varied than those expressed in their person, constitute a link between the parents' offspring and the population as a whole. For Galton, 'heredity' is neither individual nor racial, but somewhere between the two.

Galton provides a suitable summary of the significance of his physiological discussions of heredity:

> We cannot now fail to be impressed with the fallacy of reckoning inheritance in the usual way, from parents to offspring, using those words in their popular sense of visible personalities. The span of the true hereditary link connects, as I have already insisted upon, not the parent with the offspring, but the primary elements of the two, such as they existed in the newly impregnated ova, whence they move respectively developed. No valid excuse can be offered for not attending to this fact, on the grounds of our ignorance and proportionate values of the primary elements; we do not mend matters in the least, but we gratuitously add confusion to our ignorance, by dealing with hereditary facts on the plan of ordinary pedigrees – namely from the *persons* of the parents to those of their offspring. [Original emphasis][47]

This statement may appear ironic, given Galton's single-minded collection of the 'pedigrees' of famous men in both *Hereditary Genius* (1869) and in his book *English Men of Science* (1874). But it also shows the kind of reality that Galton was searching for in his interminable study of 'pedigrees' and 'descent'. Despite the physiological framework that he initially used, Galton was primarily interested in the variability of individuals in populations, to such an extent that his initial physiological speculations were to prove no more than a starting point for the major problem upon which he brought his talents to bear: the description of heredity at the populational level.

4.2 From the statistical theory of heredity to ancestral heredity

The starting point of Galton's statistical studies of heredity was a paradox based on two simple observations: (a) children resemble their parents, but are very different from each other; (b) 'exceptional' parents (in whatever terms) generally have more 'mediocre' offspring than themselves. From these two pieces of common sense, the paradox can be summarised as follows: children 'tend' to resemble not their parents but rather the typical mean of the 'race'.

This paradox runs through all Galton's writings on heredity, without exception, and lent itself to a statistical treatment at a time when the necessary mathematical tools were beginning to be widely known. Expressed in statistical terms, Galton's conception of heredity oscillated between two poles: the family and the general population, each with its own 'mean' and 'variability'. In each generation both of these poles made their effects felt: in a given family, the mean value of the children for a given character did not coincide with that of the parents (although they resemble them more than other children do), nor with that of the population as a whole, but tended towards the latter. As Galton put it in his 1865 article on hereditary talent, the children are 'meaner' than their parents. Galton initially called this centripetal tendency *reversion*, following Darwin, who, like other naturalists, had used it to describe the reappearance of ancestral characters. Galton later abandoned this term in favour of *regression*, a word that rapidly lost all hereditary connotations and is still used to describe a fundamental tool of statistics.

In this section, we shall see how Galton developed the statistical concept of heredity as a synthesis of familial variability, of social (or populational) variability and of regression, and how this concept was ultimately degraded to the notion of 'ancestral heredity'. The history of statistics will be invoked only to the extent that it relates to theories of heredity and selection. This section contains a number of contemporary figures which, in Galton's work, were often more than mere illustrations; they were the means of developing a hypothesis or even its proof. Sometimes they constituted a substitute for an overly complex mathematical proof, providing a mechanical model; sometimes, when the mathematical proof was available, they revealed its conceptual significance.

4.2.1 Statistics and heredity

Once again, we begin with Darwin, for whom the most important aspect of hereditary variation consisted of 'individual differences' that he considered differences of degree, of an 'infinitesimal' amplitude. These 'individual differences' or 'undetectable graduations' closely resembled what biometricians were later to call 'continuous variation' in a population. However, Darwin never had a clear concept of what statisticians call a 'population'. He did not use this word in the modern sense, and always spoke of plants and animals only in terms of 'races' or 'varieties'. It appears that early in his life he knew of Quételet's work 'Physique sociale' (1835),[48] but Quételet's typological vocabulary ('the average man' as a 'type') would have put him off. Indeed, the method used by Darwin in the *Origin of Species* shows that he did not really think of a 'race' as a spectrum of continuous variation. Slight variation was to be found in certain individuals who differed from the mean character and transmitted this difference to their offspring, who in turn amplified it, renewing the process, leading over time to the creation of a new variety (cf. Chapter 3). In other words, Darwin did not think in terms of the statistical distribution of an observable character in a population. That is why he always dealt with heredity and selection in terms of 'individuals', because that was all he could oppose to the 'type'. Furthermore, his correspondence with Galton during the 1870s shows that he had great difficulty in understanding arguments formulated in statistical terms.

On the other hand, right from the outset, Galton viewed Darwin's 'individual differences' in terms of language and images taken directly from Quételet's social statistics. Beginning with his 1865 text on 'hereditary talent and character', it is clear that, for Galton, 'heredity' was not to be considered only in terms of the individual, but rather in terms of the family and the race, each of which had to be understood in terms of its variability. In this sense, from the very beginning Galton had broken with the 'individualistic' conception of heredity used by Darwin. In *Natural Inheritance* (1889) he pithily summed up the theoretical choice that had underpinned his research: 'The science of heredity is concerned with fraternities and large populations rather than with individuals, and must treat them as units.'[49]

However, it was no simple matter to apply the concepts and methods of statistics to the problem of heredity. In his autobiography[50] Galton explains how he first encountered Quételet's statistical methods in 1863, two years before the publication of his first article on heredity. He had been immediately attracted to Quételet's examination of the 'law of deviation from an average', which he had discovered when reading the 1849 English translation of Quételet's 1846 book *Lettres à Son Altesse Royale le Duc Régnant de Saxe-Coburg et Gotha sur la théorie des probabilités, appliquée aux sciences morales et politiques*. What did Quételet have in mind when he applied this law?[51]

The 'law of deviation from an average' is also known as 'the law of the frequency of error', or the 'normal law', but this last term cannot be found in either Quételet's texts or Galton's early statistical writings. The term 'normal curve' only appeared somewhat later,[52] but to simplify matters we will use it throughout. This is not primarily for simplicity's sake. The 'law of deviation from the mean' has its origin in a long tradition of research on how best to estimate the position of a point on the basis of a series of observations all of which contain an element of error. This was a classic problem of early-19th-century statistics, and was of particular interest to astronomers, military engineers (artillery) and geographers (geodesy).

As is well known, Laplace developed a fundamental insight into this problem. In a series of memoirs published in the second decade of the 19th century, Laplace showed that if the measurement of a given phenomenon results from a great number of independent observations, these measures should be approximately distributed according to Gauss' law. Laplace described this distribution as the curve of 'possible error'.

In his 1846 *Lettres*, Quételet used this mathematical tool to interpret anthropometric data, thus giving it a new methodological significance, as has been pointed out by Stigler.[53] Quételet used Laplace's theorem to determine whether a series of real objects (and not mere measures) could be considered homogeneous. Laplace's theorem implied that a group of measures affected by the same major causes, and varying only in terms of many minor, accidental causes, should be distributed according to Gauss' law. Quételet's innovation was to use the Gaussian distribution as a way of detecting groups of homogeneous *objects*. He thus made explicit what had previously been merely implicit in Laplace's work: a Gaussian (or 'normal') distribution is a necessary and sufficient condition of homogeneity. The Laplace–Gauss law thus left the arcane realm of the estimation of error (in the measurement of a given object) to become a tool for detecting homogeneity in groups of real objects. In particular, it became a method for identifying 'populations' as objective entities. If, for example, the chest size or stature of soldiers was approximately distributed according to Gauss' law, this would indicate that it was a real population, within which variation was merely accidental. For Quételet, a Gaussian distribution revealed both order in apparent chaos, and also an underlying ideal type that nature tries to attain, implying that variation has no real significance. This would also explain why Darwin, if indeed he did read and understand Quételet, would hardly have been attracted by his concept of a 'population'.

Galton, however, took up Quételet's ideas. As has already been indicated, he was immediately impressed by the strange properties of the 'law of deviation from the mean' and by the possibility of applying it to a large range of anthropological, physical and social data (propensity to commit crime, for example). His admiration for this law, which reveals order in chaos, never waned. In *Natural Inheritance* (1889), he wrote:

I know of scarcely anything so apt to impress the imagination as the wonderful form of cosmic order expressed by the 'Law of Frequency of Error'. The law would have been personified by the Greeks and deified, if they had known it. It reigns with serenity and is complete self-effacement amidst the wildest confusion. The huger the mob, and the greater the anarchy, the more perfect is its sway. It is the supreme law of Unreason. Whenever a large sample of chaotic elements are taken in hand and marshalled in the order of their magnitude, an unsuspected and most beautiful form of regularity proves to have been latent all along.[54]

Galton's first reference to the 'law of deviation from an average' appeared in *Hereditary Genius* (1869). The book contains an appendix which gives a clear explanation of Quételet's method, and which accepts that a distribution with the appearance of an error frequency curve is a sufficient indication of uniformity in a population. In passing, Galton provides a totally new graphic illustration of the law. In this illustration (see Figure 2), the distribution of height in a given population is represented by a series of points, like bullet holes on a target. Following the spirit of Quételet's method, Galton explains that if two 'races' are represented in this fashion, even the untrained eye will probably recognise two separate groups. The metaphor of a target follows Quételet's idea that nature 'aims at ideal types'. But it also contains the first hint of the representation of a population as a 'cloud' of points. Given the importance that the 'cloud' image was to have in both statistics and population biology, it is worth noting where it came from. Galton's early scientific articles virtually all dealt with meteorology. A 'cloud' is a suggestive image for an aggregation that conserves its characteristics even though its constitutive elements are constantly renewed. This is also the case for a living 'population':

The vital statistics of a population are those of a vast army marching rank behind rank, across the treacherous table-board of life. Some of its members drop out of sight at every step, and a new rank is ever rising up to take the place vacated by the rank that preceded it ... The population retains its peculiarities although the elements of which it is composed are never stationary, neither are the same individuals present at any two successive epochs. In these respects, a population may be compared to a cloud that seems to repose in calm upon a mountain plateau, while a gale of wind is blowing over it. The outline of the cloud remains unchanged, although its elements are in violent movement and in a condition of perpetual destruction and renewal.[55]

Although this passage is taken from a later period of Galton's life, it explains why he represented the population as a cloud of points. A living population is not a distribution of fixed objects: it is perpetually being renewed, and the problem is to understand how its general form is maintained over the generations.

It was at this point that Galton's conceptions clashed with those of

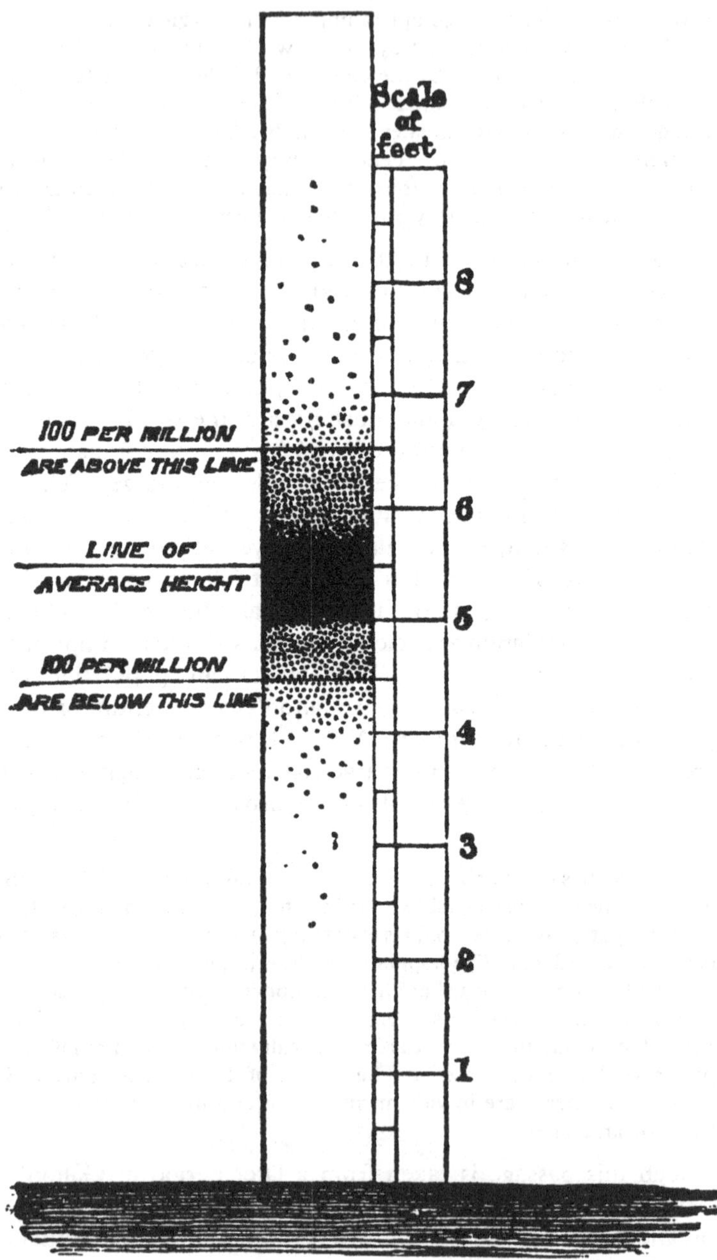

Figure 2 Illustration of the 'law of deviation from an average', from the first article in which Galton makes reference to this law (F. Galton, *Hereditary Genius*, 1869).

Quételet. The first signs of unease can be discerned in the pages of *Hereditary Genius* (1869). This book is an argument in favour of the inheritance of mental faculties in families. Quételet's curve, applied to characters such as height, suggests that populations are objectively homogeneous. In other words, differences in characters that have a Gaussian distribution are merely due to a large number of accidental and independent causes. The normal distribution thus appears to oppose even the possibility of the individual inheritance of characters. However, it is difficult to imagine Galton accepting the reduction of heredity to the accidental influence of parents on their children. After the publication of *Hereditary Genius* (1869), Galton's major theoretical objective was thus fixed: reconcile the law of deviation from the mean with heredity.

How could the phenomena of heredity be reconciled with the observation that many measurable characters found in natural populations (e.g. height) showed an approximately Gaussian distribution? Galton's solution is given in a remarkable 1877 article, 'Typical Laws of Heredity'. This paper represented a vital conceptual step forward for both mathematical statistics and population biology. Galton clearly distanced himself from Quételet and all those who had applied the law of deviation to biometrical or anthropometric data:

> Let me point out a fact which Quételet and all writers who have followed in his path have unaccountably overlooked, and which has an intimate bearing on our work to-night. It is that, although characteristics of plants and animals conform to the law, the reason of their doing so is as yet totally unexplained. The essence of the law is that differences should be wholly due to the collective actions of a host of independent *petty* influences in various combinations. [Original emphasis][56]

But of course the processes of heredity can hardly be considered '*petty* influences'. Galton drew the following conclusion, every word of which needs to be weighed carefully:

> The processes of heredity must work harmoniously with the law of deviation, and be themselves in some sense comformable to it.[57]

The realisation that there must be such a conformity 'in some sense' was the result of two complementary approaches. The first was theoretical, and consisted of a critical study of the importance of the law of deviation. The second was empirical, and took the form of his famous experiments on sweet-peas.[58] By using these two approaches, Galton altered the usual interpretations of both the normal law and heredity, and thus achieved the genuine reconciliation that he had been looking for.

The mathematical and statistical techniques used by Galton for this synthesis were typical of his ability to resolve a theoretical problem by the use of imagery, leaving the algebraic formalisation of the solution to others.[59] The problem was as follows. Laplace put forward several preconditions to the application of the law of the frequency of error. Among these, Galton noted that the factors which combine to produce the error must be (a) independent in their effects; (b) equal; (c) definable as either 'above the mean' or 'below the mean'; and (d) infinitely numerous. As he noted in 'Statistics by Intercomparison' (1875), these conditions are highly artificial,[60] in particular the first three, which have barely any relevance whatever to social and biological phenomena. It therefore follows that if social and living data do conform to the law of error, it is for reasons other than those invoked by the theory of error.[61] However, Galton could not deal mathematically with the question in Laplace's analytical terms, and his objection appeared to take the form more of a pious wish than of a rigorous proof. For this reason he immediately began to test Laplace's restrictive hypotheses by actually constructing an apparatus that would show the kind of processes that could give rise to a normal distribution.

This device was known as a quincunx. Galton mentions it for the first time in his 1877 article 'Typical Laws of Heredity', but if the text is to be believed, it had already been in use for several years.[62] For Galton, the quincunx can be thought of as a 'mechanical illustration of the cause of the curve of frequency'.[63] It consists of a vertical slim glass-fronted box. The upper part consists of a funnel into which lead shot is poured. Under the funnel a series of horizontal bars block the path of the lead shot. When a piece of lead shot hits a bar, it may equally well go to the right or the left. The shot eventually falls to the bottom of the apparatus, which is divided into a series of slots of equal width (see Figure 3). The apparatus clearly shows the effect of a binomial distribution which is similar to Quételet's conception of the error curve. If there are a large number of pieces of lead shot and a large number of rows of bars, the resultant distribution should be approximately Gaussian. The greater the number of obstacles met by the lead shot, the greater the dispersion in the slots at the bottom of the apparatus (the 'curve' will be wider). Thus the quincunx constitutes a model of the law of deviation.[64]

However, the quincunx is not only a teaching tool for convincing a relatively unmathematical public of the reality of the law of deviation. It is really an instrument for providing a very special kind of demonstration. Galton did not experiment with the quincunx, he used it as an image to devise thought experiments. The inventive stimulation provided by the quincunx is shown less by its use than by a series of drawings, in particular from the period 1876–7, when Galton sought to demonstrate that the law of deviation could be applied to a population composed of a series of heterogeneous sub-populations. The crucial passage can be found in a letter to George Darwin, written on 12 January 1877 (a reproduction and

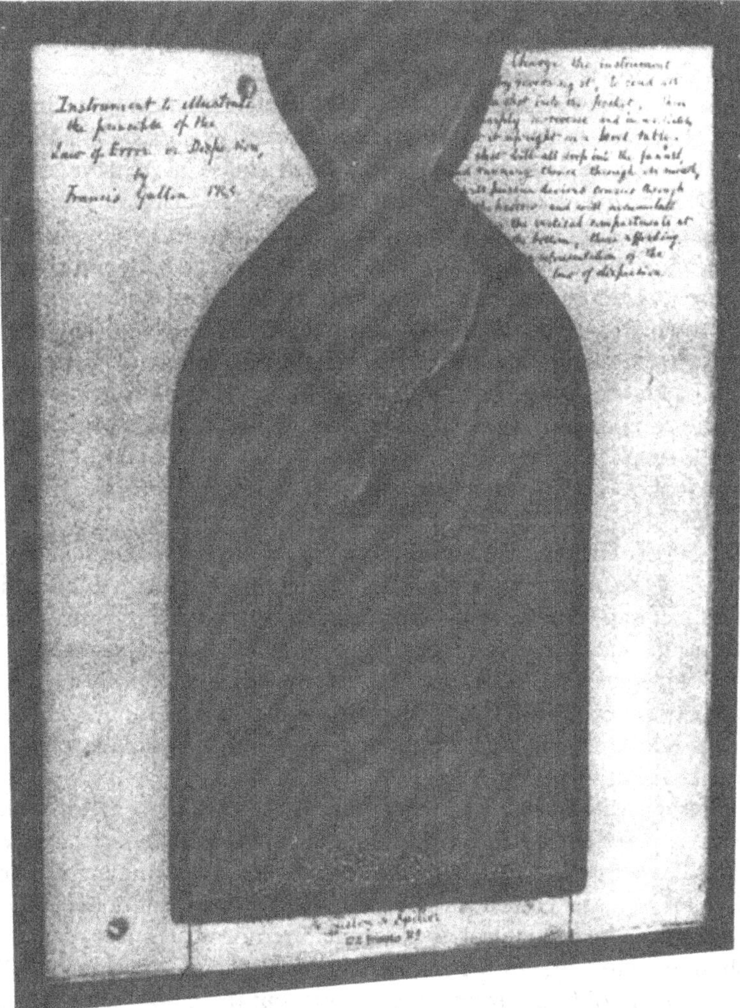

Figure 3 Original model of the quincunx, which is still on display at University College, London. The hand-written comments are Galton's. Top left: 'Instrument to illustrate the principle of the Law of Error or Dispersion, by Francis Galton F.R.S.' Top right: 'Charge the instrument by reversing it, to send all the shot into the pocket. Then sharply reverse and immediately set it upright on a level table. The shot will all drop into the funnel, and running thence through its mouth, will pursue devious courses through the harrow and will accumulate in the vertical compartments at the bottom, there affording a representation of the law of dispersion.'

commentary can be found in Karl Pearson's biography of Galton).[65] This letter is extremely interesting, because it contains drawings of something that Galton's texts had only described in words.

In these drawings (Figure 4) a new element has been introduced into the quincunx. It appears that this version only ever existed on paper, but we know that Galton was particularly skilled at making cardboard models; he may well have made a model of this version of the apparatus. In drawing II, halfway up the device there are a series of compartments that are similar to those to be found at the bottom. The bottom of each compartment consists of a trapdoor that can be opened independently. Imagine that the doors are all shut, and that a handful of shot is poured into the top of the apparatus: the resulting distribution in the middle-layer compartments would be approximately Gaussian. If the trapdoor of one of the compartments was then opened, the shot would be redistributed, and a roughly normal distribution would again be found in the bottom compartments. The same experiment can be carried out for each of the middle-layer compartments. If the number of pieces of lead shot is sufficiently large, and if the number of rows of bars is sufficiently great, the shot from each of the middle-layer compartments will be redistributed in a miniature normal distribution. Each one of these miniature distributions has its own mean value (ideally that of its middle-layer compartment), but taken as a whole, the original handful of lead shot used in the experiment will be distributed in the bottom compartments according to a normal distribution, exactly as if no middle-layer compartments had existed.[66]

Despite the fact that in explaining his experiment Galton made no reference to heredity, the implications are obvious: two successive generations correspond to two levels of compartments in a quincunx. The children of all parents sharing a given character (for example of the same height) can be thought of as the miniature normal distributions that are obtained when one of the middle-layer compartments is opened. These children are different from their parents, but the influence of the parents is shown by the children's mean value, which is not the same as those of the children of parents from another compartment (another frequency class), although in reality the mean of the children's values is not equal to the parental value (see below).

At the same time as he pursued his abstract considerations of the interpretation of the 'law of deviation' (or normal law), in 1875 Galton began an experiment on the sweet-pea (*Lathyrus odoratus*). The aim of the experiment was to try and isolate the effect of 'heredity' in a population, independently of all other considerations (in particular independently of natural selection). The character Galton chose to measure was seed

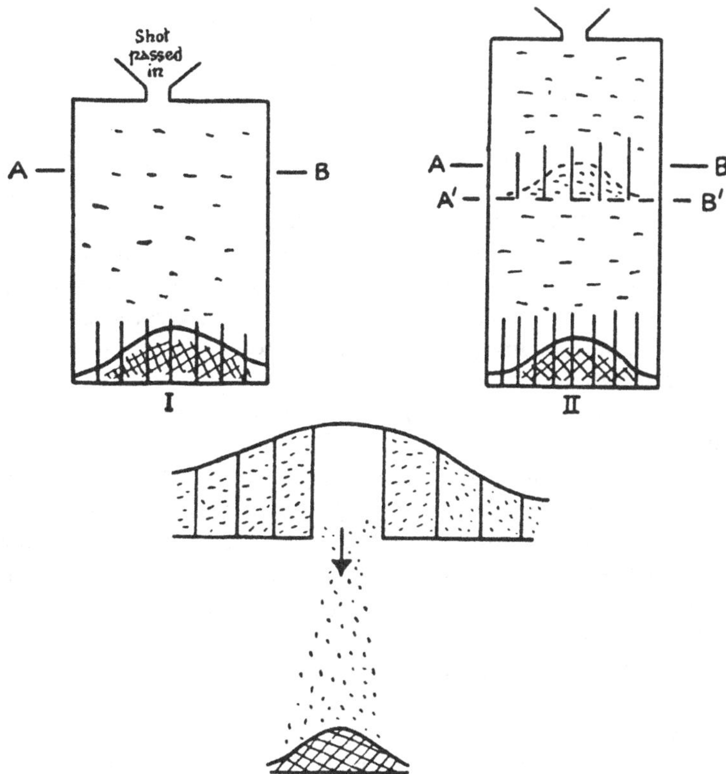

Figure 4 Schematic representation of the principle of the two-level quincunx, taken from a handwritten letter from Galton to his cousin George Darwin (1877). The figure shows how a normal distribution can be divided into several other normal distributions. This version of Galton's original (poor) drawing is by Pearson (K. Pearson, *The Life, Letters and Labours of Francis Galton*, 1914–30, vol. III B).

weight. In a large population, this character is distributed approximately according to the law of deviation. Once this had been established, Galton formed seven lots of seeds of different weights. Each lot contained identical seeds, which thus corresponded to one of the 'degrees' on the statistical scale used to order the seeds. By 'one degree of deviation', Galton meant one unit of dispersion, equivalent to 'probable error': one degree of variation in absolute value corresponds to that degree of variation that has half the values above and half below that point. Expressed in relative terms (+1°, +2°, etc.), the degree of deviation divides one of the halves of the normal curve into two equal areas. The second degree of deviation is obtained by dividing the remaining error, and so on. Figure 5 illustrates the procedure. The seven groups of seeds isolated by Galton corresponded to the seven degrees of deviation (−3° to +3°, or the classes $M - 3Q, M - 2Q, M - Q, M, M + Q, M + 2Q, M + 3Q$, where $M =$ the overall mean and $Q =$ the probable error). In this way, seven lots of seeds were constituted, each containing seven packets of 10 seeds ($7 \times 7 \times 10$), which Galton sent to various friends and colleagues so that they could grow them in different places in the United Kingdom. The aim was to compare the daughter seeds with the mother seeds. The repetition of the experiment was intended to reduce as much as possible the direct effect of external conditions and of natural selection, in order to reveal the pure effect of 'heredity'. Galton hoped to be able to interpret this effect using the model of the quincunx, and to show 'how' heredity conformed to the 'law of deviation', if this was indeed the case.

The results of this experiment can be summarised in two remarkable conclusions, presented in the same 1877 article as the quincunx model.[67] The first deals with 'family variability': the variability of the daughter seeds was the same, whatever the parental form. Furthermore, in each case the distribution of the daughter seeds followed the law of deviation, exactly like the lead shot released from a given compartment in the quincunx. Finally, the mean of each of the seven sibling categories had the same relative rank as the relevant parental strains. The second conclusion was that the offspring tended to differ from the parental type and to approach the overall mean, because the means of the sibling categories were closer together than those of the seven parental categories. Galton called this effect 'reversion', thereby expressing the fact that offspring tended to return to the 'mean ancestral type'. Expressed in a quantitative manner, 'reversion' is the product of the mean deviation of the offspring divided by the deviation of the corresponding parental type (or, and this amounts to the same thing, the mean value of the offspring for a given parental value). Galton remarked that this value was constant for all the seed groups, and thus called it the 'reversion coefficient'. The signifi-

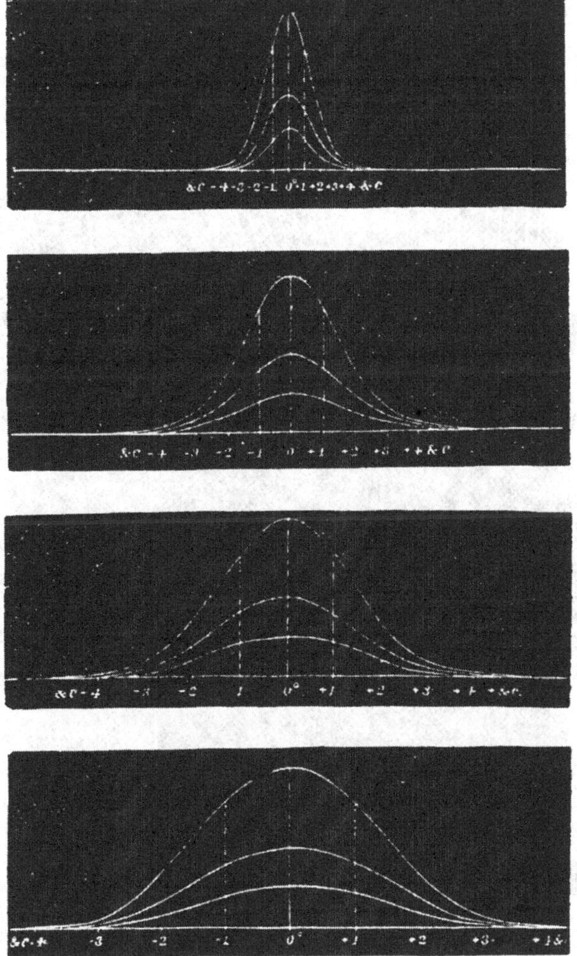

Figure 5 The 'degrees' of the frequency curve. This figure illustrates the theoretical basis of the procedure used by Galton to divide his sweet-peas according to seed weight (F. Galton, 'Typical laws of heredity', 1877).

cance of this coefficient can be seen in Figure 6: OA, OB etc. are the deviations of various parental types; OA_1, OB_1, etc. are the mean deviations of the products. The invariance of the reversion coefficient $\left(\frac{OA_1}{OA} = \frac{OB_1}{OB} = \ldots\right)$ shows that reversion follows 'the simplest law there is', a linear law. Later, Galton was to abandon the term 'reversion' for

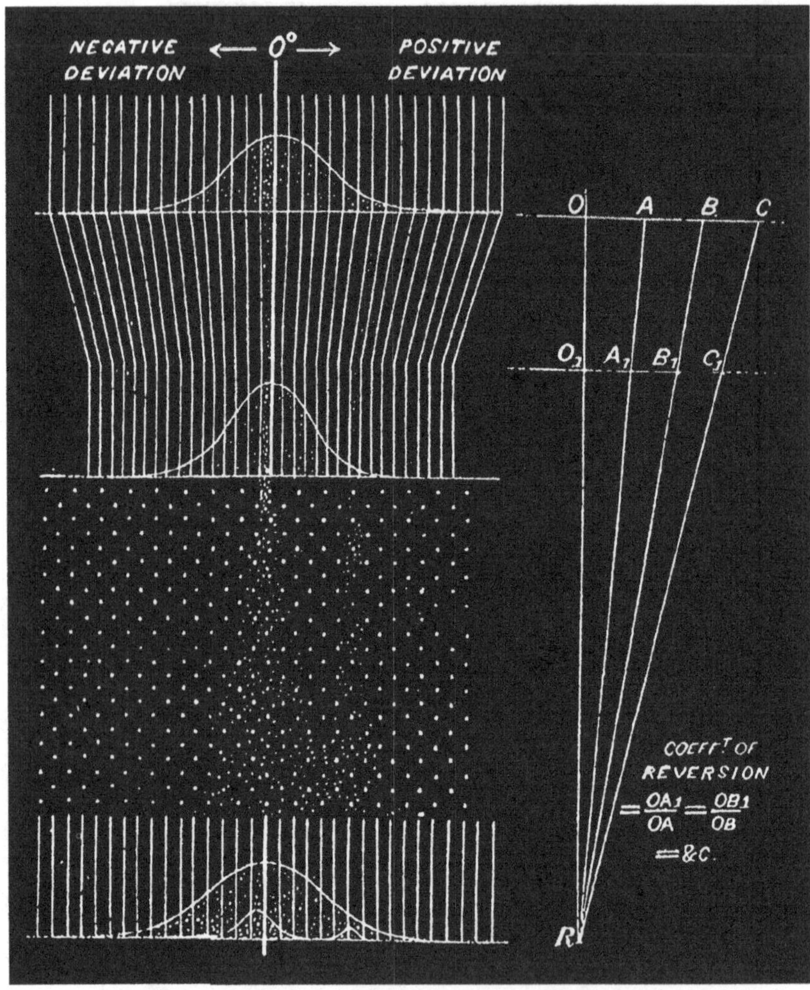

Figure 6 First appearance of the regression coefficient in the statistical literature. It is called here the '*reversion* coefficient'. For further details, see text (F. Galton, 'Typical laws of heredity', 1877).

'regression' and to speak of the 'coefficient of regression' and the 'curve of regression' in statistical terms that were increasingly separate from the problem of heredity.

I can now present an overall picture of what in 1877 Galton meant by the 'statistical theory of heredity'. The terms he used indicate without the shadow of a doubt that he hoped to resolve a problem in Darwin's writings. He explained that his sweet-pea experiment revealed the effect of 'descent' in a population.[68] If 'descent' is studied in its pure form, that is, independently of other processes such as sexual or natural selection, it is expressed by two effects: family variability and reversion. These two effects function in opposite directions: family variability leads to variation, reversion leads to homogenisation. If 'descent' (heredity) consisted only of family variability, the variation of a given race would increase with each generation.[69] If, on the other hand, reversion were the only effect of descent, populations would be homogeneous, and the only form of heredity would be racial heredity; that is, there would be no inheritance of differences. In reality, the two processes interact, producing a stable distribution of characters which is reproduced in each generation.[70] Furthermore, each of these processes conforms 'in some sense' to the law of deviation, in that they can all be expressed in terms of this law. In an appendix which was clearly written later, Galton gave an analytic treatment of the problem[71] based on a trivial use of the 'law of probable error' and of a well-known theorem of the additivity of errors (better known today as the additivity of variances, but this term only entered the statistical vocabulary in 1918).[72] A simple presentation of this algebraic demonstration shows the significance of a 'typical law of heredity'.

Galton takes as given the fact that the weight of seeds in the general population conforms to the classic Gaussian law, or the law of 'probable error'. The measure of dispersion being the 'modulus' ('c'),[73] it follows that the frequency curve of deviations from the mean is given by:

$$y = \frac{1}{c\sqrt{\pi}} \cdot e^{-x^2/c^2}$$

where x is a given variation in the measured phenomenon. This equation expresses the original state of the population. The state of the next generation can be obtained by applying (in whatever order) the effects of 'reversion' and 'familial variability'. Of course, it is assumed that neither natural selection nor assortative mating plays a role, because the aim is to describe the pure effect of 'descent'.

Reversion is expressed by a fractional coefficient that reduces variability in the overall population. This coefficient is 'r' and the variability after reversion is 'rc'. At this stage, the state of the population is given by:

$$y = \frac{1}{rc\sqrt{\pi}} \cdot e^{-x^2/r^2c^2}$$

In other words, the normal curve is narrower.

On the other hand, family variability ('v') tends to dilate the curve. The variability in the general population becomes:

$$c' = \sqrt{v^2+r^2c^2}$$

This is a straightforward application of the theorem of the additivity of the squares of the deviations. We can now express the state of the population in the offspring generation as:

$$y = \frac{1}{c'\sqrt{\pi}} \cdot e^{-x^2/c'^2}$$

In graphic terms, the normal curve has become broader.

Finally, the sweet-pea experiment showed that the overall distribution remains stable from one generation to another. Overall variability therefore also remains the same, and we can thus write:

$$c' = c = \sqrt{v^2 + r^2c^2}$$

Whence:

$$c^2 = \frac{v^2}{1 - r^2}$$

Galton thought that this final equation,[74] which he was later to call the 'law of regression', was the fundamental statistical law of heredity. It clearly expresses a simple relationship between 'racial variability' (c), 'familial variability' (v) and 'reversion' (r). It is important to note that even though it was expressed within the theoretical apparatus of the normal law, it was not deduced from this law: it is only valid under the empirically observed condition of the stability of distributions from one generation to another.

The whole conceptual schema could be presented in the form of the quincunx. Figure 6 shows in a single drawing the two statistical expressions of heredity in a given population: contraction (due to reversion) and dilation (due to familial variability). Reversion is artificially introduced into the quincunx by the use of sloped ramps. Familial variability is obtained from each of the compartments in the upper part of the apparatus, each of which represents a family. The order of the two processes does not matter, because this is not a physiological hypothesis but a statistical description. Figure 7, taken from *Natural Inheritance* (1889), summarises the three concepts that Galton hoped the quincunx would illustrate. Figure 7A shows the straightforward quincunx and the approximately normal distribution it produces. In Figure 7B, the addition of the middle-layer compartments shows the relationship between heredity and variability. In terms of the resultant distribution, there is no difference between this modified apparatus and that shown in Figure 7A, but it shows how a normal population may in reality be composed of a series of heterogeneous sub-populations, each with its own mean. If that was all heredity consisted of, each generation would be more variable than the

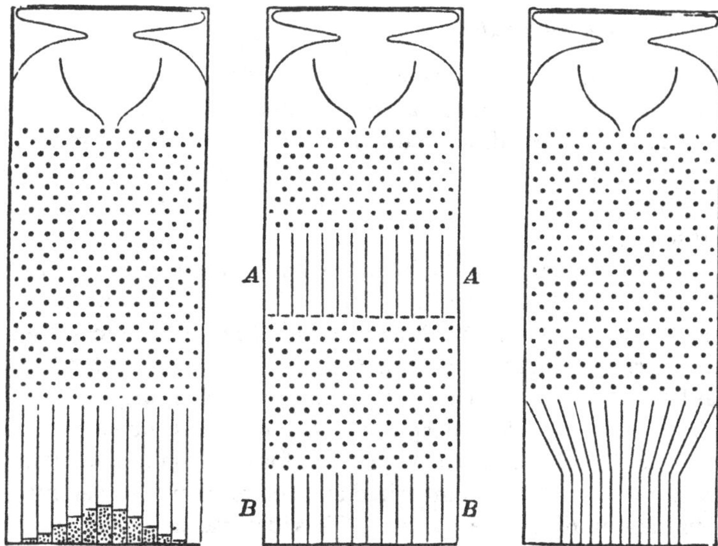

Figure 7 'Mechanical Illustration of the Cause of the Curve of Frequency' (F. Galton, *Natural Inheritance*, 1889). The apparatus on the left is the standard quincunx. It illustrates the statistical notion of the 'error curve'. The AA compartments on the modified quincunx (middle) illustrate the distribution of parents in various classes. The shot at the bottom will be slightly dispersed around the value of the compartment (children of given parents are more variable than their parents). The apparatus on the right illustrates the antagonistic process of 'regression' (children are 'meaner' than their parents).

preceding one, or in other words, the normal curve would become ever more dilated and flat. Figure 7C shows the antagonistic process of 'reversion'. It should be noted that the model does not deal with the physiological significance of these phenomena.

Whatever the inherent difficulties of the analysis, its effects on naturalists should not be underestimated, in particular on those who, following Galton, sought to integrate Darwinism and biometrical methods. Galton's statistical studies gave a quantitative and functional interpretation to concepts which in Darwin's writings remained undeveloped and even mysterious: 'individual differences', 'heredity', 'variability' and 'reversion'. Not only did Galton present a way of measuring such phenomena, he also showed that they could be viewed as complementary expressions of 'descent' considered as a 'pure' effect. Later on we will return to the fact that Galton's 1877 article was also the first in which natural selection was presented in statistical terms.

4.2.2 *Ancestral heredity*

The culmination of the statistical theory of heredity in Galton's work is well known: the notorious 'law of ancestral heredity'. The history of this law has often been described,[75] but it is such a rich source that every analysis highlights a different aspect. It is a sad story that summarises all the principal ambiguities of post-Darwinian rationalisations of heredity and selection.

The fundamental aim of the law of ancestral heredity (as Pearson was later to call it)[76] was the integration of *all* hereditary phenomena in a single conceptual framework or expression. This theoretical objective can be discerned in all of Galton's writings on heredity from 1865 to 1898. It was based on a simple, naive and somewhat archaic intuition that Galton was never to abandon, despite the failure of his increasingly complex attempts to find a statistical expression for it. This fundamental idea was that individuals not only inherited characters from their parents, but also from their more remote ancestors. This idea was developed in purely statistical terms, without any consideration of the implications for the nature of heredity. It was in these terms that certain biometricians, such as Pearson, understood and used the 'law of ancestral heredity'. Galton, however, had in mind an idea that was much more ancient: ancestral heredity led him to reintroduce the idea of heredity acting at a distance. Thus he opposed the Darwinian concept of heredity, which strictly speaking was a factor that acted directly and immediately from parents to offspring. The concept of *ancestral* heredity was in fact a way of reintroducing the idea of a genuinely racial inheritance.

Beneath the surface, the concept of ancestral heredity was linked to a critical consideration of the relevance of the Darwinian concept of natural selection. Galton had gradually become convinced that natural selection would have no effect if it acted on continuous variation that was transmitted by blending inheritance. He estimated that in the case of continuous variation the return to the mean ancestral type (as postulated in the law of 'regression') would in the long run overcome the action of selection, thus rendering impossible the gradual transformation of populations. The basic idea was that gradual selection could only temporarily alter the mean of a population: once selection was relaxed, the population could revert to its 'racial type' (or mean).

The debate around 'ancestral heredity' that took place at the end of the 19th century thus raised two fundamental problems: on the one hand, 'ancestral heredity' was presented as the logical culmination of a purely statistical, but somewhat confused, theory of heredity. On the other hand, it was clearly opposed to the Darwinian theory of natural selection – all the more remarkable because Galton had been one of its earliest proponents. The rest of this chapter deals with the first of these problems; the evolutionary aspect is dealt with in Chapter 5.

The idea of a law of ancestral heredity initially emerged in Galton's work as a development of the law of 'reversion'. How did this transition take place?

In terms of its declared aim – the establishment of a general law of heredity – Galton's 1877 article on the sweet-pea experiment had important limitations. The role of sexuality in heredity was completely ignored (sweet-peas had been chosen precisely because they were self-fertilising); the experiment was based on seeds that had been arbitrarily grouped, whereas their natural distribution was continuous; and the experiment used a species with non-overlapping generations. Galton tried to apply the method he developed for sweet-peas to a measurable human character, hoping both to generalise the law of reversion and to better understand its conceptual significance. He found that the 'reversion' to the population mean was strictly proportional to the variability.

Figure 8 shows the table of figures which was the basis of the generalisation of the law of reversion. This table, published in 1885, shows data for the height of children of parents of a given stature.[77] It enables us to understand the remarkable change of perspective that this new subject had produced in Galton's vision of heredity. The 'mid-parent' is an imaginary parent of composite sex, whose value for the character under study is the mean of the two real parental values, following transformation of the maternal value to the same scale as that of the paternal value (for example, in the case of height, the maternal value was multiplied by 1.08).[78] This technique enabled Galton to study a sexually reproducing population as though it were asexual. It is only justified if the child does not resemble one parent more than the other, that is if there is no preferential transmission depending on sex. Galton's interpretation of the data presented in this table is easy to follow. Each row of the table gives the distribution of the heights of the children of parents of a given stature (for example 72.5 inches). These distributions suggest an interpretation according to the normal law. Furthermore, the median value of each of these distributions is closer to the general mean than is the height of the mid-parents. A study of the data shows that, in all cases, the children's deviations from the mean are approximately ⅔ that of the parents' deviations, as shown in Figure 9. In this way, that 'the mean filial regression towards mediocrity was directly proportional to the parental deviation from it'[79] – a proposition that had first been demonstrated for sweet-peas in 1877 – was shown to be equally true for human stature. This character

TABLE I.

NUMBER OF ADULT CHILDREN OF VARIOUS STATURES BORN OF 205 MID-PARENTS OF VARIOUS STATURES.
(All Female heights have been multiplied by 1·08).

Heights of the Mid-parents in inches.	Heights of the Adult Children.														Total Number of		Medians.
	Below	62·2	63·2	64·2	65·2	66·2	67·2	68·2	69·2	70·2	71·2	72·2	73·2	Above	Adult Children.	Mid-parents.	
Above ..	..	..	..	..	..	..	..	..	..	..	..	1	3	..	4	5	..
72·5	..	..	..	..	..	..	..	1	2	1	2	7	2	4	19	6	72·2
71·5	..	..	..	..	1	3	4	3	5	10	4	9	2	2	43	11	69·9
70·5	1	..	1	..	1	1	3	12	18	14	7	4	3	3	68	22	69·5
69·5	..	..	1	16	4	17	27	20	33	25	20	11	4	5	183	41	68·9
68·5	1	..	7	11	16	25	31	34	48	21	18	4	3	..	219	49	68·2
67·5	..	3	5	14	15	36	38	28	38	19	11	4	..	..	211	33	67·6
66·5	..	3	3	5	2	17	17	14	13	4	..	..	..	..	78	20	67·2
65·5	1	..	9	5	7	11	11	7	7	5	2	1	..	..	66	12	66·7
64·5	1	1	4	4	1	5	5	..	2	..	..	..	..	..	23	5	65·8
Below ..	1	..	2	4	1	2	2	1	1	..	..	..	..	..	14	1	..
Totals ..	5	7	32	59	48	117	138	120	167	99	64	41	17	14	928	205	..
Medians ..	..	..	66·3	67·8	67·9	67·7	67·9	68·3	68·5	69·0	69·0	70·0	..	..	..	..	..

NOTE.—In calculating the Medians, the entries have been taken as referring to the middle of the squares in which they stand. The reason why the headings run 62·2, 63·2, &c., instead of 62·5, 63·5, &c., is that the observations are unequally distributed between 62 and 63, 63 and 64, &c., there being a strong bias in favour of integral inches. After careful consideration, I concluded that the headings, as adopted, best satisfied the conditions. This inequality was not apparent in the case of the Mid-parents.

Figure 8 Double-entry table on the basis of which Galton discovered the general statistical importance of regression (F. Galton, 'Regression towards mediocrity in hereditary stature', 1885).

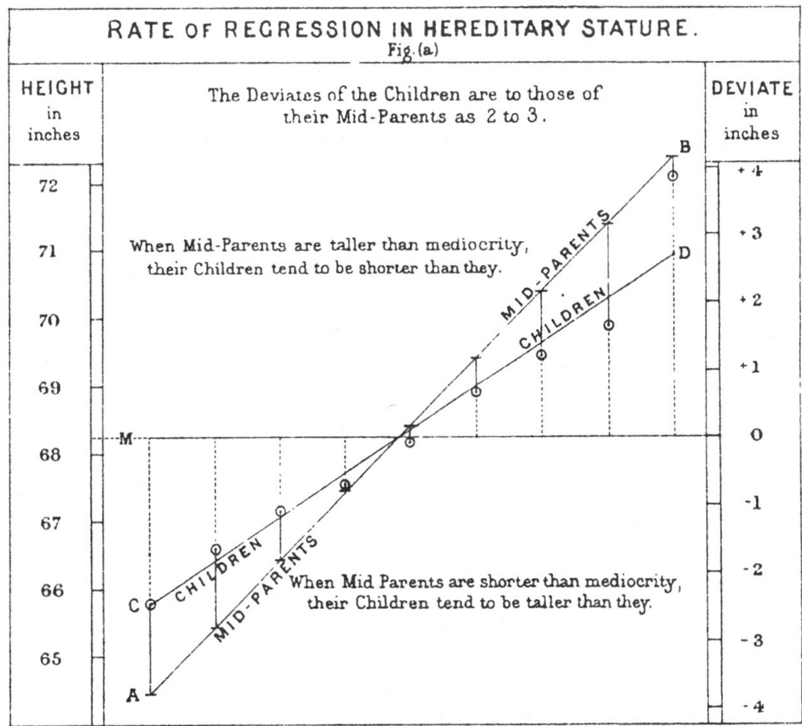

Figure 9 Diagram showing the reciprocal character of the statistical regression linking children and parents. The more the parents are 'exceptional', the more the children are (proportionately) 'mediocre'. The reciprocal is also true (F. Galton, 'Regression towards mediocrity in hereditary stature', 1885).

thus suggested the generalisation of what Galton henceforth called 'the law of regression'.[80]

Why did Galton abandon 'reversion' for 'regression'? His articles give no explicit explanation, but this change was undoubtedly due to a technical problem linked to the mathematical analysis of the coefficient of 'reversion'. The data in the table shown in Figure 9 suggest this problem quite clearly. It is extremely tempting to apply the same procedure to the *columns* as to the *rows*. The columns give the frequency distribution of the mid-parent values for children of a given height. As with the rows, these distributions are approximately normal. Furthermore, the median value of each parental distribution is closer to the overall population mean than the category of children's height to which they correspond. In other words, from a strictly statistical and descriptive point of view, the phenomenon of a return to the mean is not only observed for the children of parents of a certain stature, but also for the parents of the children of a certain height. Or, to express the matter more simply, not only do the parents always have children who are more 'mediocre' than they are, but the children also always have parents who, on average, are more 'mediocre' than *they* are (see Figure 9).[81] It would clearly be absurd to describe the second phenomenon as being a 'return' or a 'reversion': it makes no sense to talk of a return (reversion) *in the parents* of a character described *in the children*. This is doubtless the reason why Galton chose to use the conventional formula 'regression to the mean' or 'statistical regression', thus giving it a purely descriptive sense. In this sense, there is indeed a 'regression' of parental height on the height of a given group of children, and this regression can also be measured by a coefficient (in this case, of $\frac{1}{3}$). Similarly, it is possible to calculate a 'regression' coefficient for all kin relationships, in both directions. Galton, of course, did not hesitate to do precisely this.[82]

Under these conditions, the 'regression to the mean' took on a very different meaning to that of the 'law of reversion' which Galton had previously thought to be the general law of heredity. This was an empirical result, but it was closely linked to the way that the data had been classified. The result was expressed both from children towards parents and from parents towards children, from brother to brother, from uncle to nephew and from nephew to uncle. In other words, the contraction of the distribution was a result of the statistical method employed, and was not the specific product of 'heredity' or of a given kin relationship.

Galton was thus led to separate the statistical tool of regression from the theoretical context in which it had been originally conceived, and to recognise that it could be applied in all cases where there was a certain degree of correspondence between two sets of data. As he later wrote, describing the 1877 analyses of 'reversion' in the sweet-pea: 'I was then blind to what I now perceive to be the simple explanation of the phenomenon.'[83] In fact, the root of the problem was the statistical method of regression, not the theory of heredity, no matter how 'statistical' it was. The idea of a 'general theory of heredity' thus slipped through Galton's fingers even as he thought he had grasped it. 'Regression' was indeed a universal aspect of heredity, but it was applicable to such a wide range of data sets that it became impossible to consider it to be the general law of heredity. Realising that 'regression' was merely a formal instrument for measuring the degree of correlated variation of two variables, he began to look elsewhere for his law of heredity. He found it in a genuinely biological 'regression' which would be able to explain the observed statistical regression of children on parents. This was the hypothesis of 'ancestral heredity'.

In fact, Galton never used the term 'law of ancestral heredity'. Karl Pearson attributed the law to Galton, while at the same time reinterpreting it in order to give it a sense that it did not have in Galton's work.[84] Galton did however write of a 'general law of heredity' that had a 'statistical' nature and which defined 'The average Contribution of each several Ancestor to the total Heritage of the Offspring'.[85] As such, this law only appeared in 1897, but an explicit outline of it can be found in Galton's 1885 article on regression, in particular in an appendix[86] which is reproduced word-for-word in an article of 1886[87] and in a paragraph of his 1889 book on *Natural Inheritance.*[88]

We will not enter the maze of inconsistencies and modifications of the 'law' of ancestral heredity; I shall merely outline the main questions raised by its initial formulation (1885) and by its last (1897), where the concept gives rise to a 'general law of heredity'.

The primitive version of the notion of 'ancestral heredity' is given in the following terms in Galton's 1885 article on the inheritance of height in man:

> When we say that the mid-parent contributes two-thirds of his peculiarity of height to the offspring, it is supposed that nothing is known about the previous ancestor. We now see that though nothing is known, something is implied, and that something must be eliminated if we desire to know what the parental bequest, pure and simple, may amount to.[89]

Posed in this way, there were two aspects to the problem. The first concerned the statistical theory of heredity as a purely descriptive tool. In studying all the children of parents of a given height, one reasons as though all the parents of this category contributed in the same way to the character observed in the children. Galton did not think that this was true. Profoundly marked by his interminable genealogical analyses, he was convinced that knowledge of the character possessed by ancestors could improve the prediction of the expected character in the children. The concept of ancestral heredity thus tended to extend the analysis of statistical regression by including data for generations before that of the parents. The idea of a 'contribution' of previous generations to the observed character in the children thus becomes clear: Galton was in fact groping towards what was later to be called multiple regression. In this method, which Galton undoubtedly sensed without consciously developing, the regression coefficients of the children on the various generations of ancestors are not fixed *a priori,* but are merely statistical and descriptive parameters that do not have any particular causal significance. This interpretation was the only one that could give a precise operational sense to the concept of ancestral heredity, as Karl Pearson was later to point out (see below).

However, Galton's 1885 article does not refer to this purely descriptive sense of ancestral heredity. Quite the opposite. Galton thought that ancestral heredity would provide an explanation of the offspring regression observed in each generation:

The explanation of it is as follows. The child inherits partly from his parents, partly from his ancestry. Speaking generally, the further his genealogy goes back, the more numerous and varied will his ancestry become, until they cease to differ from any equally numerous sample taken at haphazard from the race at large. Their mean stature will then be the same as that of the race; in other words, it will be mediocre. Or, to put the same fact into another form, the most probable value of the mid-ancestral deviates in any remote generation is zero.[90]

This aspect of ancestral heredity serves a very different function from that of the first. It again raises the question of a genuinely racial heredity or inertia. From this point of view, offspring regression appears as an expression of a tendency that pushes individuals indefinitely back to 'the ancestral mean type', or the race type, which is confounded with the current population mean. In other words, ancestral heredity is the cause of the tendency to regress to the mean. Indeed, Galton believed that the law of the regression of offspring on parents (especially the ⅔ coefficient) could be 'deductively' derived from conjectures about the nature of ancestral heredity.[91] It should be noted that this hypothesis has as a tacit precondition that the population mean does not change. As we shall see in the next chapter, this empirically unfounded precondition led Galton to deny the efficacy of Darwinian selection.

How did Galton in 1885 calculate the 'separate contribution of each ancestor to the heritage of the offspring'? His reasoning has often been rightly presented as an example of an astonishing mathematical confusion. Galton expresses himself as though he were a solicitor reading a will: he considers the 'heritage of the offspring' as the sum of the various 'pieces of property' ('parental bequest', 'grandparental bequest' etc.).[92] Each ancestral generation transmits something, and the child's heritage is the sum total (=1) of the various inherited fractions. All these fractions obviously have to pass through the parents, who themselves contribute only partially to the child's character. Galton's calculation is based on estimated or putative regression coefficients of the various ancestral generations on each other. The unit of measurement is the deviation (D) of the mid-parent from the overall mean. The character of other generations (the offspring and the ancestral generations) is expressed as a fraction of D. The only empirical information available to Galton was the estimation of the regression of children's height on the mean height of their parents (⅔) and of the converse regression of parental means on children (⅓).[93] If the deviation of the average parent was D, the expected 'heritage' of the child could be estimated as $\frac{2}{3}$D. Furthermore, extrapolating the $\frac{1}{3}$ coefficient of the regression of the parental mean on the children to previous generations, Galton estimated that the probable deviations of the mid-ancestors of rank 2, 3, ... *n*, should be $\frac{1}{3}$D, $\frac{1}{3} \times \frac{1}{3}$D etc., or: $\frac{1}{3}, \frac{1}{9}, \ldots \frac{1}{3^n}$. The sum of the deviations of all the mid-generations contributing to the inheritance of the offspring is thus:[94]

$$D\,(1 + \tfrac{1}{3} + \tfrac{1}{9} + \text{\&c.}) = D\,(\tfrac{3}{2})$$

This calculation presupposed (wrongly) that the unknown regression coefficients could be obtained by the multiplication of those that were known. However, the value thus obtained for the 'heritage' of the offspring ($\frac{3}{2}$D) was substantially greater than the value observed for the character ($\frac{2}{3}$D). In order to bring the hypothesis into line with the facts, it was necessary to envisage a coefficient of loss at each generation, which Galton described as 'a succession tax'. If this 'tax' is the same in each generation, it must be greater than $\frac{4}{9}$, which corresponds to the factor of the sum of ancestral deviations ($\frac{3}{2}$D) and the deviation observed amongst the children ($\frac{2}{3}$D), because $\frac{3}{2} \times \frac{4}{9} = \frac{2}{3}$. This value of $\frac{4}{9}$ is the proportion in which the ancestral inheritance is transmitted from one generation to another. Another estimate, based on the hypothesis that the 'tax' on ancestral inheritance diminishes at each generation, led, via a confused and arbitrary calculation, to the value of $\frac{6}{11}$. Noting the proximity of the two values ($\frac{4}{9}$ and $\frac{6}{11}$), Galton concluded:

> The results of our two valid limiting suppositions are therefore (1) that the mid-parental deviate, pure and simple, influences the offspring to $\frac{4}{9}$ of its amount; (2) that it influences it to the $\frac{6}{11}$ of its amount. These values differ but slightly from $\frac{1}{2}$, and their mean is closely $\frac{1}{2}$, so we may fairly accept that result. Hence the influence, pure and simple, of the mid-parent may be taken as $\frac{1}{2}$, of the mid-grandparent $\frac{1}{4}$, and so on. That of the individual parent would therefore be $\frac{1}{4}$, of the individual grandparent $\frac{1}{16}$, of an individual in the next generation $\frac{1}{64}$, and so on.[95]

Neither the many inconsistencies in this argument nor the avalanche of exaggerated testamentary metaphors that accompany it will be dealt with here. The main point to note is that it is fatally flawed by the implicit precondition that the unknown regression coefficients could be obtained by the multiplication of those that were known.[96] Without this postulate, Galton could not even have *estimated* the geometrical series. And yet it is precisely this series, which Galton undoubtedly wanted to arrive at, that is the only thing to emerge clearly from this statistical confusion. As has already been indicated with regard to Galton's early physiological ideas, from his first article on heredity (1865) onwards, Galton was obsessed with the idea that the ancestral contributions were divided up according to a geometrical progression, the sum of which was equal to 1.[97]

Despite the fact that the idea of ancestral inheritance runs throughout Galton's work on heredity, it is only in its final formulation (1897) that Galton describes it as a 'general law of heredity'. In this article, the geometrical series ½, ¼, ⅛ etc. takes on the form of a proof. However, as several commentators have correctly noted,[98] two radically different formulations of the 'law' can be found in the space of a few lines.

One of the formulations is a repetition of the idea put forward in Galton's articles from the 1880s:

Let M be the mean value from which all deviations are reckoned, and let D_1, D_2, &c., be the means of all the deviations, including their signs, of the ancestors in the 1st, 2nd, &c. degrees respectively; then

$$\frac{1}{2}(M + D_1) + \frac{1}{4}(M + D_2) + \&c. = M + \frac{1}{2}D_1 + \frac{1}{4}D_2 + \&c.$$ [99]

In this equation, each of the elements expresses the constitution of the 'heritage' of the offspring. In other words, if D_0 is the deviation of an individual from the mean, it can be analysed as the sum of ancestral 'contributions':

$$D_0 = \frac{1}{2}D_1 + \frac{1}{4}D_2 + \frac{1}{8}D_3 + \ldots + \frac{1}{2^n}D_n$$

The important point in this formulation is that each mid-ancestor (mid-parent, mid-grand-parent etc.) bequeaths a fraction (½, ¼, etc.) *of its own* character (D_1, D_2...):

The occupier of each ancestral place *may* contribute something of his own personal peculiarity, apart from all others, to the heritage of the offspring. Therefore there is such a thing as an average contribution appropriate to each ancestral place, which admits of statistical valuation, however minute it may be. (Original emphasis)[100]

The other formulation of the law of ancestral heredity focuses on the child's heritage:

The two parents contribute between them on the average one-half, or (0.5) of the total heritage of the offspring; the four grand-parents, one-quarter, or $(0.5)^2$; the eight great-grandparents, one eighth, or $(0.5)^3$, and so on. The sum of the ancestral contributions is expressed by the series $\{(0.5) + (0.5)^2 + (0.5)^3, \&c.\}$, which, being equal to 1, accounts for the whole heritage.[101]

Even though Galton writes as though there were no difference between the two formulations, this second expression is completely different. It means that each mid-ancestor contributes in a declining proportion (½, ¼ etc.) to the character *of the child* (D_0 only, instead of D_1, D_2 as in the previous case). Using the same terms as above, this can be expressed as follows:

$$D_0 = \frac{1}{2}D_0 + \frac{1}{4}D_0 + \frac{1}{8}D_0 + \ldots + \frac{1}{2^n}D_0$$

These two formulations have their roots in two completely different theoretical conceptions.

The first establishes a relation between the observed characters of the child (D_0) and those of the ancestral generations (D_1, D_2...), that is, between parameters that cannot be known *a priori*. In this sense, it is similar to a general multiple regression formula:

$$y = b_1x_1 + b_2x_2 + \ldots + b_nx_n$$

where $x_1, x_2, \ldots, x_n$ are the values of given ancestors, $b_1, b_2, \ldots b_n$ are the partial regression coefficients of y on $x_1, x_2, \ldots x_n$ and where y is the predicted value for a child with a given set of ancestors.

This is Pearson's interpretation of the law of ancestral heredity, in which the 'contribution' of an ancestor takes the following operational definition: the contribution of the ancestor to the predictive formula of the probable value of an individual, given the character of his or her ancestors.[102] According to Pearson, Galton merely sought to extend the application of regression analysis in order to use available data on ancestors. There is some justification for this interpretation. For example, Galton wondered how he could use the regression coefficients of children on the other members of their cohort (who are clearly not their ancestors) in order to better predict the children's values. He also hesitated to extend the law of ancestral heredity further back than the grandparents, because of the lack of data.[103] According to this interpretation, the law of ancestral heredity merely raised methodological questions, and the terms Galton used – 'contribution', 'influence', 'legacy' and 'taxation' – were simply images used to illustrate the novel and difficult concept of 'multiple regression'. For Pearson, the only mistake Galton made was to imagine that the partial regression coefficients would be based on the series ½, ¼, ⅛... Indeed, as soon as Galton's article appeared in 1897, Pearson set about developing the equation that Galton had not been able to discover. Pearson's sophisticated analysis was based on multiple regression. Using the same data series as Galton (human height), Pearson established a series of fractional coefficients that were not the geometrical series ½, ¼, ⅛ ... and the sum of which was not 1.[104] Furthermore, this series would not necessarily be the same for every character, nor at all moments in time. For Pearson, the strength of heredity was an empirical question that could not be established *a priori*:

Galton's law makes the amount of inheritance an absolute constant for each pair of relatives. It would thus appear not to be a character of race or species, or one capable of modification by natural selection. This seems to me *a priori* to be improbable. I should imagine that greater or less inheritance of ancestral qualities might be a distinct advantage or disadvantage, and we should expect inheritance to be subject to the principle of evolution. This difficulty would be to some extent met by introducing the coefficient g, which I would propose to call the coefficient of heredity, and consider as capable of being modified with regard to both character and race.[105]

According to this interpretation, the law of ancestral heredity is merely a formal method for dealing with data on heredity. It has no physiological significance; rather, physiological hypotheses of the mechanisms of heredity would have to conform to its predictions in order to prove their validity:

The law of ancestral heredity in its most general form is not a biological hypothesis at all, it is simply the statement of a fundamental theorem in the statistical theory of correlation applied to a particular type of statistics. If statistics of heredity are themselves sound the results deduced from this theorem will remain true whatever biological theory of heredity is propounded.[106]

From 1900 onwards, this methodological interpretation of the law of ancestral heredity was to prove crucial for the biometricians who opposed the universal applicability of Mendel's laws.

However, this interpretation of the law was not truly Galton's own, as has been seen from his use of the other formulation of the law. In this formula ($D_0 = ½D_0 + ¼D_0 + \ldots + ½nD_0$), Galton is primarily interested in the constitution of the child's 'heritage'.[107] This version is both intuitively simpler and more naïve than the previous formulation. Galton simply describes each individual's constitution as being like a cake that is divided into parts. Sometimes he considers the 'legacy' of each preceding generation as a single unit, in which case the series is ½, ¼, ⅛ . . . At other points in the article he considers each individual ancestor's contribution, in which case the series is ¼, 1/16, 1/64 . . . A short note in *Nature* that appeared a year later indicates that this final interpretation won out (Figure 10). Entitled 'A Diagram of Heredity',[108] this note can be considered Galton's final public statement on heredity. The area of the square, which is taken to equal 1, represents 'the total heritage of any structure or faculty that has been bequeathed to a given individual'. The two parents ('2' and '3') 'between them contribute *on the average* half of each inherited faculty, each of them contributing one quarter of it'.[109] In the same way the grandparents together contribute a quarter of the individual's hereditary constitution, each of them 'bequeathing' on average one-eighth. The constitution of each individual is thus analysed according to the series ½ + ¼ + ⅛ + 1/16 etc., the sum of which is 1. Galton did not in fact draw the diagram: it is taken from an American horse-breeder's journal.[110]

There are many other inconsistencies and oddities in Galton's law of ancestral heredity. These will not be dealt with in detail; the following paragraphs focus on explaining how Galton got himself into this mess.

The oddities of the law of ancestral heredity can only be understood as

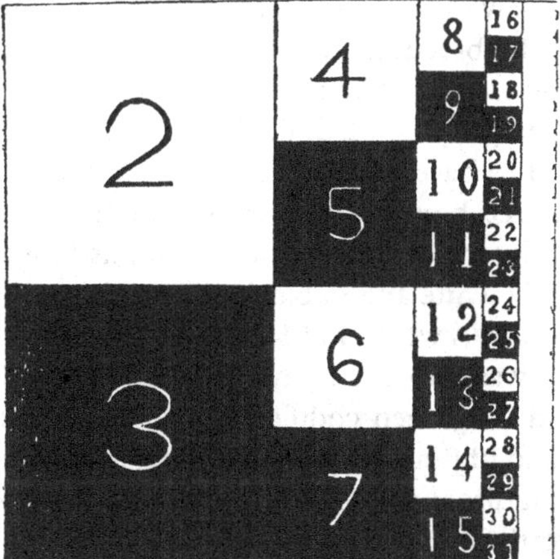

Figure 10 Schematic representation of the 'law of ancestral heredity' (F. Galton in *Nature*, vol. 57, 1898).

the result of the tension between the two different representations of heredity that Galton had always tried to focus on: statistical and physiological. The two final versions of the law correspond to these two aspects. According to the statistical formulation, information about ancestors is used to predict the character of the offspring. This version of the law thus tends to take the form of a general statistical tool, applied to data on heredity. According to the physiological formulation, the individual's germ plasm is predominant: in each generation, the fusion of the gametes involves the association of two half germ plasms, following the elimination of half of the germinal material during the formation of each gamete:

This law is strictly consonant with the observed binary subdivisions of the germ cells, and the consonant extrusion and loss of one-half of the several contributions from each of the two parents to the germ-cell of the offspring. The apparent artificiality of the law ceases on these grounds to afford cause for doubt; its close agreement with physiological phenomena ought to give a prejudice in *favour* of its truth rather than the contrary.[111]

From this physiological perspective, directly inspired by studies of cell biology and repeatedly advanced by Galton,[112] the law of ancestral heredity 'appears to be universally applicable to bisexual descent and probably beyond'.[113] Galton was convinced that the two approaches could not but agree with each other. The law of ancestral heredity was a 'statistical law',[114] but it could also be 'inferred with a considerable assurance *a priori*'.[115] These two declarations are both to be found in the same 1897 article, a few pages apart.

In fact, the origins of the law of ancestral heredity can be found in the terminology used for years by breeders in order to express the effect of interracial crosses. In these circles, it was quite usual to speak of the mixture of 'blood' and of the progressive dilution of foreign elements. A thoroughbred was thus said to have homogeneous 'blood', while a half-breed had only half the desired blood, a quarter-breed one-quarter, etc. In the same way, in human interbreeding the 'mulatto' was $\frac{1}{2}$ white and $\frac{1}{2}$ black, the 'terceron' was $\frac{3}{4}$ white and $\frac{1}{4}$ black, the 'quateron' $\frac{7}{8}$ white and $\frac{1}{8}$ black, the 'quinteron' $\frac{15}{16}$ white and $\frac{1}{16}$ black. The idea that the contribution of the 'blood' was reduced by half at each generation was thus nothing new.[116] It had long been codified and quantified, both in terms of animal breeding and of human racial discrimination. There can be no doubt that Galton first discovered the idea of the geometrical series in these conceptions. Nevertheless, the concept of *ancestral* heredity had nothing to do with the laws of hybridism. Far from being applied to examples of interbreeding, it was intended to explain hereditary effects within a given homogenous population.

Thus Galton's concept of ancestral heredity had more than one origin. In so far as it was concerned with blending inheritance,[117] it fitted well with the idea of the progressive dilution of ancestral characters. But the statistical explanation of this heredity did not require the introduction of the 'influence' of ancestors on a descendant through a geometrical series. Quite the opposite. The statistical method used suggested that such an 'influence' was merely a degree of similarity, which had only a descriptive and *a posteriori* value. On the other hand, the geometrical series lent itself to a physiological explanation, more or less directly inspired by contemporary cytology. The law of ancestral heredity, with all its obscure explanations, summarised the hesitation between a statistical and a physiological theory of heredity, that is, between a theory of heredity as measurable similarity and a theory of heredity as the material transmission of a substance ('heritage'). This hesitation was characteristic of biological thinking at the end of the 19th century and lasted until the appearance of the distinction between 'genotype' and 'phenotype'.

The fundamental conclusion of this chapter is that Galton reconciled the concepts of population and heredity. The social motivation that lay behind this work should not be ignored. Galton repeatedly produced 'scientific' justifications of individual and racial abilities and attempted to construct a 'eugenic' science that was intended to protect and improve a handful of superior lines in the human stock. However, it would be wrong for the historian to reduce the epistemological aspect of Galton's work to mere ideology. The politics of Galton and of the English biometricians have been well documented, and it would have been inappropriate to repeat this. It was more useful to examine Galton's ideas from the point of view of those statisticians and biometricians who laid down the mathematical foundations of the renewed 'Darwinism' that emerged all over the world in the first half of the 20th century. Many of them were not particularly impressed by Galton's political views, but they saw him as the inventor of the modern tools required to understand the nature of a 'population'.[118] The application of Galton's statistical tools to biological problems, in particular to heredity, has led to the development of methods that now form part of the common framework of the life sciences, and it is thus important to understand the origin of these ideas.

The nature of Galton's reconciliation of heredity and populational concepts can be understood by reference to Darwin. Darwin found it particularly difficult to understand 'variability'. Sometimes he conceived of it as 'reversion', sometimes as one of nature's capricious novelties. Darwin insisted that 'variability' was a force that had to be compatible with 'heredity', but he was unable to explain how. Galton was the first to clearly present the relationship between 'descent', 'heredity', 'variability' and 'reversion'. This detour through Galton's hereditarian ideas highlights the fate of the theory of natural selection after Darwin. Following Jenkin's (1867) objections and the publication of Darwin's *Variation of Animals and Plants under Domestication* it was clear that the theory of natural selection could not be firmly proven or refuted without clarification of the concepts of variation and heredity. As has been seen in this chapter, prior to 1900 no one explored the multiple facets of this problem more than Galton. Of course, various scientists provided more rigorous – indeed, more professional – and doubtless less ideological explanations of one or another aspect of the problem. For example, Weismann, Pflüger, Boveri, de Vries and Van Beneden all provided experimental justifications for the doctrines of the continuity and independence of the germ line and for the rejection of the inheritance of acquired characteristics. And Pearson provided a respectable mathematical demonstration of Galton's intuitions. But there can be no doubt that the general picture of 'heredity' is to be found first and foremost in Galton's work.

However, in trying to develop a rational explanation of the concept of heredity, Galton repeatedly hesitated between theoretical and statistical theories of heredity. This tension is of fundamental importance in understanding pre-Mendelian concepts of heredity. In developing a statistical theory of heredity, Galton tried to take the science of hereditary transmission out of the realm of physiological speculation; ideally, the theory would be completely free of any causal hypothesis. But unlike the biometricians, who really did adopt this approach, Galton's statistical theory of heredity had a hybrid nature. In elaborating the concept of ancestral heredity, Galton ended up encouraging archaic and confused ideas that gave credit to the concept of racial constancy and inertia.

This leads us back to the central problem of evolutionary theory. Galton's idea of heredity was thoroughly impregnated with a conservative, even decadent, vocabulary: 'return', reversion', 'regression', 'ancestral heredity'. The representation of 'heredity' in these terms inevitably tended to turn it into a force opposed to the Darwinian concept of the gradual modification of races and species. Behind the ideas of 'regression to the mean' and of 'ancestral heredity' can be discerned the outline of an open opposition to the Darwinian theory of natural selection. This is the subject of the following chapter.

5 Post-Darwinian views of selection and regression

From the broad perspective of cultural history, the end of the 19th century was marked by the eruption of a pessimistic ideology of decadence, as a reaction to the ideology of progress.[1] Obsessions with the decline of 'Western civilisation', doctrines of mental and physical 'degeneration' in psychiatry and public health, the rise and development of eugenics are all good examples of this tendency.[2] No wonder then that several major alternatives to Darwinism in this period were hostile to the idea of the unlimited progress of adaptation.[3] This chapter examines how the Darwinian tradition dealt with the theme of regression.

Among Darwin's most influential followers, two men in particular not only developed a theoretical understanding of 'regression', they also made it a central concept in their vision of natural selection: August Weismann (1834–1914) and Francis Galton (1822–1911). From both an historical and epistemological point of view, it is particularly interesting that these two Darwinian approaches to regression were completely incompatible.

Darwin considered that selection was the predominant, but not exclusive, force involved in the modification of species ('the paramount power' as he called it).[4] Weismann's and Galton's contrasting and contradictory positions on regression can only be fully understood with respect to this idea. If selection (natural or artificial) is indeed the dominant factor in modification, what happens when it ceases? Weismann argued that the inevitable result was anatomical regression, which he explained as the effect of 'panmixia'. This agreed with the idea that selection was sufficient to explain all aspects of organic evolution, and it thus tended to radicalise Darwinian positions. For this reason, the tradition founded by Weismann was called 'ultra-Darwinism' and later 'neo-Darwinism'.

In dealing with the same question, Galton also used the vocabulary of 'regression', but in a completely different way. In the absence of selection, he argued, a 'race' will return to the type that characterises it, 'type' being a statistical concept. Galton's 'statistical regression', unlike Weismann's view of the same problem, supported the idea that there were limits to the power of selection. Galton quite simply thought that natural selection as conceived by Darwin was completely ineffective: selection could have no effect on 'small individual differences', because they obey the law of 'regression' to the mean. This implies that only abrupt and discontinuous variation can be the successful target of selection, as Fleeming Jenkin had previously suggested. In other words, Galton's conception of heredity led him to reject the idea of gradual selection acting on 'minute variations'. This explains why his work was to influence both the first Mendelians (all of whom were firm mutationists and anti-Darwinians) and the orthodox Darwinian biometricians, who felt that the statistical representation of variation provided the most telling description of Darwin's 'small differences'.

For Weismann, the linked concepts of regression and 'panmixia' implied a greater fidelity to Darwin, while Galtonian regression seemed to represent an open challenge to Darwin's interpretation of natural selection. In other respects, however, Weismann's 'panmixia' and Galton's regression show an unexpected convergence. The relation between these two apparently contradictory positions represents an important moment in the history of the hypothesis of natural selection.

5.1 Weismann's panmixia: regressive evolution and natural selection

August Weismann, a major 19th-century zoologist, is known as the founder of 'neo-Darwinism', or, as it tended to be called at the turn of the century, 'ultra-Darwinism'. The spirit of this 'new Darwinism' was imbued with two radical and closely linked theses: an intransigent rejection of the inheritance of acquired characters[5] and the idea that the 'all-sufficiency of natural selection' was an explanatory principle of evolution.[6] These two theses, which were put forward in the name of parsimony, enabled Weismann to harden up the theory of heredity (rejecting both Lamarckism and, in particular, Darwin's hypothesis of pangenesis) and the theory of evolution (there are not two fundamental principles of evolution – selection and heredity – but only one: selection). The term 'ultra-Darwinism' was coined by George John Romanes in a critique of what he considered to be the unacceptable changes wrought upon Darwin's ideas by Weismann.[7] Later on, 'ultra-Darwinism' tended to be replaced by 'neo-Darwinism'.

It was in this framework that Weismann developed the idea of 'panmixia', a neologism which he often employed as an equivalent for 'degenerative evolution' or 'regression' in the anatomical sense. These terms are classically used to describe the degeneration of organs that have become non-functional (for example the reduction in size of cetacean hind limbs, wing atrophy in certain flightless birds or eye degeneration in some cave species). Weismann's choice of the term 'panmixia' to describe these phenomena could seem surprising. From an etymological point of view, the word means 'crossing of all with all', and would thus seem to refer to a mating system, in opposition to endogamy, exogamy, homogamy etc. (this is indeed its modern meaning). But this was not exactly the way Weismann employed the term. For Weismann, 'panmixia' did indeed mean 'general crossing', as the etymology requires, but the notion did not refer to the mating system, or, more precisely, not only to the mating system:

> We can describe the phenomenon which carries out the regression of a useless organ by the Greek word, which I consider to be highly appropriate, *panmixia* or *general crossing*, because this phenomenon consists essentially in the fact that we see arriving at reproduction not only those individuals in which the organ in question has reached the greatest perfection, but all individuals, irrespective of whether the organ is more or less well formed.[8]

A fundamental aspect of panmixia is thus the relaxation of selection on a given character: all individuals, irrespective of variations affecting this character, arrive at reproductive age and mate. In the long run, any variability that is not controlled by selection will lead to the degeneration, the atrophy or even the complete disappearance of the organ concerned. This series of events is clear from Weismann's first use of the term 'panmixia', in 1883:

> This suspension of the preservative influence of natural selection may be termed *Panmixia*, for *all* individuals can reproduce themselves and thus stamp their characters upon the species, and not only those which are in all respects, or in respect to some single organ, the fittest. In my opinion, the greater number of those variations which are usually attributed to the direct influence of external conditions of life, are to be ascribed to panmixia. For example, the great variability of most domesticated animals essentially depends upon this principle.[9]

For Weismann, 'panmixia' thus covers at least three things: (1) the relaxation of selection; (2) free mating; (3) anatomical regression (or 'regressive evolution'). Weismann deliberately used the word in all three senses, because he considered that they represented three links in a single causal chain. At the end of the 19th century, and for a large part of the 20th, the term was systematically used either as an abbreviation of the hypothesis that the relaxation of natural selection would lead to a 'regressive evolution',[10] or, more prudently, as a synonym for 'relaxation of selection.'

The context in which Weismann developed his interpretation of degenerative evolution by panmixia is particularly important. Panmixia emerged as part of a rich and complex reexamination of Darwin's ideas. Although not stated as such, the idea was already present in Darwin's discussions of degeneration, which put Weismann's ideas in a clearer light.

In the *Origin of Species*, Darwin explicitly envisaged the consequences of a relaxation of selection:

> In our domestic animals, if any part, or the whole animal, be neglected, and no selection be applied, that part (for instance, the comb in the Dorking fowl) or the whole breed will cease to have a uniform character. The breed will then be said to have degenerated. In rudimentary organs, and in those which have been but little specialised for any particular purpose, and perhaps in polymorphic groups, we see a nearly parallel case; for in such cases natural selection either has not or cannot have come into full play, and thus the organisation is left in a fluctuating condition.[11]

This passage appears to indicate that Darwin considered degeneration to be caused by the relaxation of selection, and the stability of a race to be due to continuous selection.[12] This impression is confirmed by the corresponding passages in the *Variation*, where Darwin goes so far as to discuss deterioration due to 'free breeding with no selection'.[13] Weismann's source of inspiration can be seen quite clearly.

Nevertheless, there are two reasons why Darwin cannot be credited with the hypothesis of panmixia. First, Darwin did not argue that the end of selection (natural or artificial) would lead to endless regression. Quite the opposite: according to Darwin, the principle of selection always functions hand in glove with the 'principle of inheritance'; selection only acts intermittently, while heredity conserves the gains of selection. In the same passages quoted above, Darwin was careful to point out that, in general, domestic races, like natural races, retain the characters they have acquired following a prolonged period of selection.[14] It is only after short periods of intense selection that the relaxation of selection has degenerative effects. During such periods, races are unstable and more easily lose the 'type' that selection has created.[15] But in general, heredity *conserves*. The other reason why panmixia cannot be attributed to Darwin is that he had a different explanation of degenerative evolution.

In certain cases, Darwin was prepared to accept the Lamarckian explanation of degeneration, in particular 'the effects of disuse':

> From the facts alluded to in the first chapter, I think there can be no doubt that use in our domestic animals strengthens and enlarges certain parts, and disuse diminishes them; and that such modifications are inherited. Under free nature . . . many animals have structures which can be best explained by the effects of disuse.[16]

For example, Darwin invokes this kind of explanation to account for the reduction in size or the complete absence of wings in large ground-feeding birds. Such birds, when they still have wings, seldom take flight except to escape danger. The ostrich, not being exposed to any danger from which it would need to escape by flight, has lost its wings, because of disuse.[17]

But Darwin also considered that many cases of anatomical regression were caused by natural selection. For example, the loss of wings in many species of littoral insects can be easily explained if those insects that remain on the ground have a greater chance of surviving than those that expose themselves to the sea winds.[18] Or again, in the case of burrowing animals such as moles, it can be imagined that the eyes not only become useless, but could also become a source of infection. Far from explaining degeneration by a *relaxation* of selection, Darwin attributed it to the joint effect of disuse and of *active* selection. According to this model, selection is not the only factor involved in evolution: it 'controls' the other factors (in particular the Lamarckian effect).[19] Darwin clearly rejected the idea of spontaneous degeneration: degenerative evolution, like all organic evolution, was in fact adaptive.

Weismann's perspective was very different. He argued that natural selection was not only a force for change, it was also a force for conservation. This position was put forward at the same time that the idea of 'panmixia' was first invoked. Natural selection in fact appears as a preliminary to 'panmixia':

> But we are here brought into contact with a very important aspect of natural selection, viz. the power of conservation exerted by it. Not only does the survival of the fittest select the best, but it also maintains it. The struggle for existence does not cease with the foundation of a new specific type, or with some perfect adaptation to the external or internal conditions of life, but it becomes, on the contrary, even more severe, so that the most minute differences of structure determine the issue between life and death.[20]

This conception of natural selection implies a radical change in its role in biological theory. For Weismann, natural selection not only modifies structures and instincts, it also, and at all times, constitutes the force that maintains them. A species that is stable at a given moment can therefore be considered a mosaic of characters, all of which are subject to the constant and normalising action of natural selection. If selection is relaxed on one of these characters, disorganisation occurs, ultimately leading to the disappearance of the organ concerned. Panmixia can be considered to be the disorder that inevitably results from the relaxation of natural selection. Weismann thus demanded both more and less of natural selection than Darwin. He did not ask it to positively explain degeneration (which he considered to be an expression of disorder) but he did require it to conserve, and thus to play the role that Darwin reserved for heredity.

Weismann first put forward the concept of panmixia at a famous lecture in Fribourg where he launched a crusade against the inheritance of acquired characteristics ('On Heredity', 1883). Because of its title and its main content, the text of this lecture has generally been considered only in terms of the history of theories of heredity. This is misleading: Weismann's argument against the inheritance of acquired characters was expressed as an argument in favour of the 'all-sufficiency' of the principle of natural selection.

In criticising the inheritance of acquired characters, the main argument attacked by Weismann was that of use and disuse; in other words, that of Lamarck: 'It seems difficult and well nigh impossible to deny the transmission of acquired characters when we remember the influence which use and disuse have exercised upon certain special organs. It is well known that Lamarck attempted to explain the structure of the organism as almost entirely due to this principle alone.'[21] Most of Weismann's lecture was devoted to refuting this argument, and in particular to showing that those effects attributed to 'use' (adaptation), like those attributed to 'disuse' (degeneration), could be more simply explained by natural selection. Weismann's discussion of 'panmixia' was directly linked to the refutation of the effect of disuse, and was clearly intended to furnish an explanation of degeneration without recourse to the Lamarckian role of disuse, in other words to the inheritance of acquired characters. Natural selection constituted a sufficient and necessary explanation of the reality of regressive evolution. Conversely, all organic stability could be explained by the constant action of natural selection.

The epistemological implications of this conceptual construction are as follows: far from rendering the study of 'heredity' autonomous, Weismann subordinated it to the principle of natural selection. Unlike Darwin, Weismann refused to consider that heredity was able to maintain characters in a group of organisms if there was no external perturbation. For Weismann, natural selection both *modifies* and *conserves*.

This makes it easier to understand the two fundamental theses of Weismann's biological philosophy – the rejection of the inheritance of acquired characters and the explanatory sufficiency of natural selection. Strictly speaking, he first expounded the thesis of the explanatory sufficiency of natural selection in 1893, in an article published in English under the title 'The All-Sufficiency of Natural Selection'. This article was a response to a polemical article by Spencer entitled 'The Inadequacy of Natural Selection', in which Spencer argued that natural selection was only an auxiliary principle of evolution, the main factor being the inheritance of acquired characters.[22] Weismann's polemic against Spencer merely confirmed what he had previously stated in his 1883 lecture 'On

Heredity': he radicalised the principle of natural selection and used it to discredit the notion of the inheritance of acquired characters.

Weismann's campaign thus took place in the context of post-Darwinian debates as to the relative roles of heredity and selection in evolutionary theory. Weismann's position in this debate was straightforward: he subordinated the theory of heredity to the theory of selection – his 'essays on heredity and natural selection' were not merely 'essays on heredity'. Weismann's position tended to clarify the biological concept of heredity, but his main motivation was to demonstrate the 'all-sufficiency' of natural selection. Nothing escapes the control of natural selection: neither life span nor adaptation nor heredity, despite the fact that all three formed part of the premises from which the hypothesis of natural selection was deduced. For Darwin, length of life, adaptation and heredity were ingredients of the concept of natural selection: the struggle for existence implied individual differences in life span, and selection presupposed both heritable differences and that these differences affected the adaptive value of organisms. Weismann admitted these Darwinian postulates in so far as they are part of a classical presentation of the principle of natural selection. But in fact his ultra-Darwinism led him to subordinate completely length of life, adaptation and heredity to natural selection. According to Weismann, natural selection determines the mean life span of organisms,[23] and it alone explains adaptation. In the case of heredity, its subordination to the principle of selection simply meant that *without selection, heredity could do nothing, not even conserve*. Such were the implications of 'panmixia.'

5.2 Galton's critique of Darwinian selection

Galton's approach was the complete opposite of Weismann's – he subordinated the theory of selection to the theory of heredity. Galton used a number of different arguments, but always in the same context of the statistical regression of offspring towards the parental means. The whole of Galton's work was characterised by the conviction, increasingly clearly stated, that Darwin's conception of selection – acting on 'small individual differences' – was incapable of permanently changing the mean type of a race. If variation is continuous, he argued, regression to the mean will dominate.

Despite the fact that the methods used by Weismann and Galton to deal with the question of regression were very different, their contemporaries must have been struck by certain remarkable similarities. Both Weismann and Galton reasoned on the basis of the hypothesis of 'free mating', and they both studied what would happen following the relaxation of selection on a given character. For Weismann, this would lead to the size of the organ regressing to a value approaching zero; for Galton it would be the population that would 'regress' to a stable mean value. The two concepts of 'regression' thus illustrated the two most extreme positions imaginable in the growing controversy over the respective weight of the 'principle of selection' and the 'principle of heredity'. Weismann's anatomical regression through 'panmixia' showed the omnipotent nature of natural selection; Galton's statistical regression showed the limits of the Darwinian hypothesis of selection and the power of heredity.

Galton's discussions of selection, despite not being the work of a real naturalist, nor even of an experimental biologist, nevertheless had an enormous influence on the way that both Darwinians and anti-Darwinians conceived of selection. As in the case of heredity, Galton's theoretical analysis of selection impressed not so much by its rigorous demonstration as by the theoretical possibilities it revealed, and by the wealth of images it introduced.

A study of Galton's writings shows that he explored three successive theoretical approaches in his critical analysis of the concept of selection, corresponding to three possible explanations of the 'regression' to the mean exhibited by races, in other words of the 'constancy' of racial types. These explanations are set out in a late text of 1894, which provides an important key to the reconstruction of Galton's conception of selectionist ideas:

> There can be no doubt as to the reality of regression. I have not only proved its existence in certain cases and measured its amount, but have shown that no race could continue constant in its characteristics unless regression existed. And again, the observed and the theoretical details of the process were found strictly to occur ... The causes why the A and B races are such definite entities may be various. In the first place each race has a solidarity due to common ancestors and frequent interbreeding. Secondly, it may be thought by some, though not by myself, to have been pruned into permanent shape by the long-continued action of natural selection. But, in addition to these, I have for some years past maintained that a third cause exists more potent than the other two, and sufficient by itself to mould a race, namely that of definite positions of organic stability ... I conceive the position of maximum stability to be the essential as well as the most potent agent in forming a typical centre, from which the individuals of the race may diverge and towards which their offspring tend on the whole to regress.[24]

Galton's three potential explanations of the stability of racial 'types' – natural selection, ancestral heredity and the existence of equilibrium points – will be dealt with in turn.

5.2.1 *1877: the first statistical account of natural selection*

Galton initially believed that it would be possible to reconcile the principle of natural selection with his statistical theory of heredity. This can be seen in his first article devoted to this subject ('Typical Laws of Heredity', 1877). It will be recalled that it was in this article that Galton first proposed the concepts of linear regression and the coefficient of regression, under the term 'reversion'. An important part of this article was devoted to natural selection, showing that it, like heredity, was compatible with the 'law of deviation' (the normal distribution). The passage in which Galton announces his intention is worth considering in some detail (stripped of all reference to any subject other than heredity, it was quoted in the preceding chapter). In fact, this text was written in such a way that the statistical treatment of heredity and that of natural selection are completely intertwined:

> Let me point out a fact which Quételet and all writers who have followed in his paths have unacceptably overlooked, and which has an intimate bearing on our work to-night. It is that, although characteristics of plants and animals conform to the law [of error], the reason of their doing so is as yet totally unexplained. The essence of the law is that differences should be wholly due to the collective actions of a host of independent *petty* influences in various combinations . . . Now the processes of heredity that limit the number of the children of one class such as giants, that diminish their resemblance to their fathers, and kill many of them, are not petty influences, but very important ones.[25]

These lines deliberately suggest that heredity, selection and reversion are biological factors that could be responsible for the fact that organisms generally show Gaussian distributions for their characters. The lines that follow confirm this impression and outline the problem that had to be resolved:

> Any selective tendency is ruin to the law of deviation, yet among the processes of heredity there is the large influence of natural selection. The conclusion is of the greatest importance to our problem. It is, that the processes of heredity must work harmoniously with the law of deviation, and be themselves in some sense conformable to it. Each of the processes must show this conformity separately, quite irrespective of the rest. It is not an admissible hypothesis that any two or more of them, such as reversion and natural selection, should follow laws so exactly inverse to one another that the one should reform what the other has deformed . . .[26]

What did it mean for natural selection to conform to the 'law of deviation'? Galton's explanation involved an ingenious analogy. In a given species, he argued, the character favoured by natural selection is not the extreme character, but that which represents the best balance between excess and absence. Thus natural selection on height never operates in favour of giants or dwarfs, but on an intermediate height. 'It is therefore not unreasonable to look at nature as a marksman, her aim being subject to the same law of deviation as that which causes the shot on a target to be

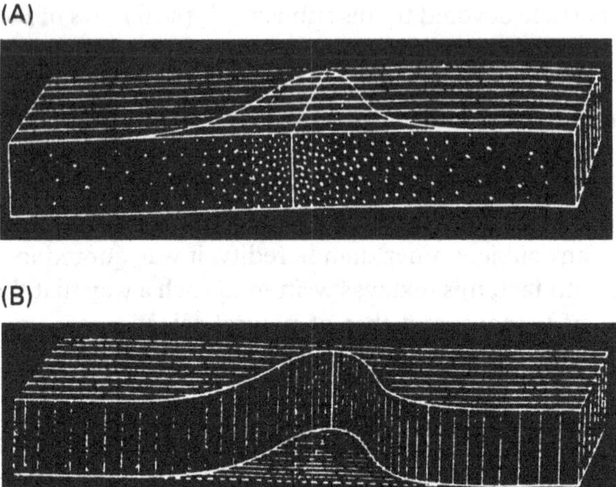

Figure 11 A gunner fires repeatedly against a rampart. In Figure 11A he is aiming at a definite line in order to create a clean cut, but in fact (Figure 11B) he makes a convex normal curve (F. Galton, 'Typical laws of heredity', 1877).

dispersed on either side of the point aimed at.'[27] Galton had already used this metaphor in *Hereditary Genius* (1869) in order to represent 'Quételet's law' as a cloud of points (compare Figure 2 with Figure 11). But here it is no longer the homogeneous population that is thus symbolised, but rather natural selection as a normalising process, operating by the elimination of extremes and the preservation of the mean character.

This representation of the process of natural selection by means of a distribution curve was the first of its kind. Galton was probably inspired by mortality curves, which had been well known since the beginning of the 19th century. To a certain extent, a curve of selective value can be considered the opposite of a mortality curve, because it represents 'the percentage of survival'.[28] Another difference from the mortality curves of the time is that the latter only showed mortality rates according to age; Galton's curves showed the distribution of survival rates as a function of any measurable character (for example, height or weight).[29] This constitutes an unmistakable statistical representation of natural selection which, it goes without saying, was to be enormously influential. Galton realised that he was introducing a subtle but decisive change into the concept of natural selection. If natural selection was described in the language of the 'law of deviation', he argued, it would become apparent that 'natural selection does not act by carving out each new generation according to a definite pattern on a Procrustean bed, irrespective of waste'.[30] In other words, natural selection should not be considered in terms of the elimination and preservation of individuals, but as a process that affects the whole population. This constitutes a clear methodological outline of the biometrical analysis of selection.

However, Galton's statistical interpretation of natural selection went further than a mere quantification of the Darwinian principle, and rapidly became a scarcely veiled attack on Darwin himself. The whole of Galton's method required that heredity, like selection (in both its elements – survival and fertility), should conform to the 'law of deviation' (the normal law).[31] Logically speaking, reversion and selection need not necessarily reinforce the racial mean.[32] For example, it is possible to imagine that 'reversion' pulls individuals towards a smaller size (that of the ancestors), while natural selection favours larger individuals. But Galton only raised this possibility in order to dismiss it. He considered that the laws of reversion and selection were '*typical* laws of heredity', that is, laws of the '*typical* character' of the population. Or again, in terms taken from Quételet, that they were laws that explained the fact that the vast majority of characters show a Gaussian distribution. If reversion and natural selection had central values that were different from that of the observed mean, then 'none of their laws [might] be strictly of the typical character'.[33] This would imply that heredity and selection play no part in the observed distribution of most characters. However, Galton believed that 'the typical laws are those which most nearly express what takes place in nature generally; they may never be exactly correct, but at the same time they will always be approximately true and always serviceable for explanation'.[34]

For this reason, Galton finally chose to present 'reversion' and 'natural selection' as two processes that 'cooperate'. Both follow the law of deviation, both reduce dispersion, both have the same central value – the mean of the preceding generation. On the basis of these hypotheses it is not difficult to imagine an analysis of the modifications induced by these two factors, because these modifications function under a symmetrical curve with a constant central value. Only the measure of dispersion would alter, being reduced in both cases. Galton outlined such an analysis and once again used the quincunx to visualise the schema he had in mind. In this illustration (Figure 12), the upper level represents the distribution of the character in the parental generation, divided into classes. At the next level, the distribution has narrowed owing to differential survival. This is represented by the fact that the bottom of the apparatus is curved, so that some of the ball bearings fall outside the central compartments. This illustrates the idea of 'percentage of survival' being variable according to the value of the character. The level immediately beneath this one has the

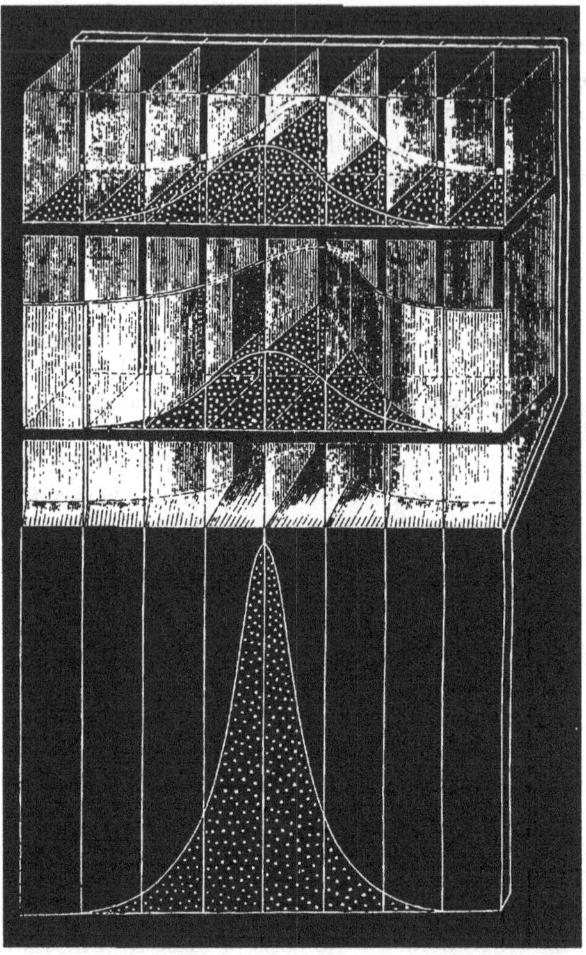

Figure 12 A mechanical model of natural selection. The upper level represents a 'normally' distributed population. The middle level represents differential survival: 'a curved partition ... separate[s] the pellets as they fall upon it, into two portions, one that runs to waste behind at the back, and another that falls to the front, and forms a new heap ... When the slide upon which it rests is removed, the pellets run down an inclined plane that directs them into a frame of uniform and shallow depth.' This lower level shows the restricted distribution of the filial population. Galton mentions that another level can be added to show differential fertility. This model can also be combined with the 'quincunx' to show variability and reversion. The modal value of the two kinds of selection and of the focus of 'reversion' do not need to be the same, but Galton thought that this was the case (F. Galton, 'Typical laws of heredity', 1877).

same shape, and shows the supplementary effect of differential 'productiveness' (fertility). The whole apparatus produces a final distribution that is narrower than that of the initial distribution and could be combined with another 'quincunx' to illustrate the dual effects of reversion and of dispersion attributable to heredity.

In an appendix to his 1877 article 'Typical Laws of Heredity', Galton presented an analysis of the problem[35] which was wholly faithful to the artificial model and can be summarised as follows:

Galton first admits that the parent population has a Gaussian distribution:

$$y = \frac{1}{c\sqrt{\pi}} \cdot e^{-x^2/c^2}$$

where x is the deviation of a given parent and c the dispersion parameter (the 'modulus', which is more or less equal to twice the 'probable error').

If the variation of y (for example height) is described by this law, the number of individuals showing a variation between x and x + dx will vary as follows:

$$e^{-x^2/c^2}dx$$

The introduction of selection implies that the survival rate of the offspring varies as a function of the parental characters (that is, their deviation from the mean). Galton assumes that the distribution of the survival rate follows a normal law: the proportion of survivors amongst offspring showing a deviation x varies according to e^{-x^2/s^2}, where s is the modulus of the normal law of selection. If it is accepted that each offspring 'absolutely resembles' its parents, the number of offspring with a deviation x will vary as follows:

$$e^{-x^2/c^2} \times e^{-x^2/s^2}$$

Or:

$$e^{-x^2(1/c^2 + 1/s^2)}$$

This simply means that the dispersion of the population has been reduced from its initial value c to a value c′, such that

$$1/c' = \sqrt{1/c^2 + 1/s^2}$$

Galton developed a similar argument for 'productiveness' (fertility), which he distinguished from 'natural selection' (differential viability).[36]

As can be seen, this formalisation of natural selection is based on the (implicit) postulate that deviations are measured with respect to the same mean value in both parents and offspring. This implies *a priori* that natural selection cannot change the mean value of the character.

The conceptual significance of Galton's schema is straightforward: he excludes those cases where selection would modify the racial mean, and thus can only be normalising or conservative. To be fair, he considers the possibility that natural selection could favour those parts of the population that do not correspond to the racial mean, but he does not seriously deal with the question, because, he argues, directional selection sooner or later leads to normalising selection.[37] In reality, directional selection would lie outside the mathematical framework presented by Galton, because this would imply that the observed distribution of characters would be asymmetric rather than normal. Pearson later pointed out that the assumption of a normal distribution for a given biological character was often an unjustified approximation, and proposed a more rigorous and complex method for adjusting the distribution curves than that of Galton. However, the key question here is not that of mathematical technique. Galton was in fact convinced that races are stable entities, and that it was legitimate to speak of them as types. It was from this point of view that he elaborated his first statistical theory of heredity and of selection, each cooperating in order to maintain the type.

It would be wrong, however, to condemn this first biometrical theorisation of natural selection as belonging epistemologically to pre-Darwinian thinking. The idea that quantitative characters have a constant mean, and the rehabilitation (against Darwin) of the vocabulary of 'type', are undoubtedly old-fashioned. But the image of natural selection as acting like a soldier firing repeatedly at a target (Figures 11A and 11B) was far from naive. In terms of its action on quantitative characters (which was, after all, Darwin's implicit assumption), selection should not be considered as acting like the Procrustes of Greek legend, who would ambush hapless travellers and make them fit the length of a bed by either stretching them on the rack if they were too small, or cutting their legs off if they were too big. Selection probably acts in a more economical fashion, carving out populations rather than exterminating individuals. The methodological debt of biometry and quantitative genetics to Galton is obvious, and was widely recognised. Furthermore, there was nothing absurd in suggesting that natural selection played a major role in stabilisation.[38] But Galton went much further, going so far as first to suggest, and then to argue, that Darwinian natural selection could not modify races, but only select between them *a posteriori*.

There is a final aspect of the 1877 paper that needs to be examined. In this article, Galton envisaged the effects of sexual selection, which he understood as a tendency to mate with a like individual (for example, a tall man marrying a tall woman).[39] This corresponds to what is called 'homogamy' in modern evolutionary theory. Such a phenomenon, argued Galton, was incompatible with the maintenance of the population *type*. This is why, in this paper, sexual selection was considered inoperative.[40] The link with eugenics is obvious: for Galton, selective marriages were the very method of eugenics. Nevertheless, this aspect of the question is not invoked in the very 'scientific' article published in *Nature*. This point will be dealt with further below (see 5.2.3 and 7.2.3).

5.2.2 Ancestral heredity, natural selection and eugenics

From 1885 onwards, 'reversion' – rebaptised 'regression' – became essentially a statistical tool, no longer considered to express the essence of heredity, which Galton now sought in terms of 'ancestral heredity'. At the same time, his criticism of Darwin's ideas became increasingly radical. The notion that natural selection 'typically' worked in the same sense as regression was replaced by the negative thesis that selection could not displace a population from its 'racial centre' for very long. Galton gave two justifications for this, one using the parastatistical language of the 'law of ancestral heredity', the other framed in qualitative terms and physiological in approach, based on the idea of states of organic stability.

As has already been seen, 'ancestral heredity' was probably one of Galton's most confused ideas. When considered from the point of view of its relation to selection, 'ancestral heredity' reveals its hidden agenda: the critique of the Darwinian concept of natural selection was closely linked with the kind of artificial selection that Galton envisaged for humans as part of his 'eugenics'.

The law of ancestral heredity is based on the idea that an individual organism inherits not only from its parents, but also from its more distant ancestors. The further back one traces a genealogy, the more the ancestral contribution tends to merge with the racial mean. Galton considered that this law explained the regression of offspring to the mean that could be observed in each generation. Each child thus carries an 'ancestral' heritage which, on average, pulls him or her back to the 'racial centre'[41] or 'typical centre'. This 'centre' can be defined as 'an ideal form, whose qualities are those of the average of all members of the race',[42] whereas 'a race is taken to mean a large body of more or less similar and related individuals, who are separated from analogous bodies by the rarity of transitional forms, and not by any sharp boundary'.[43] Galton argued that if natural selection acted on continuous variation, as Darwin had argued, it would be unable to change the 'typical centre'.

Darwinian selection could temporarily modify the mean, but as soon as it was relaxed, the population would tend to regress to its original state – that which had existed before selection began.[44] In other words, natural selection cannot fix a character mean at a new value, because ancestral heredity acts as a kind of spring that always pulls the population back to the original state of the race.

Galton's explanation was extremely artificial. He argued that if a selected sub-population deviates by a value x from the original population mean, and if selection is then relaxed, successive generations will have deviations from the original mean of rx, r^2x, r^3x etc., r being the regression coefficient. This coefficient being less than 1, the population is slowly brought back to the 'racial centre'.[45] In the vocabulary of ancestral heredity, this means that the further away one is from the last selected generation, the more the contribution of the selected ancestors diminishes and the more the 'ancestral heritage' of the race predominates.

This argument has two major flaws. First, it is incompatible with the law of ancestral heredity, according to which the parents contribute ½ to the child, grandparents ¼, and the *n*th generation $1/2^n$. An immediate implication of this law is that offspring are influenced most by their *closest* ancestors, the influence of other ancestors being more or less insignificant beyond three or four generations, as Galton had himself remarked in the 1880s. Logically speaking, this means that the trace of a selective process applied over a few generations should be greater than that remaining from the state of the population prior to selection.

The other weakness in Galton's argument is due to a completely different kind of error. Galton justified regression to the 'racial centre' by supposing the relaxation of selection. It is difficult to imagine what form regression to the mean would take following thousands or millions of generations of selection, and yet this was precisely the order of magnitude that Darwin had in mind. In such a case, what would be the 'focus of regression'[46] that populations would return to following the relaxation of selection? As Pearson pointed out in an 1896 article, 'Regression, Heredity and Panmixia', in terms of long-term selection Galton's 'regression to the mean' was no different from the limitless anatomical regression of Weismann. If this is the case, we return to the question of the origin of the 'centre of regression', and it has to be accepted that the position of this centre of regression could change over time. In other words, there could be a 'progression of the focus of regression',[47] which Pearson considered to be the necessary consequence of any directional selection. But this was precisely what Galton refused to consider. His whole argument about the relation of selection and regression was intended to show that Darwinian (i.e. gradual) selection was merely a temporary perturbation of an equilibrium that was guaranteed by regression.

Pearson also drew attention to a remarkable mathematical inconsistency in Galton's reasoning. It is possible to show that one of the consequences of the law of ancestral heredity, as conceived of by Galton, was that *there could not be* a regression to the initial population mean following a period of selection. Pearson demonstrated this in an article entitled 'On the Law of Ancestral Heredity', which carried the ironic dedication 'A New Year's Greeting to Francis Galton, January 1, 1898'.[48]

Pearson's refutation of Galton's anti-Darwinian argument takes up only a few lines. He does not deal with the various statistical refinements which he himself had introduced into the 'law'; he simply takes it as stated by Galton. Suppose that we begin with a general population, the arbitrary mean value of which – for the sake of arithmetical simplicity – is zero. In this general population, mid-parents with a deviation K from the mean are selected. The process is reiterated for n generations; selection is then relaxed. What are the values of the deviations shown by the successive generations after the relaxation of selection?

Let k_1, k_2, k_3 . . . be the deviations of the ancestral generations, k_1 being the deviation of the last selected generation, k_n being that of the first selected generation. By definition, we have:

For selected generations:

$k_1 = k_2 = k_3 = \ldots = k_n = K$ (those individuals whose deviation = K are selected)

For generations prior to selection:

$k_{n+1} = k_{n+2} = k_{n+3} = \ldots = k_\infty = 0$ (the deviation is on the average that of the general population, that is, zero)

We thus know the characters expressed by all the ancestors of the offspring obtained after *n* generations of selection. The probable deviation k_0 of this generation can be predicted by Galton's law. This states that the ancestral generations contribute their own character to the offspring in decreasing proportion (½, ¼ . . . $1/2n$), according to their distance in time. Pearson interprets this as a (simplified) multiple regression equation:

$$k_0 = \frac{1}{2}k_1 + \frac{1}{4}k_2 + \frac{1}{8}k_3 + \ldots + \frac{1}{2^n}K_n + \frac{1}{2^{n+1}}(0) + \frac{1}{2^{n+2}}(0) + \&c.$$

or:

$$k_0 = \left(\frac{1}{2} + \frac{1}{4} + \frac{1}{8} + \ldots + \frac{1}{2^n}\right)K = \left(1 - \frac{1}{2^n}\right)K$$

As an example, this would imply that after six generations of selection, the deviation of the offspring k_0 would be equal to $(1 - 1/64)K$, or 98.4 per cent of the deviation of the parents and ancestors. In other words, there is indeed the beginning of a regression to the mean of the original population, but this effect is reduced as a function of the duration of selection.

The important question, however, is what happens after selection has been relaxed. Maintaining the same conventions, numbering the generations from the first offspring following the relaxation of selection, the law of ancestral heredity enables us to deduce the deviation of the first generation after the relaxation of selection. This deviation is the fraction of K given by:

$$\frac{1}{2}\left(1-\frac{1}{2^n}\right)+\frac{1}{4}+\frac{1}{8}+\ldots+\frac{1}{2^{n+1}}$$
$$=\frac{1}{2}\left(1-\frac{1}{2^n}\right)+\frac{1}{2}\left(1-\frac{1}{2^n}\right)$$
$$=\left(1-\frac{1}{2^n}\right)$$

In the second generation after the reestablishment of panmixia, we have:

$$\frac{1}{2}\left(1-\frac{1}{2^n}\right)+\frac{1}{4}\left(1-\frac{1}{2^n}\right)+\frac{1}{8}+\frac{1}{16}+\ldots+\frac{1}{2^{n+2}}$$
$$=\frac{1}{2}\left(1-\frac{1}{2^n}\right)+\frac{1}{4}\left(1-\frac{1}{2^n}\right)+\frac{1}{4}\left(1-\frac{1}{2^n}\right)$$
$$=\left(1-\frac{1}{2^n}\right)$$

And so on.[49] The conclusion is obvious: 'The offspring will always have the same amount of the character as had the generation after selection ceased.'[50] The law of ancestral heredity, as formulated by Galton, cannot be used to discredit the Darwinian idea of selection applied to continuous variation.

Pearson in fact killed two birds with one stone: his demonstration was a major blow both to Galton's and to Weismann's conception of 'regression'. 'Panmixia' (the suspension of selection) would not imply a reversal of the process of selection. It would not lead either to regression to a stable racial centre, or to 'an indefinite distant point' (a statistical interpretation of Weismann's degenerative evolution). Panmixia is not 'regression', either in Galton's terms or in Weismann's.[51] Of course, Pearson's refutation is only valid for Galton's interpretation of the law of ancestral heredity (with its ½, ¼, ⅛ ... series for the partial coefficients of regression of offspring on ancestors), an interpretation which Pearson rejected for reasons of mathematical naiveté.

Ancestral heredity was thus a very poor argument in the critique of the Darwinian concept of natural selection. Galton was probably more or less aware of this: parallel to this argument, he set out another, very different and much more interesting – that of 'organic stability', to which we will return. However, before leaving the concept of 'ancestral heredity', it is worth examining why it was elaborated.

Galton did not believe that natural selection could gradually modify races. But he was in favour of improving human races, through a programme of selection which would be able to do for human beings, in particular as far as our moral and mental faculties were concerned, what breeders could do for domestic animals. He had confidence in the power of *artificial* selection. This is the point that reveals the coherence of his various reflections on *natural* selection. Like Darwin, Galton was convinced that artificial selection and natural selection involved similar fundamental processes. But his particular conception of artificial selection, especially his eugenics programme, enables us to understand his criticism of the Darwinian concept of natural selection, and his different interpretation of it.

Galton never argued that natural selection was unable to modify *species*. He merely refused to accept that it could modify *races*, which he considered to be stable types, recognisable by the 'normal' distribution of their characters. Galton thought that when such a distribution is found – as is generally the case – the mean is an intrinsic character of the race, which tends to be restored if it is perturbed, whatever the nature of that perturbation. Selection cannot displace the central tendency of a race, but it can act on the races themselves, which thus become the units of selection.

But how could such new 'racial centres' emerge? Galton was not very explicit on this point. Only his negative thesis is clear: they cannot arise by the gradual displacement of the normal curve. These new 'typical centres' must thus develop suddenly, with evolution taking place by leaps, natural selection only intervening afterwards: 'a race does sometimes abruptly produce individuals who have a distinctly different typical centre, in the sense in which those words were defined'.[52] But the definition of 'typical centre' given by Galton shows that this affirmation is far from unproblematic: '[The type or typical centre of a race] is to be defined as an ideal form, whose qualities are those of the average of all members of the race, or, what statistically speaking is the same thing, the average of any large and hap-hazard collection of them.'[53] In this passage, the 'typical centre' is clearly a statistical notion, and has no meaning for the individual. It is thus not clear what exactly 'individuals who have a distinctly typical centre' means. 'Individuals' or 'a group of individuals'?

This ambiguity is probably the reason why Galton so often resorted to the notion of 'sports' (Darwin's 'large anomalies')[54] or to that of 'discontinuous variation',[55] occasionally adding that a new race can only be formed on the basis of characters that do not show blending inheritance.[56] This is apparently much clearer, and is redolent of the 'mutations' conceived of by Hugo de Vries and other 'mutationists' around 1900.

But Galton's writings show that he never clearly distinguished the notion of 'sport' from that of an exceptional deviation in a normal distribution. This became obvious when he put forward the case of Inaudi, an illiterate peasant from Piedmont, as an example of a 'sport'. This case is invoked in 'Discontinuity in Evolution' (1894). Inaudi was gifted with an astonishing ability for mental arithmetic and had been tested by Galton himself. After a detailed description of Inaudi's abilities, Galton abruptly concludes with an assertion as to their heritability: 'His parents had no such power; his own remarkable gifts were therefore a sport, and let it be remembered that mental sports of this kind, however large, are nonetheless heritable.'[57] This example, with some others (e.g. rare fingerprints), is supposed to illustrate the sudden emergence of exceptional peculiarities 'without any help whatever from the process of selection'.[58] However, Galton says nothing about the kind of hereditary transmission responsible for these qualities. Are they prepotent, or do they blend? Sometimes Galton speaks of 'a slight tendency towards transmission by inheritance' (fingerprints);[59] sometimes he just says that the character is 'heritable' (mental sports).[60] It thus appears as though it was sufficient for him to be confronted with an exceptional heritable quality in order to hypothesise the emergence of a 'new racial centre'.

But if this is the case, we come back to the starting point: the children of 'exceptional' individuals, whatever their degree of 'exception', must regress to the typical mean of the race. How could they therefore found a new 'racial centre'? The solution requires another ingredient, relatively obscure from the point of view of evolutionary theory, but very clear in terms of eugenics. It will be remembered that in his 1877 article 'Typical Laws of Heredity' Galton considered that heredity could 'cooperate' with natural selection to maintain the 'typical condition' of the race. This simply meant that natural selection, to the extent that it did not affect the population mean of the character, was compatible with the 'law of deviation', in other words with the 'normal' and symmetrical distribution that can be observed for most characters. In the same text, Galton indicated that there was at least one kind of selection that could in no way accord with the 'law of deviation': sexual selection. 'In order that the law of sexual selection should co-operate with the conditions of a typical population, it is necessary that selection should be *nil*, that is, that there should not be the least tendency for tall men to marry tall women rather than short ones.'[61]

This passage is strange. Galton does not use the term 'sexual selection' in the same sense as Darwin, that is as a process of selection that depends on the advantage which certain individuals have over other individuals of the same sex and species, in exclusive relation to reproduction.[62] By 'sexual selection', Galton meant what today is called 'homogamy' or 'positive assortative mating'. In the simplest case (that envisaged by Galton), homogamy implies the choice of a sexual partner having the same character as oneself. This confusion, in which the two very different phenomena of sexual competition and homogamy were confounded under a single title, was not unusual at the end of the 19th century.[63]

This being said, it is even more curious that Galton affirmed that 'sexual selection' (that is, homogamy) was compatible with the population being maintained in a 'typical condition'. Subsequent studies of homogamy have shown that it has practically no effect on the regression of offspring on the parental values. Galton did not in fact have a statistical theory of homogamy; indeed, it was only with Karl Pearson's work at the end of the 1890s that such a theory began to be developed.[64] However, the idea that homogamy does not affect the population mean is easy to grasp: if like mates with like, it is quite reasonable to predict that the distribution of the offspring will be very close to that of the parents, perhaps with a slightly wider distribution, but certainly having the same mean. But not only did Galton not see this, he said exactly the opposite. If there is a tendency within a population for individuals to mate with similar individuals, he argued, the 'typical condition' cannot be maintained. Indeed, this assertion seems to imply that the mean value *must* change.

There is one possible explanation for this. Galton quite specifically used the term sexual *selection*, even if he used it in a different way from Darwin. By 'sexual selection', Galton did not mean an ordinary situation of homogamy. He did not envisage something like a coefficient of homogamy that would be identical for all couples, but rather something much more unusual: certain homogamous couples would have an advantage over others because of their possession of a given character. This closely resembles the objective situation created by breeders when they apply artificial selection: they do not kill the 'inappropriate' animals (this would be too costly and time-consuming), but allow only those animals with the favoured character to reproduce.

If this interpretation is valid, Galton's 'sexual selection' is difficult to understand from the point of view of evolutionary theory: it would be a mixture of homogamy and Darwinian sexual selection. It becomes utterly clear, however, in the context of the eugenic 'utopia' of which it formed a fundamental basis. As early as 1865, in an article on talent and hereditary characters, Galton had openly stated:

> No one, I think, can doubt, from the facts and analogies I have brought forward, that, if talented men were mated with talented women, of the same mental and physical characters as themselves, generation after generation, we might produce a highly-bred human race, with no more tendency to revert to meaner ancestral types than is shown by our long-established breeds of race-horses and fox-hounds.[65]

This formulation has at least the advantage of being utterly frank. Galton continually repeated and refined this position in all his writings that touched on the question of eugenics. Quite simply, it is a programme for artificial selection applied to man. The best must mate with the best, and they must do so rapidly, marrying young. In 'Kantsaywhere' (1910), Galton's unpublished utopia, candidates for citizenship have to take a general-aptitude examination. With a mark above +20, there would no limit on the number of children the candidates could have – the law and moral sentiment would encourage them to have as many as they wanted. Between +10 and +20, the moral pressure of the collectivity would limit the candidates to around three children. Between 0 and +10, they would be required to limit themselves to two children. Between 0 and −10, the law would forbid them to have more than one child, and those with a score of less than −10 would be forbidden to have any children at all.

This 'utopia' is the final caricature of the programme for the 'improvement of the human race' that Galton had continually supported from his first article on heredity in 1865 through to his death in 1911. This programme consisted of both 'positive' eugenics (social policy encouraging elites to have children) and 'negative' eugenics (legislation forbidding the mentally handicapped to marry). For many years, Galton argued that positive eugenics was the most important, and was the only form that was thoroughly moral.[66] At the end of his life, he argued the opposite, claiming that negative eugenics was 'undoubtedly the more pressing subject of the two'.[67] Whatever his changing position on this relatively minor question, he was always completely convinced that the production of a 'superior race' was quite possible.

This conviction provided the logic behind Galton's critical study of Darwinian natural selection. Galton's objective was to underline the possibility and the utility of a programme of artificial selection applied to man. In an ideological context profoundly marked by social Darwinism, by the idea that natural selection gradually and inevitably improves human populations, Galton contested the representation of selection as the gradual movement of the population mean (Figure 13). If natural selection did indeed act on continuous variation, as Darwin argued,

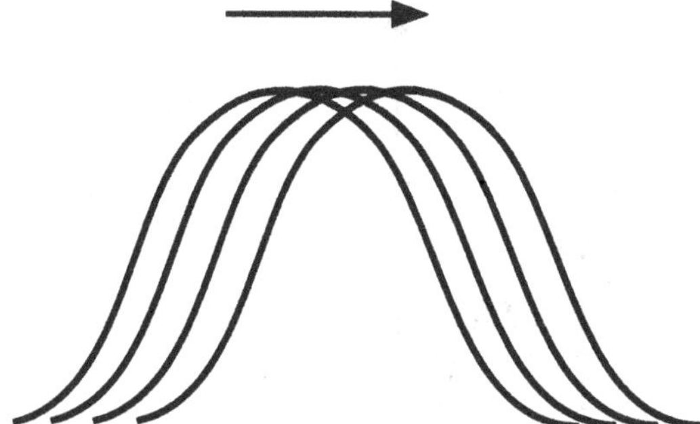

Figure 13 Illustration of the Darwinian model of selection that Galton rejected.

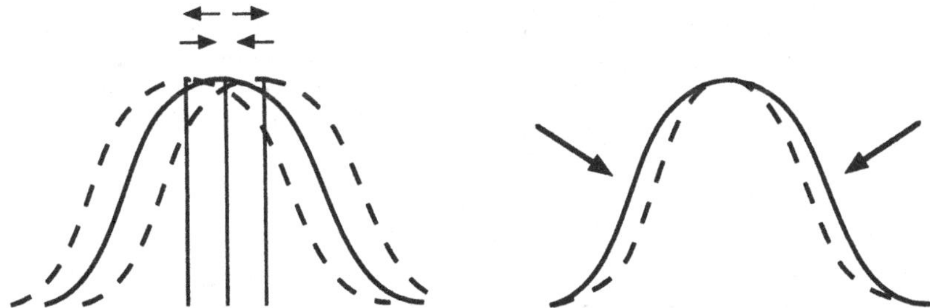

Figures 14 and 15 Illustration of the two effects of Darwinian selection that Galton accepted: temporary oscillation around an equilibrium position and normalising selection.

regression and ancestral heredity would eventually militate against the modification of the 'typical distribution'. The only conceivable change in the mean would be a temporary and minor oscillation (Figure 14). Correspondingly, the action of Darwinian natural selection is normalising: it confirms the type as long as the latter is in harmony with its environment and selection acts only on variability, which it limits (Figure 15). Under no circumstances does selection gradually move the 'racial centre'. In other words, Galton rejected the idea of selection by 'minute steps'.[68] On the other hand, he considered that artificial selection could induce the formation of a new type, on condition that exceptional individuals were sufficiently isolated (Figure 16), following the procedure carried out by breeders. Figures 13 to 16 are not original documents; they are interpretations which capture the imagery used by the post-Darwinians in their search for a 'statistical' representation of the principle of selection.

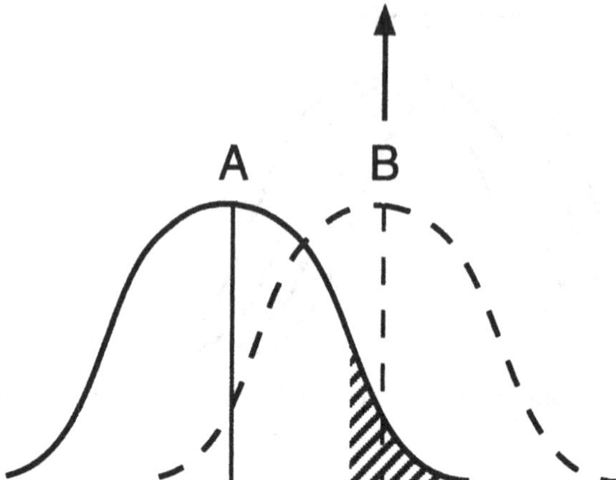

Figure 16 Representation of Galton's conception of natural selection: the rapid shift from one equilibrium position to another, on the basis of an extreme sample of the population. Of these four representations, only the third (Figure 15) had a clear statistical meaning from Galton's theoretical standpoint.

Galton's theoretical criticism of Darwin's gradual natural selection was, in practice, a justification of artificial selection applied to man. Galton liked to say that eugenics would accomplish quickly, and at low human cost, what natural selection would accomplish slowly and at a terrible price. While he did not think it possible that natural selection could alter the 'type' of a race, Galton did think that it relentlessly eliminated inferior types. In other words, he considered that natural selection operated *between* racial types.

5.2.3 *Positions of organic stability and evolutionary discontinuity*

Galton put forward another reason to justify his doubts about the Darwinian concept of gradual selection. This argument, which had a major influence on biologists in the last decade of the 19th century, was based on the belief that in nature the important changes are those that take place between stable equilibrium states. Whether it is a question of architecture, glaciers, government, crowd dynamics or any other form, some combinations are less ephemeral than others and constitute 'temporarily stable forms'.[69] This was Galton's final interpretation of 'types'.

Applied to evolution, this implies that the modification of species does not take place through the accumulation of very small differences, but by genuine leaps which involve a profound reorganisation of the organism. Galton liked to explain his point by way of a technical analogy:

> Organisms are so knit together that change in one direction involves change in many others; they may not attract attention, but they are none the less existent. Organisms are like ships of war constructed for a particular purpose in warfare, as cruisers, line of battle ships, &c., on the principle of obtaining the utmost efficiency for their special purpose. The result is a compromise between a variety of conflicting desiderata, such as cost, speed, accommodation, stability, weight of guns, thickness of armour, quick steering power, and so on. It is hardly possible in a ship of any long established type to make an improvement in any one of these respects, without a sacrifice in other directions. If the fleetness is increased, the engines must be larger, and more space must be given up to coal, and this diminishes the remaining accommodation. Evolution may produce an altogether new type of vessel that shall be more efficient than the old one, but when a particular type of vessel has become adapted to its functions through long experience it is not possible to produce a mere variety of its type that shall have increased efficiency in some one particular without detriment of the rest.[70]

This passage expresses the traditional idea of correlation, as elaborated by Cuvier.[71] Darwin gave correlation a certain importance in his theory of selection, because it enabled him to explain the fixation of certain variations that had no intrinsic value, 'correlative variations' being considered necessary consequences of any variation accumulated by natural selection.[72] However, Darwin also considered that growth constraints were to a certain extent themselves controlled by natural selection.[73]

Galton rejected this theoretical bias: not everything is possible. He explicitly argued that only a limited number of 'positions of organic stability' are available at any given time. As a result, natural selection does not favour the indefinite modification of a large number of separate traits;[74] it simply retains those equilibrium states that are in harmony with the environment.[75] Galton put forward a mechanical model of his interpretation of 'types' as 'stable equilibrium positions'. The image is that of a polyhedral stone, each face defining a stable equilibrium. However, because the faces are not all the same size or shape, some positions are more 'stable' than others. Figure 17, taken from *Natural Heredity*, illustrates the model in a two-dimensional space.

The aim of the model was to show that change can occur by 'leaps' without violating the 'law of continuity'. Let us suppose that the stone is pushed slightly. As soon as the pressure is released, the stone will resist and return to its initial position. With a stronger pressure, the stone will move further, but it will still return to its original position. However, beyond a certain point, the stone will pass to another position of stability, such that 'there is no violation of the law of continuity in the movements of the stone, though it can repose in certain widely separated positions'.[76]

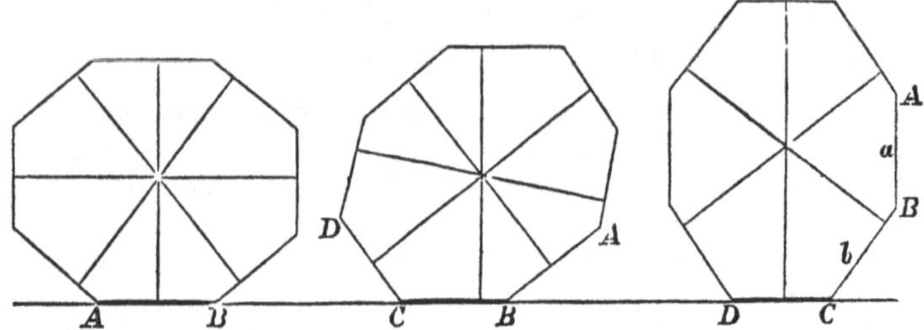

Figure 17 Galton's polygon. The equilibrium positions allow for oscillation; the transitions from one equilibrium to another are compatible with the 'principle of continuity' (F. Galton, *Natural Inheritance,* 1886). In *Hereditary Genius* (1869), Galton presents the model in three dimensions and describes it in terms of a multifaceted stone.

Through this mechanical model, Galton elaborated and popularised the notion that there are two very different modes of organic variation, subsequently known as 'continuous variation' and 'discontinuous variation' (Bateson, 1894, and de Vries, 1894).[77] The distinction was not new. Naturalists had long distinguished between variations in degree (Darwin's 'individual differences') and major anomalies or monstrosities ('sports'). Galton's contribution was to strengthen this distinction by forcefully pointing out its implications for heredity and evolution.

From the point of view of heredity, Galton opposed 'individual variations' to 'sports'. 'Individual variations' (Darwin's 'individual differences') are unstable deviations from the population mean and are the object of blending inheritance, which means that offspring are approximately intermediate between the parental values. These variations generally have a normal distribution in the population, and the offspring show a tendency to regression. Height and weight are classic examples. The second kind of variation is characterised by its stability: it is inherited in an 'exclusive' manner; in other words, the character is either inherited or not, and statistical regression of offspring towards the parental values does not apply.[78] However, Galton believed that the difference between the two kinds of variation was not absolute.[79]

In an article entitled 'Discontinuity in Evolution' (1894), Galton proposed to call these two types of variation 'divergent' and 'transilient'[80] variation. Individual variation is a temporary divergent variation, because it tends to regress in each new generation. This kind of variation has no evolutionary consequence, because, unlike 'transilient' variation, it cannot change the typical centre of a race:

No variation can establish itself unless it be of the character of a sport, that is, by a leap from one position of organic stability to another, or as we may phrase it, through *'transilient'* variation. If there be no such leap the variation is, so to speak, a mere bend or divergence from the parent form, towards which the offspring in the next generation will tend to regress; it may therefore be called a *'divergent'* variation ... I am unable to conceive the possibility of evolutionary progress except by transiliences, for, if they were merely divergences, each subsequent generation would tend to regress backwards towards the typical centre, and the advance that had been made would be temporary and could not be maintained.[81]

For Galton this discontinuist vision of evolution removes none of natural selection's pertinence, but it does change its focus. Natural selection can apply to both types of variation, but it is effective only in the case of saltational variations – those that define a new equilibrium position.

Thus Galton echoed Jenkin's criticisms of Darwin from 1867. Natural selection cannot have much effect on characters showing infinitesimal variation that are transmitted by blending inheritance. It can act on 'sports', as long as these are taken to be discrete characters, not necessarily of large size,[82] but inherited in an 'exclusive' manner. From this it follows that, strictly speaking, natural selection does not create or fashion races. It merely penalises non-adapted forms by replacing them. However, Galton did not go so far as to argue that natural selection should be considered a 'replacement rate'. This conception triumphed only during the synthesis of Mendelism and Darwinism. Nevertheless, the idea is clearly latent in Galton's work.

It would be overstating the case to identify Galton's saltational variation with the future 'mutations' of the geneticists. Apart from the fact that 'sports' were described purely phenotypically, there is a remarkable inconsistency at the heart of Galton's notion. Galton had two simultaneous and different discourses on 'sports'. On the one hand, they were seen as a sudden displacement of the 'typical centre', in Galton's statistical sense of the term.[83] From this point of view the term 'sports' cannot, strictly speaking, be applied to individuals, but only to a population, with its mean and central tendency, *and* its individual variability. The only possible rigorous interpretation of this position is that a new race is formed from an extreme sample of the general population (see Figure 16). On the other hand, Galton presented 'sports' as being those characters that are governed by exclusive inheritance, in opposition to blending inheritance. According to this definition, the coat colour of basset-hounds or eye colour in humans should be described as 'sports'. But if this is the case, there is no reason to speak of a 'typical centre', but rather of several 'types' that permanently coexist within the population.

Galton did not choose between these two very different conceptions. Sometimes saltational variation was taken to mean the formation of a new race on the basis of an exceptional sample; this representation was inherently linked to his eugenic project, and is predominant in most of his work. But Galton's later writings (from 1889 onwards) give an ever greater place to exclusive inheritance. In this respect, they influenced many of those who developed the 'mutation' theory of evolution around 1900, often in close connection with the 'mutation' theory of heredity. Bateson, de Vries and Johannsen, among others, acknowledged their conceptual debt to Galton. In fact, paralleling Galton's ambiguous view of 'sports', they acknowledged different debts under the same name. Bateson, who introduced the conventional opposition between 'continuous' and 'discontinuous' variation, understood the latter as meristic variation – that is, variation in characters that are transmitted in an all-or-none fashion.[84] De Vries, who also used the term 'discontinuous variation' in the same year that Bateson published his *Materials for the Study of Evolution* (1894), had Galton's statistical methodology in mind. For de Vries, discontinuous variation could be found in populations with a 'double-hump' distribution curve for a given quantitative character.[85] Incidentally, Weldon and Pearson, the first 'biometricians', who were later to be fiercely opposed to the mutationists, began their statistical study of evolution using Galton's idea of 'double-hump curves' (see below, Chapter 7).

Weismann and Galton were thus two major actors in the Darwinian tradition at the end of the 19th century. They both adhered to the idea that natural selection is the dominant force in evolution. They both tried to elaborate a 'hard' concept of heredity, based on the hypothesis of the continuity and independence of the germ line, and on the rejection of the inheritance of acquired characters. Thus they played similar roles in overthrowing Darwin's concept of pangenesis – it is not the organism that determines the hereditary material, but the hereditary material that determines the organism. Finally, Weismann and Galton both reelaborated the theory of selection in the context of the key problematic of 'regression'.

But the similarities end there, since the same theme served entirely different theoretical objectives in the work of these two men. Weismann's regressive evolution was the necessary consequence of an interruption of the selective process. It implied inexorable disorganisation, diminution and depreciation and that, everywhere and always, biological order depends upon natural selection. For Weismann, 'panmixia' referred to the disorganisation that came about as a result of the cessation of selection. In this theoretical context, 'regression' was an illustration of the 'all-sufficiency' of natural selection. One of the main implications of this view was that heredity is subordinated to the 'principle of evolution'. This explains why the first Mendelians at least kept their distance from Weismann's 'ultra-Darwinism' and sometimes were very critical of it, despite the fact that they agreed with its criticism of the inheritance of acquired characters. Panmixia, argued Bateson in 1894, was a 'flower of speech' that consisted of the 'belief that all distinctness is due to natural selection, and the expectation that apart from natural selection there would be a general level of confusion'.[86]

Galton's understanding of regression had a very different aim. For Galton, the principle of natural selection must be in agreement with the available knowledge of 'heredity'. According to Galton's theory of heredity, everything gravitates around statistical regression to a 'racial centre'. Amplified into the 'law of ancestral heredity', regression rapidly revealed itself to be a rehabilitation of the 'doctrine of the constancy of races' which had dominated the theory of breeding in the first half of the 19th century. Galton agreed that, on the geological scale, species were subject to endless modification. But he thought that 'race' was a stable form that could not be altered by natural selection. That is why he disagreed with Darwin's emphasis on the principle of natural selection: for Galton, selection could not act by the accumulation of extremely small differences. Galton demanded that natural selection conform to the iron law of the statistical identity of races over time,[87] which left it free either to 'cooperate' with regression, in which case it would be normalising, or to favour the emergence of a new racial centre that it would subsequently sanction but would not have created.

Galton's influence – for better *and* for worse – should not be underestimated. Of course, after Darwin, this kind of thinking about 'types' was somewhat anachronistic. The rehabilitation of typological schemes, essentially elaborated with reference to 'the race', had at least two major theoretical weaknesses: first, it was based on a flawed mathematical understanding (Pearson was quick to realise this); second, for an explicitly populational approach it contained a serious confusion between discontinuous variation and 'racial type' (a confusion that ceased only when Johannsen distinguished 'gene', 'genotype' and 'phenotype'). Furthermore, the eugenic utopia that drove Galton's hereditarian thought produced too many sinister twists in world history for it to be excused.

Nonetheless, it would be wrong to make Galton a scapegoat. Prior to the Second World War, virtually all biologists were eugenicists (historians of biology would do well to consider this fact more seriously). We will draw the curtain on this Victorian dandy by outlining the reasons for his extraordinary theoretical influence.

How could this man, obsessed by normal distributions (he ran into problems with the police for trying to verify the distribution of the length of women's skirts . . .),[88] have left such a mark on theoretical biology, when he was neither a naturalist nor a field biologist nor a 'professional' anthropologist or mathematician?

Galton's talent did not lie in a particular empirical discovery or theory, but rather in his ability to invent scientific images that oriented both his own work and that of a large number of his contemporaries in their study of previously inextricable theoretical problems. Without the founding intuitions of statistical regression and of correlation, it would be difficult to imagine the growth of biometrical studies of heredity and selection. Furthermore, the distinction between two types of variation – soon to be called 'continuous' and 'discontinuous' – prepared the ground for mutationism and, beyond that, for Mendelism. Galton thus inspired the two rival schools which at the beginning of the century confronted each other on the dual terrain of evolution and heredity. He inspired the first 'biometricians', who took up his statistical tools and used them to establish (against Galton) the possibility and the reality of natural selection as conceived of by Darwin, and at the same time opposed nascent Mendelism. But Galton's ideas also nourished the arguments of the first Mendelians, who used the same tools to try and show the inefficacy of Darwinian selection, and used the example of discontinuous variation to found a new theory of heredity.

These two chapters have sketched out the strange conceptual landscape of selection after Darwin. Darwin had rigidly associated his central hypothesis with the 'great principle of heredity' and with a certain kind of variation. By emphasising the notion of individual and immediate heredity, Darwin thought that he had freed transformationist thought from the traditional objection of the 'forces of reversion', in other words of the reversibility of evolution. By designating individual differences as the raw material of natural selection, he thought he had identified a level of reality that would enable species to show infinite variation. But those early Darwinians who studied heredity and variation were haunted precisely by the problem of regression. Weismann and Galton show us the fundamental uncertainty that marked theoretical Darwinism as the lack of an experimental science of heredity made itself increasingly felt. Wishing to preserve Darwin's heritage, Weismann radicalised the principle of selection to the point of subordinating it to the principle of heredity. Without natural selection, heredity could do nothing, not even conserve. Natural selection thus found itself in the epistemologically fragile position of having to explain everything, and of being the sole rampart against universal biological disorganisation. Galton, on the other hand, demanded that the principle of selection conform to his idea of heredity as a conservative force of regression able to oppose the modification of populations. In this view, natural selection was confined to the role of sifting 'races' that it had played no role in creating.

To summarise, post-Darwinian evolutionary theory was characterised by a profound confusion.

6 The strategy of indirect corroboration: the case of mimicry

6.1 The demonstration of natural selection: an overview

Darwin never gave a single direct proof of natural selection. This does not mean, however, that he did not prove anything. The *Origin of Species* is 'one long argument' that aims to establish 'the theory of descent with modification by natural selection.'[1] We have seen how Darwin conceived of the structure of this *'argument'*.[2] Darwin's epistemology was very explicit: he distinguished the 'mere *hypothesis*' of natural selection from the *theory* of which it was the fundamental *principle*.[3]

The hypothesis was justified by two separate approaches. Considered on its own, it was inferred as a highly 'probable' consequence of a set of empirical generalisations, about (1) the rate of reproduction of organisms; (2) the limitation of resources (the combination of the two entailing the struggle for existence); (3) the existence of variations affecting survival and reproduction; (4) the inheritance of these variations. A proof by analogy can be added to this list: the formation of domestic races by artificial selection. These circumstantial arguments imply that natural selection is not a 'principle' in the strong sense of the term that would apply to an axiomised theory, or to a theory of physics like Newtonian mechanics. The 'mere hypothesis' of natural selection is not a proposition that cannot be proved and that is only useful because of its consequences. Rather, it is a proposition that suggests the existence of something, on the basis of a number of empirical premises. In the *Origin of Species*, the 'mere hypothesis' is set out in the first five chapters. The particular meaning of the word 'hypothesis' in this context is important. Darwin's strategy was not to present those facts that could be well explained by a certain conjecture, but rather those facts that lead to a hypothesis as to the process of modification of species. The idea that natural selection is a 'mere hypothesis' suggests that it sums up a certain number of solid empirical generalisations by expressing something that they imply. Darwin's reasoning can be summarised as follows:

1. All organisms tend to increase in number according to a geometrical ratio.

2. Many more individuals are born than could possibly survive.

Hence: 3. Individuals compete for survival and reproduction.

4. For a number of characters, there is heritable variation.

5. Some of this variation affects the chances of survival and reproduction of the individual organisms.

Hence: 6. 'Every slight modification, which in any way favoured the individuals of any species, by better adapting them, would tend to be preserved'[4] (= The hypothesis of natural selection).

The other aspect of Darwin's 'long argument' is easier to understand. In the second part of the *Origin of Species* (chapters 7–13), natural selection functions as a principle that 'explains several large and independent classes of facts, such as the geological succession of organic beings, their distribution in past and present lines, and their mutual affinities and homologies'.[5] Other examples include the genesis of instincts, extinction, species divergence and rudimentary organs. These 'large bodies of facts' constitute a series of indirect proofs of the hypothesis. For Darwin, the ability of natural selection to unify and explain large and independent collections of facts justified the elevation of the hypothesis 'to the rank of a well-grounded theory'.[6]

To sum up, the *Origin* contains two levels of justification for natural selection. The 'mere hypothesis' is concerned with direct evidence for the existence of natural selection, and with the concepts required for thinking about the process. The 'well-grounded theory' is a reconstruction of natural history, in which natural selection functions as a principle that explains many kinds of facts.

At the level of the 'well-grounded theory', Darwin thought it was particularly important that the facts thus explained should be independent of each other.[7] Indeed, he thought that the power of a hypothesis could be measured by its capacity to absorb apparent anomalies and thus to transform them into proofs.[8] This is the classic strategy of validation by unification, confirmation and conversion of anomalies into resolved problems, directly inspired by the Newtonian model of experimental science. Wallace summed up this method as follows:

> There is no more convincing proof of the truth of a comprehensive theory, than its power of absorbing and finding a place for new facts and its capability of interpreting phenomena which had been previously looked upon as unaccountable anomalies. It is thus that the law of universal gravitation and the undulatory theory of light have become established and universally accepted by men of science. Fact after fact has been brought forward as being apparently inconsistent with them, and one after another these very facts have been shown to be the consequences of the laws they were at first supposed to disprove. A false theory never stands this test.[9]

Wallace went on to explain how the theory set out in the *Origin of Species* corresponded to this view: it was strengthened by its ability to integrate new problems and by its capacity to explain 'anomalies'.

In the half-century that followed the publication of the *Origin of Species*, selectionist ideas generally followed this approach. In particular, Wallace and Weismann helped to spread the image of Darwinism as a theory justified more by its explanatory power than by the rigorous elaboration of its central hypothesis. This corroborative strategy and the systematic use of the old adage *exceptio probat regulam* (which nowadays should really be translated as 'the exception *tests* the rule')[10] played an important part in the initial success of the theory of natural selection. However, it also gradually led to the panselectionist vision of natural selection which could legitimately be criticised for being merely rhetorical. All that was required was to imagine a possible adaptive significance for a character, and its existence was immediately explained by the 'principle of selection', often reduced to a 'principle of utility'.[11]

It was only towards the end of the 19th century that the problem of finding direct empirical proofs or falsifications of the hypothesis of natural selection became important. This part of our story is so rich and convoluted that it is not possible to say that the hypothesis of natural selection was demonstrated at this or that moment. In fact there were a number of key moments, but it would nevertheless be wrong to present the three decades from 1890 to 1920, during which the fate of the *hypothesis* of natural selection was decided, as a step-by-step methodological exercise that finally reached its objective. This would ignore an essential element of the crisis that shook contemporary biology to its foundations, and also some key epistemological consequences for modern evolutionists.

This can best be explained by a metaphor. It was war. In a war, you know who your enemies are, but that's about all you can be sure of. Everything else is flux. You don't always fight on the same terrain. You don't always fight with the same weapons – in fact, you often fight with arms that have been seized from the enemy, or you fight on his territory, or both. Another aspect of war – banal and tragic, but nevertheless terribly true – is that those who are the first to join up are often either dead or incapacitated when victory comes. The winners – the survivors – are no longer heroes; they no longer have any reason to fight. They may have won, but was it worth it? The true history of theoretical Darwinism in the period covered here can be glimpsed by simply saying: both sides won. This is one of the privileges of theoretical wars. Unless of course, somewhat more prosaically, both sides simply agreed to stop fighting and to forget the quarrel, if indeed they still knew what it was all about.

In the 1890s, a handful of biologists and mathematicians – in particular Walter Weldon and Karl Pearson – decided to prove the theoretical possibility and the empirical reality of natural selection. By 'natural selection' they meant exactly what Darwin had said: selection acting on continuously varying characters, producing a gradual modification of the species. They hoped that by giving a precise statistical meaning to each element of the hypothesis they would be able to prove the existence of at least one example of natural selection. These 'biometricians' (as they called themselves) thus constructed a 'mathematical theory of selection' which also turned out to be a 'mathematical theory of heredity'. But their intention, clearly stated at the outset, was to establish the possibility and reality of natural selection by 'purely statistical' means, without any causal hypothesis as to the nature of heredity. This school's debt to Galton is quite obvious even though the biometricians in fact worked hard to rehabilitate Darwinian orthodoxy against Galton.

At the same time, naturalists influenced by the success of the experimental method in physiology and embryology began the search for an experimental science of heredity and started to study discontinuous variation. These scientists – de Vries, Bateson, Davenport, Cuénot and Johannsen, among others – 'rediscovered' Mendel and, from 1900 onwards, laid the foundations of what was to be known as genetics. But as they developed the methods of this new science they became convinced that what Darwin called 'fluctuating' variation and what they called 'continuous' variation had no hereditary component, and was merely an appearance (later called 'phenotype').[12] This led to a virulent anti-Darwinism and a 'mutationist' bias: evolution takes place by leaps, and natural selection neither forms nor creates anything, but merely sifts out afterwards. This school's debt to Galton was therefore just as strong as that of the biometricians.

In both cases, concepts gave way to facts. The clash of the two schools is known as a dispute between biometricians and Mendelians. Considered from the older point of view of heredity, it was an argument between the partisans of ancestral heredity (reworked by Pearson) and the followers of Mendel. Considered from the point of view of evolution, it was an argument between orthodox Darwinians and mutationists. These two aspects were closely linked.

The biometricians rapidly lost the struggle to defend 'ancestral heredity'. As far as natural selection was concerned, things were not so clear-cut. The Mendelians did not deny the existence of natural selection; they merely rejected the idea that it acted gradually on continuous variation and that it shaped species. Nevertheless, the Mendelians were responsible for the operational definition of natural selection that is most widely used today – the diffusion of a gene in a population. Diffusion is gradual, but the variation upon which selection operates is discrete. In a way, Darwin lost and Jenkin won: natural selection came to be understood as a replacement rate.[13] However, this was merely the last of a long series of episodes. From the end of the 1910s onward, it became clear that the continuous variation analysed by the biometricians was in fact compatible with a Mendelian interpretation, on condition that the quantitative character was determined by a large number of genes. From this point onwards, the selection *of* genes could also be understood as a selection operating *on* quantitative characters. The concept of natural selection, in its Darwinian form, was rehabilitated. Darwinism thus became a dual theory, profoundly restructured by its encounter with Mendelism.

This chapter and the two that follow deal with three major strategies of justification of the principle of natural selection, which were developed in the seventy years after the publication of the *Origin of Species*: the strategy of indirect corroboration, the strategy of direct proof, and the strategy of confrontation with the theory of heredity which finally prevailed in early-20th-century Mendelism.

In the present chapter the strategy of indirect corroboration will be evoked only briefly, on the basis of one example which illustrated the explanatory power of natural selection and can be situated at the level of the 'well-grounded theory'. The *Origin of Species* had set an example by showing how natural selection unified and explained the 'various large and independent classes of facts' of natural history. After 1859, this strategy of justification took the form of a promising research programme. The principle of natural selection not only furnished a general explanatory framework for the known facts, it was also able to reveal previously unknown facts and to provide novel explanations for them. The studies of mimicry in butterflies constitute an exemplary case. Indeed, mimicry was the first application of the selectionist explanation to facts other than those presented by Darwin in the *Origin*. Given that the discovery of these facts, and the very construction of the concept of mimicry, were historically impelled by the principle of natural selection, they were soon used to illustrate the principle's heuristic power. Many similar examples can be found in the history of evolutionary biology inspired by Darwinism (generally speaking, the whole of population biology illustrates this idea). The explanatory and heuristic power of the principle of natural selection has always been, and remains, the main reason for naturalists' support for Darwinism. The example of mimicry provides a way of understanding how this strategy of justification functioned in practice.

However, the aim of the present study is not to provide an exhaustive reconstruction of the history of natural selection. Chapters 7 and 8 deal with the attempts to demonstrate what Darwin called 'the mere hypothesis'. Chapter 7 examines the intense effort deployed by the English biometrical school in the 1890s to establish the existence of natural selection, independently of any hypothesis on the physiological nature of heredity. The biometricians sought to establish a statistical methodology capable of establishing, case by case, the existence of facts of natural selection. This was a classic strategy of justification by the discovery of direct proofs.

Chapter 8 analyses the confrontation between Darwinism and Mendelism. In this context it was less a case of proving the existence of natural selection than of seeing whether and how natural selection was possible under the hypothesis of a given mechanism of hereditary transmission. There were two phases to this period. First, Mendelism found itself closely linked to the mutationist theory of evolution which appeared in the 1890s and was initially seen as a refutation of Darwin's modifying power of selection. This opening encounter concluded in the development of the first clear operational definition of selection, as a rate of replacement of genes and phenotypes. This appeared to constitute a victory for the idea that selection acts on discontinuous variation. However, in a second phase, Mendelism found itself able to explain theoretically or experimentally the fact that selection acted on (phenotypically) continuous variation. This takes the story to around 1920; by this time, confrontation had led to interpenetration or, to use a term applied in 1942 by Julian Huxley to another aspect of the story, 'synthesis'.

6.2 The strategy of indirect corroboration: mimicry

6.2.1 An exemplary case

Mimicry is an exemplary problem that shows how the hypothesis of natural selection constituted a new methodological approach. Only an outline will be presented here, although in sufficient detail to make clear the strategy of discovery opened up by the selectionist explanation.

Mimicry can be considered an 'exemplary problem' for four reasons:

1. It is the first known example of the application of the selectionist hypothesis to facts other than those invoked by Darwin in the first edition of the *Origin of Species*. Darwin subsequently entered the fray more or less enthusiastically, sometimes finding himself overtaken by events and by the implications of his theory. Thus mimicry offers the possibility of distinguishing Darwinian science from the work of Darwin himself.

2. It consists of a precise set of facts that do not involve the whole of natural history and which constitute what Darwin called 'special

difficulties' for the theory of natural selection.[14] Thus mimicry should not be given the same importance as the 'large bodies of facts' that Darwin explained by the principle of natural selection – systematic affinities, homologies etc. – which have consequences for the whole of natural history and imply that the doctrine of natural selection is a *theory* that unifies the whole of biology, perhaps the only possible such theory. In the case of mimicry, natural selection is merely a working hypothesis to be applied to a limited empirical problem. Mimicry therefore enables us to examine the explanatory power of the hypothesis in a particular case, without being drawn into a consideration of Darwin's biological philosophy. The causal chain between natural selection and the phenomena of mimicry is certainly shorter than, say, that between natural selection and 'homologies' or 'the tree of life'.

3. The selectionist explanation of mimicry is an exemplary case of the strategy of corroborative argument. For nearly 130 years, mimicry has furnished some spectacular empirical confirmations of Darwin's hypothesis, but also some unexpected twists, thus showing the heuristic power of the hypothesis of natural selection. Mimicry is one example amongst many of the capacity of selectionist thinking to generate puzzles and to overcome them.

4. Finally, mimicry has a rare advantage for a Darwinian naturalist: it has no equivalent in artificial selection. The specific reality of natural selection can thus be studied: geographical distribution, changing environment, ecological dynamics etc.

This chapter does not reconstruct the history of theories of mimicry but uses a few initial moments to reveal the strategy of indirect corroboration of the hypothesis of natural selection, and its heuristic power, through a key example.

6.2.2 *Selection and Batesian mimicry*

The term 'mimicry' was proposed in 1861 by Henry Bates[15] to describe protection by resemblance in the animal kingdom. Examples of animals that defended themselves against predators by a more or less close resemblance to their organic or inorganic environment ('homochromy') had been known for many years. Classic examples are the white fur of polar animals, the form and the colour of flatfish, or the extraordinary resemblance of stick insects to leaves or twigs. Long before Bates invented the term mimicry, a number of theories existed to explain the phenomenon. These sometimes invoked 'affinities and analogies' (later known as the theory of 'homologies'), and sometimes extraordinary coadaptation. In other words, there were formal and utilitarian explanations, but all were resolutely non-transformationist.[16]

In the first edition of the *Origin of Species*, Darwin did not deal with the phenomenon. This had no particular significance; it appears that Darwin simply did not know about the effect.[17] Furthermore, despite the fact that these phenomena had been well documented since at least the beginning of the 19th century, their study only really began after 1860 and the publication by Andrew Murray of an article on 'the disguises of nature'.[18]

However, homochromy lent itself easily to an explanation by natural selection. In October 1859 – *before* the publication of the *Origin of Species* (24 November 1859), but four months after Darwin and Wallace's communication to the Linnean Society of London – Canon H. B. Tristram wrote a memoir on the birds of North Africa and referred to Darwin and Wallace's communication in order to explain the colour of the small animals of the Sahara:

> Writing with a series of about 100 Larks or various species from the Sahara before me, I cannot help feeling convinced of the truth of the views set forth by Messrs. Darwin and Wallace in their communications to the Linnean Society ... It is hardly possible, I should think, to illustrate this theory better than by the Larks and Chats of North Africa ... In the Desert, where neither trees, brushwood, nor even undulation of surface afford the slightest protection from its foes, a modification of colour, which shall be assimilated to that of the surrounding country, is absolutely necessary. Hence, without exception, the upper plumage of every bird, whether Lark, Chat, Sulvian, or Sandgrouse, and also the fur of all the small mammals, and the skin of all the Snakes and Lizards, is one uniform isabelline or sand colour. It is very possible that some further purpose may be served by the prevailing colours, but this appears of itself a sufficient explanation ... In the struggle for life which we know to be going on among all the species, a very slight change for the better, such as improved means of escaping from its natural enemies (which would be the effect of an alteration from a conspicuous colour to one resembling the hue of the surrounding objects), would give the variety that possessed it a decided advantage over the typical or other forms of the species. Now in all creatures, from man downwards, we find a tendency to transmit individual varieties or peculiarities to the descendants. A peculiarity of colour or form soon becomes hereditary when there are no counteracting causes, either from climate or admixture of other blood. Suppose this transmitted peculiarity to continue for some generations, especially when manifest advantages arise from its possession, and the variety becomes not only a race, with its variations still more strongly imprinted upon it, but it becomes the typical form of that country. If it be objected that we see many varieties which do not become hereditary, we may reply, that these varieties having experienced changes not advantageous to their means of existence, may from that very cause become extinct.[19]

Having discussed what might be the possible significance of this hypothesis for the material he was studying, Tristram concluded that this was at least one good example of the creation of a new species from an old one 'according to natural means of selection'.[20] These statements, published prior to the appearance of the *Origin of Species*, show how quickly the selectionist explanation and all its characteristic ingredients (struggle for life between and within species, inheritance of individual peculiarities, gradual change) took root in naturalist circles. The only rider is that Tristram appears to have put forward Wallace's hypothesis (selection *between* races) rather than Darwin's (modification of races by individual selection).[21]

The example presented by Henry Walter Bates (1825–92) in 1861 to the Linnean Society on his return from an expedition to the Amazon was far more troubling.[22] Bates' memoir was devoted to the systematics of a family of butterflies called the Heliconidae. These insects live in the tropical forest and have several remarkable characteristics. They are extremely numerous (far more so than butterflies of other families); they fly in swarms; they have large, brilliantly coloured wings (bright orange or white marks on a black, blue or brown background); they generally fly slowly and they do nothing to hide themselves, particularly at night when they hang on twigs with their wings folded (the underside of the wings is also brightly coloured, which is unusual). These butterflies would therefore appear to be particularly vulnerable to predators. A large number of insectivorous birds also live in the forest; the butterflies they eat can be identified by the inedible wings that are left over and fall to the ground. These wings never belong to Heliconidae butterflies. Bates explained this observation by the bitter smell and vile taste of these insects. Their bright colours were thus interpreted as a warning signal to potential predators. Up until this point, Bates had done no more than point out a novel but nonetheless banal example of adaptation.

The difficulty was as follows. The Heliconidae butterflies share the forest with butterflies of the Pieridae family, many of which are barely different from one of the most common European butterflies, the Cabbage White (*Pieris brassicae*) – a white butterfly with small black marks at the extremity of its wings. Amongst the Pieridae butterflies observed by Bates in Amazonia were members of the genus *Leptalis*. Some of these species resemble the Cabbage White, but most of them are extraordinarily precise copies of a Heliconidae butterfly of the genus *Ithomia*. *Leptalis* mimics *Ithomia* in its wing form, colour and pattern. The resemblance is so close that even an experienced entomologist cannot distinguish the two species solely on the basis of the wings. But the two butterflies are classed as belonging to different families because their anatomy is very different – a close examination of the legs or head leaves no room for doubt, but these differences cannot be detected with the naked eye.

Bates did not find only one example of this phenomenon. He described seven mimetic species of *Leptalis* that imitated a total of six species of *Ithomia*. He also described ten other genera in which he had observed species that imitated other butterflies, noting that several species belonging to different genera would often imitate the same Heliconidae species. Bates repeatedly stated that before he formulated the hypothesis of 'mimicry', the systematics of Amazonian butterflies had appeared insoluble, because these insects did not appear to follow any of the accepted criteria for the classification of Lepidoptera.

However, this did not explain exactly what was happening. There was in fact a ready explanation for this kind of problem which was a typical example of what, since Richard Owen, naturalists have called 'analogies' – functional resemblances which show that different organisms have employed similar solutions to resolve the same adaptive problem. Analogies are opposed to 'homologies', which are similarities in *bauplan* (body-plan). For example, the similar shapes of the shark and the dolphin are analogies, because they represent similar 'technical' solutions to the same problem (fast swimming), but they are based on a very different internal anatomy. To claim that the resemblances between mimicking butterflies were analogies would be to argue that they were due to adaptative convergence. For this hypothesis to be proven unambiguously, the environmental conditions – for example the food and the climate – which might explain the phenomenon, would have to be properly described. This explanation could itself perhaps be interpreted in the Darwinian language of natural selection.[23]

However, in his 1861 paper, Bates rejected the classic explanation by 'analogy' – in other words, by similar life conditions.[24] Instead, he set out a relentless argument that led him both to admit the objective reality of 'mimicry' (one form genuinely does copy another, rather than there being convergence between the two independent forms), and to argue that natural selection was the only possible explanation of the phenomenon. For Bates, the crucial fact was that *Leptalis* (the butterfly that he took to be the mimic) showed an important intra-specific variability. Several varieties of this species could coexist in the same region; when one of them resembled *Ithomia*, it was always the most frequent. Bates also noticed that the resemblance to a given species varied from place to place, with many intermediate values between vague and perfect resemblance. This variability was found only for *Leptalis*, which thus appeared to show varying similarity to a model which was itself extremely stable. Furthermore, the *Leptalis* 'mimics' showed the same biogeographic diversity as *Ithomia*. Finally, it should be recalled that the *Leptalis* butterflies are quite edible; indeed, they are eaten when they do not resemble a

Heliconidae. All these facts, together with a number of others, led inexorably to the conclusion that this was a genuine case of 'mimicry' – a special kind of adaptation, similar to homochromy. 'I believe, therefore, that the specific mimetic analogies exhibited in connection with the Heliconidae are adaptations – phenomena of precisely the same nature as those in which insects and other beings are assimilated in superficial appearance to the vegetable or inorganic substance on which, or amongst which, they live.'[25] And, just as natural selection had been invoked by previous authors to explain homochromy, Bates argued that it was the only possible explanation for mimicry: 'This principle [which explains mimicry] can be no other than natural selection, the selecting agent being insectivorous animals, which gradually destroy those sports or variants that are not sufficiently like *Ithomiae* to deceive them.'[26]

Despite his use of the word 'sports' in this passage, Bates did not support a saltationist view of the origin of mimetic species. Quite the opposite:

> If a mimetic species varies, some of its varieties must be more and some less faithful imitations of the object mimicked. According, therefore, to the closeness of its persecution by enemies, who seek the imitator, but avoid the imitated, will be its tendency to become an exact counterfeit, the less perfect degrees of resemblance being, generation after generation, eliminated, and only the others let to propagate their kind.[27]

Bates was convinced that he had provided a genuine 'proof' of Darwin's theory of natural selection:

> Such, I conceive, is the only way in which the origin of mimetic species can be explained. I believe the case offers a most beautiful proof of the truth of the theory of natural selection. It also shows that a new adaptation, or the formation of new species, is not affected by great and sudden change, but by numerous small steps of natural variation and selection.[28]

In 1862, Darwin wrote a review of Bates' article, and did not hesitate to claim that the explanation carried the 'touch of genius'.[29] Most of the information contained in the review was subsequently incorporated into the fourth edition of the *Origin of Species*, in the chapter entitled 'Mutual Affinities of Organic Beings', complete with the remark that 'here we have an excellent illustration of natural selection'.[30] This statement is all the more striking given that the *Origin* does not contain a single direct description of a single fact of natural selection.

Although Bates had not, strictly speaking, furnished a direct proof of natural selection, he had certainly made the first attempt to 'provide' a concrete *case* of natural selection. Bates did not restrict himself to suggesting an 'application' of the theory: first he clearly identified something that could only be an adaptation (and not a homology); second, he accepted the hypothesis of natural selection because it was both impossible to explain this particular adaptation by other, more classic, factors (for example the direct action of external conditions), and because Bates clearly identified in his data certain characteristic elements of the selectionist explanation (e.g. character variability, relativity of the adaptation, the demographic dimension of the phenomenon). There remained only natural selection, which was not, strictly speaking, directly and exhaustively proven, but which was identified as the only possible explanation[31] – a formula which is characteristic of an indirect proof. The *Origin of Species* contains no similar empirical demonstration. The examples given by Darwin in the first edition of 1859 either are purely imaginary (the wolves cited in chapter 4, for example) or are analyses of 'special cases' where it is shown that the hypothesis of natural selection is *compatible* with them (e.g. the existence of sterile castes of insects). Darwin was fundamentally interested in natural history as the nucleus of a global theory of biological diversity. For many years, Bates' article constituted an exemplary model of an empirical proof of a case of natural selection. An indirect and non-quantified proof, to be sure, but a proof based on a concrete case.

The subsequent history of mimicry shows that this strategy of indirect corroboration was far more fertile than might have been expected. Some noteworthy examples follow.

Following the publication of Bates' article, the number of studies of mimicry in butterflies rapidly increased, and it became accepted as a bizarre but relatively frequent phenomenon. In many mimetic species only females show mimicry, the males tending to conserve a type that is similar to closely related non-mimetic species. This observation led to a famous argument between Wallace and Darwin. The debate began in their correspondence and then came into the open, first in an article published by Wallace in 1867,[32] then in Darwin's *The Descent of Man and Sexual Selection*, published in 1871.[33]

We have seen how Darwin postulated the existence of another form of selection, 'sexual selection', operating alongside natural selection. Unlike natural selection, sexual selection does not depend on the viability and the fecundity of individuals, but only on their greater or lesser ability to mate with the opposite sex: 'The result is not death to the unsuccessful competitor, but few or no offspring.'[34] This was why Darwin, although he confidently felt 'sexual selection' to be a natural process, always refused to make it a particular case of 'natural selection'. For Darwin, sexual selection was responsible for most of the differences between males and females that are expressed in secondary sex characters. Furthermore, for reasons that were based both on observations and on his own prejudices, Darwin thought that 'sexual selection' only affected 'ornamental' characters or, to use a more anthropomorphic term, characters involving 'beauty': 'Sexual selection implies that the more attractive individuals are preferred by the opposite sex.'[35] In other words, sexual selection is secondary and frivolous. It may greatly modify the external appearance of one of the sexes, and even, perhaps, owing to heredity and correlated growth, that of the other sex. But natural selection still controls the vital characters; in other words, it still controls adaptation. This implies that, in the final analysis, natural selection controls sexual selection, or as Darwin put it: 'sexual selection will . . . be dominated by natural selection for the welfare of the species'.[36]

Another characteristic aspect of Darwin's view of sexual selection is that it is an asymmetric process: one sex varies, the other chooses. In general, Darwin considered that males tend to vary from the species type and try to please and search out the opposite sex, while females choose, and oppose their power of careful discrimination to the indiscriminate ardour of males. This effect leads to the fact that, in general, males are more beautiful or, perhaps more precisely, 'more ornate'. Darwin thought that this rule was generally true for birds and insects, and particularly for butterflies: 'When the sexes of butterflies differ, the male as a general rule is the most beautiful, and departs from the usual type of colouring of the group to which the species belongs.'[37]

This detour was necessary for us to appreciate Darwin's perplexity in being faced with sex-limited mimicry in butterflies. If the protective function of imitation was so important for the species, why should it extend only to females? This is the question Darwin addressed to Wallace in a letter of 15 April 1868:

> And now I want to ask you a question. When female butterflies are more brilliant than their males you believe that they have in most cases, or in all cases, been rendered brilliant so as to mimic some other species, and thus escape danger. But can you account for the males not having been rendered equally brilliant and equally protected? Although it may be not for the welfare of the species that the female should be protected, yet it would be some advantage, and certainly no disadvantage, for the unfortunate male to enjoy an equal immunity from danger.[38]

Why is it then that 'the female alone had happened to vary in the right manner'?[39] In the course of his long debate with Wallace on this subject,[40] Darwin never accepted the thesis that natural selection alone could have produced such a phenomenon. His explanation (which does not go beyond a mere suggestion) was that some kind of sexual selection had to be involved. In ordinary cases of sexual selection, 'variations leading to beauty must often have occurred in the males alone, and been transmitted to that sex alone'.[41] Why should this rule not be applied to the dull males of mimetic species of butterflies? Darwin suggests that the males have not become mimetic because they would be less attractive to females. This is actually what he finally said in an addition to the second edition of *The Descent of Man*.[42] What is beautiful for man is not necessarily beautiful for a female *Leptalis*! This explanation appears somewhat rhetorical, but it shows that Darwin thought that natural selection and sexual selection were potentially antagonistic forces: in this case, sexual selection dominates, at least provisionally.

Wallace's position was very different.[43] Although he did not deny the existence of sexual selection, he rejected it as an explanation of sexual dimorphism in mimetic butterflies, arguing that natural selection could best explain the phenomenon. It is essential for the species that females, which carry the eggs and are heavier and less agile than males, should be protected. Furthermore, males are useful to the species for a shorter length of time because, once they have mated, it matters little whether or not they fall victim to predation. At this point, Wallace makes a telling generalisation. The phenomenon is exactly comparable to that in species where the female is dowdy and the male ornate, a case that is frequently found in birds and, to a lesser extent, in mammals. Here, too, natural selection in favour of females needs to be invoked: their fragility during incubation (or gestation) requires a stronger protection against predators. This does not imply that sexual selection does not take place; it does indeed intervene, and explains the males' elaborate ornamentation, but it is strictly controlled by natural selection. Wallace points out that sexual selection must be conceived of in the same way for the two sexes: in birds where the male guards the eggs, he is dowdy, whereas the females are highly ornamented. Thus, for Wallace, sexual selection always tends to differentiate the sexes, and to develop their reciprocal ornamentation and power of attraction, but always under the control of natural selection. Wallace felt able to invoke the existence of a great 'law of protective adaptation of colour and form, which appears to have checked to some extent the powerful action of sexual selection'.[44]

If Darwin and Wallace thus agreed that natural selection always ultimately controls sexual selection, they had very different explanations of sexual dimorphism in mimetic butterflies. Darwin explained mimicry by natural selection, and dimorphism by the independent action of sexual selection; Wallace accepted that sexual selection was an initial factor in differentiation, but felt that there were fundamental adaptive reasons why females were specifically protected and not males. If a female is to produce offspring, and not merely increase her chances of attracting males, it is vital that she have the time to reach the end of gestation. Males will be more exposed to predation, but their reproductive strategy ends with copulation.

In one way, the debate begun by Darwin and Wallace continues today – the contemporary scientific literature contains clear revivals of the two original antagonistic positions.[45] This debate underlines a point that is too often overlooked when epistemologists unfamiliar with evolutionary theory study selectionist explanations. There is too much of a tendency to reduce the principle of selection to a truism that can only ever be 'confirmed' by empirical study. These early debates on the relative weight of natural selection and sexual selection are convincing proof of the complexity and difficulty of the principle of selection. Without entering into the detail of a problem that has since become extremely complex,[46] it is worth pointing out a few of the theoretical difficulties that were involved in the debate between Darwin and Wallace on sexual dimorphism in mimetic butterflies.

The first problem is that of the balance between natural selection and sexual selection. The two men agreed on the need to conceptualise this balance, but they disagreed on how this should be done. This shows that the natural processes of selection (both natural selection *stricto sensu* and sexual selection) had been elaborated so as to raise questions that could not receive an *a priori* solution. A second problem was that of the two components of the 'struggle for existence' (or 'competition') – the ability to survive (viability) and the ability to reproduce (fertility). Natural selection involves both elements; sexual selection only involves the second. The subsequent history of Darwinism underlined the fact that the significance of fertility in natural selection is a fundamental problem. It is difficult to imagine how differential fertility on its own could lead to the development of adaptations, but it is clearly an integral and inevitable part of the process of natural selection, because no advantage in viability will have an evolutionary effect if it is not expressed in the form of a reproductive advantage. Finally, the debate between Darwin and Wallace led to the appearance of a third problem: does natural selection act by and for the advantage of individuals or by and for the advantage of groups? Wallace's argument shows that he considered the evolution of systems of mimetic protection in terms of the global economy of a race or of a variety: if males are less well protected, it is because they are less important to the species' economy. This postulate is not strictly necessary: an individualistic strategy could explain the same facts equally well.

Two conclusions can be drawn from these historical and epistemological remarks on Batesian mimicry. First, mimicry appeared to the naturalists of the 1860s to be the first independent experimental confirmation of the hypothesis of natural selection after the publication of the *Origin*. In this respect, it had a considerable effect on the spread of the hypothesis. Second, this remarkable phenomenon constituted a striking illustration of the idea that the Darwinian principle was not only a way of unifying natural history, but was also a working hypothesis, able to bring to light new facts and new enigmas to solve. In other words, the principle of selection was not only verifiable, it had also founded a fertile heuristic.

6.2.3 Selection and Müllerian mimicry

Another episode in the early study of mimicry highlights the strategy of indirect corroboration.

In his article, Bates mentioned in passing a curious fact: mimicry is not limited to edible butterflies that imitate inedible ones. It also affects edible butterflies (Heliconidae), suggesting that the 'models' also imitate each other: 'Not only, however, are *Heliconidae* the objects selected for imitation; some of them are themselves the imitators; in other words they counteract each other, and this to a considerable extent.'[47] This could not be explained by the hypothesis put forward by Bates. Indeed, Bates carefully avoided the problem and merely illustrated it in the magnificent plates that accompany his article. Had he pointed it out explicitly, there can be little doubt that his thesis would not have enjoyed the success it did. And yet the reciprocal mimicry of inedible species was far more impressive than the one-way mimicry of edible species. Given that this mimicry also tended to affect only the species of the same genus or of the same family, it would have been only too easy to reply to Bates that his 'mimicry' was nothing other than a case of 'affinity', or, in Darwinian language, a result of common descent. In turn, Bates would doubtless have replied that this interpretation did not agree with the facts, because there was no clear correlation between systematic affinity and reciprocal mimicry: a given species would copy one species in one locality, but not in another. But Bates had no explanation, and thus prudently restricted himself to merely repeating the fact.

The first interpretation of this curious phenomenon was made by Fritz Müller (1825–97), seventeen years later, in 1879. Müller presented his finding of 'a remarkable case of mimicry in butterflies'[48] in an article published in German, and immediately translated into English by Raphael Meldola (an openly Darwinian naturalist) on Darwin's suggestion. Müller studied a case of extreme resemblance between two species belonging to two closely related genera, *Ituna* and *Thyridia* (Danaïdae). The comparison of the two genera showed that the two species in question (*Ituna iliana* and *Thyridia megisto*) were similar not because of their common origin, but because of a convergence that happened subsequent to the separation of the two groups.[49] The individuals of the two species had wings that appeared to be virtually identical, and both possessed tufts of hairs which gave off similar unpleasant odours. In normal cases of Batesian mimicry, the mimic is always much rarer than the model; the model has the characteristic colours and forms of its family, and is inedible. But *Ituna iliena* and *Thyridia megisto* do not meet any of these criteria, except perhaps the second, which was not very important because the two species were taxonomically close.[50] But in that case, Müller asked, 'what benefit can one species derive from resembling the other, if it is protected by distastefulness?'[51] Müller's answer to his own question was to point out that the reciprocal resemblance of inedible models was advantageous because it would facilitate learning by young predators:

> Obviously, none at all if insectivorous birds, lizards, &c., have acquired by inheritance a knowledge of the species which are tasteful or distasteful to them – if an unconscious intelligence tells them what they safely devour and what they must avoid. But if each single bird has to learn this distinction by experience, a certain number of distasteful butterflies must also fall victims to the inexperience of the young enemies. Now if two distasteful species are sufficiently alike to be mistaken for one another, the experience required at the expense of one of them will likewise benefit the other; both species together will only have to contribute the same number of victims which each of them would have to furnish if they were different.[52]

Müller gave a quantitative explanation of the 'gain': it was proportional to the numbers of each of the two copies in different local populations. An interesting feature of Müller's reciprocal mimicry was that the number of species involved in such a system could be large, indeed infinite, unlike unilateral (or Batesian) mimicry, where mimicry is more efficient if there are few mimics compared to models.[53]

In his paper, Fritz Müller did not allude to natural selection. However, he was a noteworthy advocate of Darwinism in Germany,[54] and his use of the notions of 'advantage' and 'benefit' (with quantified estimations) made it clear enough in 1879 that he was working within the theoretical framework of Darwinism. Furthermore, Darwinians were quick to adopt and diffuse the theory of reciprocal mimicry as a new illustration of the fertility of the principle of natural selection.

Strangely enough, however, Müller's elegant explanation did not enjoy the same rapid success as that of Bates. Wallace took it up immediately, but Bates, together with many of those who had hailed his hypothesis, was determined not to accept it.[55] The reasons for this resistance were complex, but they can be understood if it is realised that Müller's hypothesis implied a hardening of the concept of natural selection.

By 'hardening' is meant a clearer rejection of alternative hypotheses. It will be recalled that, in their initial work on mimicry, both Bates and Wallace had justified the use of the principle of natural selection because it was necessary to reject two other hypotheses: 'affinity' and 'the similar action of external conditions' (by saying, for example, that the same food, such as a metallic salt, leads to the same wing pattern). The same approach can be found in the case of Müllerian mimicry. As Meldola pointed out, by a rather strange route Müller's hypothesis underlined the importance of acquired characters, a question that had not been clearly posed – or at least not in these terms – since the 1860s. An essential element of Müller's hypothesis was that the birds do not have an innate knowledge of which butterflies to avoid. They have to learn. However, a similar process cannot be invoked in the acquisition of mimicry by insects: form, colour and pattern are all definitively determined when the imago is formed – insects have no power to modify the external morphology of their wings by use. Thus the adoption of the hypothesis of natural selection appeared to be linked to an open rejection of the heredity of acquired characters.[56] It was no longer a question of merely saying that mimicry was not due to the 'similar action of external conditions' but of rejecting any Lamarckian effect. The hypothesis of selection was thus hardened, and was clearly opposed to the many attempts to justify adaptationist interpretations of mimicry on a Lamarckian basis.[57] This was happening at the same time as Weismann and others were trying to free Darwinism from the doctrine of the inheritance of acquired characters.

Müllerian mimicry also led to the refinement and enrichment of selectionism. New difficulties arose from the opposition of Batesian mimicry to Müllerian mimicry. In the latter case, the uniform external appearance acts as a warning signal to the predator – if it attacks, it will have an unpleasant experience. Thus Müllerian mimicry is in principle independent of the size of the mimetic populations. Things are very different in Batesian mimicry, where the mimic acts as a decoy. If the mimic becomes very abundant, there will come a point when it is no longer protected. Predators will stop avoiding it because most of the insects they encounter with that form will be edible (mimics) rather than inedible (models). Strictly speaking, therefore, selection in favour of Batesian mimicry should lead to a reduction or removal of the advantage that gave rise to it. And, in fact, Batesian species are always less numerous than their models. This apparent contradiction therefore raised the question of density-dependent selection and of its quantification. Another major difficulty was the fact that in reality there is no strict distinction between the two forms of mimicry; there is a range of degrees of palatability, ranging from absolutely edible to absolutely inedible. Mimetic populations should thus be considered as forming complex communities that cannot be studied only from an intra-specific point of view.

Today such questions have been quantitatively analysed, although not without some difficulty. But they were clearly present in the minds of the naturalists of the end of the 19th and the beginning of the 20th century, in the context of a qualitative ecology. They reveal, once again, the heuristic power of the principle of natural selection.

This closes our examination of the early theoretical understanding of mimicry. This has not been an exhaustive treatment of this fascinating episode in the history of Darwinism,[58] but merely an exemplary case of the dominant strategy of argument which, despite its weaknesses, assured the early and permanent success of Darwin's hypothesis in the scientific community. The hypothesis of natural selection was rapidly subject to independent empirical 'testing' (that is, the introduction of new 'classes of facts', besides those invoked by Darwin in his 'long argument'). These led, of course, only to indirect proofs, reinforcing the explanatory power of the hypothesis and justifying it in the same way that the multiple consequences of Newton's principles had justified their validity in 18th-century physics. Like the indirect proofs of Newtonian physics, the indirect proofs of natural selection could not be reduced to mere 'illustrations' or 'applications' of the Darwinian principle. On the contrary, the example of mimicry shows that the hypothesis of natural selection was able to generate new facts and problems. Thus the main virtue of the selectionist hypothesis at its outset consisted less in its direct verifiability than in its value as a research programme capable of enlarging the field of classic problems of natural history.

However, at the end of the 19th century, a large number of biologists began to consider that indirect proofs and heuristic fertility were insufficient. Unlike the 'principles' of the great theories of physics, natural selection could not be considered to be an unprovable proposition that was only validated by its consequences. In the argument developed by Darwin, natural selection also had the status of a proposition referring to a frequent natural process, the probable existence of which could be inferred from a certain number of empirical premises. In the chapters that follow, we will return to the question of the 'mere hypothesis', and of the strategies of proof or disproof that, at the turn of the century, explored the possibility and the reality of natural selection.

7 The search for direct proof: biometry

In the last decade of the 19th century, a small group of biologists and mathematicians, most of them English, set out to establish natural selection as a fact. Breaking with the strategy of the first Darwinians, who had justified the principle of natural selection by its explanatory power, they sought to find examples of natural selection by defining the populational parameters that would be necessary for such a proof. Walter Weldon and Karl Pearson called this quantitative method 'biometry'. This term still describes statistical methods applied to biological phenomena, but it was first used for a research programme that was intended to give a mathematical and empirical basis to Darwin's hypothesis through a statistical interpretation of selection acting on continuous variation.[1]

This episode, which lasted only from 1893 to 1901, and in particular from 1893 to 1898, is unique in the history of Darwinism. It was the first and last time that scientists carried out an exhaustive search for the facts of natural selection without considering the physiological nature of heredity. The biometricians believed it was possible to construct and test a purely statistical theory of evolution. If Mendelism had not existed, it is probable that the theory of selection would have developed along the phenomenalist lines that Weldon and Pearson intended to follow. The lightning success of the Mendelian experimental methodology from 1900 onwards made certain that this did not happen. The results of the new science rapidly rendered the biometricians' epistemological attitude completely outdated, not so much in terms of the statistical tools employed (these rapidly became widespread) but rather because the successes of experimental biology discredited the very idea that purely 'statistical', non-causal, theories could be really fruitful. However, the biometricians' attempts to find a direct confirmation of natural selection were not in vain. For many years, the quantitative studies of Weldon (England) and Bumpus (USA) remained the only direct proofs of selection in nature. Furthermore, the methodological rigour of the first biometricians played an important role in underlining the complexity of the concept of selection.

Biometrical Darwinism took two forms. First, there was the exemplary empirical approach of Weldon, who can be credited with having planned and carried out the first systematic demonstration of a fact of natural selection. Second, there were the theoretical aspects of the biometrical strategy as they appeared in the work of the mathematician Pearson. It is in this context that the limits of biometrical Darwinism will become apparent.

7.1 Biometry and the 'fact' of natural selection

7.1.1 The sources and context of Weldon's work on natural selection

The scientific itinerary of Walter Weldon (1860–1906) clearly expresses the fundamental methodological shift that took place within Darwinism – from comparative and morphological studies to statistical analysis – following the death of its founder in 1882. Weldon was trained as a zoologist, and had been inspired by the embryological implications of Darwin's ideas. From 1859 onwards, Darwin had insisted on the differences between embryonic anatomy and that of the adult organism. Darwin argued that similarities between embryos of the same class, which had been highlighted by von Baer, could best be explained by the hypothesis of common descent.[2] Taken up and developed by Fritz Müller and Ernst Haeckel, this hypothesis, known as the 'law of recapitulation', was a key focus of Darwinian investigation for several decades. As a student at Cambridge, Weldon had been exposed to Darwinian embryology through the teaching of Francis Balfour.[3] Up until 1890, Weldon's papers dealt with the embryology of various organisms, in particular of invertebrates. Throughout this period, the young zoologist had an obvious enthusiasm for Darwinism, but made no public comment on the theory of natural selection, which he accepted without discussion, being primarily concerned with its consequences for embryology.

It is interesting to see how Weldon came to the statistical study of natural selection. His published writings give no explanation for the abrupt passage from embryological monographs to his first statistical article of 1890. Fortunately, however, Karl Pearson provided an indication in the obituary he wrote for his friend in 1906,[4] in which he reproduced a note written by Weldon in 1888 which states that the evolution of new characters in the adult organism is *always* accompanied by the appearance of new characters in the larvae, and vice versa. Weldon added that he could see no clear and provable link between the two parallel modifications. This note, despite its allusive nature, is of substantial historical interest. It shows that, very early on, Weldon had broken with the dominant morphological schemes of Haeckelian Darwinism. Haeckel's law of recapitulation suggested that evolutionary novelties occur by the addition of stages in the final phases of development, implying that natural selection fundamentally acts on the adult stage of the life cycle. Against this vision of the relation between embryology and evolution, Weldon's 1888 note shows that he had been struck by what Darwin had first called 'growth correlations' and then, in the later editions of the *Origin of Species*, 'correlated variations'. For Darwin, the classic concept of correlation was not limited to synchronous variations in the organism, but also applied to successive developmental variations:

> The most obvious case is, that modifications accumulated solely for the good of the young or larva, will, it may be safely concluded, affect the structure of the adult; in the same manner as any malconformation affecting the early embryo, seriously affects the whole organisation of the adult.[5]

This passage from the first edition of the *Origin of Species* highlights the difficult problem that Weldon touched upon in 1888 and that was to become one of the major preoccupations of biometry. In order to rigorously demonstrate natural selection, the mortality rate has to be shown to depend upon variations in a given character. This supposes that individuals can be compared at different stages of their development. In nature, it is not possible to observe directly and exhaustively those individuals that survive and those that die. What can be done, however, is to compare the viability of the population at two developmental stages. A restriction of variability and a modification of the mean character are important indices of a selective process. But *a priori* it cannot be excluded that such phenomena are due to the 'law of growth' of a given character. Before selection can be unambiguously demonstrated, the law of growth for the given character has to be understood. It was almost certainly this difficult theoretical question that led Weldon to pass from morphological Darwinism to a populational and statistical Darwinism.

According to Karl Pearson, it was on reading Galton's *Natural Heredity* (1889) that Weldon realised that Darwin's grand schema was no more than a 'working hypothesis', which would have to be proved by Darwin's disciples.[6] On reading Galton, Weldon understood why ordinary embryological and morphological data were insufficient to prove the reality of natural selection. Instead of showing the degree of variation on the basis of exceptional specimens to be found in museum collections, it was necessary to measure – without any preconceptions – deviations from the populational mean type. Under these conditions, Darwin's 'small, innumerable variations' took on a precise operational definition, that of a frequency distribution.[7] Galton's work also led Weldon to realise that natural selection and heredity could be statistically analysed; if natural selection existed, it should affect the frequency distribution of the character upon which selection operated. It should be possible to measure the heritability of characters by the 'correlation coefficient' defined by Galton in 1887.[8] The biometricians continually repeated that 'the true measure of heredity is the numerical correlation between some characteristic or organ as it occurs respectively in parent and offspring'.[9] This same coefficient also permitted the quantitative measurement of what Darwin called 'growth correlations'. This term had two aspects – temporal (correlated variations of a given organ at different stages of growth) and synchronous (concomitant variations of several organs). Thus Weldon realised that Galton's statistical methods were a powerful tool for formulating and resolving problems that were crucial for Darwinism, but which could not even be expressed in the qualitative vocabulary of embryology. Weldon's view was clearly set out in a famous statement of 1893, which constitutes the founding declaration of the biometrical school:

> It cannot be too strongly urged that the problem of animal evolution is essentially a statistical problem: that before we can properly estimate the changes at present going on in a race or species we must know accurately (a) the percentage of animals which exhibit a given amount of abnormality with regard to a particular character; (b) the degree of abnormality of other organs which accompanies a given abnormality of one; (c) the difference between the death rate per cent. in animals of different degrees of abnormality with respect to any organ; (d) the abnormality of offspring in terms of the abnormality of parents, and *vice versa*. These are all questions of arithmetic; and when we know the numerical answers to these questions for a number of species we shall know the direction and the rate of change in these species at the present day – a knowledge which is the only legitimate basis for speculations as to their past history and future fate.[10]

In this passage, 'abnormality' should be taken as meaning 'variation from the mean'. All the parameters defined by Galton in his 1877 article 'The typical laws of heredity' can be seen here. But while Galton tried to make heredity, selection and organic correlation 'co-operate' in creating a stable type, Weldon saw the same tools as being theoretically neutral instruments which would permit the testing of Darwin's key hypothesis.

Before examining Weldon's empirical work in detail, the institutional context in which the biometrical project developed needs to be considered. Weldon's first statistical writings (1890–3) had an immediate and considerable impact. These studies of crustaceans marked the first application of Quételet's and Galton's statistical methods to an organism other than man. Weldon's articles dealt explicitly with correlated variations, with the homogeneity of local races and their comparison and potential divergence, and with natural selection. They convinced a large number of scientists that it was possible to put Darwinism to the test of empirical proof, using a subject much less open to polemic than man. Weldon showed that statistics could be established for populations that were neither human nor domesticated. It is hard to imagine exactly how much more difficult it was to carry out a biometrical study than an anthropometric one: despite the fact that the methods used are largely the same, shrimps and crabs do not have birth registers, nor censuses, nor health records, nor military service.

Once Weldon had set the example, things moved remarkably quickly. In 1893, Galton and Weldon founded a committee of the Royal Society of London which was intended to conduct 'statistical inquiries into the measurable characteristics of plants and animals'. Apart from its two founders, the committee initially included Francis Darwin (Charles' son), the mathematician Donald Macalister (one of Galton's collaborators), Ralph Meldola and Edward Poulton.[11] Meldola and Poulton were well known for their work on mimicry, and had the reputation of being pugnacious Darwinians. They were soon joined by Karl Pearson, a close friend of Galton (through eugenics) and of Weldon. Pearson and Weldon were colleagues at University College, London; Pearson held the chair of Applied Mathematics and Mechanics, and Weldon was Professor of Zoology. Apart from Galton, the Committee, set up on the advice of Alfred Russel Wallace, appeared to be composed of orthodox Darwinians, determined to uncover the measurable implications of the hypothesis of natural selection. In practice, from 1893 to 1897, the Committee was a centre of intense discussion of Weldon's work. In 1897 Galton proposed enlarging the circle to include partisans of the discontinuous conception of variation, in particular William Bateson, thus hoping to encourage cooperation between the two rival evolutionary schools, both of which laid claim to his work. This proved to be an illusion. Renamed in February 1897 the '*Evolution* Committee of the Royal Society', it rapidly degenerated into a series of polemics and into personal squabbling. From 1900 onwards, an intense debate about Mendelian heredity reinforced the arguments about natural selection. This dispute is known as the controversy between the 'biometricians' and the 'Mendelians',[12] although these terms have any meaning only for the period after the rediscovery of Mendel's laws in 1900, when a fundamental debate about evolution gave way to a fundamental debate about heredity. All this shows what the biometricians were doing, and why: intransigent hereditarians, they were convinced that a purely statistical approach to heredity would suffice to demonstrate the empirical validity of natural selection. For reasons that will become clear, the pre-1900 aggressive Darwinism of the biometrical school needs to be distinguished from that of its post-1900 anti-Mendelian stance.

7.1.2 *The English biometricians and stabilising selection*

Weldon set out to demonstrate the existence of cases of natural selection. The crucial epistemological question is as follows: What exactly did the biometrical school consider was required to prove natural selection?

In 1889, Weldon the embryologist, not yet turned thirty, decided to attack the statistical problem of natural selection in animal populations. His sole weapon was Galton's methodological model, which can be summed up in two points:

1. It must be possible to express the action of natural selection in terms of the 'law of error' (the normal law). The underlying idea was that natural selection acts in the same way as a marksman shooting at a target: the bullet holes (in this case the rate of selective mortality) will show a normal distribution around a central point (see Figures 2 and 11).
2. Natural selection as understood by Darwin cannot shift the race from its 'type' for any great length of time. To support this hypothesis, Galton invoked either problematic statistical arguments about 'regression' and 'ancestral heredity' or physiological arguments about 'positions of organic stability'.

The whole line of reasoning led to a rejection of the gradual modification of races by selection acting on continuous variation.

From 1890 to 1901, Weldon patiently tried to refute Galton's argument, using statistics to show that gradual selection could take place in nature. Galton's influence is clear in Weldon's first two papers on *Crangon vulgaris*, the common grey shrimp, published in 1890 and 1892. In the first[13] Weldon studied whether various physical measures (overall length, relative length of the cephalothorax and of the telson etc.) conformed to the law of error, a law which Quételet and Galton had verified only in man or in domesticated species. Weldon's study was carried out on several local populations. In each case, the 'law' was verified: the data clearly showed a bell-curve distribution. Nevertheless, each population showed different values for both the mean and the distribution, indicating that each constituted a homogeneous local race. These results had been predicted by Galton, but Weldon felt the need to discuss them in terms of natural selection rather than of 'heredity'. Weldon's writing is extremely deferential towards Galton, but under the cover of a 'confirmation' of Galton's theories Weldon in fact gives an authentically Darwinian interpretation of the data. At the beginning of the article we can read:

In his recent work on heredity,[14] Mr Galton predicted that selection would not have the effect of altering the law which expresses the frequency of occurrence of deviations from the average: so that he expected the frequency, with which deviations from the average size of an organ occurred, to obey the law of error in all cases, whether the animals observed were under the action of natural selection or not. The results of the observations here described are such as to fully justify Mr Galton's prediction.[15]

The terms are typically Galtonian, and appear to show that natural selection cannot alter the normal distribution of characters. However, Weldon's real intention becomes clear at the end of the article. Noting that the distribution curves are different in each locality, he declares that:

Since the variations observed in adult individuals depend not only on the variability of the individuals themselves (which is possibly nearly alike in all races), but also upon the selective action of the surrounding conditions – an action which must vary in intensity in different places – the result here obtained is precisely that which might be anticipated, and it is precisely that predicted by Mr Galton.[16]

This second homage to Galton is far more insidious, because it implies that the local types are determined and shaped by natural selection. While Weldon was merely making a suggestion here, and was not trying to provide an example of natural selection, it is clear from this first biometrical study that he intended to put his statistical tools at the service of Darwin's conception of natural selection.

The second article,[17] in which Weldon dealt with the problem of 'correlated variations' of several organs, marks a step back from this position and a resurgence of Galton's argument. In the *Origin of Species*, Darwin argued that such variations (for example the correlation seen in vertebrates between jaw and limb variations) 'may be mastered more or less completely by natural selection'.[18] In other words, organic correlation as such is subject to gradual modification. Galton's position, which was just as speculative as Darwin's, was the opposite: organisms can be considered technical compromises between numerous exigencies, and change in one aspect will inevitably lead to a massive change in the whole individual. This explains why Galton argued that evolution involved a discontinuous change from one form of organic stability to another. In 1892 Weldon began to subject this problem to statistical study, using the tool that Galton had called the 'correlation coefficient' (initially 'co-relation'). Like the previous article, this paper deals with five local shrimp populations (from the North and South of England, Scotland, Brittany and North Holland) and measures the correlation coefficient r for several pairs of measures (for example, overall carapace length and telson length). Weldon found that the correlation coefficients were highly variable, depending on the pairs of organs involved, but were remarkably homogeneous for a given pair of measures across the various local races. Thus the study revealed that the races had a hidden homogeneity: behind the apparent diversity of organs considered separately in local populations, the correlations of pairs of organs suggested a more profound unity of the species.

This article clearly did not constitute an empirical demonstration of natural selection (in fact Weldon did not deal with this question at all). His subsequent articles show that he had hoped to find that the correlation coefficients would reveal significant interpopulational variability. Had this been the case, it would have been possible to conclude that the growth correlation was itself under the control of natural selection, thus proving Darwin right and Galton wrong.

The opposite appearing to be the case,[19] Weldon concluded that his study confirmed one of Galton's predictions and that the correlation coefficient resolved a classic problem in morphology – the 'connection' between organs – by opening the perspective of identifying true species through the measurement of 'species-specific constants' independent of local variation.[20]

The first statistical study of 'correlated variations' thus supported Galton's 'positions of organic stability'. Both of Weldon's articles therefore gave succour to Galton's arguments.

This situation rapidly changed. From 1893 to 1901, Weldon published four remarkable papers in which he dealt head-on with each of the problems associated with a statistical demonstration of natural selection. In 1893 he published a study of correlated variations in two populations of *Carcinus moenas* (the shore crab),[21] a common crab found in shallow coastal waters throughout Europe. Easy to find, capture and rear in captivity, and with a strictly definable morphology, the shore crab lent itself to the collection of a series of precise measures from a large number of individuals. In his 1893 article, Weldon reported eleven measures of the carapace, each weighted according to overall carapace length, thus removing variation due to the size of the individual, assuming that all individuals were at the same developmental stage. The two samples (Naples and Plymouth) were composed exclusively of adult females. Figure 18 gives an idea of the measures taken by Weldon. One of these, 'frontal breadth', or the distance between the first two teeth on either side of the eyes (*CD* in the figure), is particularly important in both the 1893 article and in those that followed.

In general, the study of *Carcinus moenas* confirmed the conclusions of Weldon's work on the grey shrimp. On the one hand, virtually all the characters measured (ten out of eleven) showed a normal distribution for the two populations, from which Weldon inferred that they were homo-

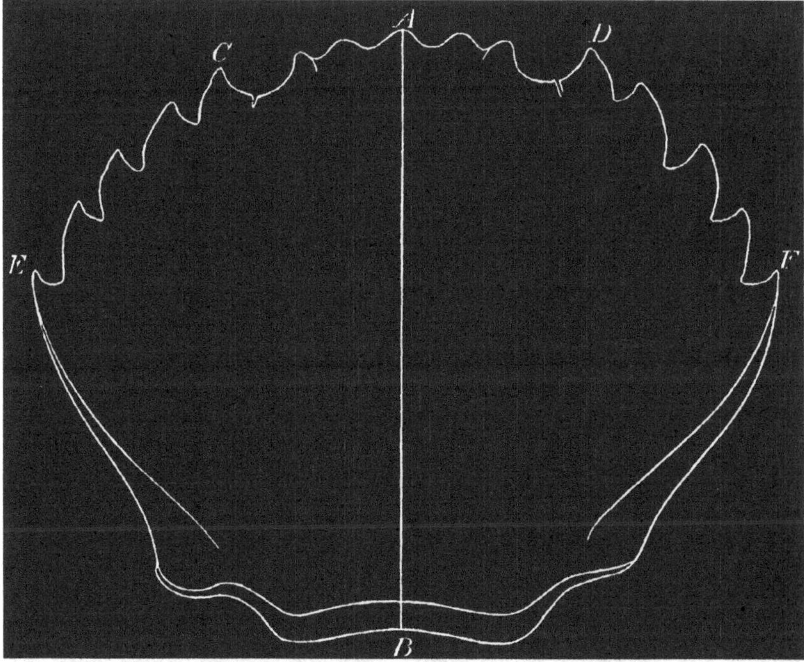

Figure 18 Measures taken by Weldon on the carapace of the shore crab. 'Frontal breadth' (the distance between the eyes) is the distance *CD* (W. F. R. Weldon, 'On certain correlated variations in *Carcinus moenas*', 1893).

geneous (Figure 19). On the other hand, he found that the correlation coefficients for twenty-three pairs of characters were approximately the same for the two samples. Up to this point, the paper contained nothing new. However, Weldon also observed that in the Neapolitan population, one character – 'frontal breadth', marked *CD* in Figure 18 – had a non-Gaussian, clearly skewed distribution. This kind of distribution could not be analysed using Galton's statistical methods. Weldon therefore turned to Karl Pearson for advice, asking him whether it was possible to analyse the skewed distribution as though it represented the superposition of two normal distributions. Pearson solved the problem and showed that the frequency distribution observed by Weldon could indeed be resolved into two normal distributions with different means and dispersions. The result of this mathematical 'dissection' can be seen in the striking diagram reproduced in Figure 20. The inevitable conclusion was that 'the female of *Carcinus moenas* [was] slightly dimorphic with respect to frontal breadth'.[22] In other words, the study had proved

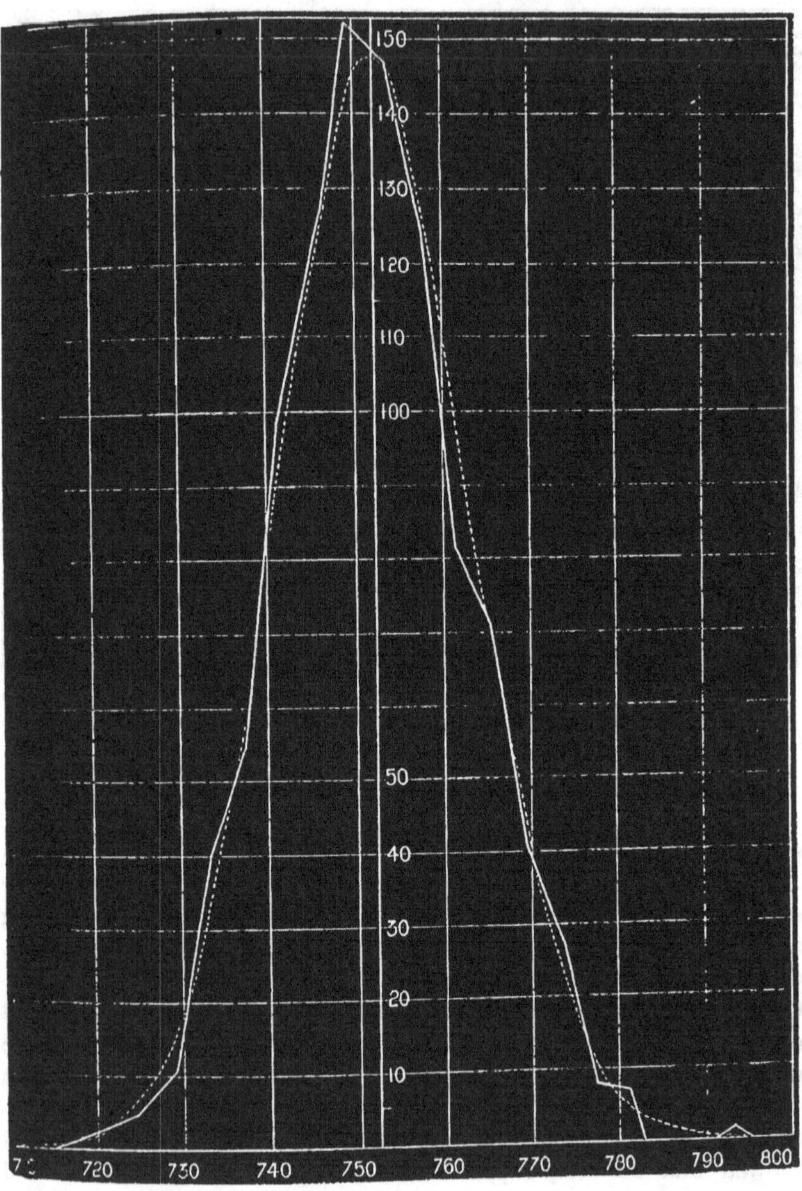

Figure 19 One of many normal distributions identified by Weldon in his studies of natural populations of *Carcinus moenas*. This example is the antero-lateral margin. The dotted line represents the corrected observed curve (W. F. R. Weldon, 'On certain correlated variations in *Carcinus moenas*', 1893).

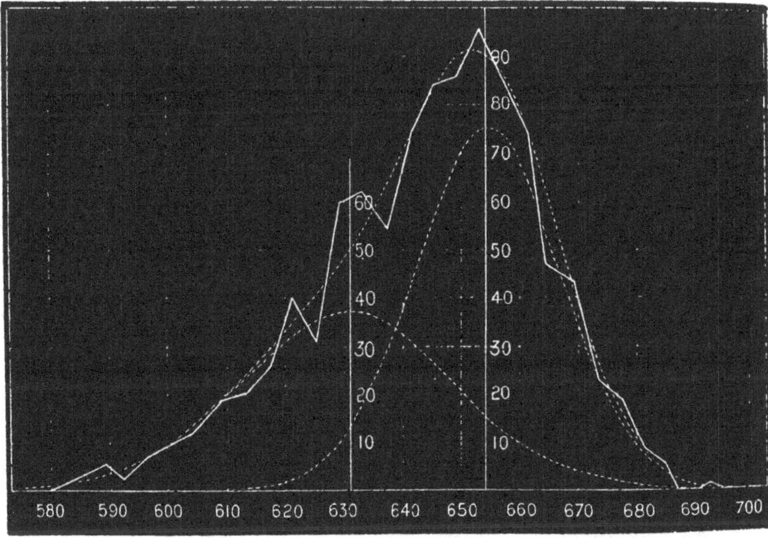

Figure 20 Distribution of frontal breadth in shore crabs from the Bay of Naples. The dotted lines are Pearson's 'dissection' (W. F. R. Weldon, 'On certain correlated variations in *Carcinus moenas*', 1893).

the coexistence of two 'races', of two distinct homogeneous populations in the same locality.

With characteristic prudence, Weldon did not draw any other conclusions in this paper. But the mathematical study in which Pearson resolved the problem posed by his naturalist colleague is particularly revealing.[23] This article marked the beginning of a close collaboration between the two men and gives an indication of Weldon's intentions. Formally speaking, Pearson's long monograph goes way beyond the boundaries of evolution. It presents a method for analysing 'abnormal'[24] – non-Gaussian – frequency distributions. Such distributions are found for all sorts of measures. Since the work of Lexis (1877), infantile mortality curves were known to provide a classic example. (Indeed, Lexis appears to have originated the term 'normality' to describe Gaussian curves, adult mortality being 'normal' and following a Gaussian distribution, whereas infantile mortality, which shows a highly skewed distribution, is 'abnormal'.)[25] 'Abnormal' curves were also well known in meteorology (barometric frequency curves) and in economics, to which Pearson had made an important contribution through his study of the distribution of interest rates. However, the first question that strikes a statistician on studying a non-Gaussian distribution is whether the measures taken from the population are homogeneous, that is, are all affected by the same kind of causes.

Pearson set out three alternatives:

- Skewed curves that correspond to a binomial distribution in which the parameters *p* and *q* (the probabilities for and against a given event) are not equal. Such curves are perfectly compatible with a homogeneous material; infantile mortality curves are a good example. Pearson only deals with this possibility in passing.[26]
- Skewed curves that result from the superposition of two or more normal curves with different means, modes, standard deviations and areas.
- 'Symmetrical abnormal curves, which are compounded of two or more normal curves' with similar means and modes, but with different standard deviations and areas (or, inversely, with similar standard deviations and areas, but different means and modes).

Most of the article is devoted to the last two situations. Pearson sets out an elegant method for 'dissecting' frequency distributions, based on the consideration of 'moments', a concept which he introduced in this article. Without going into the detail of this complex argument, the fundamental elements can be summarised as follows. When a variable x takes values $X_1, X_2 \ldots X_n$, the 'moment' to the power $r = \overline{\Sigma^r} = \Sigma X^r/N$. These moments can be expressed in terms of the mean. The 'centred moments' to the power $r = (m_r = \Sigma(X - \overline{X})^r/N)$. The first-order moment is the mean; the second-order centred moment is a measure of dispersion (since Fisher, this is called the 'variance', but Pearson did not use this term); the third- and fourth-order moments describe the degree of skew and the height of the curve. The significance of the higher-order moments is less intuitive. Pearson's 'dissection' test is based on a comparison of the moments: if a skewed curve is produced by the superposition of two normal curves, the first five moments of the skewed curve will be equal to the sum of the corresponding moments of the two normal curves, expressed in terms of the mean of the skewed curve. In the case of a symmetrical curve, the first six moments are involved. In practice, dissection requires the resolution of ninth-degree equations, which makes the whole operation somewhat long-winded. Nevertheless, from a purely algorithmic point of view, Pearson's 1894 article remains a key moment in the history of statistics.

Pearson felt that this work marked his passage into the world of Darwinism, for which he had previously expressed little interest. The title of the article is quite explicit – 'Contributions to the Mathematical Theory of Evolution'. This was the first of eighteen papers, all of which had the same basic title. This series, which stretched from 1894 to 1912, provided the mathematical bases for much of 20th-century biometrical statistics.

The evolutionary implication of this first study of the 'mathematical theory of evolution' was as follows: by applying his method for the dissection of non-normal distribution curves to data such as Weldon's, Pearson hoped to catch evolution in the act. If a homogeneous population could be dissected into two sub-populations that were heterogeneous for a given character, it was reasonable to think that Darwin's 'divergence of character' had finally been observed. This concept, defined in chapter 4 of the *Origin of Species*, contains the idea of the differentiation of one race into many as an effect of natural selection. Weldon's Neapolitan crabs clearly suggested such an interpretation.

Pearson summed up the procedure used to interpret the results of the theoretical dissection of the frequency curve as follows:

> (α) If the first dissection is possible and the second is not, a real evolution is going on; (β) if the first dissection is possible and the second is possible, and both groups give sensibly the same percentage, we have a mixture of two heterogeneous materials and no true evolution, unless the organs be so closely allied that one must vary directly with the other (*e.g.*, length of right and left legs); (γ) both dissections are possible, but give groups with different percentages; we have *both* organs evolving at the same time.[27]

Applied to the Neapolitan crabs, this method would favour the first hypothesis. The distribution of the 'frontal breadth' character is skewed and can be dissected into two normal curves with different means, standard deviations and areas. Pearson concluded that a 'real evolution' was taking place in the Neapolitan crabs.

This method is extremely ingenious, because it provides a way of distinguishing two cases with very different meanings: two races cohabiting in the same locality is not the same thing as a single race that is differentiating. Pearson and Weldon considered that they had come extremely close to a direct observation of natural selection. However, the meaning of their data was ambiguous. For Pearson's conclusion to be true (and thus for a 'real evolution' to be taking place), replicate observations would have to be made and the distributions would have to change over time. Furthermore, if 'evolution' there was, its nature was not at all clear. Was it a genuine case of differentiation produced by disruptive natural selection? Or was it a case of a demographic change in two coexisting forms in a given locality? In the first case, the skewed curve would be the product of the superposition of two curves that were moving away from each other (Figure 21). In the second case, it would result from the superposition of two curves with changed areas (Figure 22). The problem was that there was an ambiguity at the heart of Pearson's interpretation: the observed result could equally imply an *effect* of natural selection (divergence) or a *process* of natural selection (replacement of one form by another). The first possibility is typically Darwinian, the second

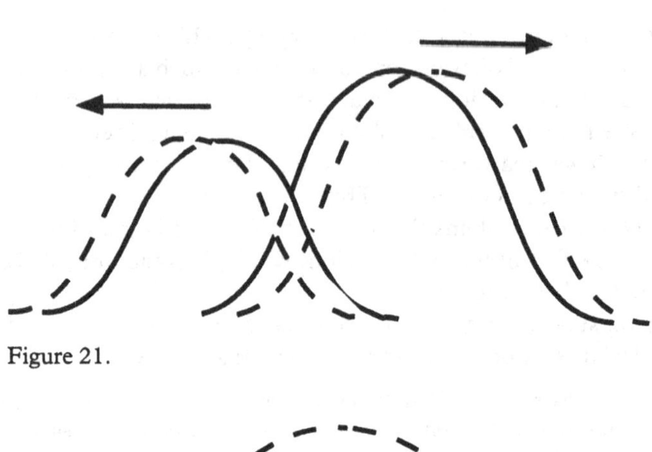

Figure 21.

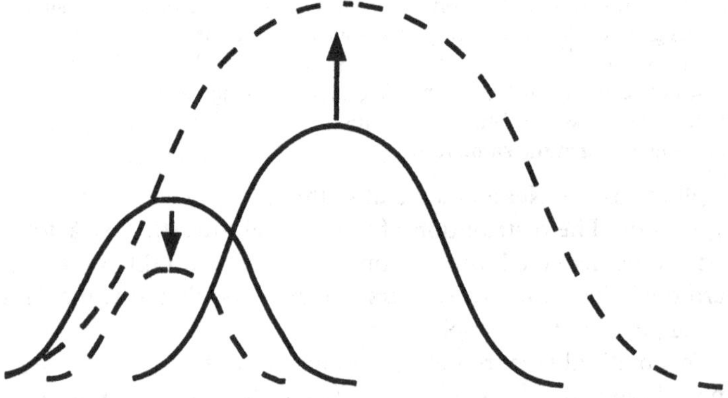

Figure 22.

is typically Galtonian. That is no doubt why Weldon, in his own article, merely concluded that there was a dimorphism in the female Neapolitan crabs, and did not try to interpret this result.[28]

We know that for a certain time Weldon continued to believe that selection could be shown to exist by identifying cases of dimorphism using biometrical methods, and by analysing them using purely mathematical methods, resolving skewed curves into Gaussian ones. This procedure involved a representation of natural selection that was extremely close to that of Galton: the rapid appearance of a new 'racial centre' (or mean type) with its own Gaussian distribution, and the division of the race into two, with selection acting by competition between the two races.

Immediately after the publication of the 1893 article on correlated variations in the shore crab, Weldon and Galton joined forces at the Royal Society to create the 'Committee for Conducting Statistical Studies on the Measurable Characteristics of Plants and Animals' referred to above. The initial objective of the Committee was to determine the exact statistical meaning of the hypothesis of natural selection. One of the first tasks given to the Committee was to apply Pearson's method for resolving skewed distributions into Gaussian curves to a material other than the crab. With this in mind, Weldon began a massive study of herrings. The observed populations had a clearly skewed distribution, and Weldon thought that, with Pearson's aid, he would be able to repeat the demonstration of dimorphism that had been so successful with the crabs from the Bay of Naples. The project was a failure. The skewed curve resisted Pearson's attempts at dissection. It raised particularly difficult mathematical problems, and its biological meaning was unclear, because, as Pearson pointed out, the skewed distribution was in fact compatible with a homogeneous population.[29] The study was presented to the Committee, but was not published.[30] Shaken and depressed by this failure, Weldon gave up the idea of detecting natural selection through dimorphism, and turned to a different strategy that was both more direct and more Darwinian in its inspiration.

The new strategy appeared in a famous article presented in 1895 to the Royal Society under the title 'An Attempt to Measure the Death-Rate Due to the Selective Destruction of *Carcinus moenas* with Respect to a Particular Dimension'.[31] This study has often been rightly cited as the first direct empirical proof of a selective process taking place in nature. It deserves particular attention, more for the theoretical problems that it raised than for what it proved. Weldon's great merit, as his friend Pearson pointed out, was to have set out all the problems that have to be dealt with biometrically before natural selection can progress from a hypothesis to a law.[32] But Weldon's study also shows what a purely 'statistical' approach to natural selection means, or could have meant.

In an appendix, Weldon clearly indicated that his aim was to develop a method that would permit a choice between the two competing conceptions of natural selection. The first conception was that originally defended by Darwin and Wallace:

> The view originally put forward by Darwin and Wallace is that specific modification is at least a gradual process, resulting from 'the accumulation of innumerable slight variations, each good for the original possessor' ('Origin of Species', chap. xv). This view rests on the assumption that each of those small differences which are to be observed among a group of individuals belonging to the same species has generally some effect upon the chance of life.[33]

The second conception was that the effect of small variations was negligible and that the modification of species essentially takes place through large but rare variations ('sports').[34] According to this view, variations such as those described by the normal distribution curve have little or no effect on survival.

The bulk of Weldon's article was a specific example to show that continuous variation is exposed to destructive selection. Once again Weldon studied *Carcinus moenas*, but this time he measured only one character: frontal breadth, the character that had enabled him to detect a dimorphism in the Bay of Naples population. This time, however, he studied only the Plymouth population, which had shown a typically Gaussian distribution for frontal breadth. In other words, unlike the Neapolitan population, the Plymouth adult population was homogeneous. Weldon tried to discover whether the adult distribution was the result of a selective destruction that affected the crabs during their development, by comparing the variability of the population at different developmental stages. As the age of the crabs was not directly known, carapace length was taken as a reasonable estimate of age, and the measures of frontal breadth were expressed relative to total carapace length. Weldon was interested in the dispersion of the measures taken, and in particular whether the dispersion narrowed with age. Around 8000 crabs were measured, divided into thirty-six age (size) classes, each composed of around 200 individuals. Calculation of the 'quartile-deviation' (Weldon's measure of dispersal) for these age-groups led to the following conclusion: in the youngest individuals the distribution was significantly narrower than in the adults; the distribution then grew larger with increasing age (size), became significantly greater than in adults, and finally narrowed again in the ultimate stages of growth. In other words, the variability of the population initially increased, and then decreased as the population approached adulthood.[35]

In this study, demographic parameters were not measured directly – conclusions were drawn on the basis of statistical measures taken from crabs caught on the beach and classed according to their size. Weldon interpreted the increase in variability with size by invoking the Darwinian principle that 'many variations appear at a late period of development'.[36] The later restriction of variability was attributed to selective mortality, which was more intense because the animals showed a greater deviation from the mean adult type.

The empirical argument contained in this article was based simply on this kind of data. Weldon did not discuss the possible adaptive significance of such a narrowing of variation. He did, however, set out a long mathematical interpretation to account for the phenomenon, were it to be due to selective elimination. The idealised diagram reproduced in Figure 23 shows his reasoning, which impressed many of his contemporaries. The diagram represents what happens in a population of crabs between

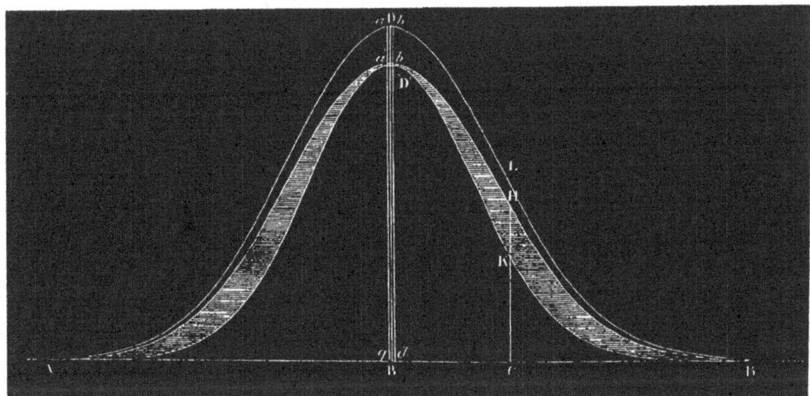

Figure 23 Illustration by Weldon of the two effects of random elimination (L → M) and selective elimination (M → K). Note that in this model the mean is not changed (W. F. R. Weldon, 'An attempt to measure the death-rate due to selective destruction of *Carcinus moenas* with respect to a particular dimension', 1895).

the moment of growth when variability is at its greatest (c_1 being the quartile-deviation at this point) and the adult stage, when variability is restricted (c_2 being the quartile-deviation of the adult population). The values of c_1 and c_2 are known empirically. In the sample studied by Weldon, the young crabs showed a maximum variability for a carapace width of 12.5 mm, and the dispersion of the frontal breadth was c_1 = 10.79 mm; adult crabs (14.5 mm and greater) had a dispersion c_2 = 9.96 mm. Between these two developmental stages, an unknown number of crabs die. This mortality can be attributed *a priori* to two causes: some crabs – probably a large number of them – die for reasons unconnected with the character under study (frontal breadth). Because it strikes indiscriminately, this non-selective mortality does not alter the crabs' variability but only their number. This phenomenon is represented by the outer curves, which have the same quartile-deviation c_1 (but different areas). Selective mortality, on the other hand, implies a reduction of variability and affects the shape of the curve. The middle curve has a different form from the other two. This is the distribution curve for the adult crabs (quartile-deviation c_2).

It is relatively easy to calculate the total cost to the population of selective mortality. Let k_1 be the median ordinate value of the distribution prior to non-selective elimination (the outer curve) and k_2 the median ordinate value following non-selective elimination. The population size is given by $k_2 \cdot c_1 \cdot \sqrt{\pi}$ because the distribution is Gaussian. Following selective elimination, the ordinate median value of the curve is still k_2

(Weldon's example is one in which selective elimination acts against values that deviate from the mean; this is not necessary, but it makes calculation easier). Selective elimination leads to a narrowing of the distribution, which is now given by c_2. The population size of the Plymouth sample was thus reduced by (10.79 − 9.96)/9.96 = 7.7%. This corresponds to the overall cost of selection for the population. For individuals, the selective mortality rate varies according to the character: it is zero for those individuals having the mean value. In principle, one can calculate the exponential function that describes the variation of the mortality rate for a given deviation. Weldon did this, using a method similar to that put forward by Galton in 1877, in an article discussed previously.[37]

This was the first attempt to measure a process of natural selection directly. With this series of studies, the history of the concept of natural selection reaches a key epistemological point: natural selection was supposed to appear as a pure and simple fact, free of any biological theory, merely requiring a statistical method in order to be revealed. Weldon summed up this theoretical position as follows:

> By purely statistical methods, without making any assumption as to the functional importance of the frontal breadth, the time of life at which natural selection must be assumed to act, if it acts at all, has been determined, and the selective death-rate has been exhibited as a function of the abnormality, while a numerical estimate which is at least of the same order as the amount of the selective destruction has been obtained.[38]

A bit further on, he makes clear that a complete treatment of the problem would mean taking into account the inheritance of the character. But a 'theory of the mechanism' was no more necessary than in the case of selective mortality.[39] A 'purely statistical' approach to heredity, based on correlation, would suffice. In this way the 'direction and rate of evolution' could be determined 'without introducing any theory of the physiological function of the organs investigated. The advantage of eliminating from the problem of evolution ideas which must often, from the nature of the case, rest chiefly upon guess-work, need hardly be insisted upon,' argued Weldon.[40]

These words need to be placed in their historical perspective. In 1895, a thirty-five-year-old morphologist, won over to Galton's statistical methods, used these techniques to rehabilitate, against their inventor, the Darwinian thesis that selection acts gradually on continuous variation. In so doing, Weldon took up arms against two rival schools. First he distanced himself from the 'orthodox' Darwinians, for whom the recognition of the reality of natural selection was based on the qualitative identification of an adaptive advantage produced by a given character. In the 1890s the leader of this 'adaptationist' Darwinism was none other than Alfred Russel Wallace, tireless advocate of an interpretation of selection in terms of the 'principle of utility'.[41] In 1895 Weldon did not explicitly cite Wallace, but his target was quite clear: 'knowing that a given deviation from the mean character is associated with a greater or less percentage death-rate in the animals possessing it, the importance of such a deviation can be estimated without the necessity of inquiring how that increase or decrease in the death-rate is brought about, so that all ideas of "functional adaptation" become unnecessary'.[42]

At the same time, Weldon found himself opposed to the experimental biologists, who argued that a physiological theory of heredity was a precondition for any theory of evolution. For Weldon, it did not matter whether evolution took place gradually or by leaps as a consequence of sports, as Bateson had argued in 1894.[43] Natural selection could be studied by statistical methods, independently of any theory of the 'mechanism' of heredity. Weldon devoted half of his monograph to showing how his method could equally be applied to 'exceptional variations' or 'sports'. Weldon listed five hundred 'exceptional' variations in a particular character (the row of marginal teeth along the edge of the shell) that he had noted amongst his 8000 crabs. At this level of observation, 'exceptional' variations appear in a new light: they were doubtless 'exceptional' in that they did not fit into a normal distribution, but, with care, they could be found in sufficient numbers to permit a statistical analysis.

Weldon's work was thus part and parcel of a very special theoretical climate – it followed four decades of utilitarian rhetoric that had gradually transformed the Darwinian revolution into a kind of non-religious natural theology. His study of selective mortality in crabs came at a time when the physiological theory of variation and heredity that Darwinism needed so much was not expected to be developed for many years, if at all. In this dual context, the biometrical enterprise consisted of the construction of a phenomenalist theory of selection and of heredity that considered itself free of any causal prejudices.

This last point appears clearly in a debate between Weldon and E. R. Lankester that erupted in the pages of the scientific journal *Nature* shortly after the publication of the article on selective mortality in *Carcinus moenas*.[44] Lankester accepted that Weldon had established a link between the variation of a certain dimension of the crab (frontal breadth) and that of the mortality rate in young individuals; Weldon's method allowed him to predict the probability of survival of a young animal for a given character. But, argued Lankester, such an induction did not allow Weldon to state that the variations in the measured dimension were the *cause* of the mortality rate. To the extent that Weldon did not put forward any adaptive or functional explanation, nothing ruled out the possibility that the observed character was merely the concomitant of another, unknown, character that was the real cause of the change in the viability of the animal.

Weldon's answer, while it did little to further the theory of selection, gives an excellent indication of the epistemological atmosphere in which he worked. Weldon replied that he considered causality in the same way as Hume did. A cause is never anything other than 'an object, followed by another, and whose appearance conveys the thought of that other', such that, in practice, science never attains a perfect knowledge of 'the circumstance in the cause, which gives it a connection with the effect'. This reply, littered with quotations from Hume's *Inquiry Concerning Human Understanding,* was not merely academic. Weldon and Pearson were convinced that the classic category of causality was too narrow, and that in the human and biological sciences it was necessary to replace it by the notion of statistical correlation, which is in some ways a generalisation of the concept of causality. Correlation, they argued, was the concept and the tool that would make it possible to develop a mathematical theory of both biology and the humanities.[45] In hoping to discover a 'purely statistical' theory of natural selection and of heredity, the biometricians were not so naive as to believe that this was the only possible theory, but they were sure that other theories would have to be compatible with this notion. At worst, biometry appeared to be the only possible waiting strategy; at best, the only possible way of handling scientifically a domain in which the 'law of chance' seemed to express the intrinsic nature of events.

However, even as a demonstration of the empirical reality of natural selection, Weldon's article was open to three serious criticisms.

The first has already been referred to – Weldon described a selective mortality rate which he associated with a precise character (carapace frontal breadth), but he was not able to say how this character affected the individual's chance of survival. As a result, his study did not, strictly speaking, provide a direct proof of natural selection. Despite his 'philosophical' response to Lankester, Weldon was well aware of this problem. Immediately after the publication of the 1895 article, he set out to identify the physiological factor that was *associated* with frontal breadth and that might affect the crabs' ability to survive. But this implied that the reality of natural selection could not be established by 'purely statistical' means, without any causal hypothesis. We will return to this point.

The second criticism was that the observed narrowing of variation could be due to a 'law of growth' and selection. There are many characters that are highly variable at a given stage of development but that converge to a typical state at adulthood. Weldon had dealt with this objection in an appendix to the article, clearly influenced by the bitter discussions within the Committee. In this appendix, the paper's conclusion is rewritten in the following terms:

In order to estimate the effect of small variations upon the chance of survival, in a given species, it is necessary to measure *first*, the percentage of young animals exhibiting this variation; *secondly*, the percentage of adults in which it is present. If the percentage of adults exhibiting the variation is less than the percentage of young, then a certain percentage of young animals has either lost the character during growth or has been destroyed. The law of growth being ascertained, the rate of destruction may be measured.[46]

This merely acknowledges the problem rather than giving an answer. Weldon recognised that the observed phenomenon could be explained in two totally different ways, each equally compatible with the data. This objection may seem strange to today's reader, imbued with the modern distinction between phenotype and genotype. A 'Mendelised' Darwinism would study the changes within a population from generation to generation, and its fundamental question would be: Is a given gene conserved in the population, and in what proportion? The question of the developmental stage at which the hereditary variation is expressed would be interesting, but secondary. But in the non-Mendelian biology of pre-1900, however, this latter question was fundamental. In dealing only with what was subsequently to be called the 'phenotype', the temporal description of variation had to take into account the developmental course of the character. Several years later, Weldon felt this criticism was sufficiently important to cause him to write a long paper that effectively constituted a direct reply (see below).

The third criticism was as follows. Even if Weldon had indeed revealed an authentic case of selective mortality, he had only found an example of conservative selection, a selection that confirmed the grouping of the population around an unaltered mean (today we would say a case of stabilising selection). Once again, therefore, the monograph supported Galton's ideas. Indeed, the statistical method employed by Weldon is more or less that put forward by Galton in his 1877 article 'The Typical Laws of Heredity' (see Chapter 5). The diagram reproduced in Figure 23 shows that the method used in Weldon's 1895 article is only valid if, at all ages, the variations are symmetrically distributed around the mean. Thus Weldon had not demonstrated the modification of a species by natural selection – the main sense of the Darwinian concept of selection. Once again, Weldon was well aware of this limitation:

It is necessary for the employment of this method that the variations should be distributed on each side of the mean with sensible symmetry, and that the position of minimum selective destruction should be sensibly coincident with the mean of the whole system. Such statistical information as is at present available leads to the belief that these conditions may be expected to hold for a large number of species, which are sensibly in equilibrium with their present surroundings, so that their mean character is sensibly the best, and the change of mean from generation to generation is at least very small. They cannot be expected to hold in cases of rapid change such as those induced by rapid migration or other phenomena resulting in a rapid change of environment. For the investigation of such rapid change, it would be necessary to treat the more general case, in which the chances of deviations of opposite sign are not sensibly symmetrical, and in which the mean is not necessarily the position of maximum destruction. The treatment of this case requires the help of a professional mathematician.[47]

In this extremely lucid commentary, Weldon underlines the limits of his study of selective mortality in the crab:

- The study had only demonstrated the existence of stabilising selection; in other words, the only kind of 'Darwinian' selection accepted by Galton.
- For many species (if not most) this is the only kind of selection that one can expect to observe.
- The statistical demonstration of a case of selection that significantly modifies the type would be extremely complex; the article gives no hint of such a demonstration.

This implies that in 1895 there was still no direct proof of natural selection as an agent that modifies species. The statistical study of selective mortality in crabs could comfort only those who doubted the very possibility of gradual modification by natural selection.

In this respect, it is extremely interesting to compare Weldon's monograph with that of H. C. Bumpus, published three years later. This article dealt with the 'elimination of the unfit' in the domestic sparrow (*Passer domesticus*), which had recently been introduced into the USA.[48] For many years this article has been cited as the only direct proof of an example of natural selection, apart from Weldon's work. Bumpus' paper employed a very different methodology from Weldon's.

Bumpus was an American naturalist, renowned for his studies of variation in birds. In 1898, he declared his intention of providing an empirical proof of the existence of natural selection. Despite the fact that he and his contemporaries continually used the Darwinian principle, he argued, 'we forget we are really using a hypothesis that still remains unproved and that specific examples of destruction of animals of known physical disability are very infrequent'.[49] The article is based on quantitative data, but these are taken from a very different kind of example from that presented by Weldon. Bumpus studied a single event – selective death in a group of sparrows as a consequence of an exceptionally violent snowstorm on 1 February 1898. The conditions under which Bumpus collected the data are not described; he merely states that at the end of the storm, 136 sparrows were brought in a state of shock to the Brown University Anatomy Laboratory. 'Seventy-two of these birds survived; sixty-four perished.'[50] The entire monograph is devoted to a precise description of each of the birds according to ten morphological criteria: size, weight, beak length, femur length etc. Bumpus did not bother himself with the cumbersome statistical apparatus of the biometricians; he provided individual measures, calculated the mean and standard deviation and made an extremely clear qualitative interpretation of the data.

Bumpus shows that the elimination of 64 birds cannot be considered 'accidental', because those that died were structurally different from those that survived. For all the measures taken, those birds that died were generally the most extreme individuals, both the largest and the smallest, those with the longest and the shortest wings, and so on. It was as if natural selection had tried to restore an 'ideal type':

> *Natural selection is most destructive of those birds which have departed from the ideal type, and its activity raises the general standard of excellence by favoring those birds which approach the structural ideal.* [Original emphasis. This is the only sentence in italics in the whole article][51]

Bumpus was not very clear about what he considered a 'structural ideal' to be. It seems that it was nothing other than the ordinary sparrow, such as it was prior to its introduction into America:

> In an earlier lecture, on the 'Variations and Mutations of the Introduced Sparrow', facts are adduced which, it was claimed, were sufficient to show that the English sparrow, since its introduction into this country, has found life so easy that the operation of natural selection has been practically suspended, and that the American type consequently has become degenerate. No active agent had eliminated anomalies, until they had become over four times as numerous as in England. When calling attention to the occurrence of these variations, and to the fact that they were an indication of the absence of an active eliminative factor, I little thought that within a few months I might witness the action of an eliminating factor that would test the structural qualifications of *all* the birds: destroy those which had departed unduly from the ideal type, and thus raise the general standard of excellence.[52]

Bumpus made an extremely interesting remark. He noted that if he no longer reasoned on the basis of the mean, but of individuals, it was frequently the case that a bird with an exceptional character survived, as long as several large variations were not present in the same individual – no bird showing more than three exceptional characters survived the storm. Hence the conclusion that natural selection did not in and of itself favour a given mean character, but rather a complex of characters, a 'type': 'It is the *type* that nature favours [original emphasis].[53]

At the heart of Bumpus' 1898 article, as in Weldon's 1895 monograph, we find Galton's conception that when natural selection acts on continuous variation it favours the conservation of the 'type'. It is remarkable that more than half a century after the first formulation of the hypothesis of natural selection in Darwin's 1842 'Sketch', the two first direct experimental confirmations of the hypothesis dealt with cases of 'conservative' selection.[54] Indeed, Bumpus wrote of 'a general tendency toward conservatism on the part of the survivors'.[55] Neither Weldon nor Bumpus had therefore proved what Darwin really had in mind: the modification of a local race.

Karl Pearson gave clear statistical expression to the difference between the two possible forms of 'natural selection' that a naturalist could encounter.[56] If natural selection is a modification of the composition of a population solely due to differential mortality, 'secular natural selection' has to be distinguished from 'periodic natural selection'. The opposition of 'secular' and 'periodic' was borrowed from the language of physics and astronomy. From the work of Laplace onwards, astronomers have spoken of secular and periodic variations in the movement of the planets. A secular change is a unidirectional, unilinear change that takes place over a long period of time. A periodic change, as the term suggests, is cyclical (for example, the perturbation of the earth's orbit when in conjunction with Jupiter). By 'secular natural selection', Pearson meant a modification of the mean and standard deviation of the curve of variation from one adult generation to the next. By 'periodic natural selection', he meant a process that was repeated in each generation, and would be expressed by a modification of the standard deviation (and, perhaps, of the mean) at several steps of development *in the same generation*, but which would not modify the state of the adult population compared to that of the previous generation. Periodic selection is thus a process that, in each generation, maintains the mean adult type. Quite clearly, this kind of selection could not be a factor of 'progressive' evolution.

7.1.3 Weldon and directional selection

Only too aware of the limits of his 1895 monograph, Weldon soon began to look for an example in which natural selection would be the only possible explanation of a morphological change. Luckily for him, the shore crab offered the possibility of directly observing an example of directional evolution. As we will see, however, the demonstration of the process was to require an important change in Weldon's methodology.

In 1898, Weldon presented the results of a new study of *Carcinus moenas*,[57] in which he analysed the distribution of frontal breadth in adults from the Plymouth Sound population between 1893 and 1898. He found that frontal breadth had systematically shrunk in adults, for all size classes, and at the same period of the year. This was possibly a sign of what Pearson had called 'secular selection'. To be certain, however, Weldon had to show that the observed gradual reduction in the parameter really was due to selective mortality.

In his 1895 article, Weldon had indicated that a 'statistical' proof of a case of innovative selection would involve analysing skewed distribution curves and comparing them at successive points in development, all of which he thought would present substantial difficulties.[58] In 1898, this purely statistical procedure gave way to another approach, which brought the naturalist close to traditional experimental methods. Instead of looking for an exponential function that would express the mortality rate as a function of the character studied, Weldon sought a possible cause of an eliminating selection that acted against those crabs with the largest frontal breadth. This question inevitably led him to study those factors in the environment that had changed, to put forward a functional hypothesis, and to design experiments to test the hypothesis in artificial conditions.

In terms of an environmental factor that might have led to the change in the size of the crabs, Weldon noted that a large dike had recently been built, leading to a partial closure of Plymouth Bay and to changes in the local currents. Because of this change, the kaolin of the granite moors around Plymouth tended to accumulate in suspension in the shallow coastal waters. Furthermore, the urban population of Plymouth had increased considerably in the second half of the century, and substantial quantities of waste had been ejected via the sewers at various points along the coast. The water in the bay had thus become increasingly muddy, making existence difficult for coastal species, some of which had disappeared. These facts suggested that the reduction in the frontal breadth of the crabs might be due to increased quantities of mud.

To test this hypothesis, Weldon carried out an experiment. He placed several hundred crabs in a large container holding seawater and fine-ground white clay which was continually agitated so that it would remain in suspension. After a certain amount of time, some crabs died. They were separated from the living animals and a distribution curve was established for the frontal breadth of the two samples. When repeated several times over, the experiment gave results of the kind shown in Figure 24: the survivors had a narrower frontal breadth than the dead crabs. Repeated using mud taken from the coast, the experiment gave the same result. Weldon's conclusion was that the suspended mud led to a selective mortality, and that this was the cause of the observed change in mean frontal breadth:

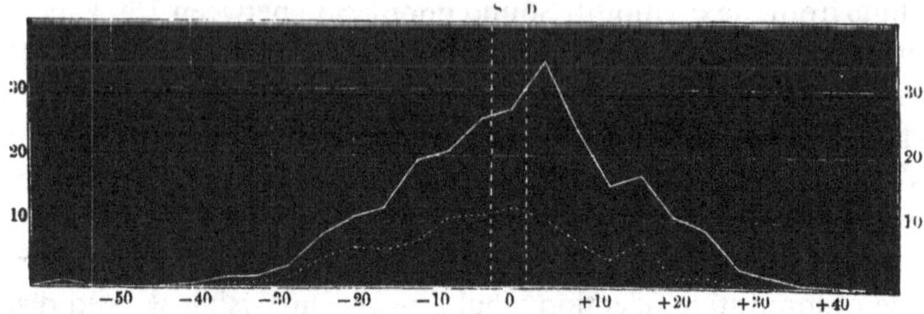

Figure 24 Effect of kaolin on frontal width in crabs, under artificial conditions. The solid line corresponds to the original population distribution in the aquarium. The dotted line shows the distribution after exposure to kaolin. Weldon's aquarium appears to represent a kind of 'population cage' thirty years before they were developed. (W. F. R. Weldon, Presidential address to the Zoological Section of the British Association, 1898).

I see no shadow of reason for refusing to believe that the action of mud upon the beach is the same as that in an experimental aquarium; and if we believe this, I see no escape from the conclusion that we have here a case of Natural Selection acting with great rapidity because of the rapidity with which the conditions of life are changing.[59]

This conclusion was also tested experimentally. Weldon took young crabs and kept them in a protected environment, that is, in perfectly clear seawater, to see what would happen after moulting in artificial conditions. If the mud had indeed induced the selective mortality, the distribution of frontal breadth for a given size of crab should not be the same in the protected experimental population and in a natural sample. As crabs do not moult at the same time, they had to be reared separately, in individual containers, and observed separately. After much effort, Weldon succeeded in obtaining 527 moulted animals, each in its own jar equipped with a double siphon. The experiment showed that the experimental sample had a slightly larger mean frontal breadth than wild crabs of the same size. On the basis of these results, Weldon finally allowed himself a functional interpretation:

Of course, if the observed change in frontal breadth is really the result of selection, we ought to try to show the process by which this selection is effected. This process seems to be largely associated with the way in which crabs filter the water entering their gill-chambers. The gills of a crab which has died during an experiment with china clay are covered with fine white mud, which is not found in the gills of the survivors. In at least ninety per cent. of the cases this difference is very striking; and the same difference is found between the dead and the survivors in experiments with mud.[60]

Strangely enough, despite its exemplary rigour and lucidity, this study did not have the same impact as the 1895 article that had described an example of 'periodic' (or stabilising) selection by purely statistical means. This may have been due to the circumstances surrounding its publication. The article did not appear in one of the major scientific journals of the time, but rather – without a title – in the transactions of the British Association for the Advancement of Science, where it appeared in the form of an address by the president of the Zoology section, who at the time was none other than Weldon himself.

It further appears that Weldon was unhappy with the study, despite the considerable effort he had put into setting up his 'experimental crabbery'. However ingenious the experimental procedure had been, Weldon considered that he had not completely proved an example of natural selection. A complete demonstration would, he felt, have to be fundamentally 'statistical'. Based only on the observation of distribution curves of animals in nature, at different ages and in different generations, such a demonstration should ideally lead to the estimation of a 'rate of selective mortality' in natural conditions. More precisely, it would involve the calculation of a selective mortality rate (and of the cost for the whole population of the selective process) and of an algebraic function describing the variation of the mortality rate as a function of the variation of the character affecting viability. By following this method, the simple fact of natural selection would have emerged out of the data, without any auxiliary hypothesis. It was because this purely 'statistical' approach was mathematically and materially too complex that Weldon turned to a classical experimental method, with all that implied in terms of causal hypotheses and laboratory procedures. It was also for this reason that Weldon felt he had not fulfilled the programme he had adopted before the 'statistical' committee of the Royal Society in 1895: the 1898 article did not give the correlation between parents and offspring, nor did it give the correlation between the measured character and the character that was directly involved in the selective process (the crab's filtration apparatus). Finally, it did not provide an estimate of the 'rate of selective mortality'. In other words, 'secular' (i.e. directional) selection was not really 'proved' as a fact of nature.

In 1901 Weldon published his last paper on natural selection.[61] Purely biometrical in scope, it dealt with a small gastropod, *Clausilia laminata*, which has a shell with a large number of spirals around a straight axis. Weldon studied the variability of one of the parameters that determine the shell's spiral – the radius of the peripheral spiral for a given degree of coiling. This parameter had two advantages. Apart from the fact that it was sufficient to determine the form of the shell, it also revealed, in the adult, the previous steps of development (the adult shell, constructed by a series of consecutive additions, provides a summary of individual growth). Weldon found a new case of stabilising selection, using a method similar to that in his 1895 study of selective mortality in *Carcinus moenas*. He compared the variability of young shells with those of adults, and found that the mean was conserved but variability was reduced. This time he gave a direct reply to the criticism of the 'law of growth' that had been raised in 1895. The species he had chosen allowed him to recognise the precise stage of development of the surviving adults, thus permitting a comparison of the variability of the survivors with that of the general population at all stages of growth. This comparison showed that the variability of those animals *that survived into adulthood* was narrower than that of the general population. It could thus be rigorously concluded that the change in variability with age was a consequence of selective mortality and not of some 'growth correlation'.

Through this elegant study Weldon provided an empirical answer to one of the strongest methodological criticisms made of the Darwinian hypothesis of natural selection at the end of the 19th century.

Weldon's final study was interesting for another reason. In sampling shells from different soil depths, he systematically found the same phenomenon: there was a close similarity in populations of *Clausilia laminata* separated in time. It seemed reasonable to conclude that the constancy of this species was due to 'periodic' selection. In other words, biometrical observations appeared to confirm the thesis that Weismann had advanced on the basis of purely theoretical considerations – natural selection has a decisive role not only in the modification of species, but also in the conservation of their characteristics.

A final word on Weldon. His work provides an exemplary illustration of that phase of Darwinism that I have described as the strategy of direct proof. Weldon and the biometricians thought they could legitimate the principle of natural selection by giving it the status of a 'fact'. With this in mind, the statistical analysis of variation appeared as a way of avoiding most causal hypotheses, particularly those related to the physiological nature of heredity and the mechanisms of adaptation. Their hope, of course, was that by accumulating comparable data on the correlation between 'mortality rate' and variation, they could make natural selection a law on the basis of strictly inductive inference.

The project did not last long, but it profoundly marked the biological community, which showed great ambivalence towards the biometricians for reasons that can be summed up as follows.

The biometricians provoked opposition for epistemological reasons. They not only favoured a particular biological hypothesis, they also fought aggressively to impose both a methodology and a conception of science. For the biometricians, statistics was more than just another tool; it was the required form for scientific data confronted with the 'laws of chance'. From an epistemological point of view, they were open partisans of a phenomenalism that found its inspiration in the work of David Hume and especially of Ernst Mach. Pearson devoted entire books, such as *The Grammar of Science* (1892), to arguing that scientific theories do not provide any information about reality beyond that furnished by sensory data, and are simply statements that permit the probabilistic prediction of one kind of phenomenon on the basis of another. From this point of view, statistics, and in particular the concept of correlation, not only show how data should be collected, but also how they should be interpreted. This purely inductive vision of the discovery and foundation of theories must have been highly provocative for most turn-of-the-century biologists. The methodological rigour of biometry appeared old-fashioned faced with the successes of experimental biology in fields such as heredity, causal embryology, biochemistry or animal physiology. In these fields statistics was at best merely one tool amongst many, and at worst an attitude that effectively neutralised experimental investigation.

This having been said, Weldon must be given due credit for his work. Despite the fact that, in his ceaseless search for the *fact* of natural selection, Weldon did not contribute to the development of the *theory*, it would nevertheless be wrong to think that his difficult project of direct proof was completely useless. As Pearson rightly noted in his funeral eulogy for Weldon,[62] he was the first to pose clearly the problems that have to be resolved in order to define the existence of selection in nature: How can a deviation be measured, and in what units? How should the age of the animals, the length of adult life and the overlap of generations be dealt with? How can random elimination be distinguished from selective elimination, and from an effect of growth? Has selection taken place between birth and adulthood? In practice, all these questions are extremely difficult to answer, especially in natural conditions. Weldon's rigour in formulating these questions, and the methods he developed to respond to them, profoundly marked 20th-century biologists.

Furthermore, Weldon's empirical results were far from trivial. It was no mean feat to prove an example of natural selection – for over half a century, roughly up until the 1950s, Weldon's studies remained the only examples of direct proofs of natural selection in nature. Bumpus' 1898 paper is the only other comparable work, but, unlike Weldon, Bumpus did not study a particular *population*. Neither was it trivial to show that the morphological stability of a given species could be due to stabilising selection, or that directional selection, when it occurred, could be extremely rapid (as was the case for the Plymouth crabs). One of

Weldon's oft-forgotten lessons is that such assertions are always contingent, and really require an empirical demonstration.

Finally, all Weldon's work on natural selection was characterised by a major limitation: heredity had no place in his studies. In all his 'programmatic' declarations[63] he affirmed that the rigorous proof of natural selection would require a 'correlation' between the parental character and that of the offspring, but he never met this requirement. Furthermore, it can be seriously doubted whether this way of viewing heredity was useful for the theory of natural selection. This point will be dealt with further in the light of some theoretical aspects of biometrical conceptions of selection.

7.2 Theoretical aspects of biometrical Darwinism: Pearson

For better *and* for worse, Karl Pearson (1857–1936) embodied the theoretical heart of biometrical Darwinism. Pearson was a mathematician by training and by profession, and for twenty-seven years held the chair of Applied Mathematics and Mechanics at University College, London. Pearson's early career was characterised by a series of epistemological, political and historical essays. On reading Galton's *Natural Heredity* (1889), Pearson developed three ideas that were to mark the rest of his work: (1) the use of statistics permits the study of an immense range of problems in the human and life sciences; (2) Darwinism is of fundamental importance in both these domains; (3) social problems must be viewed from the standpoint of eugenics.

Throughout Pearson's work, these three points can be found intertwined with his view of scientific methodology (he had studied the neo-Kantian and phenomenalist circles that were prevalent in the Germany of the 1880s) and with his political convictions (he was a Fabian socialist and an enthusiastic supporter of women's suffrage). All this led to an inimitable style in which mathematical esoterism and Darwinian preaching were never completely independent of a bio-social project which, in his final years, led him to express his sympathy for German National Socialism. Pearson was one of those people who, despite their influence in the sphere of methodology, are driven by their obstinacy into scientific dead-ends.

This is not the place to retrace Pearson's intellectual biography. That would entail telling the stories of mathematical statistics and of eugenics, in both of which Pearson played an essential role in the early years of the century.[64] It will suffice to note that Pearson's evolutionary thinking had this dual framework. Most of his eighteen articles published between 1894 and 1912 under the collective title 'Contributions to the Mathematical Theory of Evolution' (or 'Mathematical Contributions to the Theory of Evolution') in fact deal with the methodology of statistical inference and of anthropometry. Only those aspects of his work that had a direct impact on the theory of natural selection will be examined here.

Pearson's contribution can be summarised in three points:

- A statistical conceptualisation of the key concepts of Darwinism
- A fundamental vision of the relation between heredity and selection
- A reelaboration of the concept of fitness for reasons that were initially linked to eugenics

Strictly speaking, these three aspects cannot be separated – statistics, the theory of heredity, and eugenics form the omnipresent three-dimensional framework in which Pearson deployed his evolutionary ideas. Nevertheless, each of these dimensions had specific consequences for Pearson's work and ideas.

Pearson's statistical reinterpretation of Darwin's concepts is one of the foundation-stones of the 20th-century mathematical theory of evolution. According to Sewall Wright (who can scarcely be accused of complacency with regard to Pearson's political incorrectness), Pearson's work constituted its 'biometric foundation'. Pearson's conception of the relation between heredity and selection, and in particular what he called the 'fundamental theorem in selection', is of substantial epistemological interest because it reveals the intrinsic limits of biometrical Darwinism. Finally, Pearson's critical examination of the components of fitness, despite being rooted in a thoroughly eugenicist framework, was an ambiguous, but fundamental, conceptual advance for the theory of selection.

While this approach brings nothing new to Pearson's biography, it clarifies the role played by this brilliant and slightly *passé* figure in the Darwinian tradition at a critical moment in its development – the period immediately prior to its confrontation with Mendelism.

7.2.1 The statistical reinterpretation of Darwinian concepts

One of Pearson's most remarkable statistical contributions was his analysis of frequency distribution curves by the method of moments. This analysis gave a mathematical status to asymmetric curves at a time when all distributions were supposed to be normal. This problem has already been referred to in the context of Weldon's work; here we will examine its impact on the study of evolution. Formally speaking, the analysis developed by Pearson was completely independent of the question of evolution, and really has its place in a general history of applied mathematics.[65] However, this technique was developed in a very specific context. It will be recalled that Galton, following Quételet, considered the 'normality' of a frequency distribution to be an index of the homogeneity of the observed population. Any non-normally distributed population was heterogeneous, especially a population with a skewed distribution. Galton did not know how to analyse such distributions mathematically, although he suspected that they were in fact composed of several normal distributions. While this representation is at least plausible in the case of multimodal curves (i.e. curves with several 'humps'), this had to be proved, and it also had to be borne in mind that non-normal curves do not always have this form.

The Darwinian aspect of this mathematical problem is simple to grasp. Darwin's model of selection implies that it is the population (the 'variety' or 'race') that is gradually transformed. *A priori* it is possible to conceive of this process according to two different statistical approaches. Either the population is not 'homogeneous', and two or more 'forms' coexist, in which case the population distribution should be considered a composite of several distributions, each having its own parameters. The gradual evolution of the 'race' would imply that one of the forms grows in number, to the detriment of the others (see Figures 16 and 22 above). Or there is a single homogeneous population, the distribution of which cannot be dissected into several elementary curves. In this case, evolution is described by the modification of the parameters that characterise the population (principally the mean and the dispersion – see Figures 13 and 15). The gradual divergence of two populations represents a complex development of this view (see Figure 20).

In 1891, when Weldon noted that the Neapolitan crabs had a bimodal distribution (see Figure 20 above), he put forward the reasonable hypothesis that this distribution could be resolved into two normal curves. As his correspondence with Galton shows,[66] his aim was indeed to find the first direct proof of natural selection, but this selection as seen from Galton's standpoint – the appearance of a new 'racial centre' substituting itself for the original form.

The numerical data had to be shown to be mathematically compatible with a composite distribution. As explained in section 7.1, Weldon, newly appointed Professor of Zoology at University College, asked Pearson to examine the problem theoretically. Pearson responded enthusiastically and wrote the first of his 'Contributions to the Mathematical Theory of Evolution' (1894).[67] This was the article dealing with the dissection of frequency curves analysed above. Another article rapidly followed, this time dealing with skewed distributions in general.[68]

Taken together, these two somewhat esoteric articles had a subtle but decisive impact on the development of a statistical theory of evolution. Pearson first showed that it was possible for *asymmetric* distributions to correspond to thoroughly 'homogeneous' populations, and that in fact this would quite often be the case. He also showed that, inversely, a symmetric distribution could be the result of the superposition of several different distributions. Pearson thus put an end to the myth of the 'normal curve' as either a necessary or a sufficient condition to describe a homogeneous population. It was an illusion to believe, like Galton, that a skewed distribution was a definite indication of a heterogeneous population, either because two populations were diverging or because they coexisted in the same locality.

Pearson's second point, which was extremely important in the intellectual climate of the 1890s, dealt with regression. It will be recalled that on the basis of this statistical concept, Galton had denied the possibility of a population mean's gradual shifting because of natural selection. Pearson rapidly revealed the methodological error at the heart of this argument.[69] Mathematically speaking, the 'regression of the offspring on the parents', despite its name, did not take place on the parental mean, but on the general offspring population mean. For (statistical) 'regression' to produce a genuine return to the parental population type, the distributions of the two successive generations would have to have the same mean. This is not necessarily the case; only case-by-case studies can determine whether or not the parental population mean is the same as that of the offspring. Where the parental population is subject to selection, this is extremely unlikely – a powerful 'periodic' (stabilising) selection (acting for example on the early steps of development) would have to counterbalance the directional selection on the parents.[70] In other words, there is no basis for the idea that 'regression to the mean' could be opposed to the action of directional selection. Statistical regression is thus merely a predictive tool, not an evolutionary force. Selection and statistical regression do not have the same epistemological status – they should not be considered as antagonistic forces with one continually undoing what the other has done. Statistical regression is not Weismann's 'panmixia'; it is not a 'reversal of natural selection'.[71] Shortly after Galton's death, Pearson summarised this episode in a declaration in which criticism discreetly rubs shoulders with homage:

> It is a remarkable fact that many biologists have accepted Galton's theory of regression, without seeing that there is no regression whatever on Galton's hypothesis after the first generation, at which selection is stopped! This misunderstanding arises from the common belief that a word carries its own definition and that it is not useful to study its algebraical significance.[72]

Another of Pearson's remarkable contributions was his precise statistical characterisation of the vocabulary of selection. Pearson set out the fundamental definitions that were subsequently taken up by the genetic theory of selection. His intention was to describe all the parameters that had to be measured in order to submit the hypothesis of natural selection to empirical tests, and to give it an unambiguous quantitative content.

Any evolutionary biologist studying selection from a statistical point of view has to dissociate selective mortality from random mortality. The very concept of natural selection implies that death does not strike at random, and that the chances of death and survival are linked to variations in certain characters. In 1877, Galton had proposed a formalisation of the problem by arguing that the rate of survival for a normally distributed character should itself be normally distributed (this implies that a selective process, unlike random mortality, alters the dispersion of the population). It was precisely this working hypothesis that Weldon had adopted and illustrated in his 1895 study of selective mortality rates in the shore crab (see Figure 23). Pearson generalised this approach, freeing it from all reference to the normal law, and introduced a new idea that was to be of fundamental importance for 20th-century evolutionary biology. He showed that non-selective mortality, like selective mortality, should produce a significant change in the characters of a population. The smaller the population, the more random sampling should produce changes in the group mean: 'Isolation of a few individuals who form a random sample may produce very sensible modifications of race characters.'[73]

This position can be recognised as the source of what population geneticists would later call 'random drift' (today generally known as 'genetic drift'). Pearson is not talking about the fixation or random elimination of 'genes' – his theory is purely statistical. However, he did quickly realise that stochastic phenomena could be of fundamental importance for the theory of selection. He spoke of 'random selection' to describe random sampling, thus indicating that this was a genuine factor in evolution, interacting with 'natural selection'.[74] This idea found little support at the time, but there can be no doubt that it inspired the first population geneticists.

Pearson's most spectacular contribution was his clarification of the modes of selection. In 1896, in a long paper entitled 'Regression, Heredity and Panmixia', Pearson imposed a conception of selection that was to last for many years.[75] In this article, he distinguished three modes of selection in nature: 'natural selection', 'sexual selection' and 'reproductive selection'. In itself, this three-way division was not new. The distinction between 'natural selection' and 'sexual selection' was fundamental to Darwin's work. The distinction between evolution by selective mortality and evolution by differential fertility ('productivity')[76] was not new, either. With great foresight, Darwin had argued that viability and fertility were two fully functional elements of 'natural selection'. However, turn-of-the-century Darwinians commonly considered that natural selection could only act by deciding between life or death, and thus could only be measured by a mortality rate (see section 7.2.3).

Pearson's observations brought an undeniable methodological clarification to each of these three types of selection.

The distinction between the two modalities of 'natural selection' (selection by differential mortality and selection by differential fertility) has already been discussed. Natural selection can be 'secular', modifying the mean over generations; it may also be 'periodic', preserving the adult mean and acting on the mean or dispersion of the population at different stages of growth. This distinction today goes under the terms 'directional selection' and 'stabilising selection'.

The distinction between the two heterogeneous modalities of 'sexual selection' was more subtle. Darwin had defined sexual selection as follows:

> This form of selection depends, not on a struggle for existence in relation to other organic beings or to external conditions, but on a struggle between the individuals of one sex, generally the males, for the possession of the other sex. The result is not death to the unsuccessful competitor, but few or no offspring.[77]

Darwin's definition was straightforward – certain individuals have a greater chance of reproducing because of an advantage they have in terms of access to members of the opposite sex. Such a phenomenon constituted a clear example of 'selection', but after Darwin's death the concept became less precise. Galton, for example, used 'sexual selection' to describe the preference for a partner of the same type. Pearson, on the other hand, made a distinction between 'preferential mating' and 'assortative mating'.[78] Preferential mating is expressed by the preference of one sex taken as a whole for partners with given characteristics. Statistically speaking, this preference can be measured by a difference in the mean and dispersion between the distribution of the general population of one sex and the distribution of those who found a sexual partner. For example, in a human population this would be revealed by the fact that married women would not have the same height distribution as the general population of women. 'Assortative mating', on the other hand, involves individual preference for a partner having a given character, as a function of one's own character. Assortative mating can be measured mathematically by the correlation coefficient between couples' characteristics. Pearson noted that these two modalities of 'sexual selection' referred to two fundamentally different phenomena that could act in opposite directions. Preferential mating should in principle lead to a modification of the population mean (this would happen, for example, if all females had a preference for tall males), while assortative mating would on the contrary maintain the population mean, although perhaps broadening the distribution slightly (this would happen, for example, if tall people mated with tall people, small people with small people etc.). In today's terminology, 'sexual selection' would generally be restricted to what Pearson called 'preferential mating' (or 'tribal taste'), and 'homogamy' or 'assortative mating' would be used for what he also called 'assortative mating' or 'individual taste'. Whatever the term, it is interesting that Pearson subsumed the two processes under the category 'sexual selection'. This tells us a great deal about his confusion with regard to the subject.

At this stage a semantic point needs to be made. The distinction between sexual selection in the strict sense of the term and homogamy (or assortative mating) has always been a moot point. The problem is apparently linked to the connotations of the word 'selection' in 'sexual selection'. Darwin provided a way of understanding this difficulty in one of his philosophical discussions of 'natural selection'. In reply to those who complained that the expression was naively anthropomorphic, Darwin argued in the third edition of the *Origin of Species* (1861) that the concept excluded all reference to choice. Natural selection is not a result of 'choice', and, strictly speaking, nature does not 'act'. It is thus only 'by metaphor' that a term implying both action and choice is used in biological theory.[79] This observation is difficult to apply to the case of sexual selection, where there are indeed individuals who recognise and choose their mate. It does not matter whether these processes are voluntary and conscious, as their human meanings would imply – forms of sexual selection can easily be imagined in plants. The important point is that, in sexual selection, discrimination involves an individual biological agent who succeeds or fails in the search for a sexual partner. The difficulty in distinguishing between sexual selection and homogamy is obvious. In both cases, there is indeed a 'choice'. This choice is not of the same kind, but it is still a 'choice', and no amount of terminological wriggling can get round this fundamental problem. In contemporary theoretical writings the expression 'mate choice' is generally used to describe the totality of phenomena involved in homogamy. Strictly speaking, sexual selection could equally be described in the same way, as it was in the 19th century. Pearson's distinction between 'preferential mating' (sexual selection *stricto sensu*) and 'assortative mating' (homogamy) thus touched on an important and difficult point.

As well as 'natural selection' and 'sexual selection', Pearson also recognised a third form of selection in nature, 'reproductive selection', which he defined by the correlation between fertility and the variation of a given character. Against Galton, and against the conviction of most other eugenicists (amongst whom he was a major propagandist), Pearson gradually became convinced of the inseparability of the processes of 'natural selection' and 'reproductive selection'. The circumstances in which he restored this fundamental idea of Darwin's will be discussed in section 7.2.3.

Thus Pearson contributed to the clarification of the statistical presentation of Darwinian categories. In his analysis of distribution curves, of regression and of the modalities of selection, he demystified the concepts that had been forged by Galton, putting them in their proper place as descriptive tools that were *a priori* neutral in terms of an evaluation of Darwin's concept of the modification of species.

7.2.2 *Heredity and selection in biometrical theory: Pearson's 'fundamental theorem in selection'*

This section deals with the central theoretical point of biometrical Darwinism: the relation between heredity and selection.

It has already been shown how, according to Weldon's perspective, the exhaustive proof of an example of natural selection required, amongst other things, the measurement of the correlation between parents and offspring for the character in question, this correlation being considered a measure of the intensity of heredity. Leaving aside the question of a physiological theory of heredity, the reason for this requirement is easy to understand. If the correlation between the parental character and that of the offspring were zero, selection could not alter anything in the population, because the composition of the offspring population would be independent of that of the parents who survived and reproduced.

Nevertheless, it was in Pearson's work that the link between the statistical theories of heredity and selection appeared most clearly. His major 1896 paper – 'Regression, Heredity and Panmixia' – had three simultaneous objectives. In the first place, it reexamined the general mathematical theory of correlation. In 1888, Galton had introduced the term 'correlation coefficient'.[80] The theory was in fact much older; it had been developed in the middle of the 19th century by the French mathematician Bravais, in the context of the study of errors in the determination of a point in astronomy.[81] Bravais's analysis was formally more rigorous and more general than that of Galton, because it dealt with the correlation of three variables, and not only of two. In his 1896 monograph, Pearson took up Bravais's analysis and generalised it to *n* variables.[82] He further showed that the theory of multiple correlation was of fundamental interest for the analysis of heredity. Finally, he extended the application of the theory of multiple correlation to selection.

These points clarify the enigmatic discussion at the end of the article. In the final paragraph, Pearson formulated a 'fundamental theorem in [or of] selection'.[83] The theorem, or rather the problem to which it was intended to respond, was introduced by Pearson as follows:

> *Fundamental Theorem in Selection.*– The general theory of correlation shows us that, taking $p + 1$ correlated organs, if we select p of them of definite dimensions, the remaining organ will follow a normal law of distribution, of which the standard-deviation and mean can be determined. Now, in the problem of natural selection, we do not select absolutely definite dimensions, and the p organs selected may be specially correlated together in selection, in a manner totally different from their 'natural' correlation or correlation of birth. We, therefore, require a generalised investigation of the following kind: Given $p + 1$ normally correlated organs, p out of these organs are selected in the following manner: each organ is selected normally *round a given mean*, and the p selected organs, pair and pair, are correlated in any arbitrary manner. What will be the nature of the distribution of the remaining $(p + 1)^{th}$ organ?
>
> Geometrically, in p-dimensional space we have a correlation surface of the pth order among the p organs, and out of this, with any origin we please, we cut an arbitrary correlation surface of the pth order – of course, of smaller dimensions – the problem is to find the distribution of the $(p + 1)^{th}$ organ related to this arbitrary surface cut out of what we may term the natural surface.
>
> If the p organs are organs of ancestry – as many as we please – and the $(p + 1)^{th}$ organ that of a descendant, we have here the general problem of natural selection modified by inheritance. [Original emphasis][84]

Although this passage could have been less esoteric, it sums up the spirit of the biometrical theory of selection. What exactly was the problem that Pearson wished to deal with?

The passage becomes much clearer if '$p + 1$ correlated organs' is fully understood. By 'organ' Pearson referred to any measurable character, such as height. Second, the expression 'correlated organs' does not refer to anatomically distinct organs, but to the same character measured in different generations of the same population. This is what is meant by 'organs of ancestry', in other words the character state in the p generations of parents. Pearson may have chosen the letter p to signify 'parental generation', but this is mere supposition. Whatever the case, the '$(p + 1)^{th}$ organ' refers to the character state in the population in generation $p + 1$.[85]

'The p selected organs, pair and pair, are correlated in any arbitrary manner.' This means that the correlation coefficients between the character state in a given generation and that in all other generations are known. In fact, Pearson was mainly interested in the correlation of the final generation $(p + 1)$ with each of the parental generations. This suggests that he wanted to pose the problem of selection in the language of ancestral heredity.

The term 'law of ancestral heredity' does not appear in this article (it was not used until 1898).[86] But it was in this 1896 paper that the multiple-regression equations which Pearson considered to be the essence of the law of ancestral heredity were first applied to the problem of heredity.[87] In the simplest example, the equations take the following form:

$$x_0 = b_1x_1 + b_2x_2 + b_3x_3 + \&c$$

where $x_1, x_2, x_3 \ldots$ is the deviation from the mean of the character in a given generation and $b_1, b_2, b_3 \ldots$ the regression coefficient of the character on the parental generations (b_1 for the parents, b_2 for the grandparents etc.).

The same equation can be written in an equivalent form, using correlation coefficients:

$$x_0 = r_1 \cdot \frac{s_0}{s_1} \cdot x_1 + r_2 \cdot \frac{s_0}{s_2} \cdot x_2 + r_3 \cdot \frac{s_0}{s_3} \cdot x_3 + \ldots$$

where $r_1, r_2, r_3 \ldots$ are the correlation coefficients between a given generation and the previous generations, and $s_1, s_2, s_3 \ldots$ are the standard deviations.

This equation resembles a formula predicting the character of a descendant on the basis of the character of its ancestors. In practice, its use requires knowledge of the distribution of the characters in the ancestral generations, which in 1896 was not true for any empirical case whatsoever.

Pearson's question is as follows: how would this predictive formula change if, in each generation, only a fraction of the population were to reproduce? Technically speaking, Pearson is asking about the consequences of an arbitrary truncation of the 'correlation surface' that describes all the correlations between the various generations. A large part of the article is in fact purely mathematical – Pearson presents and develops the then novel theory of multiple correlation; he thought it was possible to formalise it by using it to explain the totality of facts of heredity and selection.

Pearson's theoretical project was to absorb the problem of selection into the problem of heredity, which would itself be absorbed into the theory of correlation. His stated hope was that statistics would suffice to define simultaneously both the 'law' of heredity and the 'fundamental theorem' of selection. Pearson's incontestable mathematical achievement was his treatment of multiple correlation. The law of ancestral heredity was an application of this, namely a multiple-correlation equation connecting a given generation with all its ancestors for a given character. This equation can be represented geometrically as a hypersurface (of $p + 1$ dimensions). The 'fundamental theorem in selection' describes the results of truncating such a surface, which will depend on the distribution of the character in the population. In broad terms, the 'fundamental theorem' had to solve the following problem: how does selection carve out the correlation surface connecting offspring and ancestors? The theorem makes it possible, in principle, to predict the character of offspring, given an exhaustive knowledge of the character's distribution among ancestors.

It could be argued that Pearson succeeded in his ambitious programme, because he developed a general application for his multiple-regression equations. However, the successes of this approach were also its failures. Pearson's formal approach was so general that it was compatible, *a priori*, with a number of entirely different scenarios. This appears clearly in the final pages of the article,[88] when Pearson wonders whether the 'focus of regression' of a population remains stable or shifts during selection. If it is stable, a population that has been subject to selection will eventually return to its pre-selection mean once selection has been relaxed. This, of course, was Galton's model. Darwin's conception of natural selection, however, implied that the focus of regression could 'progress'. While Pearson was undoubtedly more sympathetic to the latter hypothesis, he was obliged to admit that his 'fundamental formulas' could not differentiate the two hypotheses, but merely expressed both of them. He concluded that only experimentation and observation could decide whether the focus of regression moved or not, and if so, at what speed.[89]

By any account, this was a meagre theoretical result. In the final analysis, Pearson's 'fundamental formulas' of heredity and selection were revealed for what they were – an elegant way of summarising the endless data of as yet non-existent genealogical and demographic tables. They were also a way of avoiding taking any theoretical risks whatsoever. Pearson's 'fundamental theorem' – unlike that of Fisher, which was to be developed in the 1930s – was the opposite of a theoretical principle. It was an instrument for reading data tables, and a passive instrument at that. Even the most esoteric mathematics could not hide this epistemological bias.

For Pearson, however, this epistemological bias was a positive advantage. A self-proclaimed follower of Ernst Mach, Pearson thought that scientific laws and theories are *summaries*, valid only as *descriptions* and not as explanations of phenomena: '... science is a conceptual *description* and classification of our perceptions, a theory of symbols which economises thought. It is not an explanation of anything. It is not a *plan* which lies in phenomena themselves.'[90]

It was from this point of view that Pearson approached the Darwinian theory of selection. At the end of the 19th century, no theory was more 'conjectural' and 'explanatory' than that of Darwin. Pearson forced himself to convert the Darwinian edifice into a 'descriptive theory': 'As I interpret that theory, it is truly scientific for the very reason that it does not attempt to explain anything. It takes the facts of life as we perceive them in a brief formula, involving such conceptions as "variation", "inheritance", "natural selection" and "sexual selection".'[91] The law of ancestral heredity and the 'fundamental theorem' were obvious illustrations of this epistemological programme.[92] Incidentally, it is quite rare for scientists to conform so consistently with their own conception of science.

The weakness of Pearson's theory of selection is in fact due to the conception of heredity on which it was based. For the biometricians, heredity was never anything other than a correlation coefficient: 'the true measure of heredity is the numerical correlation between some characteristic or organ as it occurs respectively in parent and offspring'.[93] One of the major problems of a theory of heredity which has this position as its sole basis is that the correlation coefficients between parents and offspring are extremely variable, depending on both the character studied and on the species.

As J. B. S. Haldane noted with irony,[94] anybody interested in the theory of selection could still use this statistical tool to give a clear answer to a certain kind of question. For example: Imagine that in a given human population adults less than 1.6 m tall are stopped from reproducing. What will be the mean height of the children in the next generation? The way to answer this kind of question is very simple: consult the existing tables of height distribution, remove all those parents with a height of less than 1.6 m, and calculate the mean height of those that remain. One could proceed slightly faster by reasoning on the basis of the correlation coefficient between parents and children, which in fact summarises the data present in the tables. One could equally pose the same question in a somewhat more sophisticated manner, for example, by saying that the parents would be 'selected in a normal manner around a given mean'. But in principle, the answer will be of the same kind, because, once again, the correlation is known empirically. Things get a bit more complicated if selection is maintained over several generations – the correlation of the offspring with their ancestors more distant than their parents has to be known. Without such data, multiple-regression equations such as those set out by Pearson in 1896 remain stuck in an empty formalism.

Pearson tried to give a respectable and rigorous mathematical form to both the law of ancestral heredity and the fundamental theorem of selection. In practice, he overburdened them both with an incredibly complicated arsenal of coefficients, which tended to fuse the two formulae. On the one hand, the fundamental theorem of selection was a complex development of the law of ancestral heredity. This is why Pearson tried to construct the law in such a way that, in its most general form, it included the process of selection. This required an astute choice of symbols, and the addition to each of the terms of the multiple correlation equation of a coefficient that 'summarises' the selective effects in each generation. On the other hand, Pearson admitted that the strength of heredity (what today would be called 'heritability') varied in intensity, and was ultimately modulated by natural selection. This led to the introduction of 'coefficients of heredity' which, in the law of ancestral heredity, reflected this subordination of the principle of heredity to the principle of evolution.[95] But this spiral of increasingly abstruse mathematics could not hide the method's fundamental flaw. In the absence of a hypothesis on the nature of heredity, Pearson could scarcely hope to give his 'heredity coefficients' any other empirical significance than that of a summary of sundry evolutionary factors. Hence his paradoxical emphasis on dissociating the effects of heredity from those of natural selection, sexual selection and homogamy, and expressing the 'influence' of these factors in 'constants of heredity'.[96]

This underlines the fundamental limit of the biometricians' Darwinism: they believed that there could be a 'purely statistical' theory of heredity. The critical point was not their statistical bias – by trying to define the quantitative criteria for the identification of natural selection, sexual selection, assortative mating and many other factors in evolution, they undoubtedly carried out useful work. But when they dealt with heredity, their method was a dead-end. Reduced to a mere index of resemblance with no causal significance, heredity was both the privileged target and the blind spot of the biometrical evolutionary problematic. It was a 'privileged target' because heredity was an exemplary model of an epistemology that rejected causality in the name of correlation, and a 'blind spot' because in the absence of any physical significance (a correlation coefficient is non-dimensional), heredity became a magnitude that was suspected of merely 'summarising' various factors.

With biometry, Darwinism paradoxically reached the critical point of its history. Thanks to the biometricians, natural selection appeared clearly as a 'fact'. But because the physiological basis of heredity remained unknown, the biometricians reduced heredity to a mere method of regrouping data, thus threatening the theory of selection with ceasing to be a theory at all. This, of course, was not Pearson's view. He thought that the law of ancestral heredity – precisely because it embraced all the phenomena of heredity – had a place in the biological sciences which was analogous to that of the principle of gravitation in Newtonian physics: 'If Darwinian evolution be natural selection combined with heredity, then the single statement which embraces the whole field of heredity must prove almost as epoch-making to the biologist as the law of gravitation to the astronomer.'[97]

In fact, the methodological strategy of biometrical Darwinism can be taken as a measure of the conceptual confusion that was typical of turn-of-the-century biology. By continually repeating that heredity could only be represented by a contingent correlation coefficient that would vary between organs and between species, the biometricians gave succour to the idea that evolution is the fundamental principle of biology, and that heredity is a force with a variable intensity, subject to the principle of evolution. Pearson himself had said as much in 1898: 'I should imagine that greater or less inheritance of ancestral qualities might be a distinct advantage or disadvantage, and we should expect inheritance to be subject to the principle of evolution.'[98]

From this standpoint, it is quite reasonable to imagine that all sorts of factors (homogamy, environment and, of course, natural selection) 'influence' heredity and that they appear in the coefficients that empirically measure heredity. Two years later, with the rediscovery of Mendel's laws, a completely different vision of heredity would dominate – the idea of 'heredity as force' was replaced by the concept of 'heredity as particle'. As a result, priorities were reversed: the 'principle of evolution', and in particular its Darwinian form, tended to become a subsidiary element of the science of heredity.

7.2.3 Fitness: theory and practice

A final aspect of biometry deserves special discussion: that dealing with what modern evolutionary theory calls the 'components' of natural selection. Today it is quite a commonplace to say that the general state of an organism can be broken down into two components – viability and fertility. In principle, both these factors are measured by demographic parameters, and their product defines that character which since Fisher (1930b) has been called 'fitness'.[99] This word – which has as one of its oldest meanings 'adaptation' – came into common usage at the end of the 19th century as an abbreviation for 'fitness to survive'. This expression was itself nothing more than a reversal of the term that Spencer had proposed as a replacement for 'natural selection' – 'survival of the fittest'.

At the turn of the century, the inclusion of fertility in the concept of 'fitness', and thus in the definition of natural selection, was by no means self-evident. It was widely thought that fitness to survive could be recognised and measured only by longevity, which was in turn taken to be an index of adaptation to the environment. It was in fact relatively common to distinguish 'natural selection' as a process acting only through differential mortality, from differential 'productivity' or 'fertility'. It is therefore particularly interesting to trace the passage of the concept of natural selection from a vision strictly related to differential mortality, to the modern conception which links natural selection to overall reproductive capacity. The biometricians, and in particular Pearson, played an important role in the elaboration of the modern concept of fitness, which took place completely independently of the Mendelian revolution. The metamorphosis of this concept followed a rather bizarre route, much of which has been forgotten by modern scientists. The contemporary scientific literature shows that the first generations of Darwinians had great difficulty in imagining that fertility was a component of 'fitness', and thus a decisive element in the definition of the process of natural selection. In fact, the incorporation of fertility into the concept of selection had far-reaching implications for powerful and deeply rooted aspects of social reality. It is particularly significant that it was one of the most zealous eugenicists – Pearson – who finally decided to incorporate fertility into the quantitative definition of natural selection.

Before analysing Pearson's position, we need to remember Darwin's view of the question, which was in fact less straightforward than most of his readers believed. Darwin's contemporaries tended to be struck by formulations such as: '... natural selection acts by life and death, – by the preservation of individuals with any favourable variation, and by the destruction of those with any unfavourable deviation of structure'.[100] Or again by the following formulation, which replaced the one above in the fifth edition of the *Origin of Species* (1869): '... natural selection acts by life and death, – by the survival of the fittest, and by the destruction of the less well fitted individuals'.[101]

These passages suggest that natural selection is a factor in the modification of species which acts only through death or through the greater or lesser longevity of individuals, which amounts to the same thing. However, this was only one aspect of the process that Darwin had in mind. When Darwin carefully defined his key concepts, he expressed himself somewhat differently. For example, when he introduced the concept of competition or 'the struggle for existence', he made it clear that he included '*not only* the life of the individual, *but* success in leaving progeny' (added emphasis).[102] In the definition of natural selection given in Chapter 4, Darwin used a symmetrical formulation:

can we doubt ... that individuals having any advantage, however slight, over others, would have the best chance of surviving *and* of procreating their kind? On the other hand, we may feel that any variation in the least degree injurious would be rigidly destroyed. This preservation of favourable individual differences and variation, and the destruction of those which are injurious, I have called Natural Selection. [Added emphasis][103]

This is quite clear: Darwin included reproduction in his definition of competition and of natural selection.

It could be objected that these passages do not in fact have the meaning that a contemporary evolutionist would give them. Perhaps Darwin simply meant that individuals only propagate their type under the (obvious) condition of having 'survived'? If this is the case, the passages quoted above would not imply any distinction whatsoever between the probability of survival (viability) and the probability of reproduction (fertility), and Darwin would not have thought that differential viability and fertility were forces that might not coincide in the process of selection.

This is an important point. In fact, relatively few of Darwin's writings imply that the dynamics of selection are the result of the superposition of at least two distinct forces. There is, however, one passage that shows that Darwin was at least conscious of the problem, an addition to the sixth edition of the *Origin of Species* (1872):

A critic has lately insisted, with some parade of mathematical accuracy, that longevity is a great advantage to all species, so that he who believes in natural selection 'must arrange his genealogical tree' in such a manner that all the descendants have longer lives than their progenitors! Cannot our critic conceive that a biennial plant or one of the lower animals might range into a cold climate and perish there every winter; and yet, owing to advantages gained through natural selection, survive from year to year by means of its seeds or ova? Mr E. Ray Lankester has recently discussed this subject, and he concludes, as far as its extreme complexity allows him to form a judgement, that longevity is generally related to the standard of each species in the scale of organisation, as well as to the amount of expenditure in reproduction and general activity. And these conditions have, it is probable, been largely determined through natural selection.[104]

These lines are remarkable, and strikingly modern. Darwin was aware that the relation between fertility and longevity is complex; he conceived of it in terms of an investment; finally, he clearly stated that fertility is as important as individual longevity in the strategy of 'survival' and suggested that the relation between longevity and fertility should be considered as an equilibrium subject to variation and exposed to the action of natural selection.

There is another factor that emphasises the importance of the distinction between viability and differential fertility in Darwin's thinking. This can be found in the notion of sexual selection as a process distinct from that of 'natural selection'. In the case of sexual selection, change occurs only because of differences in the reproductive performance of individuals. This is the clearest possible illustration of the fact that the two components of natural selection – viability and differential fertility – are not identical.

How could the first Darwinians have a different understanding of the concept of natural selection and argue that it only produced change through death, without any role for fertility? This complex relation probably has more to do with the history of ideology than with the epistemological history of theories; I will merely outline how, immediately following the publication of the *Origin of Species*, the problem of fertility had an effect on the theory of natural selection. This will also constitute a suitable introduction to Pearson's work on the question.

It is very difficult to measure fertility precisely. For many years, the only useful available data were demographic. One 'fact' in particular haunted 19th-century demographers – the 'lower' classes and races tended to be more prolific. The poor, said the demographers, marry earlier and, being feckless, have a large number of children. The 'upper' classes, on the other hand, being prudent and wishing to give their offspring the best possible education, limit their fertility. As to geniuses, Galton noted with alarm that they showed a marked tendency not to marry or leave offspring.[105] Hence the fear that the civilised nations would eventually be submerged by the proletariat, giving rise to 'retrograde evolution'. If the population level is maintained mainly by the 'lower classes', argued Galton, then 'a large element of degradation is inseparably connected with those other elements which tend to ameliorate the race'.[106] Such statements, despite being heavily marked by contemporary ideology, or, more precisely, because they were so marked, could only encourage the idea that differences in fertility were not related to natural selection. This led some authors to simply reverse Darwin's argument, as, for example, in the following passage by Greg (1868), cited by Darwin in *The Descent of Man* (1871):

> The careless, squalid, unaspiring Irishman multiplies like rabbits: the frugal, foreseeing, self-respecting, ambitious Scot, stern in his morality, spiritual in his faith, sagacious and disciplined in his intelligence, passes his best years in struggle and celibacy, marries late, and leaves few behind him. Given a land originally peopled by a thousand Saxons and a thousand Celts – and in a dozen generations five-

sixths of the population would be Celts, but five-sixths of the property, of the power, of the intellect, would belong to the one-sixth of Saxons that remained. In the eternal 'struggle for existence', it would be the inferior and less favoured race that had prevailed – and prevailed not by virtue of its good qualities but of its faults.[107]

Similar passages abound in the anthropological literature of the time. In some Parisian circles it was a commonplace to say that the French population was becoming Breton, so fertile were the Bretons: 'Mr Francis Galton tells me that he was recently informed by credible medical authorities in Paris that the French population is becoming Breton, owing to the fact that this element of the population does not limit its fertility to anything like the same extent as other elements.'[108]

It is thus relatively easy to understand how and why fertility was excluded from the principle of natural selection. Human populations, and in particular the 'civilised nations', were – it was argued – exempt from the action of natural selection. They were in the state that Weismann described as 'panmixia'. In panmixia, natural selection is suspended, and all individuals arrive at reproductive age. Uncontrolled by selective mortality, indiscriminate fertility leads to the proliferation of the unfit, and thus to 'regressive evolution'. The behaviour of social classes was thought to aggravate this tendency – for Galton and the eugenicists it was quite obvious that the 'upper' classes limit the number of children they have in order to rear them better and thus give them a better chance of triumphing in social competition. The lower classes do exactly the opposite. The result is that a relentless mechanism is set in motion that will lead to the decay of civilisation. This will happen because of an increasingly bitter class struggle and the fact that natural selection becomes ever weaker in the course of human history.

This was the speculative schema that formed the basis of Victorian eugenics. The eugenic programme was fundamentally a project of artificial selection with the aim of braking the retrogression of 'civilisation' by acting on the factor that natural selection no longer holds in check: human fertility. It did this in two ways: either by 'increasing the productivity of the best stock' (positive eugenics) or by 'repressing the productivity of the worst' (negative eugenics). These are Galton's formulations, from a 1901 lecture on 'the possible improvement of the human breed'.[109]

This position is clearly cast, and logically so, in the language of breeders. In breeding, 'selection' does not involve killing undesired animals, but rather stopping them from breeding, and favouring the reproduction of those that are considered to be the best. Artificial selection is typically selection by reproduction. It was this kind of selection that the eugenicists proposed to apply to man, in order to overcome the weaknesses of a natural selection which, it was thought, was no longer effective. When this vision does not spiral into genocidal madness, it can be characterised as a middle-class ideology[110] based on the belief that it is possible to 'improve the human race' by sterilising the insane and by giving decent people welfare payments to encourage them to have children, to keep themselves clean and to send their infants to school.

To be sure, all this is neither the best nor the purest epistemology. But this aspect of the history of the concept of natural selection cannot simply be ignored. Darwin had constructed a concept that implied a subtle co-operation between death and reproduction. The first generation of Darwinians had adopted a model of evolution by selective mortality, effectively ignoring the role of fertility. There were, no doubt, theoretical difficulties in understanding how fertility and mortality interacted in natural selection. Nevertheless, it was probably demographic reasons of the kind dealt with above that initially led to the exclusion of fertility (or 'productivity') from the concept of natural selection. Darwin was quite clear about this. When, in *The Descent of Man*, he mentioned the work of Galton and others on the 'retrograde march' of the civilised nations that were allegedly undermined by the higher fertility of the poorest classes, Darwin took the trouble of going back to the demographic sources and of reinterpreting them. Thus, in the same chapter of the *Descent*, he noted that infantile mortality in the first five years of life in towns was twice that in the countryside. He also remarked that the mortality of women who marry very young was also unusually high. These facts, together with others, led him to argue that the famous retrograde march would not go unopposed – the higher fertility of the poor could after all be interpreted as a strategy to meet the terrible infantile mortality that existed in the towns, amongst both rich and poor, but more amongst the poor.[111] This response not only reveals a vision of the masses that was slightly different from that of his cousin Francis Galton, it was also a discreet but firm reminder of the richness of selectionist reasoning.

These aspects of the ideological context explain why the first quantitative studies, far from measuring natural selection in terms of 'differential reproduction', on the contrary took great care to separate its effects from those produced by differences in fecundity. Galton's 1877 paper, in which he put forward the first expression of the process of natural selection in statistical terms, was analysed above (see section 5.2.2).[112] It is worth noting here that in this remarkable document, which served as the basis of Weldon's and Pearson's experimental and theoretical studies, Galton presented natural selection as a process that modifies the frequency curve of a population character which is heritable and which can be measured by a 'percentage survival' that varies as a function of the character. He further indicated that the effect of differential fertility (or 'productivity') could be expressed by an algebraic function that had the same form as that which expressed natural selection.[113] This formal similarity was due to the fact that the two processes could both be measured by 'percentage survival':

> Natural selection is measured by the *percentage of survival* among individuals born with like characteristics. Productiveness is measured by the average number of children from all parents who have like characteristics, but it may be physiologically looked upon as the *percentage of survival* of a vast and unknown number of possible embryos, producible by such parents. [Added emphasis][114]

This formulation might appear to be very close to an operational definition of natural selection integrating the chances of death and of reproduction. In fact, the opposite is the case. Not for a single moment did Galton envisage using the term 'natural selection' to describe a process that would involve fertility. If he was struck by the formal symmetry, it was because this approach enabled him to understand how fertility could oppose the action of natural selection. Within this framework, 'fitness' cannot be considered an overall demographic index. In fact, this term was absent from ordinary Darwinian literature, and in particular from biometrical writings, up until the end of the 1890s. In the major studies of Weldon (1893–8) and Bumpus (1898) dealt with earlier in this chapter, the only measure discussed was differential mortality.

These points clarify the strange line of thought adopted at the end of the 1890s by Pearson, that indefatigable advocate of both orthodox Darwinism and Galtonian eugenics, on the relation between 'reproductive selection' and 'natural selection'.

There can be no doubt that Pearson's interest in the role of fertility in evolution was motivated by strong support for eugenics. This can be seen in most of his theoretical articles that deal with the problem, in particular the long series of 'Contribution[s] to the Mathematical Theory of Evolution' (1894–1912). However, a theoretical position cannot be reduced to the motives that inspired it. It was thus Pearson, the most eugenicist of the biometricians, who reconciled the concept of natural selection with that of fertility, and who for the first time proposed a usage of the term 'fitness' that was close to the modern sense of the word (as a global demographic index, including longevity and reproductive performance).

Pearson did not suddenly stumble upon this integrated concept of natural selection, which he knew to be Darwin's original position, and which could not easily be reconciled with eugenics. It has already been noted how, in his major work of 1896, he introduced the concept of 'reproductive selection' as a factor of evolution distinct from 'natural selection' and 'sexual selection'.[115]

What exactly did 'reproductive selection' imply? First and foremost, it implied that the degree of fertility is like any other heritable character. According to the biometrical method, the empirical confirmation of this hypothesis required the study of individual genealogies sampled from homogeneous populations, and the calculation of correlations for fertility between offspring, parents and various progenitors. Pearson tried to carry out just such a study on the rare data available – books of genealogy for the English country gentry[116] and the registers of racehorse stables.[117]

This kind of investigation was unprecedented and turned out to be very complicated, owing to a large number of statistical problems. Pearson found very weak parent–offspring correlations, generally less than 0.1. However, this did not stop him from coming to a clear conclusion, in the most fully developed of his articles on the question: 'The investigations of this memoir have been to some extent obscure and difficult, but the general result is beyond question. *Fertility and fecundity, as shown by investigations on mankind and on thoroughbred horses, are inherited characters*' (original emphasis).[118]

The hypothesis of reproductive selection required that variations in fertility should be correlated with those of a given character. In the absence of such a correlation, differences in fertility could not have any effect on evolution. Pearson's empirical studies of this question remained very fragmentary, but he did carry out some, having noticed that in certain populations of plants fertility was not uniformly distributed among individuals sharing a given character, and that maximal fertility generally corresponded to the modal value of the character.[119]

These empirical procedures were open to a number of methodological criticisms, but they were nevertheless unique and clearly showed Pearson's theoretical orientation. He was attempting to apply to differential fertility the same method that Weldon had applied to the 'rate of selective mortality'. This implied studying a homogeneous population[120] and showing that the differences in fertility were able to maintain or modify the frequency distribution for a given character within a population.

At the same time as he tried to test the hypothesis of 'reproductive selection', Pearson was only too well aware of the fundamental theoretical implications this concept posed for Darwinism. The key part of his work on this question was concentrated in three short years, from 1897 to 1900. After this period, Pearson threw himself into the battle against the first Mendelians. The concept and the term 'reproductive selection' had been introduced into Pearson's 1896 study 'Regression, Heredity and

Panmixia'. Various documents show that Galton had expressed important reservations with regard to the expression 'reproductive selection', to which he preferred 'proliferal selection' or 'genetic selection', probably because the term 'reproductive selection' appeared to interfere with what he called 'sexual selection' (see Chapter 5). In deference to Galton, Pearson often used the term 'genetic selection' at the same time as 'reproductive selection'. He did not renounce his term,[121] because it seemed to be quite appropriate for what he had in mind – a process that could gradually modify the frequency distribution of a character in a population, owing to a correlation between the character and 'reproductivity', that is, the mean number of offspring per individual showing that particular character.

In 1897, in a book entitled *The Chances of Death*,[122] Pearson dealt with the theoretical implications of reproductive selection. The terms he used show that he knew he was dealing with a fundamental aspect of the theory of selection. He declared that if fecundity was heritable, and if it was correlated with a given character, it would have to be recognised as 'a *vera causa* of progressive evolution'.[123] The term *vera causa* – the real and not fictitious cause – is that used by Darwin in his search for the fundamental factors of the modification of species. By using this term, imbued with the Newtonian ideal of science, Pearson put reproductive selection on the same footing as natural selection, suggesting that reproductive selection should be considered a major factor in the *natural* modification of species.

This assertion was founded on a fairly obvious statistical point: a population cannot be stable if the modal value of a given character and that of the associated fertility do not coincide.[124] If they do not coincide, the distribution of the character would change in such a way that the most fertile individuals would also be the most 'typical'. In other words, as well as 'natural selection' and 'sexual selection' there exists another factor involved in evolutionary change. Even in the absence of the two classic modes of selection, it is possible to conceive of a gradual and non-random evolution. Pearson thought that if 'reproductive selection' existed, it would have to be considered one of the most powerful and rapid forces of modification. He did not fail to mention in passing that reproductive selection would constitute an exact, natural analogue of human artificial selection.[125] He further argued that if empirical data confirmed the heritability of fertility and its correlation with character distributions, the hypothesis of a progressive selection by differential reproduction would inevitably be confirmed.

However, Pearson's theoretical intention was not to introduce a new factor of evolution, distinct from natural selection. His post-1897 writings show that he was in fact moving towards a revision of the generally accepted conception of natural selection. In 1898, in a long paper devoted to 'reproductive or genetic selection', Pearson argued that fertility was a character that deserved a very specific analysis from the point of view of the theory of natural selection. From a Darwinian perspective it was quite reasonable to imagine that reproductive performance was the product of evolution by natural selection, each species investing in reproduction according to its own mode of life. But – added Pearson – fertility was precisely not a character like any other:

> The problem of whether fertility is or is not inherited is one of very far reaching consequences. It stands on an entirely different footing to the question of inheritance of other characters. That any other organ or character is inherited, provided that inheritance is not stronger for one value of the organ or character than another, is perfectly consistent with the organic stability of a community of individuals. That fertility should be inherited is not consistent with the stability of such a community, unless there be a differential death-rate, more intense for the offspring of the more fertile, *i.e.*, unless natural selection or other [*sic*] factor of evolution holds reproductive selection in check.[126]

In other words, fertility cannot be thought of simply as a character that is affected by natural selection. Rather, it needs to be considered, if not one of selection's components (Pearson did not argue this in 1898), then at least a factor that perturbs the fundamental mechanism of natural selection. It was for this reason that, two years later, in the second edition of *The Grammar of Science*, Pearson's argument was hardened in the direction of a necessary collaboration of mortality and fertility in all natural selective processes: 'I do assert that progressive changes in living forms can only be looked upon as the product of the action and reaction of natural selection and reproductive selection.'[127] At this point, Pearson was in fact on the threshold of a complete reorganisation of the concept of natural selection, which he was to carry out in two ways.

First, there was a shift in meaning. This was particularly obvious in the case of the word 'fitness', which by the end of the 19th century had imposed itself among the Darwinians because of the success of Spencer's expression 'the survival of the fittest'.[128] According to the traditional point of view, the 'fittest' was the organism best adapted to the environment, 'fitness' connoting vigour, health and the totality of capacities that permit a given individual to 'survive' more or less well than others. Pearson had underlined that this definition was insufficient and that it was necessary to accept that the 'fittest' was also 'the most fertile', or, more precisely, whatever constituted the best compromise between individual adaptation and productivity. Pearson called this 'net fertility' – the gross fertility (the effective number of offspring at birth) minus all those who do not arrive at adulthood.[129] The concepts of fertility and fitness thus become indistinguishable – true 'fitness' involves fertility, and true fertility involves the life expectancy of the offspring:

> ... we must look upon fertility not as merely associated with purely reproductive characters, but consider, at any rate, *net* fertility as closely allied to health, fitness, and strength in the whole complex of organs and characters which form the individual. Thus it seems possible to understand, even if the point still wants quantitative demonstration, how the modal character tends to become associated with the character of maximum net fertility. When the environment by natural selection produces a given type, with that type it ultimately associates the maximum fertility ... Health, strength, and fertility are functions we may suppose of the fitness under a given environment of organs in the individual environment.[130]

This passage contains the first presentation of the concept of fitness as a necessary synthesis of individual vigour and fertility, both depending on the interaction of living beings with their environment. Once this semantic step had been taken, the road was open to a fully demographic concept of natural selection. Pearson, however, went no further than this verbal change. It could be argued that with Pearson there was a symbolic rupture with the 'sacrificial' conception of selection (selection as a summary of mortality), leading to the 'productivist' conception of selection (selection as a summary of reproduction). This latter concept of selection has characterised the whole of 20th-century Darwinian evolutionary biology and, as we have seen, in fact involved a return to Darwin's original positions.

However, Pearson's modification was not merely terminological. By including fertility as part of natural selection, Pearson introduced the idea that natural selection should not be considered an omnipresent infallible force controlling all characters, but rather the effect of a force that was always in unstable equilibrium with other forces. It was reasonable to argue that reproductive selection and selection by mortality were more or less closely linked. This explains why Pearson sought empirical indices of a correlation between fertility and longevity: 'In the course of our investigations we have seen that the relationship between fertility and duration of life does not cease with the fecund period. We thus reach the important result that characters which build up a constitution fittest to survive are also characters which encourage its fertility.'[131] But he also indicated that the relationship between the two kinds of selection should be considered a perpetually oscillating 'balance'.[132] Mortality and fertility are functions of the environment, and it is unlikely that environmental changes will affect both factors in a perfectly coordinated manner.[133] Whatever the case, Pearson's references to this problem are never more than allusions. However, the principle had been laid down: the theory of selection had to be recast in terms of a plurality of interacting forces.

Finally, what was the relation between 'reproductive selection' and Pearson's eugenics? It has already been pointed out that Pearson constantly alternated between a desire to rehabilitate orthodox Darwinism against Galton, and a eugenicist proselytism that claimed to follow Galton. In fact, Pearson's attitude towards reproductive selection changed, depending on whether the context in which he was writing was mainly 'Darwinian' or 'eugenicist'. In *The Grammar of Science*, a few pages after the passage where the concepts of fitness and differential fertility were literally fused, Pearson pursues his point in a political comment that is absolutely unambiguous:

> I am prepared to maintain that the middle classes (owing to their long period of selection and selective mating) produce relatively to the working classes a vastly greater proportion of ability; it is not the want of education, it is the want of stock which is at the basis of this difference. A healthy society would have its maximum fertility in this class, and recruit the artizan class from the middle class rather than *vice versa*. But what do we actually find? – A growing decrease in the birth-rate of the middle-classes; a strong movement for restraint of fertility and limitation of the family, touching only the intellectual classes and the aristocracy of the handworkers! . . . The dearth of ability at a time of crisis is the worst ill that can happen to a people. Sitting quietly at home without external struggle a nation may degenerate and collapse, simply because it has given full play to genetic selection and not bred from its best. From the standpoint of the patriot, no less than from that of the evolutionist, differential fertility is momentous . . . Our social instincts have reduced to a minimum the action of autogeneric selection within the community, they must now lead us to consciously provide against the worst effects of genetic selection, – a survival of the most fertile, when the most fertile are not socially the fittest.[134]

The expression 'genetic selection' in this passage is strictly synonymous with 'reproductive selection'. Pearson sometimes used the term to keep Galton happy, the latter considering 'reproductive selection' to be ambiguous.[135] But this is merely a detail. This passage clearly shows where Pearson found the inspiration for his concept of a tension between selection by differential mortality and 'reproductive' selection. Pearson was indeed convinced that his comprehensive concept of fitness did not apply to civilised man: '. . . if fertility be inheritable or correlated to inheritable characteristics, then in the case of civilised man material selection at present would appear to be quite secondary to reproductive selection as a factor of progressive evolution'.[136]

If Pearson contributed to the renewal of the naturalist concept of selection, it was because of the highly ambiguous blessing of his eugenicist ideology. How could Pearson reconcile his vision of natural selection as a progressive factor of evolution associating viability and reproduction with the use of these same concepts in his decadent eugenicist propaganda?

Pearson was probably aware of the more or less untenable nature of the positions held by the eugenicists with regard to the theory of natural selection. He doubtless considered that the eugenicist cause would be strengthened if it were not seen to be based on a 'scientific' misunderstanding. The concept of natural selection thus had to be clarified to reinforce the case for eugenics. The theoretical basis of eugenics was the thesis that natural selection no longer guaranteed evolutionary progress in civilised man. Pearson thought he had shown why this was the case. In nature, selection involves the cooperation of two causes of change – differential viability and differential fertility. No progress is possible if these two factors are not at least partially correlated. But in civilised man, viability and fecundity are no longer linked, because of the special competition that structures 'civilised' societies (the 'upper' classes limit the number of children they have, whereas the 'lower' classes do not). At the same time, Pearson argued that there were hereditary differences of ability between the social classes. The conclusion of this series of arguments can easily be predicted – the reason why 'natural selection' no longer plays a role in man is because its mechanism no longer functions. Naked eugenicist ideology thus went hand in hand with the theoretical reelaboration of the concept of natural selection.

Biometry was a complex episode, and its fascination for historians of science is easy to understand. The biometricians can be credited with having defined the parameters that have to be measured in order to establish natural selection as a fact. It could be argued that this was a stupid and worthless exercise. However, in science the search for facts and examples is never vain. Who would reproach Foucault for having sought a direct proof of the rotation of the earth three hundred years after Copernicus? The major epistemological weakness of the biometricians lay in their intransigent phenomenalist epistemology. Fuelled by their enthusiasm, this method led them to believe that the theory of natural selection could be constructed without any biological hypotheses whatsoever, in particular without a hypothesis about the physiological functioning of heredity. This explains why no sooner had they demonstrated (or believed they had demonstrated) cases of natural selection than they launched into a controversy with the first Mendelians. As to the eugenicist sympathies of several – if not most – biometricians, these need to be taken into account. But it would be too easy to argue that the biometricians did 'bad science' because of their 'bad ideology'. In fact, at the turn of the century, most statisticians and biologists of heredity adhered to an ideology of degeneration and of social control by eugenics. The Mendelians were no better in this respect.[137]

The analysis presented here has deliberately concentrated on a hitherto neglected theoretical effect of the eugenicist atmosphere on the problematics of natural selection. By repeating over and over again that in humans natural selection was suspended, by arguing that the civilised nations would degenerate owing to the proliferation of the unfit, the eugenicists finished by focusing the attention of evolutionists on the idea that naturalists' conception of selection should include differential fertility. The result was ambiguous in its motivations but was fundamental for the development of Darwinism, because it brought to light a lesson that was explicitly contained in Darwin's founding writings. Of course, this result could have been reached by a different route, but this is what actually happened. Modern evolutionary biologists may wish they had more respectable ancestors, but this is not the case.

8 Establishing the possibility of natural selection: the confrontation of Darwinism and Mendelism

If there is a key event in the history of Darwinism, it must be its confrontation with Mendelism. It was through its contact with the new science of heredity that the theory of selection became truly intelligible. The confirmation of natural selection by consequence or direct proof was replaced by a study of the objective conditions that could make the process possible. By clarifying the experimental meaning of concepts such as variation and heredity, Mendelian genetics transformed Darwin's prerequisites into an experimental and theoretical foundation upon which the concept of selection could be reconstructed. The hypothesis of selection lost some of its generality, but it gained in empirical plausibility.

The marriage of the two approaches was not all plain sailing. The experimental science of heredity which developed around 1900 initially seemed to have given the *coup de grâce* to Darwinian selection. And while this same science was eventually to appear as a complement of Darwinism, this was only at the price of a subtle but decisive modification of the principle of selection.

This chapter deals with some of the most famous episodes in the history of 20th-century biology – those relating to the early days of genetics – but only in so far as they relate to the theory of selection. It does not constitute a systematic history of the origins and developments of Mendelian genetics.

When discussing Mendelism, we need to distinguish between the principles set out by Mendel in 1865 and the theoretical context in which these principles were rediscovered and exploited from 1900 onwards. Practically all of the first Mendelians were part of the 'mutationist' school – in 1901 Hugo de Vries, one of the rediscoverers of Mendel's laws, had given this name to his theory of the origin of species.[1] This evolutionary theory, which owed a great deal to Galton, was set out at the beginning of the 1890s. For the mutationists, only 'discontinuous' variation (Bateson, 1894) was the source of evolutionary change. Thus they were vigorously opposed to the Darwinians, even before the rediscovery of Mendel's laws in 1900. When the new science of heredity, soon baptised 'genetics', was organised around these laws, it initially appeared to favour the mutationists' hypotheses because it implied a particulate conception of heredity. In the early years of the 20th century it is thus sometimes difficult to distinguish the positions and analyses of 'mutationism' from those of 'Mendelism'. As will become clear, however, the two concepts have to be separated for reasons that are not merely chronological. The mutationists' criticisms of natural selection will therefore be dealt with separately from the early effects of Mendelism on the theory of selection.

8.1 Mutationist criticisms of selection

The biometricians had been unable to respond to at least one of the objections raised by Fleeming Jenkin – natural selection has only a limited amount of variation to act upon. Jenkin justified his criticism by basing himself on the experience of breeders:

> We all believe that a breeder, starting business with a considerable stock of average horses, could, by selection, in a very few generations, obtain horses able to run much faster than any of their sires and dams ... But would not the difference in speed between each successive generation be less and less? ... The rate of variation in a given direction is not constant, is not erratic; it is a constantly diminishing rate, tending therefore to a limit.[2]

Darwin's only response to this objection was to insist on the amplitude of the transformations made by breeders and horticulturists in domesticated species. But he had no direct experimental evidence of the availability of new variation at the end of a process of selection.

Forty years on, Jenkin's objection still carried weight, but this time against the biometricians, who were remarkably silent on the question. Weldon did not include the problem of the limits of available variation in his empirical demonstration of natural selection. As for Pearson, his mathematical models of selection accepted without justification that when selection takes place the variability of the offspring in each generation is sufficient to shift the population mean indefinitely. Pearson certainly did not state that such an effect was physiologically possible (this would be nonsensical), but his formal description of selection was constructed as though there were no intrinsic limit to the power of selection.

The question of the limits of available variation was nevertheless fundamental for the hypothesis of natural selection, and lent itself to a direct experimental approach. It was also the main source of 'mutationist' theories. The fundamental concepts of mutationism – in particular 'mutation' and 'pure line', but also key concepts of the future science of genetics such as the distinction between 'genotype' and 'phenotype' – were constructed as part of an attempt to refute the Darwinian principle of selection by showing that the nature of variation was incompatible with gradual and indefinitely effective selection. The names of two botanists are closely associated with this project: Hugo de Vries (1848–1935) and Wilhelm Johannsen (1857–1927), both of whom played a major role in the emergence of Mendelian genetics. Their positions can be taken as a summary of the mutationist critique of Darwinism.

8.1.1 The mutationist theory of the origin of species (de Vries)

In 1889 – the same year that Galton published *Natural Inheritance* – Hugo de Vries published a brilliant essay entitled *Intracellular Pangenesis*.[3] This was one of the most vigorous and imaginative discussions of heredity produced during the pre-Mendelian period. As the title suggests, it was aimed against the Darwinian hypothesis of pangenesis, but the thesis it puts forward with regard to heredity led de Vries to question Darwin's model of selection as well.

According to de Vries, Darwin's 'provisional hypothesis of pangenesis' was composed of two distinct propositions:

1. In every germ-cell (egg-cell, pollen-grain, bud, etc.) the individual hereditary qualities of the whole organism are represented by definite material particles. These multiply by division and are transmitted during cell-division from the mother-cell to the daughter-cell.

2. In addition, all the cells of the body, at different stages of their development, throw off particles; these flow to germ-cells, and transmit to them the qualities of the organism, which they are possibly lacking.[4]

De Vries rejected the second proposition, which he called the 'transportation hypothesis', but accepted the first. His own theoretical intentions flowed from this position:

My problem in the following pages will be to work out the fundamental thought of pangenesis independently of the transportation hypothesis . . . I shall be guided by the thought that the physiology of heredity, and especially the facts of variation and of atavism, indicate the phenomena which are to be explained, while microscopic investigation of cell-division and fertilization will teach us the morphological substratum of those processes.[5]

Because he rejected Darwin's proposed origin for the hereditary particles – budding – de Vries refused to use the Darwinian term 'gemmules', preferring instead his own term 'pangens', which he described as the 'special material bearers for the various hereditary characters'.[6] For de Vries, there were a limited number of distinct types of pangens, which corresponded specifically to definite characters in the organism. De Vries considered them 'single factors more or less independent of each other':[7]

'These factors are the units which the science of heredity has to investigate. Just as physics and chemistry go back to molecules and atoms, the biological sciences have to penetrate to these units in order to explain, by means of their combinations, the phenomena of the living world.'[8]

Once the pangen had been defined, de Vries distinguished two radically different types of variability. The first was 'fluctuating variability', which he attributed to an alteration of the numerical relationship between different kinds of pangens,[9] each of which could be present in a varying number of copies in the nucleus of each cell. De Vries thought that this number depended upon the environmental conditions to which individuals were exposed.[10] For de Vries, fluctuating variability was of no consequence for evolution. He went on to describe a second type of variability: 'some pangens may change their nature more or less in successive generations or, in other words, new kinds of pangens may develop from those already existing'.[11] Only this variability could explain the appearance of genuinely novel characters, and also the origin of varieties and species.[12]

The distinction between fluctuating variation in the *number* of pangens and the constitutive variation of their *type* was the starting point for de Vries' book *The Mutation Theory*, which was openly opposed to the classic Darwinian representation of the origin of species. Between *Intracellular Pangenesis* (1889) and the first volume of *The Mutation Theory* (1901), de Vries' work followed two lines. The better known is hybridisation – by crossing varieties and species that were as different as possible, de Vries hoped to prove the existence of the pangens he had postulated in 1889. In 1900 this led him to 'rediscover' Mendel's laws of segregation.[13] In the terminology of *Intracellular Pangenesis*, this discovery implied abandoning the theory of multiple copies of pangens, and accepting the view that each type of pangen existed in two copies and two copies only in the cells of each individual, but in two possible alternative forms. His other line of research was directly linked to a critique of the Darwinian principle of selection. This will be more fully analysed here.

During the 1890s, de Vries discovered Galton's statistical studies of heredity and became particularly enthusiastic about the idea that asymmetric curves are indicators of heterogeneous populations. In an 1894 article on artificial selection in plants – 'Galtonian demi-curves as an indication of discontinuous variation'[14] – de Vries showed it was possible to use selection to separate a population having an asymmetric, more or less bimodal curve into two normally distributed populations with different means. De Vries argued that his selection experiment had produced a new racial centre that defined a population discontinuous with the overall population, and potentially able to replace it. Like Weldon's

1893 paper on the shore crab, this study was influenced by the Galtonian model of selection, in which one homogeneous population replaces another. But de Vries' approach was different in that it was wholly based on the experimental method. Instead of carrying out a theoretical dissection of the observed frequency curves, as Weldon and Pearson had done, de Vries showed that it was possible in practice to separate the population into two homogeneous and discontinuous races. This illustrated the idea that selection – artificial or natural – does not modify a population, but can only separate competing discontinuous forms.

At the same time, de Vries set out to show that selection has no effect on 'fluctuating variability', which he agreed showed a Gaussian distribution. In *The Mutation Theory*, he devoted a whole chapter, with the provocative title 'Selection Alone Does Not Lead to the Origin of New Species',[15] to this project. This chapter consists mainly of a compilation of the registers of plant-breeders (of cereals and various other plants). De Vries recognised that selection had accomplished major quantitative modifications in plants, for example in doubling the sugar content of sugar-beet, but he argued that this improvement rapidly reached an upper limit and that it was maintained only as long as selection continued to act.[16] As soon as selection was suspended, he said, plants tended to revert to the pre-selection mean of the general population. To justify this assertion, de Vries reported his own experiments on selection in maize (1886–94), the results of which were so ambiguous that they provoked a polemic with the biometricians, which has been analysed in some detail by William Provine.[17]

From his work on hybridisation, monstrosities and artificial selection, de Vries developed a general conception of the origin of species, according to which species emerge during periods of rapid and major variation, known as 'mutations'. These periods are exceptional, and most species are 'immutable' for most of the time. Indeed, they generally show a continuously distributed variability. But this variability, which Darwin thought was made up of 'individual differences', cannot serve as the basis of the modification of species: 'According to the Mutation theory, individual variation has nothing to do with the origin of species. This form of variation, as I hope to show, cannot even by the most rigid and sustained selection lead to a genuine overstepping of the limits of the species and still less to the origin of new and constant characters.'[18] Thus the Darwinian thesis of the gradual and indefinite transformation of species by selection had to be rejected. However, de Vries did not wholly reject the concept of natural selection, which he considered to act *post hoc* by eliminating non-adapted forms – selection thus had a fundamental regulating role, but was not involved in the origin of species. Natural selection transforms nothing; it merely provides an *a posteriori* check on brusque and massive variations which it plays no role in creating:

> The struggle for existence ... may refer to two entirely different things. On the one hand the struggle takes place between the individuals of one and the same elementary species, on the other between the various species themselves. The former is a struggle between fluctuations, the latter between mutations. In the former case those that survive are the individuals which find conditions favorable to them ... It is by this process that local races arise, and by it that acclimatization is rendered possible. If the new conditions of life are relaxed, the adapted race reverts to the form from which it sprang.
>
> The natural selection of newly arisen elementary species in the struggle for existence is an entirely different matter. They arise suddenly and without any obvious cause; they increase and multiply because the new characters are inherited. According as the young or the parent form is better fitted to the environment will the one or the other of them survive. Species no more *arise* as the result of this struggle for existence, than they do as the result of the struggle between the variants of one and the same type – though for different reasons in the two cases. In order that species may engage in competition with one another it is evidently an essential condition that they should already be in existence; the struggle only decides which of them shall survive and which shall disappear.
>
> ... In a word, from the standpoint of the theory of mutation it is clear that the role played by natural selection in the origin of species is a destructive, and not a constructive one.[19]

This is the vision of evolution that de Vries called the 'mutation theory'. The term 'mutation' was in fact ambiguous. In one sense, it meant the sudden occurrence of a new hereditary unit or 'pangen', and in this respect was part of the vocabulary of variation.[20] But de Vries thought that this type of variation was so rare and important in its effects that 'mutation' should also be used to describe the very process by which new species were formed: 'The name I propose to give to this "species-forming" variability is Mutability – a term in general use before Darwin's time.'[21] Thus 'mutation' was not only a concept in the theory of heredity and of variation, it was also the name of a theory of evolution. Despite the fact that de Vries did not wholly reject natural selection, his theory was completely anti-Darwinian, since it implied that species are formed suddenly and abruptly by 'saltation',[22] in a single generation, in individuals that immediately take on the status of 'elementary species'. According to this view, natural selection has no role in the *formation* of species, but only in maintaining them.

The 'mutation theory' was not entirely speculative – it was in part based on observations of a plant he had studied for many years, *Oenothera lamarckiana*. This story is well known to geneticists, but it is worth repeating. In 1886, de Vries noticed a large number of specimens of *Oenothera lamarckiana* in an overgrown field. This plant is an American species of evening primrose with large yellow, white or purple flowers that is sometimes cultivated in European gardens. Among these plants, de Vries noticed two unusual varieties that differed from the wild type by a number of characters. Once transferred into an experimental garden, the two varieties proved to be perfectly stable. More important, the two varieties, together with five others, appeared regularly in the cultivated wild stock – of 50,000 artificially cultivated plants, 800 showed one of these sudden variations, and furthermore produced stable offspring, with no reversion to the species type. De Vries decided to call this phenomenon 'mutation' – the brutal appearance of a genuine 'elementary species'. However, *Oenothera lamarckiana* remained the only example of evidence for the 'mutation theory' – despite a great deal of effort, de Vries never found another species that showed the same phenomenon. Nevertheless, he did not consider this to be a fundamental obstacle. To the extent that 'mutation' was the process by which new species were formed, it was reasonable to think that it would be relatively infrequent: species are generally stable, and only occasionally and exceptionally enter into periods of 'mutability'. During such periods, mutations are extremely frequent and are oriented in precise directions. According to this understanding, not only does selection not have a constructive role, but even the initiative and the rhythm of change are due to mutation. Mutation pressure is thus the fundamental factor of evolution.[23] However, de Vries also forcefully asserted that mutations take place in all possible directions, and he rejected any 'tendency to vary in a definite direction'. Thus, despite his opposition to the Darwinian hypothesis of gradual selection, he paid homage to Darwin for having shown that the evolution of the natural world could be explained without recourse to supernatural agents:

> Whether this selection takes place between individuals, as DARWIN and WALLACE thought, or whether it decides between the existence of whole species, as I think; it is the possibility of existence under given external conditions which determines whether a new form shall survive or not.[24]

In other words, while selection has no formative role, it does nevertheless have a regulating role in the emergence of species. It plays no role in their *origin*, but is decisive for their *survival*.

This view of evolution enjoyed widespread rapid success in the early years of the 20th century. At first most of the pioneers of Mendelism (Darbishire, Lock, Punnett, Goldschmidt, Johannsen, Cuénot, Castle, Morgan) more or less accepted the 'mutationist' vision of evolution, which had the overwhelming and unique advantage of showing that heredity could finally be experimentally investigated. Furthermore, 'mutationism' did not have the old-fashioned and confused philosophical implications of Lamarckism or orthogenesis, nor did it deny the existence of natural selection, but merely rejected the formative role assigned to selection by Darwin. It was the success of the mutationists' positions that provoked so many proclamations of the 'death' of Darwinism.[25]

8.1.2 'Pure lines' and selection (Johannsen)

De Vries' theories were fairly speculative and highly vulnerable to the objection that they were, for the most part, based on observations of a single species. The most developed form of the mutationist critique of the principle of selection came about as a result of experimental studies of 'pure lines' carried out in the early years of the century. These studies had such an impact that in a very short time the destinies of the theories of selection and of heredity became indistinguishable.

The fundamental importance of 'pure lines' in the history of genetics is well known, but they were not only important for Mendelism. The 1903 study in which Johannsen introduced the idea of 'pure line'[26] needs to be understood in context. Johannsen, a Danish botanist, began his work in 1901 in the wake of the rediscovery of Mendel's laws and the success of de Vries' mutation theory. This was a key moment in the opening phase of 'genetics'. But Johannsen's work on pure lines was not 'Mendelian' either in its methods or its objective. His experiments did not involve crosses, had no direct relation to the laws of hybridisation and were based on the variability of quantitative characters – the kind of variation that interested the biometricians.

Johannsen's aim was in fact to study the nature of selection, using an experimental test to decide between its Galtonian and Darwinian interpretations. This approach had three objectives:

- To test the validity of Galton's law of regression to the mean
- To determine if the products of a selective process are stable (Darwin) or if they return to the pre-selection population mean (Galton)
- To determine whether or not there is a definite limit beyond which selection cannot act

In fact, these three points deal with Jenkin's three objections of 1867, which were subsequently developed from a statistical point of view by Galton and from an experimental point of view by de Vries. Johannsen's study appeared to be a crucial experiment that would clarify the common conceptual field of heredity and selection. Although Johannsen's method was biometric (measuring and comparing the mean and the dispersion of a quantitative character), his 1903 monograph was the starting point for the subsequent distinction between 'genotype' and 'phenotype' (Johannsen, 1909) that was to be fundamental for the history of genetics.

This study of 'pure lines' thus had an exceptional theoretical background – rooted in Galton's statistical methodology, but swept along by the Mendelian revolution, it was crucial both for the theory of selection and for the theory of heredity.

Johannsen's 1903 monograph dealt with the common bean, *Phaseolus vulgaris*, in particular the 'Princess' variety which was commercially available in large quantities. Johannsen's primary intention was to test the validity of Galton's law. He had great admiration for Galton, to whom the study was dedicated. It will be remembered that Galton had discovered the law of regression (or 'reversion') on sweet-peas, another self-pollinating plant belonging to the Papilionaceae. Like Galton, Johannsen studied seed weight. After buying a sample of 16,000 seeds collected in 1900, he first showed that the weight of these grains was normally distributed. He then kept some and grouped them in weight classes. These seeds were planted in 1901, and the seeds produced by the resulting plants were collected and classically studied by the law of regression to the mean, with the samples being analysed *en masse*.

Apart from verifying Galton's law, Johannsen also tried to clarify the concept of 'pure lines'. By 'pure line', Johannsen meant all the offspring produced by a single self-fertilised seed. This 1903 study should not be interpreted in the Mendelian language of 'homozygosity' (Bateson and Saunders, 1902),[27] since weight is not a character that behaves in a Mendelian fashion. From a stock of commercially bought seeds (1900 harvest), nineteen pure lines of beans were set up in 1901 from 19 individual seeds. The experiment lasted three generations; the strain seeds (1900 harvest) were called 'grand-parental' seeds. In the summer of 1901 Johannsen obtained a first generation from these individual seeds, which he classed according to weight within each pure line. These 'parental' seeds (1901 crop) were planted in 1902, producing a generation of 'daughter-seeds' (1902 crop).

Considered as a whole, the daughter generation showed a Gaussian distribution, as Quételet and Galton would have predicted from a 'homogeneous' population (Figure 25). However, the experimental procedure enabled Johannsen to detect whether the apparent homogeneity was real. The results of this experiment were presented in a series of twenty tables, each one corresponding to a 'pure line' (A, B, C... T). Figure 26 gives a translation of one of these tables. This shows, for example, that a 'grandparent' harvested in 1900 and weighing 950 mg, in 1901 produced 'parental' seeds with a mean weight of 520 mg, which in 1902 gave 'daughter-seeds' classed according to the weight of the parents (strictly speaking, this should be the *parent*, because the mode of reproduction is uniparental).

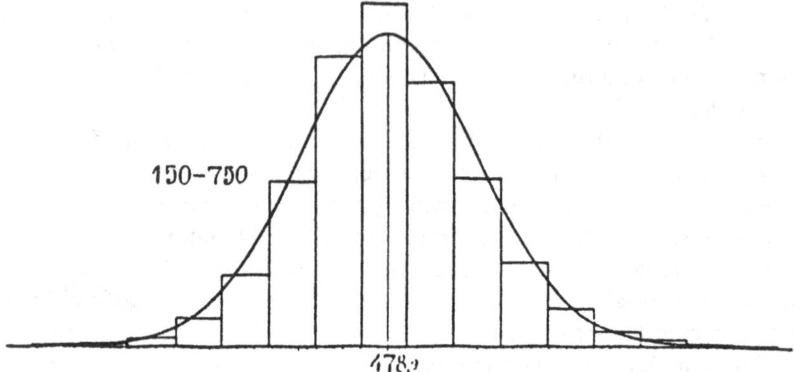

Figure 25 Variation of the weight of 5,494 seeds in the 1902 harvest (W. L. Johannsen, 1903).

The implications of the data contained in this table are striking. In the daughter generation, seed weight fluctuates around the mean of this generation (in this example, 557.19 mg), independently of the variations of the parental seeds. There is no correlation between the deviations of the daughter seeds from *their* mean and that of the parental seeds from *their* mean. This suggests that the variations within a pure line are wholly random from the point of view of heredity, and that the 'pure lines' are homogeneous in their hereditary constitution, or 'stirp', to use Galton's expression (as Johannsen did).[28] If this observation is correct, pure lines are not only 'pure' in their genealogy, but also in their heritable 'type'. This implies that Galton's conception of 'type' should be abandoned – 'type' does not express itself in the mean of the general population, which is heterogeneous, but in the juxtaposed 'pure lines'.

Johannsen interpreted his experiment in terms of both Galton's 'regression' statistic and the modifying power of selection. The superposition of these two interpretative frameworks led to an entirely novel articulation of the concepts of heredity and selection.

At first glance, the data lent themselves to an interpretation in the Galtonian language of regression. Johannsen noted that as long as the general population was studied *en masse*, the law of regression to the mean was verified. For example, if 25 very big seeds and 25 very small seeds are planted, the seeds of the next generation will, on average, be less extreme than their parents; in other words, there has been a partial regression to the mean. But things are very different from the point of view of pure lines. Within each pure line, there is a *complete* regression of the daughter seeds towards the line's type. Put another way, within a pure line, the correlation between parents and offspring is zero (in statistical jargon, the

Mean weight of grand-parent seeds (1900): 950 Mean weight of parental seeds (1901): 520			
Weight classes of parental seeds (1901)	Descendants (1902)		
	Mean weight	Number of seeds	Standard deviation
350-400	572	86	85.1
450-500	535	118	89.7
500-550	570	77	105.4
550-600	565	72	98.4
600-650	566	48	65.7
650-700	555	74	98.4
All lines pooled	557.19	475	93.0

Figure 26 Variation of daughter seeds in one of Johannsen's pure lines ('line B') (W. L. Johannsen, 1903; my translation). The paper contains nineteen similar tables, each of which corresponds to a 'pure line'. The key figures to compare are the mean weight of the parent seeds (520 mg) and the weight of the daughter seeds as a function of the parental weight class. There is clearly no correlation between the weights of parents and offspring.

regression coefficient of the offspring on the parents is zero). In a given generation, the daughter seeds 'regress' completely, and not partially, towards the mean of their line. Within a pure line, tall parents do not have particularly tall offspring, nor do short parents have particularly short offspring. Observed within-line variation is thus not attributable to hereditary constitution, but to environmental differences, such as the position of the seed in the pod, the position of the pod on the stem, the position of the plant in the field etc. To summarise, in a 'pure line', individuals regress only to the type of their line, and not at all to the general population mean.[29]

This has an immediate implication for the theory of selection. Johannsen had in fact shown two things. On the one hand, selection can be effective at the population level: the offspring of a given seed is stable, and does not show a return to the mean of the population from which the seed was taken. This contradicted Galton's position. But on the other hand, selection is completely ineffective within a 'pure line'. In such lines, there can be no progressive change – the extreme parental classes can, on average, reproduce only the type of the line, and not their own character. The efficacy of selection is intrinsically linked to the possibility of 'purifying' the population, and thus cannot continue indefinitely: 'And it will now be easily understood that the action of selection cannot be carried beyond fixed limits – it must indeed cease when the purification, the isolation of the particular most strongly deviating line, practically speaking, is carried to completion.'[30] This contradicts Darwin, or at least the classic biometric interpretation of Darwin (Pearson).

Figure 27 summarises Johannsen's main conclusion. In the upper part of the figure (A), the curve represents the distribution of the seeds in the initial sample. The lower parts show respectively the effect of selection *of* a single pure line (B) and *in* a pure line (C and D). The selection of the largest seeds (C) has the same effect as the selection of the smallest seeds (D). The same procedure was carried out for all nineteen pure lines, and, within each pure line, for each weight class of 'parent' seeds. The elegance and the effectiveness of Johannsen's study comes from the fact that in a single protocol it groups together:

- A biometric representation of a population
- A conception of heredity (the article prefigures the distinction between phenotype and genotype)
- A clear representation of the efficacy *and* the limits of selection

What were the fundamental conceptual consequences of the theory of 'pure lines'? As has already been suggested, this theory was characteristic of a period in which many biologists believed that with the birth of genetics, all the problems of the causal theory of evolution would be solved *ipso*

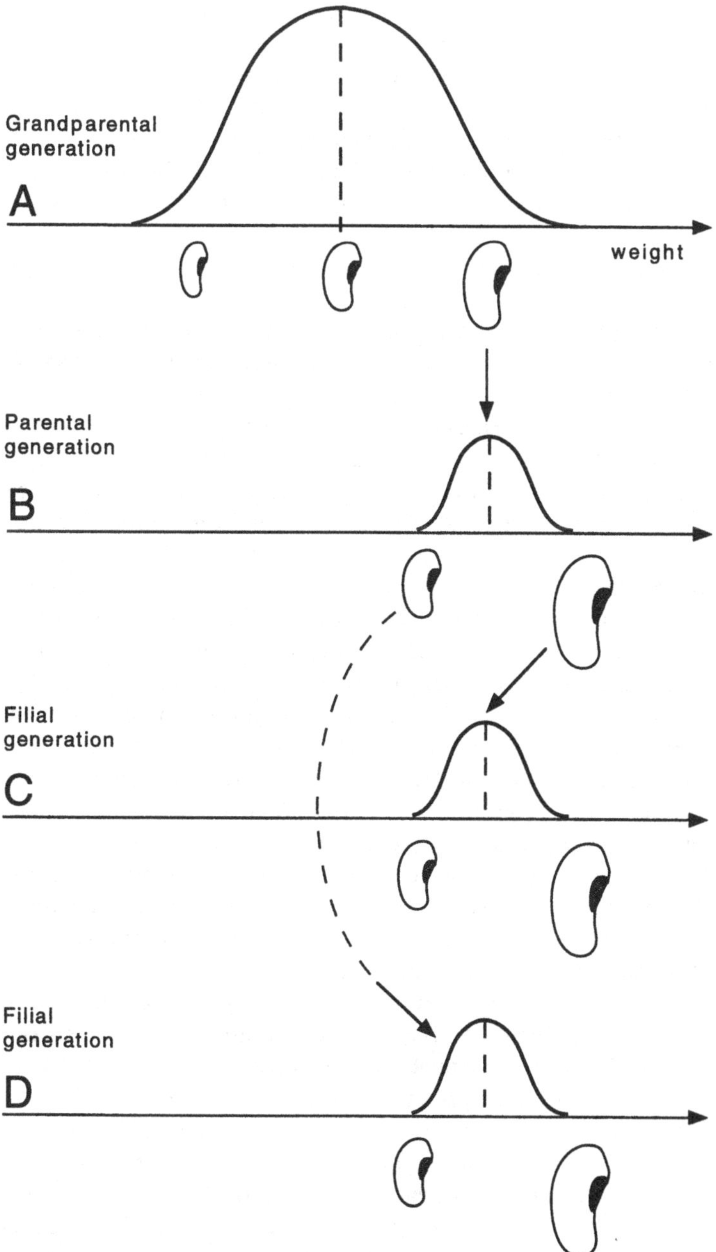

Figure 27 Schematic illustration of Johannsen's analysis of 'pure lines' of beans (*Phaseolus vulgaris*).

facto. Johannsen's work contains the most explicit and well-worked-out version of this conception, to the extent that the theory of 'pure lines' is at root both a theory of heredity and a theory of selection. Although this syncretic approach was not entirely justified, it sufficiently marked 20th-century biology for these aspects to be worth examining.

What were the implications of 'pure lines' for the science that Bateson was soon to call genetics?[31] Johannsen's 1903 work was clearly the starting point of, and the fundamental justification for, the distinction between phenotype and genotype, which the author was to propose in 1909.

The distant origin of this distinction can be found in the old opposition between 'constitution' and external appearance. In the second half of the 19th century, first Galton and then Weismann had given this opposition an anatomical interpretation. Galton distinguished the 'stirp' from the 'person'. The stirp (from the Latin *stirps* – root) is the totality of 'latent' characters that are transmitted by the germ cells, while the 'patent' characters are those that are expressed in the differentiated tissues of the visible person and which 'represent' the stirp, in the sense that an elected assembly represents a group.[32] Weismann preferred to speak of the '*germen*' and the '*soma*', the *germen* being the temporary carrier of the independent and potentially immortal qualities of the *soma*.[33] In his *Intracellular Pangenesis*, de Vries had argued that the hereditary units ('pangens') were present in the nuclei of all cells of the organism. 'Stirp', '*germen*', 'pangens' – these were all ways of isolating 'heredity' by giving it an anatomical location. Nevertheless, this representation of heredity had its limits – the localisation of heredity was of no use for determining which part of an observed character is heritable and which is not. This is why Galton and the biometric school developed a purely phenomenalist theory of heredity, based on the comparison of characters that could be measured in related individuals. In this theoretical context, the analysis of regression and the calculation of correlations *was* the representation of heredity. The advantage of this approach was that, despite the absence of an understanding of the physiological nature of heredity, it was nevertheless possible to speculate about the 'heredity' of a character. On the other hand, this statistical approach was severely limited by the fact that it presented heredity as a force of variable intensity, depending on the character and the situation. The biometric literature of the time is full of figures dealing with the theoretical significance of the comparative strength of 'nature' and 'nurture'.[34] At the end of the 19th century the representation of heredity was thus divided into an anatomical concept that identified heredity by a number of simple anatomical oppositions (*germen*/*soma*, nucleus/cytoplasm), which failed to come to grips with 'heredity', and a phenomenalist concept based on the measurement of the intensity of heredity, which left its physiological nature unclear.

This conceptual configuration was completely subverted by Johannsen's work on 'pure lines'. On the one hand Johannsen's methodology was biometric – he analysed the 'pure lines' in terms of distribution curves, means, standard deviations. But on the other hand the study led not to a measure of heredity, but to a dual understanding which was not anatomical, unlike the classic opposition between soma and germ line. Johannsen opposed the 'type' of the line and the 'fluctuations of the line', which were not two separate things, but two aspects of the same reality which six years later he was to call 'genotype' and 'phenotype' (note the continued use of 'type'). This is all the more important because the *non-Mendelian* analysis of pure lines led to a fundamental clarification of Mendelism.

Once again, the starting point has to be Galton's understanding of 'type'. For Galton, the type was the mean of the general population, the 'racial centre' towards which individuals tend, on average, to return. Johannsen transferred the notion of the 'type' of the general population to the pure lines that constitute the population. But in these pure lines, 'regression' to type is not partial, but complete. Statistically speaking, the *within-line* parent–offspring correlation is zero. In other words, there is a 'fluctuation' in the line that has no hereditary significance. Furthermore, the absolute mean value of a line is not constant from one generation to the next for conjunctural reasons, but the differences between lines are nevertheless maintained. This leads to a paradoxical notion of 'type' – the type of a pure line has a biometric expression in its frequency distributions, but these distributions do not express the whole truth but are merely its phenomenal manifestation. This was the origin of the notion of 'phenotype'. In the 1909 paper in which he introduced the term, Johannsen explicitly emphasised its descriptive and biometrical significance:

> The 'type' in Quételet's sense is a superficial phenomenon which can be deceptive. Only further study will decide whether only one or, as more often the case, several biologically different types are present [in a population]. Therefore I have designated a statistical, *i.e.* purely descriptively established type, as an 'appearance type' [*Erscheinungstypus*], a phenotype. Phenotypes are measurable realities, just what can be observed as characteristic, in variation distributions of the 'typical' measurement, the center around which the variants group themselves. Through the term phenotype the necessary reservation is made, that the appearance itself permits no further conclusion to be drawn. A given phenotype may be the expression of a biological unit, but it does not need to be. The phenotypes recognized in nature by statistical investigation are in most cases expressions of such a unity.[35]

The concept of 'genotype' had a Mendelian aspect which was absent from the 1903 study of pure lines. Some time after the publication of this article, Johannsen tried to extend its conclusions to varieties showing crossing reproduction, in a paper entitled 'Does Hybridization Increase Fluctuating Variability?'[36] Hybridisers had long known that when two varieties or two species are crossed, the first filial generation (F_1) is homogeneous, while the following generation is much more variable. Johannsen's theory of pure lines was thus open to the objection that it was not valid for hybrid and panmictic populations. In this second study Johannsen observed that in artificially crossed pure lines of beans, the weight of first-generation hybrids was no more variable than that of the two parental populations. If it was accepted that a quantitative character such as weight was defined by an unknown but definite number of Mendelian characters, 'type' could be understood as a definite Mendelian combination. In the first hybrid generation, one and only one 'typical combination' appears. In the following generation, the segregation of independent factors produces a whole series of new combinatory 'types'.

At this point in the argument, one could ask whether anything remained of the notion of 'type'. Johannsen could not fail to be aware of one of Mendel's remarks in his 1865 paper, which the latter thought to be one of the most important consequences of his 'law of development of hybrids'. Mendel noticed that in a hybrid population, the number of possible forms increased with the number of 'different traits'. If n was the number of segregating characters, the total number of possible forms was 3^n. For seven characters, for example, there are 2187 possible 'forms'. As a result, Mendel concluded that even in a small population there was a high probability that the individuals would all be different, and that it should be expected that in each generation new forms would arise that had been absent from previous generations.[37] Although Johannsen nevertheless retained the language of 'type', he did so in order to express the idea that, whatever the number of 'types' in the population, the opposition of 'fluctuating variability' and 'difference of type' remained valid. It was in this way that the idea of 'genotype' as a combination of genes was constructed. However, the term did not appear until 1909, when Johannsen wrote: 'This constitution we designate by the word genotype.'[38]

To summarise, the distinction between phenotype and genotype appears as the effect of a synthesis between biometric theoretical analyses and the Mendelian description of heredity. Phenotype is indeed a 'type' in the sense used by the biometricians – it can be revealed by statistical analysis. Genotype, however, is a hypothesis concerning the structure of hereditary constitution which could be revealed only by a hybridisation procedure. These two concepts are clearly complementary, but there is a fundamental difference between them. Because it is a statistical entity, a phenotype can only be properly defined at the level of a population. A genotype, on the other hand, is the factorial structure of a zygote; several individuals or whole populations can have the 'same genotype', but a genotype is not a statistical entity. It can certainly be inferred from data and by statistical methods, but it is not defined by statistical parameters.[39]

Now that I have sketched out the context, it is easier to understand the implications of the 'theory of pure lines' for the new science of heredity. In 1903 this theory did not have a Mendelian method. But by audaciously extrapolating his conclusions to 'Mendelian populations' (in other words hybrids), Johannsen laid the basis of the distinction between phenotype and genotype, that is of the philosophy of genetics. Any study of heredity would henceforth be obliged to specify whether its object was apprehended by purely descriptive statistics or by a causal hypothesis concerning the structure of hereditary determinants.

What were the effects and consequences of 'pure lines' for early-20th-century evolutionary biology?

Johannsen had clearly sided with de Vries and the partisans of the 'mutation theory'. In the most convincing way possible, the experiments on pure lines illustrated the idea that selection was not a force of transformation, and that it should only be understood as an *a posteriori* selection of novelties that it did not construct: 'in the cases that I have examined, selection only acts in populations to the extent that it chooses specimens of already-existing types'.[40] Selection creates nothing, it only isolates. It does not create, it does not transform, it only eliminates. This argument was characteristic of all the 'mutationists'.

This point is often misunderstood. In one way, the idea was already present in Darwin, who always said that selection did not create variation, but presupposed it. How then could the mutationists believe that they had sounded the death-knell of Darwinism? The important theoretical point here is the nature of the variation that is exposed to selection. Darwin considered that selection acted on slight individual differences, or 'continuous variation' (to use a later term). Because experiments showed that this kind of variation was always present, in all species and for most characters, this implied that although selection did not create variation, it could continually act on species and transform and mould them. It was precisely this central notion of Darwinism that the mutationists rejected by pointing out that because 'continuous' or 'fluctuating variation' was non-heritable, it could not be used by selection to form species. Selection was thus seen as a force that was strictly limited by the available discontinuous variation.

Twentieth-century Darwinism was to be reconstructed on the basis of this critique. To use a term frequently encountered in popular presentations, neo-Darwinism really is a theory of 'mutation-selection'. All the great Darwinians of the 20th century, from Ronald Fisher to Ernst Mayr, have repeated, following de Vries, that evolution is a two-stage process, in which, in the last analysis, the rhythm of selection depends on the rhythm of mutations. Nevertheless, this doctrine was not clear initially, for the simple reason that the concept of mutation itself raised major experimental problems. The mutationist critics of selection had an Achilles' heel – the very concept which they wielded so effectively against Darwinism.

The theory of pure lines is a useful backdrop for understanding this important point. In 1903 Johannsen concluded that selection could only choose between and not transform his pure lines of beans. On this basis he felt able to affirm that de Vries' 'mutations' were the only means by which evolution could go beyond the 'limits' that the selection of 'fluctuating' variations was unable to surpass. However, the concept of mutation had as yet no operational definition.

The Mendelian approach, apparently so favourable to the first mutationists, was soon to fatally perturb their evolutionary scenario. Quite rapidly, the suspicion arose that 'mutations' were merely rare combinations of Mendelian factors. Pearson explicitly formulated this objection in a text written in the same year that Johannsen published his monograph on pure lines. Let us suppose, said Pearson, that the individuals in a hybrid population differ by ten Mendelian characters. The probability of having one 'purely allogenic' individual (homozygous at all ten loci) is $(\frac{1}{4})^{10}$, in other words less than one in a million: 'it will be clearly seen that the rarity of some of the more exceptional normal constitutions may easily lead to their being looked upon as "mutations", even if they appear in the offspring of a population many generations removed from hybridisation'.[41] In fact, in a Mendelian model 'there is no room for the appearance of "mutations", although certain variations with very small frequency would be extremely rare in a limited population'.[42] Apart from the word 'mutation', this observation merely repeats the remark of Mendel referred to above. Pearson, however, pushed this Mendelian logic in the direction of biometric positions. As a statistician, he thought it obvious that the 'law of development of hybrids' was a binomial law, and when applied to a large number of characters this law of distribution could be considered to be a continuous distribution. As long as there were fewer combinations realised in a given generation than the total number of possible combinations, this system would constitute a continual and virtually unlimited source of variability which was completely compatible with the Darwinian model of gradual selection.

Johannsen was apparently unaware of this possibility when he proposed to extend the conclusions drawn from pure lines to hybrid populations. In his 1907 paper on hybridisation and fluctuating variability[43] he affirmed that selection could do nothing more than choose between types of combination. Claiming that all possible Mendelian combinations were present in the second generation of hybrids, he inferred that the theoretical situation was exactly the same in the case of pure lines, and that selection could never go beyond combinatory 'types', thus confirming mutation as the only source of progressive change. The key point here is the postulate that all possible combinations were present in the second generation. However, it only requires the number of Mendelian factors to be large, or the population size to be small, for this to be false. It was perfectly possible to imagine that the number of Mendelian factors was so large that all their possible combinations could not be explored in a finite amount of time (for example, the age of the earth). It could easily be argued that Mendelian combinations had, on their own, furnished all the variability necessary for evolution since the origins of life. Of course, this was utterly speculative, but no less so than the concept of 'mutation' at the time.

It is scarcely surprising that some Mendelians pushed Mendelism towards positions that were radically opposed to those of mutationism. The most extreme example was J. P. Lotsy, who in 1916 proposed that the diversity of life was a result of hybridisation. Lotsy was a Dutch botanist who had made his reputation by developing Mendelian analyses of complex cases of hybridisation. In his book *Evolution by Means of Hybridization*,[44] Lotsy argued that not only species, but also the higher taxa (including classes and orders), had their origins in repeated hybridisation events. Apart from Mendelian combination, this model required that the history of life began with several primitive forms, and not one. In the preface, Lotsy dealt lightly with this implication of the theory of evolution by hybridisation:

> [The author] is f. i. very little impressed by the kind of criticism which calls it 'inconceivable' 'verging on the absurd' etc., to believe that crossing can ever have been the underlying cause of the origin of new species, from authors who firmly believe that the origin of new species should be ascribed to some kind of variability; because it seems to him 'absurd' that those who advocate the origin of new species from a *single* ancestor one, should reproach an author who defends such an origin from *two* ancestral species, of stating an 'inconceivable' opinion. [Original emphasis; 'f. i.' is presumably an abbreviation of 'for instance'][45]

Lotsy's model also implied that evolution was not so much a 'progression' as a 'succession' of forms, gradually exhausting an immense but finite field of possible combinations.[46] As Peter Bowler has rightly noted,[47] the most surprising thing is not that this kind of speculation existed, but rather that criticisms of it were so rare.

William Bateson, the most active and determined of the early Mendelians, had a vision of evolution that was not very different from that of Lotsy. Bateson interpreted dominance and recessivity according to a theory that has been classically called a 'presence–absence theory'. According to this theory, a recessive gene is in fact a gene that has disappeared. One of the consequences of this interpretation was the rejection of the idea of mutation as a specific alteration of a gene – a gene can remain or can disappear, but it does not change. From this theoretical starting point, Bateson elaborated a general representation of evolution as consisting of successive and irreversible losses of genes. As he also had to explain the appearance of new *characters*, Bateson postulated the existence of inhibitor genes that affected other genes. Each time an inhibitor gene disappeared, hitherto latent genes were expressed and produced the appearance of an evolutionary novelty. This implied that the original germ plasm contained all that was necessary for the evolution of life.[48]

This evolutionary theory, a product of one of the greatest names in the early history of genetics, is at least as disconcerting as that of Lotsy. Both theories date from the middle of the 1910s, around fifteen years after the rediscovery of Mendel's laws. Both are marked by an explicitly preformist and degenerative bias, and both are formulated in Mendelian language. Their influence should not be exaggerated, but their very existence shows that Mendelism was compatible with interpretations of evolution that were in complete opposition to mutationist theories. They also indicate that in the first two decades of the 20th century, mutationism's theoretical position was sufficiently ambiguous to turn into its opposite. It was thus a relatively weak theoretical opponent, despite the arrogance of its early adepts and the rapidity with which they proclaimed the decline and death of Darwinism.

The concept of 'mutation' was soon shaken by the very data that constituted its sole experimental basis. As we have seen, from the end of the first decade of the century, many geneticists suspected that the *Oenothera lamarckiana* 'mutations' were nothing other than permanent hybrids, which would easily explain their relatively high frequency. Bateson was the first to voice his suspicions,[49] which were to be confirmed in the 1910s by the work of Davis, Renner and Muller – some of the most elegant studies in the early history of genetics. Most of the *Oenothera* 'mutations' were in fact due to relatively rare recombinations of certain heterozygotes, which produced a stable form because of homozygote lethality.[50] Inevitably, the concept of mutation was profoundly shaken by these results, at least until Muller's work in the 1920s gave a clear experimental meaning to the idea by distinguishing between events produced by genetic alteration (mutations *stricto sensu*) and those produced by recombination.[51]

It is thus scarcely surprising that once the empirical meaning of the concept of 'mutation' had been clarified, it was easily absorbed by the theory of natural selection. The origin of the profound reorganisation of the theory of selection that took place after 1900 can in fact be found in the specific rationality of Mendelism. The following section examines the fundamental conceptual effects of the Mendelian model of heredity on Darwinism.

8.2 The effects of Mendelism on the theory of selection

We have arrived at the decisive step. Mendelian genetics provided evolutionary biologists with a framework within which selection could take on a clear methodological meaning. It gave them a way of representing the mode of action of selection and of understanding its effectiveness in both natural and artificial situations. Although natural selection was successfully reconstructed on this new basis, it did not emerge unscathed. The following section analyses the development and configuration of the new form of selection.

Although this study is not a history of genetics, Mendel has to occupy an important place in our story. We will first deal with the thorny question of the relation of Mendel to Mendelism. We will then analyse Mendelism's effect on Darwinian evolutionary theory.

8.2.1 What was Mendelism?

This is a complex question. Mendel was only retrospectively identified as having discovered the physiological theory of heredity. In 1900–3, the pioneers of genetics were able to bring their theoretical premises into focus by reading Mendel's 1865 paper on plant hybridisation. The early phases of Mendelism raise a problem that is more or less the opposite of the one that characterises the history of Darwinism. The principle of natural selection has a name and an author (no doubt about that), and it is quite clear what it is supposed to explain. If the principle of selection presents a problem, it is at the level of its operational meaning and its empirical reality.

In the case of Mendelism, things are exactly the opposite. Mendel's operational schema was wholly clear and reproducible and, right from the outset, had a phenomenological impact. But it is hard to give it a name. Mendel himself spoke of a 'law' which he called the 'law of the formation and development of hybrids'.[52] In this law, which had an algebraic form, the essential conceptual element is not clear – is it the disjunction of gametic types, the hypothesis of 'dominance' or the independent segregation of characters? It is no accident that textbooks generally refer to 'Mendel's *laws*', although Mendel only ever spoke of *a* law. After an initial hesitation, the first Mendelians rapidly agreed on the 'law of gametic purity', called by Bateson the 'essential part of the discovery [of Mendel]'[53] and by Castle 'Mendel's law of heredity'.[54] However, while this law exists operationally in Mendel's writings, it is not at all clear that it had the same conceptual significance for Mendel as for the Mendelians, nor that it was intended to explain the same thing. The history of Mendelism thus confronts us with a genuine problem of anticipation. Were Mendelian principles really 'rediscovered', as was argued by all those who laid down the foundations of genetics in the 1900s, and if so, in what sense?

As has already been emphasised, this section deals with Mendelism's central proposition, because that enables us to understand the renewal of Darwinism. We will first analyse the form this proposition took in the writings of the first Mendelians, in other words those who raised 'Mendel's law' to the status of a fundamental law of the science of heredity. We will then see to what extent it did indeed exist in Mendel's own writings.

8.2.1.1 Mendelism and Mendelian genetics. The history of the 'rediscovery' of Mendel's principles is well known.[55] In March 1900, Hugo de Vries sent a manuscript in German on the 'law of hybrid segregation' to the German Botanical Society.[56] The paper was published on 25 April 1900 and makes explicit reference to Mendel. De Vries had made a French summary, which was published a month earlier, on 26 March 1900, in *Comptes Rendus Hebdomadaires des Séances de l'Académie des Sciences*.[57] This text, entitled 'Sur la loi de disjonction des hybrides' (On the law of hybrid disjunction), did not refer to Mendel. On 22 April, the botanist Carl Correns (1863–1933), having learnt of de Vries' French summary, rapidly wrote a summary of his own work on hybridisation, underlining the parallels of his and de Vries' findings with those of Mendel, and sent the article to the German Botanical Society under the title 'G. Mendel's Regel über das Verhalten der Nachkommenschaft der Rassenbastarde' (G. Mendel's law concerning the behaviour of varietal hybrids). This paper appeared in May 1900.[58] Finally, in June, a young Austrian lecturer called Erich von Tschermak (1871–1962) sent, again to the German Botanical Society, an article on artificial crossing of *Pisum sativum*, which contains results that are comparable to those found by

Mendel using the same material.[59] The article – a summary of a doctoral dissertation that had been presented in January 1900 – was published in July, Tschermak having previously published the whole of his thesis in May 1900.

So in the same year, in the same journal, three botanists 'rediscovered', or claimed to have rediscovered, the Mendelian interpretation of hybridisation. Much has been written about their independent intellectual itineraries, and indeed it does seem probable that they each learnt separately of Mendel's work. On the other hand, the coincidence of publication, or rather the succession of publications, in the same German journal, was probably not a chance event. The fact that each of the botanists put so much emphasis on Mendel seems to have been part of an attempt to limit the impact of the work of the other two – the term 'rediscovery' was not coined by historians, but by Correns himself, who, in his 1900 note, explained not only that he had 'come upon the same results as de Vries' on peas,[60] but had also 'given exactly the same explanation' as that furnished by the abbé Gregor Mendel in 1866.[61]

Two points with regard to the content of the three papers in which Mendel was 'rediscovered': the first is purely nominal, while the second deals with the nature of the 'rediscovery'. All three studies, like those of Mendel, dealt with hybridisation. Neither de Vries nor Correns used the word 'heredity'; both spoke of 'traits,' 'characters', 'factors' and 'predispositions' (*Anlagen*). Despite this vocabulary (obviously close to de Vries' 'pangens'), and the fact that they proposed the same law as Mendel, neither de Vries nor Correns suggested that a general theory of heredity could be constructed on the basis of their data. Tschermak did use the language of heredity, but, as Robert Olby has noted, the manner in which he expressed himself was typical of pre-Mendelian representations of heredity, and the claim to 'rediscovery' is highly contestable in his case.[62]

Thus the 'rediscovery' of Mendel in 1900 was not the foundation of a new science of heredity. Leaving aside the case of Tschermak, de Vries and Correns 'merely' rediscovered a way of explaining hybrid series. Certainly, they both stated that the Mendelian law of segregation was generally applicable, although in 1900 they differed on the extent – de Vries thought it universal,[63] while Correns vigorously disagreed.[64] But neither of them took the question any further than the crossing of hybrids. Their subsequent evolution confirms that neither of them thought that he had found a general law of heredity. Correns devoted most of his energy to studying cytoplasmic heredity, while de Vries, who did not even mention Mendel in his *Mutation Theory* (1901–3), ended up arguing that this law, which did not fit the data of his beloved *Oenothera*, was in fact an exception to the general rule of crossing.[65]

The second point raised by the 'rediscovery' of Mendel is: What exactly did de Vries and Correns 'rediscover'? They both spoke of Mendel's law, following Mendel's 1865 article in which he writes in general terms of a 'law of the formation and development of hybrids'. De Vries first called it 'la loi de disjonction des hybrides' (the law of disjunction of hybrids), then opted for the German term '*Spaltungsgesetz*' (law of splitting or division), which was translated into English as 'segregation'. For both de Vries and Correns, the fundamental postulate of this law was that the characters that differentiate hybrid forms can be analysed in terms of independent *pairs.*[66] Each pair is composed of 'antagonistic' characters, the antagonism signifying that while the two characters are both present in the organism (in this case a hybrid), they cannot coexist in the gametes, and thus must separate when the gametes are formed. In other words, the germ cells produced by hybrids are not themselves hybrid. De Vries explained this hypothesis very clearly: '*The pollen grains and ovules of monohybrids are not hybrids* but belong exclusively to one or the other of the two parental types.'[67] The law of segregation is synonymous with what William Bateson, a year later, called the law of 'gametic purity':

> The essential part of the discovery is the evidence that the germ cells or gametes produced by crossbred organisms in respect of given characters may be of the pure parental types and incapable of transmitting the opposite character . . . there may be, in short, perfect or almost perfect discontinuity between these germs in respect of one of each pair of opposite characters.[68]

This is the central proposition of Mendelism and forms the basis of the factorial combinations with which we are so familiar. This also explains why, on reading Mendel, de Vries, Correns or any didactic presentation of what was later to be called Mendelian genetics, one has the impression that all three authors were espousing exactly the same doctrine.

However, it is not enough to say that the central proposition of Mendelism was also that of the 'Mendelian science of heredity' (or, as it was called from 1905 onwards, genetics). While it was a necessary methodological ingredient, it is not sufficient to define 'genetics'. Robert Olby has emphasised that the central proposition of the Mendelian theory of *heredity*, or genetics (as against Mendel's own ideas), was that all the inherited characters of sexual organisms are determined by a finite number of units present in two copies in each individual which are able to take alternative forms ('alleles').[69] The central supposition of *genetics* is thus that all inherited characters are determined by one or more pairs of factors that are not necessarily identical. It is not surprising that there is no contemporary reference for this statement – it is a heuristic proposition, developed between 1901 and 1903, a kind of gamble upon which the

Mendelian science of heredity was built. The methodological and conceptual configuration that this proposition contained was that genetics was both a generalisation and an interpretation of the 'Mendelian' understanding of hybridisation.

1. *The generalisation of the Mendelian schema.* Mendelism was generalised from a schema of hybridisation into a schema of heredity. This conceptual transition has often been expressed by saying that what Mendel had conceived for interracial heredity, genetics extended to intra-racial heredity. As the number of verifications of the law of segregation increased, the first Mendelians rapidly arrived at the idea that the same individuals could be hybrids for one character and pure for another, in such a way that the old terminology of 'hybrid' and 'thoroughbred' became obsolete. For example, this change appears clearly in the following passage from Cuénot written in 1902:

> The word hybrid and its opposite, thoroughbred, now have a very precise meaning; first of all, these two terms only have a value with regard to a given character; in practice, one always specifies or implies that an individual is hybrid or thoroughbred for a given character; thus the son of a blond Frenchman and a brunette Italian woman is a hybrid for hair colour, but he is a thoroughbred from the point of view of the white race, to which both parents belong.[70]

In the same year, Bateson suggested a new vocabulary. He proposed the term 'allelomorph' to describe the character unit that could exist in the form of an antagonistic pair, and to use 'heterozygote' and 'homozygote' instead of 'hybrid' and 'pure type'.[71] Despite the existence of alternatives, these terms were rapidly adopted because they expressed a clear break with the racial representation of heredity. The main connotations of 'heterozygote' and 'homozygote' are cytological and not taxonomic. Furthermore, they directly evoke the most important aspect of Mendel's view of hybridisation – the idea that factors or predispositions (*Anlagen*) separate into their pure state in the gametes and combine in the fertilised egg.

2. *An interpretation of the Mendelian schema.* In itself, Mendel's understanding of segregation has an essentially formal significance. It does not imply that for a given character, all *individuals* have a pair of hereditary elements (irrespective of how these elements are conceived – germs, chemical particles or 'predispositions' of an unknown physiological nature). Segregation, which de Vries and Correns rediscovered in Mendel, merely implies that a *cross* should be described in terms of a pair of antagonistic characters. If the individual is hybrid, the two characters must coexist in that individual as two distinct elements, one of which will generally be 'latent' or 'recessive' and the other observable: 'In the

hybrid', wrote de Vries, 'the two antagonistic characters lie next to each other like dispositions (*Anlagen*).'[72] But neither de Vries nor Correns (nor before them, Mendel) said that an individual with a pure form of a given character carried two identical *Anlagen*. They did indeed argue that the zygote of a pure parental form is produced by two gametes that are identical for the given character, but nothing in these early texts suggests that once the zygote has been formed something like two particles remain in existence. In *Intracellular Pangenesis*, de Vries accepted that there could be a variable number of copies of the same pangen in the cells of different individuals. In other words, the law of segregation does not require *a priori* that two (identical or different) elements separate in the gametes and that two elements subsequently fuse in the zygote. It merely requires that *if the individual is a hybrid*, then the *Anlagen* act separately (and thus, in particular, that they do not fuse). The symbolism used by Mendel and by all the Mendelians in the period 1900–2 supports this interpretation. A pure form was not represented by 'AA' but merely by 'A', while a hybrid, by definition, was represented by 'Aa'. This symbolism reflects the constraint of the algorithm: de Vries, Correns, Cuénot and Bateson – following Mendel – initially speculated only on the frequency of 'pure' and 'hybrid' forms and never on that of the 'elements' or 'factors' (which today we would call 'genes').

The greatest difference between Mendelian genetics and Mendelism in general was the hypothesis that each hereditary character has a corresponding structure in each individual. It is hard to say exactly where and when this thesis came to the fore. To a certain extent, the notions of chromatic reduction (Weismann), of the continuity of chromosomal lines in somatic cells (Boveri) and of the localisation of the hereditary units in the nucleus (de Vries) had paved the way for the idea. But it was only in 1902 that the hypothesis of a strict correspondence between the cytological structure of the nucleus and the formal structure of Mendelian crosses was put forward by a twenty-five-year old American student, Walter Sutton (1877–1916). This well-known passage is worth rereading:

> I may finally call attention to the probability that the association of paternal and maternal chromosomes in pairs and their subsequent separation during the reducing division as indicated above may constitute the physical basis of the Mendelian law of heredity.[73]

Once this equivalence had been proposed, the Mendelian theory of heredity appeared in its fully recognisable form. The law of segregation becomes the statistical expression of a process of physical division and combination; the alleles become parts of chromosomes; heterozygosity *and* homozygosity become words for a real structural parity; finally the distinction between *germen* and *soma* can be interpreted in terms of the dissociation between a structure (the future genotype) and its variable manifestation (the phenotype). This explains why, after 1902, the Mendelians gradually modified their symbolic conventions, representing homozygotes by their germ structure (e.g. AA) rather than by their appearance (A).

This sheds some light on the historical form of turn-of-the-century Mendelism. Mendelism first appeared as a formal predictive schema for the descent of hybrids. In the early 1900s this was called the 'law of segregation', the 'law of gametic purity' or 'Mendel's law', and according to the elegant formulation of William Castle was based on the idea that the formation of gametes is the inverse of fertilisation.[74] If the gametes that unite to form the zygote are different, the organism produced by the zygote will in turn form different gametes. The simplicity and the elegance of this hypothesis flow from the fact that the number of hypotheses on the structure of the germ plasm is reduced to a minimum. In particular, the hypothesis did not require that for every character there were, in each individual, two determining factors. It simply required that hybrids, and only hybrids, be in one way or another 'double' in their germ-plasm line. This is what de Vries and Correns 'rediscovered' in 1900, without suggesting that it was generally applicable. They limited its application to 'true hybrids'[75] and to situations where one of the characters is completely dominant.[76] A mere two years later, not only did these restrictions appear unnecessary, but the Mendelian schema was presented as a central proposition of a general theory of heredity.

8.2.1.2 Undefinable Mendel: neither geneticist nor Darwinian. After this clarification of the historical meaning of Mendelism, it will be easier to understand some aspects of Mendel's work that are important for the present study. It has often been noted that the theoretical horizons of Mendel's original writings were greater than those of the first Mendelians. It is thus important to return to the sources of Mendelism in order to understand better its confrontation with Darwinism. The following section will deal with only two points: (1) What exactly did Mendel discover? (2) What was the explicit implication of that discovery for the transformation of species?

8.2.1.2.1 Mendel's law. The date and the title of Mendel's first paper are significant. His 'Versuche über Pflanzen-Hybriden' ('Experiments on vegetable hybrids') was dated 1865, the same year that Galton published his first article on heredity.[77] Although in this paper Mendel did not use this term (in German *Erblichkeit* or *Vererbung*), he did frequently use such terms as reversion, disposition (*Anlage*),[78] constancy of type, characters, factors, elements of germ and pollen cells. This vocabulary is characteristic of all those – horticulturists, breeders, physicians and cytologists – who were interested in what was beginning to be called 'heredity'. Mendel's correspondence with Nägeli shows that one of the main objectives of his 1865 paper was to abandon the vague terms with which hybridisers described the relationships between different members of hybrid lines. Evoking Gärtner's experiments on peas, for example, Mendel wrote: 'Statements like: "Some individuals showed closer resemblance to the maternal, others to the paternal type", or "the progeny had reverted to the type of the original maternal ancestor", etc., are too general, too vague, to furnish a basis for sound judgement.'[79] This shows that if Mendel did not have the ambition of constructing a general theory of 'heredity', he at least sought to clarify the concept in the case of hybridisation.

This said, what did Mendel in fact discover? We are all taught that Mendel formulated several laws, their number and names varying according to the textbook. Classically, there are said to be three laws: 'the law of uniformity of the first generation', 'the law of segregation' (or of 'gametic purity' or 'disjunction') and 'the law of independent segregation of characters' (or 'the law of independent assortment'), to which at the beginning of the century was (wrongly) added a 'law of dominance'. All these 'laws' have a close relation with the content of Mendel's 1865 paper; indeed, the first three correspond to things that Mendel said. As far as dominance is concerned, Mendel introduced the term (together with 'recessivity') but he did not formulate a universal law of dominance.[80]

This traditional textbook presentation does not do justice to Mendel's theoretical ambition. He did not have in mind several *laws*, but *a* law. More precisely, a 'generally applicable law of the formation and development of hybrids'.[81]

The important term in this formulation is 'development'. Mendel had elaborated a law of serial development in the mathematical sense of the term, based on the strict and simple principles of combinatory analysis. Formally speaking, this law states that any hybrid for a given character produces an offspring distributed according to definite proportions. If the pure parental forms are A and a, and the hybrid Aa, the offspring of the hybrid will be distributed according to the ratio 1A : 2Aa : 1a. (These symbolic conventions are those used by Mendel himself.) The law supposes the existence of a 'pair of differing traits' and permits, under certain restrictive hypotheses (in particular that of the constancy of such traits),[82] the deduction of the 'series' of distributions over an infinite number of generations. Given that Mendel reasoned on the basis of the simplest possible system of reproduction (self-fertilisation), the iterative procedure producing the series was relatively simple – in each generation the pure types (A and a) reproduce according to type, and the remaining hybrids split according to the formula 1A:2Aa:1a. Figure 28 reproduces the table of offspring ratios that Mendel established on the basis of this hypothesis.[83]

Mendel extended this law to several independent characters – explicitly for three characters, and implicitly for *n* characters, merely indicating the number of constant forms in the hybrid offspring:

> If *n* designates the number of characteristic differences in the two parental plants, then 3^n is the number of terms in the combination series, 4^n the number of individuals that belong to the series, and 2^n the number of combinations that remain constant. For instance, when the parental types differ in four traits the series contains $3^4 = 81$ terms, $4^4 = 256$ individuals, and $2^4 = 16$ constant forms; stated differently, among each 256 offspring of hybrids there are 81 different combinations, 16 of which are constant. [In modern terms, the 'constant combinations' are the homozygous crosses. It should be remembered that this is a self-fertilising system][84]

Throughout the paper, Mendel frequently uses the expression 'combination series' and presents his 'law of development' (of hybrids) as a 'law of combination of differing traits'.[85] J. Jindra has drawn attention to the probable source of Mendel's idea of ordering the offspring of hybrids according to an algebraic series. Mendel had taken Ettinghausen's course at the Physical Institute of Vienna – Ettinghausen was the author of *Combinatory Analysis* (*Die Combinatorische Analyse*, 1826). It appears that Mendel's series were a direct application of a method of serial development presented by Ettinghausen under the title 'variation'.[86]

What was the epistemological status of the algebraic law of development of hybrids? Once again, Mendel's 1865 paper is quite explicit. In the first part,[87] Mendel proceeds phenomenologically, reasoning on the basis of traits, and showing that the results of his crossing experiments could reasonably be interpreted in the formal language of combinatory analysis. At this stage of his study, no biological hypothesis is advanced, neither about the physical significance of the 'traits' nor about the process underlying the formal combination. In particular, there are no references to 'elements', nor *Anlagen*, nor reproductive cells. The 'law of combination' is established independently of any physiological hypothesis, and is discussed initially only in the phenomenal language of the 'trait' (*Merkmal*). The insistence with which Mendel used this word – 'visible index' or 'apparent character' – is clearly no accident.

In a letter to Nägeli dated 18 April 1867, Mendel emphasised the phenomenalist meaning of that part of the paper in which he established

Generation	A	*Aa*	*a*	Expressed in terms of ratios A : *Aa* : *a*
1	1	2	1	1 : 2 : 1
2	6	4	6	3 : 2 : 3
3	28	8	28	7 : 2 : 7
4	120	16	120	15 : 2 : 15
5	496	32	496	31 : 2 : 31
n				$2^n - 1 : 2 : 2^n - 1$

Figure 28 Translation of the table in which Mendel expresses the distribution of forms in the offspring of a hybrid for a pair of characters ('Versuche über Pflanzen-Hybriden', 1865; reproduced from C. Stern and E. R. Sherwood (eds.), *The Origin of Genetics: A Mendel Source Book*, 1966).

the algebraic law of the 'development' of hybrids. Nägeli had made the following remark: 'You should regard the numerical expressions as being only empirical, because they cannot be proved rational.'[88] Mendel replied as follows:

> My experiments with single traits all lead to the same result: that from the seeds of hybrids, plants are obtained half of which in turn carry the hybrid trait (A*a*), the other half, however, receive the parental traits A and *a* in equal amounts. Thus, on the average, among four plants two have the hybrid trait A*a*, one the parental trait A, and the other the parental trait *a*. Therefore 2A*a* + A + *a* or A + 2A*a* + *a* is *the empirical simple series* for two differing traits. *Likewise it is shown in an empirical manner* that, if two or three differing traits are combined in the hybrid, the series is a combination of two or three simple series. Up to this point I don't believe I can be accused of having left the realm of experimentation. [Added emphasis][89]

The empirical part of the paper is followed by a typically explanatory part, which Mendel entitled 'the reproductive cells of hybrids'.[90] The explanation involves interpreting the 'combinations of traits' as being the result of the 'association' (or 'union') of reproductive cells, each of which exclusively carries one 'trait' or the other.[91] This explains why Mendel speaks of the 'internal make-up of the individual forms',[92] the 'identical factors ... acting together in the production of constant forms',[93] the reproductive cells' 'potential for creating identical individuals'.[94] Mendel did not merely suggest this interpretation in the language of cellular theory, he also 'tested' it by crosses between hybrids and parental forms (what today would be called a 'backcross').[95] Then, and only then, did he present a schema showing the correspondence between the 'combination series' of the 'traits' and the union of the gametes during fertilisation.

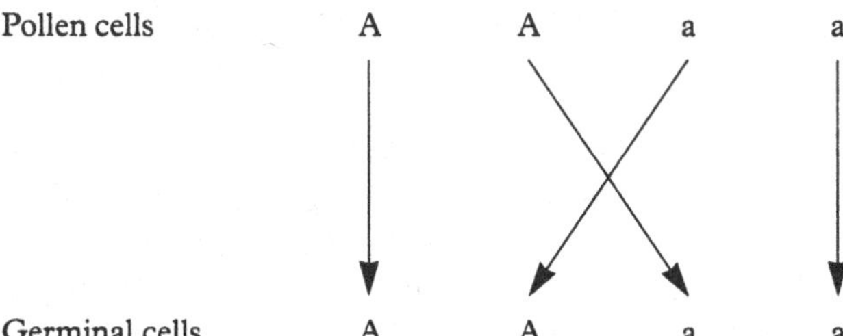

This needs to be examined in detail. In the following extract, all the symbols are those in Mendel's original text, and all emphasis is Mendel's. 'Germinal cell' means 'egg cell':

> According to the laws of probability, in an average of many cases it will always happen that every pollen form A and *a* will unite equally often with every germinal-cell form A and *a*; therefore, in fertilization, one of the two pollen cells A will meet a germinal cell A, the other a germinal cell *a*, and equally, one pollen cell *a* will become associated with a germinal cell A, the other with *a*.
>
> The result of fertilisation can be visualized by writing the designations for associated germinal cells and pollen cells in the form of fractions, pollen cells above the line, germinal cells below. In the case under discussion one obtains:
>
> $$\frac{A}{A} + \frac{A}{a} + \frac{a}{A} + \frac{a}{a}$$
>
> In the first and fourth terms germinal and pollen cells are alike; therefore the products of their association must be constant, namely A and *a*; in the second and the third, however, a union of the two differing parental traits takes place again, therefore the forms arising from such fertilisations are absolutely identical with the hybrid from which they derive. *Thus, repeated hybridization takes place.* The striking phenomenon, that hybrids are able to produce, in addition to the two parental types, progeny that resemble themselves is thus explained: $\frac{A}{a}$ and $\frac{a}{A}$ both give the same association, A*a*, since, as mentioned earlier, it makes no difference to the consequence of fertilisation which of the two traits belongs to the pollen and which to the germinal cell. Therefore
>
> $$\frac{A}{A} + \frac{A}{a} + \frac{a}{A} + \frac{a}{a} = A + 2Aa + a$$[96]

This extract shows the similarity between Mendel's view and that which was to impose itself in genetics. It is easy to understand why whole generations of geneticists and historians have believed that the fundamental doctrine of classical genetics (the idea that all characters in an individual are determined by two hereditary elements) was present in Mendel's work.

Nevertheless, this belief is most probably mistaken. The conceptual relation between Mendel's schema and genetics can be summarised in three points:

1. Despite the fact that he did not use the term, Mendel originated the concept of 'gametic purity' – the idea that in a hybrid, the reproductive cells retain and transmit only one of the two traits for a given character. The main implication of this concept is that the offspring does not express the 'weight of its ancestry'.[97] If, for instance, two gametes which are similar for a given character associate, the zygote will be completely free of the hybrid 'taint' of the parents. As Bateson rightly argued, this is the essential theoretical point that Mendel discovered.[98] Mendel was clear that the purity of the traits within the reproductive cells was intrinsically linked to the twin postulates of the constancy and segregation of traits:

> The course of development [of a hybrid series] consists simply in this; that in each generation the two parental traits appear, separated and unchanged, and there is nothing to indicate that one of them has either inherited or taken over anything from the other.[99]

2. Mendel did not clearly anticipate the distinction between 'genotype' and 'phenotype'. This can be seen in the symbolism reproduced above. Mendel sums up his explanation of the empirically observed law of development in the following formula:

$$\frac{A}{A}+\frac{A}{a}+\frac{a}{A}+\frac{a}{a}=A+2Aa+a$$

If we accept for a moment that the left-hand side of the formula is a 'genotype', as has classically been argued, what is the right-hand side of the formula? It is not a 'phenotype' because, under the conditions of perfect dominance that Mendel adopts throughout his paper, the classes are in fact 3A + *a*.

The only possible interpretation is that the left-hand side of the formula gives the structure of the possible pairs of gametes, while the right-hand side gives the constitution of the zygotes. This constitution takes into account only the type of elements involved in trait determination, and not their number. Mendel's model could thus imply, among other things, that two identical *Anlagen* fuse into one (or behave as one).

3. Mendel did not put forward the doctrine that each character of a sexual organism corresponds to one (or more) pair(s) of hereditary determinants. There is no proof that Mendel thought that observable characters were determined by two permanent and potentially identical hereditary units. This has been shown by J. Heimans and Robert Olby,

who put forward two decisive arguments on this point. Heimans[100] drew attention to a curious feature of some of Mendel's unpublished manuscripts, which show that Mendel, reasoning on the basis of three mutually exclusive traits in the gametes, believed that *all three* could coexist in the same hybrid. In modern terms, this would mean that an individual could possess three alleles. This severely weakens the proposition that Mendel had a bifactorial concept of inherited characters.

Mendel did not argue that the zygote always possessed two copies of each factor, whether or not these two copies are different. He merely affirmed that in hybrid organisms, the 'compromise' between the 'antagonistic elements' was temporary and ended when the reproductive cells were formed.[101] As Robert Olby has pointed out,[102] there is a passage in Mendel's 1865 paper that suggests that the bifactorial conception of hereditary constitution is not only absent, but is in fact incompatible with Mendel's model. This passage is extremely interesting, because it is the only point at which Mendel speculates on what might happen to the 'elements' that determine the 'traits' during the organism's life cycle. The translation is by Robert Olby:

> In the formation of these cells all the elements present participate in a perfectly free and equal arrangement, *whereby only the differing elements are mutually exclusive* [wobei nur die Differierenden sich gegenseitig ausschliessen]. [Added emphasis][103]

This passage indicates at least one thing: Mendel did not think that the separation of identical elements was necessary, but only that different elements had to separate. This passage also suggests that, for each character, each reproductive cell contains more than two 'elements' or 'factors'. If the organism is not hybrid for a given character, two identical 'elements' do not necessarily separate; they *may* stay together during the process of the formation of two gametes from a previous cell.

Whatever the case, for Mendel, as for the first Mendelians, the understanding of hybridisation is based on the notion of pairs of *distinctive traits*, not on the notion of heredity as a *bifactorial determination* of observable characters. Mendelian combination does indeed require a principle of duality, but in its physical aspect this duality is limited to the reproductive phase – to that point when an observer of the 1860s could have apprehended the organism as a 'double'. The segregation postulated by Mendel concerns only the typical qualities carried by the gametes, and not the pairs of particles.

This difference between Mendel's Mendelism and that of genetics does not alter the predictive power of the law of segregation. Had the bifactorial hypothesis and its chromosomal interpretation turned out to be false,

Mendel's schema would not have been affected. It could still have laid the basis of a formal, empirically testable, science of heredity. Perhaps Mendel's genius lay in the fact that, unlike many others, he refrained from speculating about the structure of the germ plasm, and merely used the available cytological data.

If it is difficult to see in Mendel's actual writings anything other than the bifactorial conception of heredity, this is doubtless because of the power of the chromosomal imagery that we retrospectively project onto the texts. The chromosomal theory of heredity postulates that hereditary characters are carried by two continuous lines of filaments; once we have this physical model in mind, it is very difficult to imagine that Mendelian combination was not defined in terms of pairs of *things* with an individual existence in time and space. But Mendel did not have such an image. His vision of the germ plasm was certainly particulate, like that of Darwin and Galton, or, later, of de Vries. But for Mendel, as for Darwin, Galton and de Vries, *the number of particles of the same type* was not important. The law of combination was defined on the basis of types, not of individual particles.

8.2.1.2.2 Mendel and the transformation of species. What was the implication for evolution of Mendel's discovery? This is in fact quite explicit in his writings.

Mendel did not merely evoke 'evolution' and the 'transformation of species' – they were the context in which his experiments on hybridisation took on their full meaning. The proof can be found in the introduction and conclusions to Mendel's 1865 paper. In the introduction, Mendel uses the following terms to describe his ambition of finding a general law of development of hybrids: 'It requires a good deal of courage indeed to undertake such a far-reaching task; however, this seems to be the one correct way of finally reaching the solution to a question whose significance for the evolutionary history of organic forms must not be underestimated.'[104]

This opening declaration says nothing about how the problem of hybridisation affects the theory of evolution. Mendel explains this point in the conclusion, where he discusses at some length the results of hybridisers (in particular Gärtner) on the '*transformation of one species into another by artificial fertilisation*' (original emphasis).[105] The classic form of such experiments was simplicity itself – species A was crossed with species B, then the resultant hybrid was crossed with species B, then the descendants closest to B were crossed with species B, and so on, in the hope of obtaining individuals with an offspring that would consistently show type B, without any reversion to type A. In reality, these experiments were somewhat disappointing, because the hybridiser always found a tendency to reversion, however many crosses had been made with species B.

Mendel states that 'if . . . the development of forms proceeded in these experiments in a manner similar to that of *Pisum*, then the entire process of transformation would have a rather simple explanation'.[106] From the first hybrid generation onward, a definable proportion of individuals will be strictly identical to A or B. This proportion depends on the number of characters by which the two species differ. If they only differ by one character, ¼ of individuals will be pure type B (and the same for A). For two characters, the proportion will be $1/16$; for three, $1/64$; for n, $1/2^n$. This fraction corresponds to the individuals produced by the pollen and germ cells which would be type B for all characters.

Mendel's 'combination series' also enabled him to explain why the results of experiments on complete and definitive 'transformation' were so disappointing. In such experiments, the two species differ for many and often for very many characters. This implies that the proportion of favourable combinations will be very low. For example, for seven different traits – the number of characters that Mendel studied in peas – the probability of the reappearance of a pure parental form in the offspring of a hybrid is 1 in 16,384. As Mendel remarked with regard to this example, if the population is limited, there is a strong chance that the desired form will not appear.[107]

Mendel developed this point further, arguing that his combinatory model explains the extraordinary variability of cultivated plants, which is not only or not even principally due to 'changes in conditions':

> . . . nothing justifies the assumption that the tendency to form varieties is so extraordinarily increased that species soon lose all stability and their progeny diverge into an infinite number of extremely variable forms. If the change in living conditions were the sole cause of variability one could expect that those cultivated plants that have been grown through centuries under almost identical conditions should have regained stability. This is not known to be the case, for it is precisely among them that not only the most different but also the most variable forms are found . . . It remains more than probable that a factor that so far has received little attention is involved in the variability of cultivated plants. Various experiences force us to accept the opinion that our cultivated plants, with few exceptions, are *members of different hybrid series* whose development along regular lines is altered and retarded by frequent intra-specific crosses.[108]

Given Darwin's perplexity when faced with the great variability of domestic species, this passage is astonishing.

Mendel did not restrict his discussion of the origin of variability to domestic species – he clearly stated that the laws of hybridisation are probably the same for all plants, and probably for all organisms.[109]

Furthermore, in the final lines of his paper, the allusion to a 'continuous evolution of plant forms'[110] suggests that while Mendel did not intend to construct a theory of evolution, he nevertheless clearly saw the consequences of his ideas for such a theory. In 1865, six years after the publication of the *Origin of Species*, a hypothesis on the possible origin of variability was inevitably of major importance, given that Darwin had admitted his complete ignorance of the question.

While Mendel's monograph was clearly centred on evolutionary preoccupations, it is nevertheless difficult to discern Mendel's view of his relationship with Darwin. At the beginning of the 20th century, some geneticists (e.g. Bateson) argued that Mendel carried out his experiments on peas against the background of a disagreement with Darwin's evolutionary ideas.[111] This makes no sense for purely chronological reasons: Mendel began his experiments on peas in 1857, two years before the publication of the *Origin of Species*. Furthermore, from a conceptual point of view, the 'disagreement' mentioned by Bateson is contradicted by Mendel's own testimony. In one of his letters to Nägeli, he underlines an interesting point of theoretical convergence between his own hypothesis of the independence of traits in hybrid combinations and the comparable Darwinian idea of the independent variation of characters:

> Darwin and Virchow have pointed to the high degree of independence that is typical for individual characters and whole groups of characters in animals and plants. The behaviour of plant hybrids indisputably furnishes an important proof of the correctness of this point of view.[112]

It is significant that in his only direct allusion to Darwin, Mendel mentioned his hypothesis of the discontinuity of characters as a 'proof' of one of Darwin's crucial doctrines (selection operates on more or less independent characters).

It would, however, be unreasonable to push the parallel too far. It is virtually certain that Darwin did not know of Mendel's work.[113] Furthermore, it is probable that had he read it and understood it, he would have been very suspicious of a doctrine that he had firmly condemned in principle in *The Variation of Animals and Plants* (1868) – the doctrine that assigned the origin of variability to crossing, and only to crossing. Such a doctrine, he explained, implies that all new variation is an effect of reversion.[114] From Mendel's point of view, this doctrine was not wholly satisfying. Mendel seems to have oscillated between two theoretical positions – on the one hand, the old Linnean and fixist idea that hybrids delimit the ultimate point of possible transformation between species (which has as a corollary that there is no real transformation of species); on the other hand, the idea that the possible number of combinations of characters might be so great that there would be, in practice, a sufficient source of variation for a continuous and indefinite evolution.

This as about as far as the study Mendel's vague evolutionary vision can go. What were the conceptual effects of Mendelism on the theory of selection in the period 1900–20?

8.2.2 *From Mendelian genetics to Darwinism: the new shape of selection*

We now possess the necessary contextual and epistemological tools to interpret the most important event in the history of Darwinism: the Mendelian reconstruction of the principle of selection. The preceding pages focused on the complexity of the conceptual constellation that was called 'Mendelism' in the early 1900s. For contemporary biologists of the time, it was not at all clear that Mendelism was merely a 'theory of heredity'. For most of the pioneers of Mendelism, the new experimental science of heredity was closely associated with a new non-Darwinian and mutationist theory of evolution. But it was precisely the Mendelian science of heredity and mutation that turned out to provide the firmest experimental basis for Darwinian evolutionary theory.

The specific effect of the Mendelian theory of heredity on the theory of evolution and the unique coherence of the new paradigm of selection which developed in the first two decades of the 20th century are of decisive importance. Mendelian genetics had three major effects on the Darwinian theory of evolution:

- It clarified the relation between descent and heredity.
- It developed a vision of heredity that implied that populations were structurally stable from the standpoint of heredity. In other words, Mendelian genetics swept away the notion of heredity as a 'force' or a 'tendency' that varies in intensity.
- This inertial aspect of Mendelian heredity led to an understanding of the stability of the changes obtained by selection, and to a vision of the kinetic description of selection (i.e. a predictive approach to the rhythm of evolution with various factors altering gene frequencies – mutation, selection, sampling effect etc.).

These three theoretical effects were interlinked and marked the end of the obscure problematic of 'reversion' in evolutionary theory.

8.2.2.1 The clarification of the relation between descent and heredity. In 19th-century biology the concepts of descent and heredity were intertwined. This can be seen with particular clarity in Darwin, but he was by no means alone in this. This association between descent and heredity had its origins in the old representation of 'blood' that flows in 'lines'. After Darwin, the appearance of the notion of the continuity and independence of the germ line, far from dissolving this association, merely reinforced it. We have seen how for Galton (and for Weismann)[115] this doctrine went hand in hand with the concept of 'ancestral heredity' – in other words, with the idea that ancestry or descent was essential to the concept of heredity.

Mendelian methodology led to an unequivocal break with this view of heredity. The nature of this rupture can be precisely defined by confronting the starting points of the biometricians and of the Mendelians. The biometricians developed the most radical and rigorous conception of heredity as a 'weight' of ancestry. This point has already been dealt with at some length – Galton, Weldon and Pearson constructed a theory of heredity on the basis of the quantification of the resemblance between related individuals as a function of their ancestry. According to this theory, each ancestor contributes a certain part to the germinal constitution of an individual. The culminating point of the theory was thus a 'law of ancestral heredity'. As we have seen, in its most developed and respectable form (Pearson), this law consisted of a multiple-regression equation of the offspring on ancestors with varying degrees of distance. It is a purely phenomenal law, neither containing nor implying any hypothesis as to the mechanism of heredity. But it is impossible not to see this law as a declaration that heredity should be considered in the language of descent, of genealogy, of pedigree, of lineage.

The theoretical bias of the Mendelians was completely different. They were interested in the genetic structure of the individual, and more precisely in the composition of the zygote. Once established for the two parents, this structure determines the nature and the ratios of the offspring. If, for example, an 'aa' individual is crossed with another 'aa' individual, their offspring will be exclusively 'aa', whatever the 'ancestral heredity' of each parent. The hereditary factor is thus all that counts, not the origin of this factor; the 'determinant' counts, not the 'ancestry'. From this standpoint, pedigree is only an instrument of analysis and has no implication for heredity.

The methodological opposition between the biometrical and the Mendelian approaches can be formulated another way. Both schools made use of probabilistic and statistical methods, but these tools did not have the same meaning in the two cases. The biometricians emphasised the importance of statistics as a tool that made it possible to use the maximum amount of empirical information with the minimum number of causal hypotheses. This is the real meaning of the law of ancestral heredity: it renders the prediction of the character of the offspring more precise by integrating as much data as possible on the character of various ancestors. Thus in 1903, Pearson, following Galton, could still declare that 'a knowledge of the characters of the parents does not accurately define the character in the offspring'.[116] Similarly Weldon, criticising the premises of the first Mendelians, objected that they deprived themselves of a key predictive instrument, particularly in the case of characters showing continuous variation:

> The fundamental mistake which vitiates all work based upon Mendel's method is the neglect of ancestry, and the attempt to regard the whole effect upon offspring, produced by a particular parent, as due to the existence in the parent of particular structural characters ... Not only the parents themselves, but their race, that is their ancestry, must be taken into account before the result of pairing them can be predicted.[117]

According to this perspective, statistics were used to overcome the fact that there was no causal hypothesis as to the mechanism of heredity.

Mendelism proceeded by exactly the opposite method. Once the parental genotypes are known, the probability of the genotypes of the next generation is deduced. The probability is thus incorporated into the mechanism. On average, the Aa × Aa cross gives 1AA:2Aa:1aa, because each of the parents necessarily produces gametes in the proportion 1a:1A. The predictive schema is thoroughly causal and deterministic. Unlike Galton and Pearson's empirical laws of correlation, Mendel's law of segregation is not *statistically based on* experimental evidence but is rather *statistically confirmed by* experimental evidence.

This methodological opposition is explicitly expressed in the symbolic conventions used by the two schools. When the biometricians spoke of successive generations, they generally used the letter P (for parent) – P_n ... P_3, P_2, P_1 are the various ancestral generations which converge on an offspring F (filial generation). This symbolism shows that the biometricians worked backwards towards the ancestors (see Figure 29). The Mendelians, on the other hand, worked forwards to the offspring. That is why they had only one parental generation (P) and many filial generations (F_1, F_2, F_3...). The Mendelians rarely went any further than two or three generations, because their objective was to reconstruct the structure of the P generation by observing how it split or could be analysed in its offspring. These two positions can easily be represented schematically (see Figures 29 and 30).

To summarise, the Mendelians were not interested in the fate of individuals in a given line, but in the deduction of their genetic structure. Once this structure had been revealed, the genealogy of the individuals concerned was of no interest. The Mendelians rejected ancestrality ('descent' in Darwinian terms) as an element in the concept of heredity. For them, heredity was a question of structure, not of lineage.

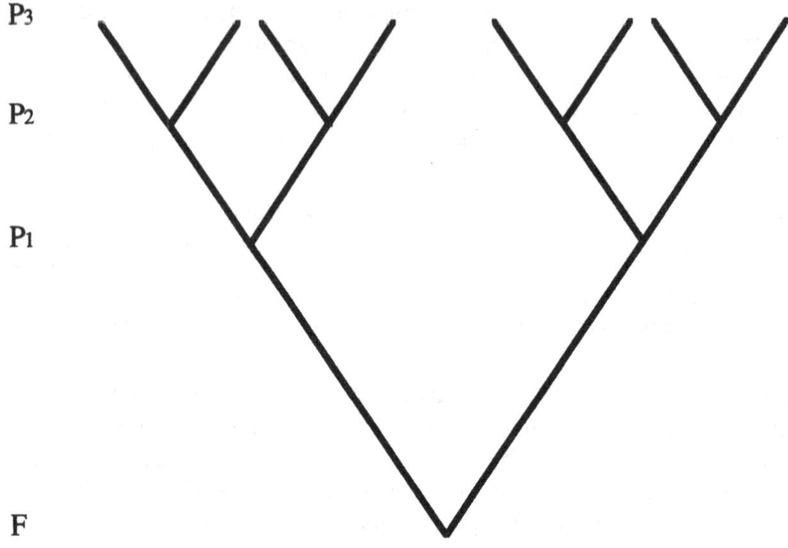

Figure 29 Biometrical representation of generations.

This is no doubt one of the reasons why Mendelism initially appeared to be violently opposed to Darwinism. By rejecting 'ancestral' heredity, Mendelism put itself forward as a theory of heredity freed from the theory of descent. However, this antagonism was to reveal itself as relatively superficial, and in reality the Mendelian vision of heredity supported Darwinian evolutionism, because the doctrine of gametic purity destroyed two interconnected beliefs that had continually undermined and confused the principle of selection.

The first of these beliefs, very widespread among breeders, was that the 'force of heredity' increased the longer the character had been transmitted in the race. Darwin had vigorously opposed this idea, but without any empirical proof apart from breeders' reports.[118]

The second belief was that races could always revert to previously possessed characters, and that they tended to do so spontaneously. This fear of 'regression' and the differing ways in which it was expressed in evolutionary theory towards the end of the 19th century have been dealt with in previous chapters.

To these two beliefs, Mendelism opposed the following position: if a character behaves according to the law of segregation (or the law of gametic 'purity', which amounts to the same thing), there is no reason to think of 'heredity' either in terms of a theorem about forces, or in the language of reversion. Heredity should not be viewed as a force, an energy or

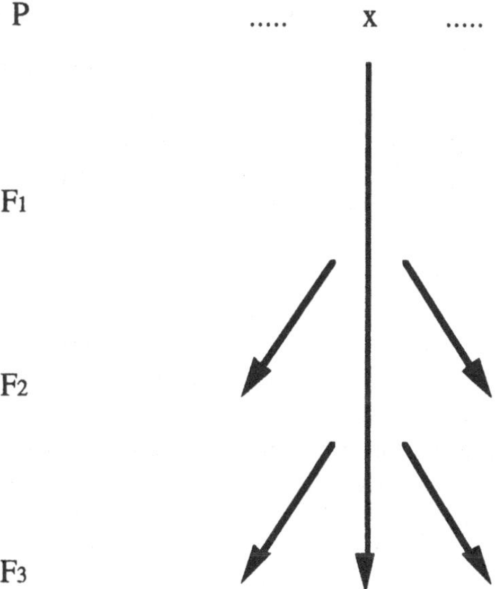

Figure 30 Mendelian representation of generations.

a tendency, but as a structure. If selection interacts with heredity, it is not as an antagonistic power, but as a force that can modify a structure.

However, this fundamental consideration only takes on its full meaning under two conditions: first, the properties of the structure of Mendelian heredity have to be studied at the population level, because selection modifies only the structure of a population, and not that of an individual zygote; secondly, and correlatively, there has to be a factor of change in the elements of structure (mutation). These two developments, which were crucial in the reconciliation of Darwinism and Mendelism, will be dealt with separately.

8.2.2.2 Panmictic equilibrium (the Hardy–Weinberg law, 1908). The year 1908 was the key year in the confrontation of Mendelism and Darwinism. That year, two researchers – Wilhelm Weinberg (1862–1937), a German obstetrician, and Godfrey Hardy (1877–1947), a brilliant English mathematician – independently established the law that was to be the starting point of population genetics.[119] Known as the 'Hardy–Weinberg law', or as 'the law of panmictic equilibrium', this law, which can be simply demonstrated, is a direct consequence of Mendel's law of segregation. It states that in a randomly mating population, the frequency distribution of Mendelian combinations for a given character (locus) is stable, whatever the frequencies of the Mendelian characters (genes). The law supposes that we can abstract from all evolutionary forces (deterministic or stochastic), such as selection, mutation or sampling error. This is, of course, an ideal situation which can never be realised, but it provides a model that can be used to evaluate the effects of various evolutionary factors. The fundamental theoretical interest of the Hardy–Weinberg law is that it isolates the simple effect of Mendelian heredity in a randomly mating sexual population.

If there is something surprising about the discovery of the Hardy–Weinberg equilibrium, it is that it did not happen earlier. Mendel could easily have derived it, as could several geneticists and mathematicians who studied the law of hybrid segregation and its implications at the population level. If it took a mathematician and a physician, neither of whom was an evolutionist, to discover the law, that is probably because evolutionary biologists' theoretical attention was drawn elsewhere. The antecedents of the Hardy–Weinberg law are thus particularly interesting.

Mendel's work itself contains the first step towards the Hardy–Weinberg equilibrium, in the shape of a reservation he made about the classic doctrine of reversion held by hybridisers. In 1865, Mendel showed, on the basis of his law, that the destiny of a mono-hybrid population (that is, hybrid for one pair of characters) could be calculated for as many generations as one wanted. The table summing up this calculation is reproduced above (see Figure 28). The renewed splitting of A*a* hybrids into A + 2A*a* + *a* offspring leads to a progressive reduction in the proportion of A*a* hybrids. After *n* generations, the population has the following structure: $(2^n - 1)\mathrm{A} : 2\mathrm{A}a : (2^n - 1)a$. Mendel discussed this result by saying that his law confirmed the idea of Gärtner, Kölreuter and many other hybridisers who argued that 'hybrids have a tendency to return to the parental forms'.[120] Mendel thought this remark sufficiently important to repeat in the conclusion to his paper.[121] He thus had in mind a law that would be able to describe rigorously the causes and the rhythm of the 'reversion' of hybrid forms to pure types – a law showing why the hybrid form is unstable, and why reversion to pure type is a necessity. Seen from this angle, the Mendelian 'law of development of hybrids' parallels the evolutionary thematic of reversion and atavism.

It could be objected that Mendel's law of disjunction was not a law of reversion because it was based on data from self-pollinating plants (only the first generation was produced by an artificially fertilised cross). If Mendel had extrapolated his calculations to a situation of panmixia, using simple mathematics, he could have noticed that the population would remain constant. And yet Mendel did not envisage this possibility.

There is one aspect of Mendel's research that illustrates his theoretical interest in a law of reversion. After his experiment on peas, Mendel looked for another plant that would verify his law of development of hybrids. He rapidly found a certain number, but he also found some species, such as *Hieracium* (hawkweed), that stubbornly resisted all attempts at combinatory analysis. *Hieracium* shows a remarkable constancy of hybrid forms. We now know that this plant is occasionally apomictic (characterised by reproduction without fusion of the gametes). Mendel, of course, did not know this, and his persistence in studying *Hieracium* after the publication of his 1865 paper[122] verged on the tragic. Nägeli's role in this episode has often been highlighted: he encouraged Mendel to work on *Hieracium* with the obvious intention of convincing him that his laws of hybridisation were not as universal as he claimed. This is undeniably an aspect of the story, but Mendel's misadventures with *Hieracium* also reveal his own theoretical intentions. His 1869 paper contains the following telling lines:

> The question of the origin of the numerous and constant intermediate forms has recently acquired no small interest since a famous *Hieracium* specialist has, in the spirit of the Darwinian teaching, defended the view that these forms are to be regarded as [arising] from the transmutation of lost or still-existing species.[123]

This suggests that by sticking to *Hieracium*, which presented a series of key problems, Mendel hoped to show that what the 'Darwinians' called the transmutation of species could be reduced, in one way or another, to the combination of hybrids, and perhaps thus to a dynamic of reversion.

However, this is mere speculation. The important point is that in his 1865 paper, Mendel had no sooner indicated that his law explained the reversion of hybrids towards parental types than he made an important reservation – although hybrids reduce in proportion in each generation, 'they can never disappear entirely'.[124] In theory at least, any hybrid in a population gives rise to both pure and hybrid offspring. As a result, even in the limited conditions of self-pollination, Mendelian rationality suggests that, in principle, a residual variability will persist indefinitely. This is the conceptual root of the Hardy–Weinberg law.

After the rediscovery of Mendel's law in 1900, biologists rapidly sought to predict its consequences for a panmictic population. In 1902 the statistician G. Udny Yule was the first clearly to pose the question: 'what, exactly, happens if the two races *A* and *a* are left to themselves to intercross freely *as if they were one race*?'[125] Yule shows quite simply that if the two initial populations 'A' and 'a' are of the same size, the random crossing of individuals will indefinitely reproduce the same population structure, generation after generation (that is, 1A: 2Aa: 1a). Yule reasoned on the basis of the total dominance of one character, which implies that the population remains in a steady state, with a 3 : 1 ratio of dominants : recessives. But he did not consider what would happen if 'A' and 'a' were initially in unequal proportions.

The same observation applies to Pearson's study of Mendelian heredity, published in 1903. Using more sophisticated formal techniques than Yule, and reasoning independently of the hypothesis of dominance, Pearson showed that the stability of the population in panmixia was a necessary consequence of the Mendelian law of segregation: 'with random cross-fertilisation there is no disappearance of any class whatever in the offspring of the hybrids, but each class continues to be reproduced in the same proportions'.[126] Retrospectively, this statement seems so clear that Pearson has sometimes been credited with the discovery of the Hardy–Weinberg law. However, Pearson – like Yule – reasoned on the basis of the highly specific case in which the two Mendelian factors were initially present in strictly equal frequencies, and did not generalise the case to other allelic frequencies.

The circumstances in which the mathematician Hardy came to formulate the law that bears his name are particularly interesting. Hardy reports that in 1908 the geneticist R. C. Punnett asked him to solve the problem of the evolution of Mendelian proportions in a free-crossing population. Punnett had his doubts about Yule's statement that any dominant Mendelian factor introduced into a population would establish itself at a frequency such that the ratio of (phenotypically) dominant forms to recessive forms would be 3 : 1. If this assertion were true, argued Punnett, it would signify, for example, that brachydactyly, which is a dominant character in humans, should, over time and in the absence of countervailing forces (selection), come to affect three-quarters of the human population.[127] If there was any doubt about the matter, this anecdote confirms that in 1902–3 neither Yule nor Pearson had set out the law of panmictic equilibrium.

Hardy did not have much difficulty in solving the problem. In a few lines, and simply using 'the mathematics of the multiplication-table', he showed that whatever the initial frequencies of the forms AA, Aa and aa, a stable equilibrium should be established from the second generation.[128] Applying his model to the case of a single AA homozygote introduced into a homogeneous aa population, he showed that a theoretical equilibrium is immediately established, which will indefinitely maintain the new character in constant proportions, irrespective of whether the character is dominant or recessive. Hence the following unambiguous declaration: 'In a word, there is not the slightest foundation for the idea that a dominant character should show a tendency to spread over a whole population, or that a recessive should tend to die out.'[129]

In other words, the probabilistic genetic structure of a panmictic population is constant as long as the only factors affecting that population are those of Mendelian inheritance. The same law had been discovered a few months earlier by the German physician Wilhelm Weinberg, but this co-discovery went unnoticed for many years.[130] It should also be noted that in 1903 the American geneticist W. E. Castle had also arrived, empirically, at the conclusion that the Mendelian structure of a polymorphic race is stable in the absence of altering factors such as selection.[131]

The fundamental theoretical importance of the Hardy–Weinberg law for the study of evolution cannot be overestimated. This law of equilibrium sounded the death-knell of the thematic of reversion, in whatever form. Once the principle of panmictic equilibrium had been established, and under the condition that all inherited transmission ultimately obeys a Mendelian logic, heredity could no longer be viewed either as a force of reversion to 'type' (e.g. Galton) or as a principle incapable, on its own, of conserving the gains of progressive evolution (e.g. Weismann). The Hardy–Weinberg law provided the causal science of evolution with something like the equivalent of the principle of inertia. It implied that any panmictic Mendelian population, as long as it is not subject to the action of an evolutionary force, conserves the same genetic structure indefinitely. It also implied that if a given factor (selection, mutation, random variations) changed the structure of the population, the effect of the perturbation would be maintained once the cause of the change had disappeared. As Hardy rightly noted in his 1908 article, the 'stability' of Mendelian proportions should not be understood as a law of constancy, but as a characteristic of a structure that conserves modifications: 'The sense in which the distribution $p_1:2q_1:r_1$ is "stable" is this, that if we allow for the effect of causal deviations in any subsequent generation, we should, according to theory, obtain in the next generation a new "stable" distribution differing but slightly from the original distribution.'[132]

The Hardy–Weinberg law looks like a principle of inertia that could be the starting point for a mechanics of evolution. Of course, it is an idealised simplification – in practice, all populations are subject to one evolutionary force or another. Even if it is admitted that a population is completely isolated and protected from any selective or mutational process, its genetic structure would only be truly constant if there were no random variation. This condition supposes that the population size should be infinitely large, which obviously does not hold in reality. Furthermore, the Hardy–Weinberg law has meaning only for sexually reproducing species, where reproduction involves meiosis. The principle of panmictic equilibrium thus applies to only a limited part of the history of life. Nevertheless, these restrictions are not fundamentally different from those of the physical sciences, whose principles are also simplifying abstractions that are valid only under certain spatio-temporal conditions.

Once the principle of panmictic equilibrium is clear, it is easy to understand that evolutionary theory had to make a series of conceptual leaps, even though, of course, these changes did not occur overnight. They can be summed up as follows:

1. The model of panmictic equilibrium put an end both to quarrels over the primacy of this or that 'factor of evolution' and to arguments about the priority of the 'principle of heredity' over the 'principle of evolution' (or vice versa). The Hardy–Weinberg law provided a reference model that made possible the evaluation, quantification and detection of the effects of various 'factors of evolution'. Mutation and selection were thus soon to appear no longer as alternative explanations, but as forces that interacted in the dynamic of genetic frequencies.

2. The model further put a stop to the wave of confused speculation about 'panmixia' and 'regression'. According to the particulate and combinatory logic of Mendelian genetics, panmixia no longer denoted regression but represented the ideal stationary state of a sexually reproducing population unaffected by any evolutionary force. In the Mendelian lottery, panmixia (or random mating) was formally equivalent to the random fusion of gametes, and thus to the random formation of pairs of alleles. The panmictic model[133] (which it would be more correct to call 'pangamic') incorporates the idea of a Mendelian population as a 'gene pool'[134] to which all individuals contribute equally. Each 'factor of evolution' can thus be defined by its effect on the homogeneous field of genetic and genotypic frequencies, as measured by the 'deviation from panmixia' that it produces. This situation is analogous to the heterogeneous forces of physics which express themselves in the homogeneous language of deviations from an ideal inertial trajectory.

3. Unlike the biometrical interpretation of evolution, which was wholly turned towards the deduction of the present on the basis of the 'ancestral past', the statistics of population genetics were aimed at providing a prediction of the future of a population on the basis of its initial state. This did not involve a passive reinterpretation of data, but rather the calculation of the kinetics that predicted the rhythm of change under different hypotheses. From the outset this approach was genuinely experimental, because it was based on causal hypotheses and was open to correction by experience.

8.2.2.3 The new shape of selection. In its Mendelised version, the theory of selection has two fundamental aspects:

- It takes the form of a kinetic prediction of the speed with which a character spreads or regresses in a population under the effect of selection.

- It is linked to a theory of mutation, because if selection is merely the diffusion of a new allele in a population, the new allele is not itself created by gradual selection.

These aspects of the new theory of selection were successively constructed during the first two decades of the 20th century.

The idea of a predictive kinetic description of selection was not new. In his criticisms of Darwin, Fleeming Jenkin concluded that there was at least one theoretical case in which the effectiveness of selection was clear – the hypothesis according to which selection acted on 'preponderantly' (dominantly) transmitted sports. In this case, if the variation is in the slightest bit advantageous, it should indeed invade the population and become fixed. Jenkin pointed out that the development of a mathematical theory of this hypothesis was complex but possible. This was more or less the situation in which the Mendelians found themselves in the early 1900s, but with two important differences. First, most of them, being heavily involved in the mutationist polemic against Darwinism, had simply not thought of reconstructing the concept of selection on this new basis. Second, the Mendelian view of heredity was substantially more subtle than that evoked by Jenkin. If characters are determined by pairs of factors, and if, furthermore, dominance and recessivity exist, the effect of selection is not immediately apparent.

The American geneticist Castle was the first to attempt to construct a kinetic description of selection on the basis of the Mendelian hypothesis. He did this in an article entitled 'The Laws of Heredity of Galton and Mendel, and Some Laws Governing Race Improvement by Selection', published in 1903. This article went unnoticed by most scientists, perhaps because Castle dealt with the question only from the point of view of animal husbandry and artificial selection. Castle's work was encouraged by the exceptional research climate that reigned in American agronomic institutes at the beginning of the century,[135] in which discussions of Mendelism and the factors of evolution developed in an operational and pragmatic perspective. U.S. plant- and animal-breeders were among the first to become aware of Mendelian principles. They wanted to know whether 'recessive' defects could be eliminated, or whether they would continually reappear, whatever was done. By showing that there were two ways in which a breeder could succeed in obtaining a race of 'pure dominants', Castle gave an elegant and pragmatic answer to this question.

The first way of purifying the race consisted of crossing an individual that shows the 'dominant' character (in 1903 the word 'phenotype' was not yet in use) with a recessive individual. If all the offspring were themselves dominant, there was a good chance that the genitor was also

RESULTS OF SELECTION FOR THE DOMINANT CHARACTER A IN THE VARIOUS GENERATIONS FOLLOWING A CROSS BETWEEN A PURE A AND A PURE B.

Generation.	Parents.	Offspring.	Per cent A or $A(B)$.
1	$A + B$	$A(B)$	100
2	$A(B)$	$A + 2A(B) + B$	75
3	$A + 2A(B)$	$4A + 4A(B) + B$	88.8
4	$A + A(B)$	$9A + 6A(B) + B$	93.7
5	$3A + 2A(B)$	$16A + 8A(B) + B$	96
6	$2A + A(B)$	$25A + 10A(B) + B$	97.2
7	$5A + 2A(B)$	$36A + 12A(B) + B$	98
8	$3A + A(B)$	$49A + 14A(B) + B$	98.4

Figure 31 'Results of selection for the dominant character *A* in the various generations following a cross between a pure *A* and a pure *B*.' (W. E. Castle, 'The laws of heredity of Galton and Mendel, and some laws governing race improvement by selection, 1903).

dominant. A race of such individuals would thus be purified of undesirable recessive characters. Castle noted in passing that such a result would not be possible if Galton's law of ancestral heredity (the 'atavism' noted by breeders) were true.

However, it would be both time-consuming and costly for a breeder to proceed in this way, because it would involve sacrificing a large proportion of existing individuals. The key question is whether the systematic elimination of recessives has a chance of purifying the race. It was in answering this question that Castle elaborated, for the first time, a kinetic description of selection under the Mendelian hypothesis.

Castle's reasoning was extremely simple. In a totally hybrid population with two alleles 'A' and 'B', if A is dominant over B the hybrid type can be given as 'A(B)'. Crossed to each other, such hybrids give 'A' dominant homozygotes, 'A(B)' dominant heterozygotes and 'B' recessives. Again, these symbols preceded the standard conventions introduced later for the designation of genotypes and phenotypes. Castle simply asked: What would be the result of random crossing between individuals, under the hypothesis that all recessives are eliminated in each generation? The table reproduced in Figure 31 gives the results of this calculation over eight generations. The proportions obtained in each generation are exactly those predicted six years later by the Hardy–Weinberg law (that is, if the frequencies of alleles A and B are p and q in the parents, the genotypic frequencies in the offspring are $p^2 : 2pq : q^2$). Castle indicated in particular that if selection is suspended in a population, its expression (or 'degree of purity') will stabilise at the level attained at that moment.[136] This demonstration clearly does not have the generality of those of Hardy and Weinberg in 1908, because it is based only on a case-by-case numerical verification. But it is also apparent that, in 1903, Castle saw a major implication of the Mendelian 'combination series' for evolutionary theory. He further showed that it did not matter if dominance was not absolute, and that this was merely an accessory and not a necessary aspect of Mendel's hypothesis.

The table in Figure 31 shows that the selection of dominant phenotypes leads to a progressive purification and that the genetic structure changes, even though the breeder only ever selects individuals with the same appearance – in each generation, the proportion of pure dominants ('A') increases. Correlatively, the overall frequency of individuals with a 'dominant' appearance increases, while individuals with a recessive appearance decrease in frequency. Castle summarised these results in a diagram (Figure 32). This diagram, obtained by elementary arithmetic, is the first known representation of a kinetic description of selection.

Castle's model lent support to the mutationist ideas of de Vries and Johannsen. He did not present selection as a process that could produce a race, but merely as a means of improving already existing races by the diffusion and fixation of a Mendelian factor. Shortly afterwards, Castle adhered to the mutationist theory of evolution, holding that Mendelian factors do not originate in selection but in sudden large-scale variations, in other words, in 'mutations'.[137] This episode shows that, very early on, there was a possible reconciliation between the theory of selection and mutationism – in its principles, the formal framework laid down by Castle in 1903–5 did not differ from the position that was later to prevail in the shape of neo-Darwinism.

However, this synthesis only took place some years later, and at least until the middle of the 1910s mutationism and Darwinism appeared to be incompatible theoretical positions. The leading mutationists of the early years of the century (e.g. de Vries, Bateson, Darbishire, Lock, Goldschmidt, Johannsen, Cuénot, Morgan) thought that their theory disqualified the Darwinian conception of selection from playing a role in forming species and races. Reciprocally, the rare orthodox Darwinians of the time did not fail to ask where the 'discontinuities' so dear to the geneticists came from. Weldon, for example, in a remarkable article published in

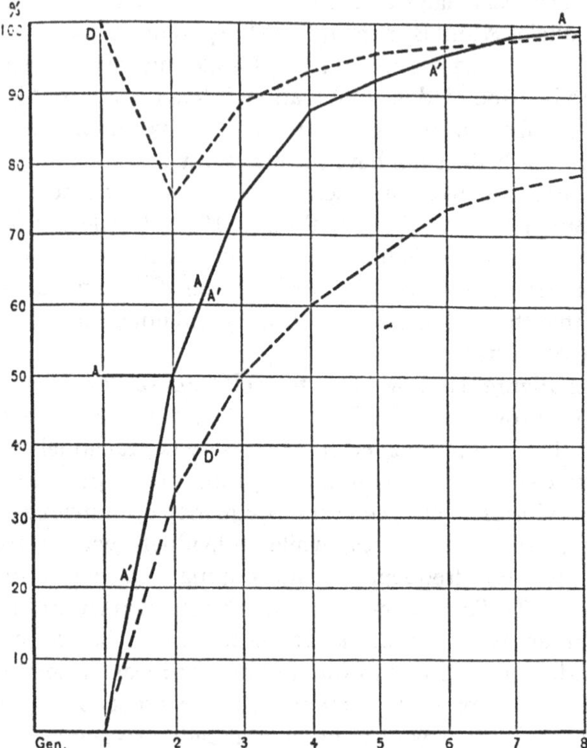

A, Rate of Race Improvement by Elimination, Dominance being Alternative.
D, Same, Dominance being Uniform.
A', Chances in 100 of obtaining a Pure Individual, Dominance being Alternative.
D', Same, Dominance being Uniform.

Figure 32 Effect of eliminating selection on recessives in a monohybrid population. Abscissa: generations. Ordinate: percentage of dominants. The *D* curve is a graphic representation of the table given in Figure 31. In modern genetic language (which Castle did not use), it gives the proportion of the dominant phenotype at each generation; in Castle's terms: the 'per cent of A or A(B)'. In generation 1 (F_1), there is no double recessive, and consequently the percentage of the dominant phenotype is 100%. The *A* curve represents another table, based on a curious hypothesis, 'alternative dominance' (where 'dominance alternates between the two characters A and B', a case which Castle thought was common). The *D'* and *A'* curves represent the 'chances in 100 of obtaining a pure individual [of type A]' under the two hypotheses of 'uniform' and 'alternative' dominance, respectively. In modern terms, these two curves represent the rate of progression of the genotype AA (W. E. Castle, 'The laws of heredity of Galton and Mendel, and some laws governing race improvement by selection', 1903).

1902, showed that the various varieties (and in particular the peas) on which de Vries and Bateson verified Mendel's laws were in fact the result of a long process of selection on characters that showed continuous variation.[138] Indeed, Mendel's own experiments had begun with an initial phase of selection of the characters that were then used to establish the law of development of hybrids.

In retrospect, the reasons why the fusion of the theories of mutation, of Mendelian heredity and of selection took so long to appear and even longer to win out are relatively simple. Once again, it has to be remembered that at the beginning of the century categories such as 'mutation' or 'Mendelian factor' were extremely unstable. First, it was not at all obvious that 'mutations' were alterations in Mendelian factors or genes. On the contrary, the scientific literature of 1900–10 shows that, for most biologists, this was only one possibility among many. Furthermore, the first mutationists had a tendency to think that developmental constraints oriented mutations in a certain direction. As Peter Bowler has shown,[139] mutationist conceptions were often marked by orthogenetic positions, including the idea that mutation pressure pushes the species in a given direction, whether the character is useful or not. This explains why the mutationists did not recognise natural selection as having the role Darwin ascribed to it – natural selection could accelerate transformation in the case of an advantageous character, or cause its extinction if the character was clearly non-adaptive, but it could not be considered to be a factor that shaped species. This vision of evolution can be found, for example, in an article written by Morgan in 1909 for a popular science journal. In this paper, which has the striking title 'For Darwin', Morgan – who was not yet convinced of the Mendelian theses – took up a key idea of de Vries' and argued that there were small hereditary but discontinuous variations, which he called 'definite variations'. These variations did not go beyond the limits of the species, but would spontaneously recur in many individuals. This led Morgan to an unusual argument that placed him midway between the mutationist and Darwinian conceptions of evolution:

> Any new, definite character will gradually appear in all the individuals whether it is useful or not. If it is useful it may sooner implant itself on the race than if it is indifferent; for more individuals may survive that possess it, than of those without it. It will spread faster, but in any case it will come in the long run. Thus we see that it spreads, not because it is advantageous, but because it is a definite variation.[140]

Reading these lines, one senses that the reconciliation between the theories of selection and of mutation was not far off.

This was to happen in the 1910s, on two complementary bases: the experimental theory of mutation, and an understanding of the formal consequences of Mendelian genetics, and in particular of the Hardy–Weinberg equilibrium, for the theory of selection. These two developments were independent, but as they converged they forged the new paradigm of natural selection.

The experimental theory of mutation was principally the work of the school of Morgan, and later of Muller, in the period 1910–20. Morgan, very much attached to mutationist conceptions, was only converted to Mendelism in 1910, when results from his laboratory on the numerous mutations that occurred in laboratory strains of *Drosophila* convinced him that these 'mutations' could best be interpreted by a Mendelian analysis. The importance of the experimental work carried out by Morgan and his students (in particular by Alfred Sturtevant, Calvin Bridges and Hermann Muller) for the reinterpretation of the basic concepts of Darwinism cannot be overestimated.[141] By accumulating over two hundred 'mutations' in the same species in the space of a few years, and by showing that these could be analysed as alterations of Mendelian genes, Morgan literally dissolved the idea that mutation creates species.

Without entering into the detail of this extraordinary work, those experimental advances of the Morgan school that had a direct impact on the theory of evolution can be summed up as follows:

1. Mutations consist fundamentally of alterations of genes, and are transmitted in a Mendelian fashion. In a given species mutations can take on a wide variety of forms more or less independently affecting virtually all characters, can combine with each other, and can occasionally survive and persist.

2. Mutations can have major effects, but they can also be at the limit of detectability. In other words, the size of the effect is entirely secondary from a genetic point of view: both 'small' and 'large' mutations are local alterations of chromosomes.

3. For a given character, mutations can occur in all directions; they are not 'directed'.

4. Mutation should be clearly distinguished from recombination.

This last problem was resolved by Muller when in 1922 he showed that mutation is a local change in a gene, thus putting an end to the idea that modifications in the hereditary material (and thus mutations) could be explained by mere recombination effects. This study proved that mutations were genuine novelties.[142]

All of these positions, progressively extended to many characters and many species, led to mutations of Mendelian genes being understood as the ultimate, if not unique, source of the inherited variability of species. The new theory of mutation marked the triumph of the discontinuist representation of variation, but also contained several decisive elements of the Darwinian doctrine of variation – the representation of the organism as a mosaic of more or less independent characters, the smallness and non-directionality of hereditary variation, and the fact that this variation could not be reduced to a combinatory series. These doctrines constituted decisive experimental arguments in favour of a return to the Darwinian fold.

In parallel to these experimental breakthroughs, the biologists of the 1910s came to understand the extraordinary effectiveness of selection under the Mendelian hypothesis. The new form of selection appeared in its clearest form in a book by R. C. Punnett, *Mimicry in Butterflies* (1915).[143]

Punnett hoped to show that mimetic adaptations in butterflies had not been gradually constructed in the course of evolution, but had appeared suddenly as major mutations. In his book he rejected any role for natural selection in the shaping of mimetic forms and limited its action to the diffusion of forms that had appeared spontaneously within the population.[144] This argument was typical of the mutationist discourse popularised by de Vries and Johannsen in the early years of the century. However, despite – and perhaps because of – this, the main interest of Punnett's book is that it elaborated, with a novel rigour, the notion of selection as a dynamic of replacement.

In chapter 8 of his book, Punnett examined what would happen in a population subject to selection for a variation transmitted in a Mendelian fashion.[145] He first takes on board Hardy's law, which shows that there are 'positions of equilibrium' for the distribution of genotypes, whatever the frequencies of the Mendelian factors that compose them. He then asks at what 'rate' the population could pass from one equilibrium to another if either the dominant or recessive form were to have a measurable selective advantage. What would happen if the selective advantage were 1 per cent, 10 per cent etc. in favour of the dominant form? It is obvious that the advantaged form would tend to replace the other, but could it replace it completely, and according to what law?

In an appendix, Punnett published a remarkable table by the mathematician H. T. J. Norton. The manuscripts in which Norton worked out the data in the table have not been found; thus we only know the results of his work (summarised in the table) and not the analytical procedure he used. The algorithm he employed seems to have been similar to that used ten years later by J. B. S. Haldane in his first 'Mathematical Theory of Natural and Artificial Selection' article.[146] Whatever the case, Norton's work can be seen as the first attempt to formalise a kinetic description of natural selection on the basis of the Mendelian hypothesis. The table, which is well known (see Figure 33), explicitly shows that, even with very

Percentage of total population formed by old variety	Percentage of total population formed by the hybrids	Percentage of total population formed by the new variety	Number of generations taken to pass from one position to another as indicated in the percentages of different individuals in left-hand column							
			A. Where the new variety is dominant				B. Where the new variety is recessive			
			$\frac{100}{50}$	$\frac{100}{75}$	$\frac{100}{90}$	$\frac{100}{99}$	$\frac{100}{50}$	$\frac{100}{75}$	$\frac{100}{90}$	$\frac{100}{99}$
99·9	·09	·000								
98·0	1·96	·008	4	10	28	300	1920	5740	17,200	189,092
90·7	9·0	·03	2	5	15	165	85	250	744	8,160
69·0	27·7	2·8	2	4	14	153	18	51	149	1,615
44·4	44·4	11·1	2	4	12	121	5	13	36	389
25·	50·	25·	2	4	12	119	2	6	16	169
11·1	44·4	44·4	4	8	18	171	2	4	11	118
2·8	27·7	69·0	10	17	40	393	2	4	11	120
·03	9·0	90·7	36	68	166	1,632	2	6	14	152
·008	1·96	98·0	170	333	827	8,243	2	6	16	165
·000	·09	99·9	3840	7653	19,111	191,002	4	10	28	299

Figure 33 Estimate of the speed of change in a Mendelian population subject to selection. The first three columns give the distributions of the three genotypes (e.g. AA, Aa, aa) in the population. For example, the third line gives 90.7 + 9.0 + 0.3 = 100%. The other columns show the number of generations required to pass from one distribution to another, for different selection intensities, and under complete dominance or complete recessivity (H. T. J. Norton, in R. C. Punnett, *Mimicry in Butterflies*, 1915, appendix 1).

weak selection intensities (around 1 per cent), there is complete replacement of one form by another, whether the advantaged form is dominant or recessive. In his commentary, Punnett insists on the fact that the replacement process will carry on to its logical conclusion, and that it is rapid, even for very weak selection intensities and recessive characters:

> A selective advantage of 10% operating against the recessives will reduce their numbers in 70 generations from nearly one-half of the population to less than one-fortieth ... If the selective rate is diminished from 10% to 1% the number of generations necessary for bringing out the same change is nearly 700 instead of 70 – roughly ten times as great. Even so, and one can hardly speak of a 1% selective rate as a stringent one, it is remarkable in how brief a space of time a form which is discriminated against, even lightly, is bound to disappear. Evolution, in so far as it consists of the supplanting of one form by another, may be a very much more rapid process than has hitherto been suspected.[147]

Punnett also examines rates of selection that were far weaker: even if selection intensity is of the order of 10^{-6}, the process is still effective on the geological time-scale. While this analysis was primarily statistical, it clearly indicated that there could no longer be any doubt as to the efficacy of natural selection in the context of Mendelian heredity.

Last and most important, Punnett showed the main methodological consequence of this view. Mendelism provides a simple and precise way of measuring natural selection: 'The development of Mendelian studies has given us a method, rough perhaps but the best yet found, of testing for the presence, and of measuring the intensity, of natural selection.'[148] In a Mendelian context, the intensity of selection is in fact measured by a simple replacement rate (thus by a number) and not, as in Weldon's biometrical method, by an algebraic function that integrates the selective value and the value of a character showing continuous variation. Punnett put forward a case study to test this method: the replacement of the light form of the silver-birch moth (*Amphydasys betularia*, or, as it is known today, *Biston betularia*) by a melanic form that behaves dominantly. This was the beginning of a story that has become a legend, which will be dealt with in Chapter 10.

It does not matter that Punnett's intentions were anti-Darwinian (he hoped to show the role of major mutations). The formal approach he put forward applies equally well to 'small' or 'big' mutations – in each case a replacement rate is worked out and measured in the same way. Punnett's thought-provoking table convinced a large number of biologists that Mendelism and Darwinism were compatible. If small-scale mutations could simultaneously affect a number of characters, selection could play an active role in shaping organic forms. At around the same time Morganian genetics proved the existence of small non-directed spontaneous mutations. From a methodological point of view, the gulf between mutationism and Darwinism became negligible. In 1916, Morgan embraced the theory of selection and renounced the idea that a mutation could establish itself in a population merely by spontaneously recurring.[149] This was a year after the publication of both Punnett's book and his own magnum opus, *The Mechanism of Mendelian Heredity* (1915). It can reasonably be argued that this was the moment when the new shape of selection became a paradigm.

By the end of the 1910s, therefore, Darwinism, far from being incompatible with the new theories of heredity and of variation, had been revitalised by them. The main effect of Mendelism was to free the theory of evolution from the polymorphous concept of reversion, and to permit a predictive description of the kinetics of selection. As to the theory of mutation, reduced from being a doctrine of speciation to the more humble rank of a doctrine of variation, it ceased to be opposed to selection and became its complement. After 1920 it was a commonplace to say that mutation and selection involved two different levels of the evolutionary process – one was the origin of inherited variation, the other was the origin of adaptations.

In a way, this language was the same as that used by Darwin to describe the relation between variability and selection – Darwin had always argued that selection did not create variability, but used it to form adaptations. However, the specificities of the new paradigm need to be emphasised. In the genetic theory of selection, the process described is that of replacement. This process has a beginning and an end: a mutant allele appears, is diffused and becomes fixed. This conception substantially strengthened the atomistic tendencies that were already perceptible in Darwin's philosophy of evolution, despite his gradualism. As has been repeatedly pointed out in the preceding chapters, Darwin saw natural selection as carrying out the diffusion and elimination of 'variations' rather than of 'individuals'. Nevertheless, his hypothesis of pangenesis involved a conception of individual variation as not only an 'individual difference' but also a continual *process* of change. In the new genetic theory of selection, however, the ontology of selection became explicitly and above all else particulate.

To sum up, with the advent of Mendelism the objections raised in 1867 against the theory of selection by Fleeming Jenkin finally found a response. In a way, Mendelism proved Jenkin right, to the extent that it bound the representation of the selective process to a particulate hypothesis of heredity. But the Mendelised theory of selection also proved Darwin right, because it showed that selection acts on an n-dimensional field of polymorphic loci, and thus cannot be reduced (as the first mutationists had argued) to the survival of exceptional major anomalies or to the spontaneous appearance of new forms that immediately take the shape of species.

8.3 Genetics comes to grips with continuous variation: the rehabilitation of orthodox Darwinism

Beyond its reconciliation with mutationism, the Darwinian theory of selection won out in another, more literal, way. The characteristic aspect of the Darwinian concept of selection was its continuist starting point. For Darwin, natural selection acted fundamentally on *infinitesimally small* differences, and its effect is a *gradual* transformation of species. This in fact prevented Darwin's hypothesis from becoming a mere tautology – by adopting the motto 'Natura non facit saltum' (Nature makes no leaps),[150] Darwin not only marked out his philosophical ground, he also took a massive risk. Whether or not selection acts effectively on continuous variation can only be answered by empirical investigation. The mutationists thus initially believed that they had genuinely refuted Darwin. As we have seen, their fundamental argument was that selection had no effect on 'fluctuating' variability, but only on discontinuous variability – in other words, on 'mutations', which alone are inherited. The corollary of this argument was that selection (both artificial and natural) is not a process of formation, but only of substitution.

As has been seen, the mutationists were right in a way: they proved that, genetically speaking, selection is a process of replacement. The experimental science of heredity imposed this doctrine, even if it was qualified by the fact that 'genetic' mutation is more subtle than the old idea of the sudden creation of a species.

This, however, is merely one side of the turbulent relation between Darwinism and Mendelian genetics. After the rediscovery of Mendel's laws, it rapidly became apparent that the Darwinian paradigm had been buried a trifle too hastily, and that it showed stubborn signs of being alive and kicking. The opposition between non-hereditary fluctuating variation and hereditary variation due to 'mutation' turned out to be a fiction. It was no longer sufficient to reformulate the principle of selection; it was necessary to understand in genetic terms the reality of the precise process described by Darwin.

A detailed examination of the way in which quantitative genetics forged its methods would be outside the scope of the present study. Instead, some seminal moments which highlight the genetic rehabilitation of original Darwinism will be outlined.

The work of the American geneticist William E. Castle (1867–1962) constituted the most vigorous reaction to mutationist dogmatism. Castle's early support for Mendelism has already been mentioned, as has his equally precocious support for the 'mutation theory of organic evolution' and in particular the idea that selection does not strictly speaking produce anything, but merely improves races by the diffusion of favorable factors.[151] In 1903, Castle also put forward the first elaboration of a general law of panmictic equilibrium and the first attempt at defining a kinetic description of selection under the hypothesis of Mendelian heredity.

Of all the founders of Mendelian genetics, Castle alone dared both to question the central dogma of Mendelism and to undertake the rehabilitation of the principle of selection in its most orthodox Darwinian form, in the name of 'genetics'. In Castle's mind, these two theses were intimately linked.

In his first Mendelian writings, Castle questioned the universality of the central dogma of Mendelism – the law of 'gametic purity'. In 1903 he suggested that Mendelian factors do not necessarily segregate, but could sometimes associate in the gametes to form a 'mosaic character'.[152] Initially a minor reservation, this assertion gradually grew into a general doctrine, known as the 'theory of gametic contamination'.[153] This stated that, despite the fact that Mendelian characters are independent units in sexual reproduction, they should not be conceived of as solid phases, strictly separated in the zygote germ plasm, but rather like differently coloured layers of wax superimposed on a single base (the chromosomes). When the layers separate, they may accidentally carry with them fragments of another layer.[154] Expressed less metaphorically, this implies that Mendelian determinants can modify each other: 'Gametic purity is not absolute, even in sharply alternative inheritance.'[155]

Apart from this difference with the central dogma of Mendelian genetics, Castle came to strongly oppose the mutationist postulate that 'fluctuating' variability is not inherited. In a famous set of experiments on rats carried out between 1906 and 1919, he tried to show that Mendelian characters were not as constant as was thought, that they showed heritable fluctuations, and that they were exposed to the effective action of gradual and continuous selection. This typically Darwinian approach, coming from a famous geneticist and a rigorous and capable experimentalist, took the form of a direct criticism of the doctrine of the constancy of 'pure lines'. For many years, Castle thought he had shown that this doctrine was false, and that the principle of selection in its most literal Darwinian form had to be rehabilitated.

These two examples of dissent were of fundamental importance to the theories of both heredity and evolution. With regard to the critique of the notion of gametic purity and to the idea of a fluctuating variation affecting Mendelian characters, the experimental results ultimately proved Castle wrong. In both cases, however, the phenomena on which he focused attention could not be ignored, and they had the salutary effect of shaking the saltationist, pre-formist and anti-Darwinian dogmatism of early-20th-century geneticists to its foundations. The result was the emergence of a genuinely different view of genetics, and in particular a different view of the genetic significance of evolution.

Castle's selection experiments were mainly done on rats. Begun in 1906, they lasted until 1919 and involved fifty thousand animals. They have been summarised many times in both the scientific literature and in textbooks, and they are well known to historians of science.[156] In the first three decades of the 20th century they constituted a major reference for anyone interested in the theory of selection. The key points will be outlined here.

Initially, this was a classic Mendelian study of variation in the coat colour of rats. Like many other rodents, rats show a marked discontinuity between those that have a uniform grey coat (normally the case) and those that are hooded or piebald. These last two forms both have black heads and a black streak of varying length along the back, the rest of the body being white. Together with one of his students, H. MacCurdy, Castle first showed that the 'hooded' character behaved as a recessive Mendelian factor with regard to the grey coat colour, which is dominant.[157]

Castle also observed that piebald rats showed important variations in coat pattern. As these variations did not appear to behave in a Mendelian fashion, he initially thought that they were not hereditary, and that only the alternative grey/piebald could be analysed genetically. He began a selection experiment within a pure line of piebald rats, genetically characterised as double recessives for this character. The experiment involved the systematic selection of both the lightest and the darkest animals. Two strains were thus created, the character of which was to be accentuated through selection. According to Mendelian theory, the two strains should have conserved their mean characters, with the same 'fluctuating' variation, and produced the same statistical regression. However, to Castle's great surprise, the selection experiment produced some astonishing and intriguing results. Over the years, Castle and his co-workers were able to constitute a series of lines showing varying degrees of blackness, from rats that were entirely black to rats that had only black heads, including all possible intermediate variants. Furthermore, all the lines were stable. Figure 34 is taken from an article published in 1914 by Castle and J. C. Phillips, and sums up the whole

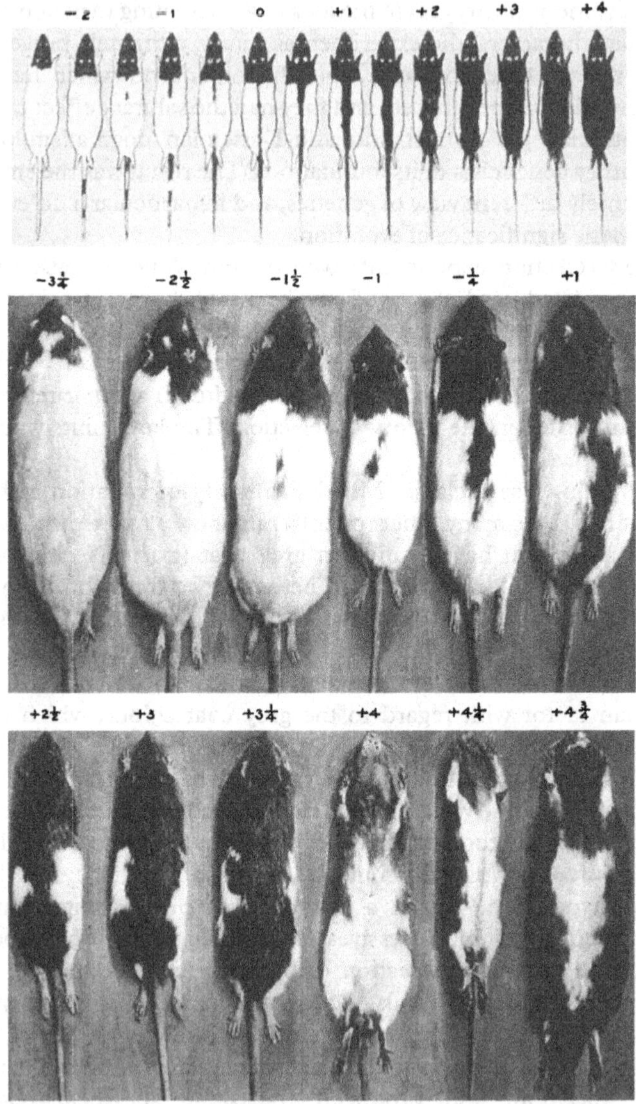

Figure 34 Arbitrary scale of the ‘hooded’ character in rats selected by Castle and Phillips. Below: corresponding photographs. The last three animals are shown on their backs (W. E. Castle and J. C. Phillips, ‘Piebald rats and selection’, 1914, pl. 1).

selection experiment.[158] The lines marked '+' were produced by selection for more extensive marking, those marked '−' were produced by selecting for less extensive marking. The conclusions of the study were straightforward: despite the fact that they came from a single strain of pure recessives, the various selected lines showed no spontaneous regression to the initial type; in fact, biometrical analysis showed that selection had not only changed the mean of the variability curve, but also its amplitude – the observed modifications went far beyond the variation found in the initial stock. And finally, when the direction of selection was reversed, the time taken for reversion to the initial type was directly proportional to the duration of selection.

This experiment showed that selection could act effectively on 'fluctuating' variability. Furthermore, because Castle was working on a phenotype that he supposed was dependent on only one Mendelian character, his selection experiments appeared to go against the results of Johannsen on 'pure lines'. In fact, the lines of hooded rats were neither 'pure' nor 'constant'. Castle interpreted his results by arguing that selection had modified the Mendelian character itself, by acting on its 'fluctuations'. This interpretation was linked to a critique of the law of gametic purity – the central dogma of Mendelian genetics. Throughout the experiment, Castle had carried out all sorts of crosses between the selected strains, with a control strain and with various other varieties. For example, the cross of the '+' strain with wild grey rats produced hybrids from which it was possible to re-extract double-recessive piebalds. These double recessives were less pigmented than those of the '+' strain, which Castle interpreted as being due to a 'contamination' of the transformed Mendelian character. From this followed the idea that the gene (and thus the genotype) was unstable, and that this instability could give rise to a process of selection. Thus there was an explicit link between Castle's dissent over genetics and his support for Darwinism. As the title of his 1914 summary article makes clear, the same set of experiments had the aim of testing both 'the effectiveness of selection' and 'the theory of gametic purity in Mendelian crosses'.[159]

Castle proposed that genes as such could be exposed to selection because of the small fluctuations that affected them. The fate of Castle's idea is well known. Various geneticists (MacDowell, Pearl, Sturtevant, Wright, Pictet) showed that what Castle had taken to be homogeneous genetic material was in fact heterogeneous. Castle had selected not one Mendelian character, but several, which meant that his work was just another example of 'multiple factors'.[160] Attention was also drawn to the fact that the selection of hooded rats tended towards a ceiling, corresponding to the uniformity of genotypes. Finally, sophisticated crossing procedures showed that the allele that Castle thought had been altered was in fact constant throughout the experiment, and that the hypothesis that slightly different alleles had appeared during the selection experiment should be rejected. To sum up, the genetic interpretation of Castle's experiments turned into a victory for Johannsen's thesis: selection can only purify genotypes that are mixed in a population; it cannot modify genes, and thus cannot modify genotypes.

However, this victory for genetic rationality did not leave the mutationist representation of evolution unscathed. Castle's selection experiments, properly interpreted, convinced many scientists that the moment had come for a genuine synthesis of genetics and Darwinism. The founders of population genetics, and in particular Sewall Wright (who had been Castle's assistant) and Fisher, said that Castle's data had made such a synthesis a necessity.[161]

First, these experiments rendered untenable the sharp opposition of discrete (or inherited) variability and quantitative or fluctuating variability. If continuous variation is taken to be the macroscopic effect of a polygenic determinism, it is open to the action of natural selection. Second, the example of hooded rats showed that a given Mendelian factor, unaltered in its structure, could be 'modified' in its effects owing to the genetic context. This was the case, for example, for the recessive gene isolated by Castle in his rats, which sometimes gave a black coat and sometimes a coat that was virtually white. At the end of the 1920s, the term 'modifier' was coined to describe a gene that could modify the expression of another gene. The concept of modifier genes joined the already established idea of epistasis (genetic interaction). With the development of experimental studies, these two concepts were to have a major theoretical importance. If it was possible for a given allele to determine opposite characters because of different genetic contexts, genes should no longer be viewed as simply and directly determining characters, and selection should not be seen as merely sifting out particles. If the same gene could be dominant in one context and recessive in another, if it could change its expression depending on the sex of the individual, if it could sometimes be lethal and sometimes advantageous, then the theory of selection could not be reduced to a mere kinetic description of the replacement of hereditary particles.

Thus the rehabilitation of the Darwinian principle of selection, far from simplifying this principle, opened a particularly complex theoretical and experimental field of investigation – population genetics. This involved the reconciliation of two competing approaches to heredity – biometry and Mendelism. The history of this fusion will be only briefly outlined here.[162]

The fundamental fate of continuous variation under the Mendelian hypotheses could easily be foreseen. For many years, the biometricians had admitted the theoretical possibility that the physiological mechanism of heredity might be particulate. This was the hypothesis put forward by Galton, who had noted in *Natural Inheritance* that characters such as height must depend on a large number of more elementary characters (for example the sizes of different bones). Galton pointed out that even if these characters behaved physiologically as independent and discrete particles, the wide range of possible combinations would produce the effect of continuous variation and of a heredity that appeared to be 'blending'.[163] This proposition was based on a well-established statistical tradition, that of Laplace's interpretation of Gauss' law: if a large number of independent causes are involved in a measurable effect, the frequency distribution of the measures can be thought to be no different from the normal law.

When Mendel's laws were rediscovered, the biometricians applied this old maxim. In 1902, Udny Yule noted with regard to 'unit-characters', or Mendelian factors:

> Surely it would be a very moderate estimate that the number of [Mendelian] units could not be less than 50? Yet this would suffice to give ... *over a thousand-million million types*! Even then if the variations of 'units' do take place by discrete steps *only* (which is unproven), discontinuous variation must merge insensibly into continuous variation simply owing to the compound nature of the majority of characters with which one deals. There does not seem any escape from this conclusion.[164]

A year later, Karl Pearson made a first mathematical analysis of this suggestion, in an article entitled 'On a Generalised Theory of Alternative Inheritance, with Special Reference to Mendel's Laws'.[165] In this paper, Pearson showed that the Mendelian hypothesis was quite capable of explaining continuous inherited variation. If it was accepted that a relatively important number of loci are involved additively in the inheritance of a quantitative character, the major laws of heredity established statistically by Galton and biometry would apply. In particular, Pearson showed how the Mendelian approach was compatible with an analysis of the phenomena of heredity in terms of statistical regression and ancestral inheritance. In fact, this can be deduced from the multifactorial hypothesis. It could be argued that this astonishing article, in which Pearson also outlined the law of panmictic equilibrium, was the founding document of quantitative genetics. However, Pearson noted that Mendelian predictions did not agree with the correlations observed for the height of related human beings, in particular between collaterals (siblings, aunts and uncles, cousins ...). As Yule was later to remark, this disagreement flowed from the fact that Pearson assumed complete dominance. If there was partial dominance, the predicted and observed correlations between close relatives were in fact very similar.[166]

The fundamental doctrines of quantitative genetics were developed early in the century, long before the publication of Fisher's canonical article of 1918, which is often credited with having laid the foundations of the discipline. The biometrical literature shows that Pearson and a number of others quickly recognised the formal possibilities provided by Mendelism for the study of continuous variation. If these early studies did not become widely known, this was because of Pearson's continuing unconditional epistemological hostility with regard to Mendelism.

However, the impact of the first hesitant steps of theoretical quantitative genetics on early-20th-century biologists should not be exaggerated. It was not the mathematical formalism that convinced scientists of the plausibility of the multifactorial hypothesis, but the experimental studies of Nilsson-Ehle (1909–11) and of Tammes (1911). These studies investigated the genetic determinism of various quantitative characters in plant organs: stem or grain length in wheat and oats (Nilsson-Ehle)[167] and colour intensity in flax flowers (Tammes).[168] These two botanists had postulated that such characters were determined by a small number of Mendelian factors, each allele acting additively on the intensity of the character. In these studies, they reasoned as though there were two alleles at each locus that acted as '+' or '−' factors. For example, if a character such as the length of an organ is determined by two loci, the degrees of the character would be determined by the number of 'doses' of the '+ gene' (aabb = 0; Aabb or aaBb = 1; AAbb or AaBb or aaBB = 2 etc.). The hypothesis may seem simplistic, but it led to rich and testable predictions. It was this kind of study which, long before Fisher's modelling, convinced scientists of the ability of Mendelian genetics to explain and absorb quantitative variation.

This brief glance at the origins of quantitative genetics (or biometry) will have to suffice. However, a further point needs to be made: it is an oversimplification to state that the 1910s were marked by opposition between discrete (hereditary) variation and fluctuating (non-hereditary) variation. The physiology of hereditary transmission required that hereditary determinants should have a discrete location and organisation. But continuous variation was no less 'hereditary' than discontinuous variation, and could, in the final analysis, be interpreted by the Mendelian hypothesis. From this point onwards, there was no obstacle in the way of a rehabilitation of the principle of selection in the very form in which it had been conceived by Darwin – a principle of the construction of forms by 'adding up deviations so slight as to be hardly or not at all appreciable by the human eye'.[169]

A dual theory of selection thus emerged from the confrontation of Darwinism with Mendelian genetics. Initially, Mendelian formalism and the mutation theory imposed a vision of selection as a process of gene replacement in a population. This language provided the principle of selection with the operational intelligibility that it had previously lacked. As J. B. S. Haldane liked to point out, once absorbed into population genetics the theory of selection took on the form of a 'mechanics' of evolution. Given the forces (selection intensity, mutation pressure, sampling error) and the initial state (the genetic structure of the population at a given moment), it is possible to predict the probable state of the system in all future generations. Such a theory is indifferent as to the size of the variation – it matters little whether the mutated allele has a 'large' or a 'small' effect.

But side by side with this kinetic description of gene replacement, Mendelism also had to give way to a more literally Darwinian selection, the selection of quantitative variability. With hindsight, we can say that this kind of selection posed (and still poses) the most difficult and most fundamental theoretical problems. If it is true that the genetic theory of selection is still expressed in the intuitively simple language of gene frequencies, it would nevertheless be naive to believe that it can be reduced to a description of the rate of accumulation of favourable mutations. The great challenge for 20th-century Darwinism has been to cope with systems of interaction in which genes acquire a selective value.

The Darwinism that emerged from Mendelism was not simplified but pluralised. On the one hand, it contained a kinetic description of genetic change, formalisable and highly predictive – a 'stoichiometry of evolution', to use Lotka's expression.[170] On the other, it implied a difficult but fundamental analysis of the dynamics of the modification of quantitative characters, which, strictly speaking, are the only units that have selective value.

Part 3

The genetic theory of selection

By the end of the 1910s, the long opening crisis of the Darwinian hypothesis of natural selection was well and truly over. Although it had certainly not been adopted as a paradigm by most evolutionists, the 'hypothesis' was theoretically possible, in that it could be shown to be compatible with the data and doctrines of the new science of variation and heredity (genetics). A new Darwinism, freed from a series of major problems, could now develop. The origin of the term 'neo-Darwinism' has already been discussed.[1] Together with 'ultra-Darwinism', it was first applied to the evolutionist conceptions of August Weismann, which differed from Darwin's by two radical and intimately linked doctrines that were put forward in the name of parsimony: (1) a total rejection of the inheritance of acquired characters; (2) the 'all-sufficiency' of natural selection for the theory of evolution.

Throughout the century, Weismann's two principles have grown in influence. Whenever Darwinism is said to be in crisis (this happens quite regularly), one of Weismann's two historic postulates is certain to be the focus of debate. There are two cast-iron ways of shaking contemporary evolutionary orthodoxy – argue that individuals can change the direction of evolution (e.g. through ethological or embryological factors) or criticise panselectionist arguments (e.g. natural selection does not explain extinction, or divergence, or some other effect). 'Neo-Darwinism' thus has a real historical and conceptual coherence, going beyond the contexts in which it was developed (the 'ultra-Darwinism' of the beginning of the century, population genetics, the 'synthetic' theory of evolution . . .).[2]

Nevertheless, as a category, 'neo-Darwinism' does not sufficiently take into account the theoretical matrix that produced 20th-century evolutionary biology. Theodosius Dobzhansky, in *Genetics and the Origin of Species* (1937), gave the clearest description of the central methodological postulate of the new Darwinism: 'Since evolution is a change in the genetic composition of populations, the mechanisms of evolution constitute problems of population genetics.'[3] In other words, population genetics provides a homogeneous and sufficient methodological framework for any causal science of evolution; there is no need for any other, argued Dobzhansky: 'Experience seems to show . . . that there is no way toward an understanding of the mechanism of macro-evolutionary change, which requires time on a geological scale, other than through a full comprehension of the micro-evolutionary processes observable within the span of a human lifetime.'[4] Such unsubtle statements have the great merit of clearly defining the conceptual framework within which the debates and theories of contemporary 'neo-Darwinism' need to be understood.

Twentieth-century Darwinism has long been thought to be a more complex and subtle beast than 'population genetics'. After all, does it not describe itself as a 'synthesis'? The importance of its massive empirical harvest is patent. The

'Modern synthesis' is a great book of nature, containing many marvels, and an infinite number of fascinating empirical chapters; in this respect, its history is indeed extremely complex.[5] But there can be no doubt about the homogeneity of belief and of language that is shared by the whole of evolutionary biology. The most striking fact is that a whole generation of biologists from a wide range of disciplines, all trained in Mendelian logic, gave up metaphysical armchair theorising and interminable quarrels over 'the principal factor of evolution' and were converted to the pacifying idiom of genetics and populational thinking. For this reason, the following chapters concentrate on the central theoretical beliefs of the New Synthesis rather than providing a detailed description of the disciplines that participated in it.

This final part contains an historico-critical examination of Mendelised 'neo-Darwinism'. This neo-Darwinism was fundamentally a neo-Mendelism.[6] Recast in the formal language of population genetics, the hypothesis of natural selection took on two unique epistemological features. First, it found itself absorbed into a theoretical apparatus that was not Darwinian, and the models of which could equally well serve completely different evolutionary scenarios. Theoretical population genetics emerged as the transcendental topic of the science of evolution under the Mendelian hypothesis. In this formal context, natural selection was no longer a principle, nor even a probability, but rather a parameter that interacts with a number of others within a homogeneous theoretical field open to many other evolutionary scenarios. Population genetics was a formulation of the conditions under which natural selection could be the driving force of adaptive evolution (or the 'paramount power' of evolution, as Darwin put it). This relativisation of selection highlights the second remarkable epistemological aspect of modern discussions of the Darwinian hypothesis. As a formal predictive science able to set out conditions under which selection *could* be the major factor of evolution, population genetics enabled the emergence of a genuine theory *of* natural selection. Not a theory of its far-off consequences (as was the case with the vast 'argument' that Darwin called the 'theory of natural selection' in the *Origin of Species*), but a formal, predictive theory of natural selection as such. In this final part these assertions will be investigated by examining first the rationality of the genetic theory of selection and then the question of its empirical content.

There are several reasons why the chapters that follow do not contain an exhaustive history of seventy years of neo-Darwinian thought. From the 1920s onwards the crisis of Darwinism had largely been resolved; once Mendelism and the mutation theory had been given an experimental basis, the road was finally open for a systematic exploration of the theoretical possibilities opened by the hypothesis. The paradigm set out, the time had come to exploit it. While previous chapters necessarily examined the founding crisis of Darwinism in some detail, the key question here will be to draw up a balance-sheet.

9 The place of selection in theoretical population genetics

The attraction of population genetics lies in its formalism. Because of its enthusiasm for the possible rather than the real, it sometimes appears as though it could do without empirical data altogether. In practice, population genetics does not tell us much about the way in which evolution has taken place. In the form in which it developed in the 1920s and 1930s, and which it largely retains today, it can be considered the transcendental topic of the science of evolution. That is, it sets out and constitutes the boundaries of what can be discussed in modern evolutionary theory. Natural selection is no longer a fundamental principle but a parameter which measures one of many 'forces' of change. Despite the open bias of its founders, theoretical population genetics is not in itself 'Darwinian'. If it nonetheless appears as the modern theoretical version of Darwinism, this is because it formalised the conditions that would enable selection to be a key factor in evolution.

9.1 Founding models: population genetics has its first debate

The fundamental themes of population genetics were constructed through mathematical models that the vast majority of biologists (if indeed they did any more than simply talk about them) found utterly impenetrable. For the most part, the models of population genetics were developed in the space of a few years, from the middle of the 1920s to the beginning of the 1930s, by three legendary scientists: R. A. Fisher (1890–1962), J. B. S. Haldane (1892–1964) and S. Wright (1889–1988).[1] These models, despite their rigid formalism, quickly led to major conceptual changes within the evolutionary community. We need to distinguish 'commonsense' effects from the genuinely visionary debates that launched the new discipline.

The commonsense effects of these models rapidly made themselves felt, mainly because Fisher, Haldane, Wright and a handful of others (E. B. Ford, J. Huxley, T. Dobzhansky) were able to use simple language to express the conceptual implications of their formal models.[2] By the end of the 1930s, many evolutionary biologists accepted the lessons of a discipline which, for the most part, they simply did not understand.

These lessons were directly expressed and easily understood because they represented a clear response to the doctrinaire positions that had dominated half a century of post-Darwinian debate. Despite their virulent conflicts, the supporters of neo-Lamarckism, orthogenesis, mutationism and ultra-Darwinism all believed in the existence of an ultimate 'factor of evolution' which would enable biology to go beyond existing evolutionary antinomies.

Population genetic reasoning dissolved this mirage, together with the eclectic agnosticism that it naturally provoked. In populational genetics, evolutionary space is defined by a field of gene frequencies. This field has a limited number of dimensions (although the actual number is not known in practice). A certain number of 'forces' can modify the observed frequency distributions – recurrent mutations, migration, natural selection, sexual selection, mating systems and sampling variation. The effects of these factors are expressed in a homogeneous language – the coefficients that measure the various forces are merely exit and entry rates. In the probabilistic evolution of the 'genetic structure of a population', the qualitatively heterogeneous causes of change can be expressed in a homogeneous parameter space composed of different classes of factors – ecological (natural selection), geographical (migration), cytogenetic (mutation), ethological (sexual selection, mating system, generation structure) and stochastic (sampling effect or 'random genetic drift'). These factors express themselves as commensurable effects, in the same way that the heterogeneous forces of physics express themselves in the homogeneous language of motion. In such a parameter space, it is possible to make *a priori* deductions as to the weight of these factors and their interaction. A typical question would be: What is the equilibrium frequency with a given rate of mutation, a given rate of migration, a given selection pressure and a given probability of random extinction? As in physics, the underlying mathematical theory rapidly becomes extremely complex. In 1924, in the founding document of the new mathematical theory of selection, Haldane noted that the problems raised by this theory required the use of hitherto unexplored mathematical functions: '. . . the mathematical problems raised in the more complicated cases to be dealt with in subsequent papers seem to be as formidable as any in mathematical physics'.[3]

Despite these difficulties, the theoretical genetics of populations immediately led to striking and novel conclusions, which both excited the evolutionary biologists of the 1930s and invaded the popular literature.

The preceding chapter evoked the famous table of 1915 in which H.T. J. Norton first formalised the evolution of a gene frequency for various selection rates. The table (Figure 33) shows that a very weak selective advantage (around 1 per cent) is sufficient to diffuse and fix an allele in relatively few generations on an evolutionary time-scale. This table gave a precise, testable and striking meaning to the Darwinian idea that 'infinitesimal' advantages could, over time, have major effects.

At the same time, experimental genetics was accumulating data on spontaneous mutation rates. Morgan and Muller showed that in general these rates are lower than 10^{-5}, and often very much lower. This provided decisive evidence against orthogenesis, and in particular against the version advocated by the first 'mutationists'. In genetic terms, orthogenesis implied repeated mutations occurring in the same direction and the idea that 'mutation pressure', on its own, can modify species. *A priori*, this was not absurd. But the highest known rates of mutation found in nature were still several orders of magnitude lower than the lowest selective advantages that could be detected in the laboratory. The same argument could be used against neo-Lamarckism. Following the failure of Weismann's experiment on mutilations, turn-of-the-century Lamarckians argued that somatic effects accumulated very slowly in the germ cells. Nevertheless, in this case too, even very weak selective effects would still be stronger. These arguments, originally put forward by R. A. Fisher in 1930,[4] were repeated in the 1930s and 1940s by E. B. Ford, J. B. S. Haldane and J. Huxley in a series of widely read books.[5] As William Provine has pointed out, through its emphasis on parsimony, population genetics made its main doctrinal competitors outdated.[6]

However, it would be wrong to think that the new discipline simply gave its seal of approval to Darwinism. In fact, it provided a new way of thinking about the principle of selection, freeing it from the crushing weight of its self-evidence. For example, the models showed that in small populations selection was effectively counterbalanced by the random fixation of alleles ('genetic drift'); or again, that an unfavourable, potentially lethal, recessive mutation would stabilise itself at a low frequency, but would not completely disappear.

Thus the effect of population genetics on evolutionary common sense was to present a way of overcoming the qualitative heterogeneity of causal factors through a homogeneous language of effects. Population genetics made the major doctrinal quarrels of the 19th century completely outdated. It put an end both to the mirage of identifying a single great principle of evolution and to agnostic eclecticism. Finally, and above all, it constructed *a priori* a field of possibilities. The rest of this chapter deals with this final point.

The common project of the founders of theoretical population genetics was the construction of a statistical mechanics of genetic variation. The theoretical problem can be expressed as follows: given the existence of a certain number of forces acting on the field of genetic variation (mutation, migration, selection, mating system, linkage, various random variations etc.), how can models be created that will predict the frequency distributions of genes at several thousand loci within species of varying sizes, and over many, many generations?

Faced with this extremely difficult question, the founders of population genetics did two things. First, they built simplified models with a strong predictive capacity. These models generally dealt with a single locus, or sometimes two, and assumed very restrictive hypotheses – constant selection rates, constant population size, selective value independent of frequency and density, etc. Even with such simple hypotheses, formalisation presented major mathematical problems. Nevertheless, Fisher, Haldane and Wright were able to elaborate algorithms that could be used and applied to experimental or natural populations.

From this pragmatic point of view, Haldane's contribution was the most remarkable. He concentrated on the question of modelling what could be really measured in natural or experimental conditions. Thus Haldane, more than Fisher or Wright, was interested in very high selection rates – not because he thought that such rates were more probable for most loci, nor because he thought they were more pertinent for explaining evolution, but simply because they were the only ones that could be measured. This empirical orientation of population genetics will be dealt with in the next chapter.

Beyond their simplified models, the founders of population genetics also had in mind the question of the overall evolution of the genome. If modelling a single locus is difficult, representing the total cloud of genes in a population, although not impossible, does not lend itself to a highly predictive model and requires something akin to a biological philosophy. Fisher, Wright and Haldane realised that if they tried to reason on the basis of the totality of the genome, they would inevitably lose in information content what they gained in generality. Their discussions of this question need to be understood as a kind of fundamental meditation on the possible scenarios of evolution.

The initial debate between the founders of population genetics will first be presented somewhat summarily; a more technical analysis will be outlined in the subsequent sections.

The key debate took place between Fisher and Wright, between an elementarist or 'genic' approach to evolution (Fisher) and a more 'organismic' approach (Wright). For Fisher the fundamental question was the destiny of evolutionary 'atoms' – genes. The ideal that the science of evolution should aim towards, he felt, was the kinetic theory of gases. This image can be found throughout his work, from his early writings (not yet mathematical) through to his final studies.[7] Wright's 'organismic' approach – the term is his own[8] – can be characterised by the postulate that gene interaction is the fundamental aspect of evolution.

Fisher reasoned on the basis of very large, virtually infinite, panmictic populations. This enabled him to consider that the probability of random fixation of an allele was virtually zero. However, Fisher did not ignore the mathematical theory of random fixation (what today would be called 'random genetic drift') – he was in fact the first to attempt to theorise it, at a time when it was called 'the Hagedoorn effect'.[9] In 1921 A. L. and A. C. Hagedoorn had suggested that 'random extinction' of alleles constituted a major factor of evolution, independent of any determinist factor (in particular mutation and selection). A year later, Fisher put forward the first model of this hypothesis,[10] with the sole empirical precondition that the population should be infinitely large. If Fisher relativised the effects of random variations on genetic structure, this was partly because he thought that the majority of species have *de facto* a very large population size and thus form a panmictic continuum. It was also because this idealised situation (the phrase is Fisher's) enabled him to construct a theory which, although not deterministic, was at least directional in its analysis of evolution. According to Fisher, evolutionary 'progress' is a necessity, in the sense that fitness is a magnitude that is necessarily maximised.[11]

Sewall Wright took more or less the opposite theoretical tack. Impressed by the heterogeneity of the different strains of guinea-pigs that he had initially studied, Wright thought that in nature as in captivity, restricting the population size had drastic consequences.[12] He thus reasoned on the basis of populations that were divided into sub-units of limited size in which stochastic effects cannot be ignored. Any small population will tend to become inbred, in proportion to the time for which it is isolated, and will thus tend to have a homogeneous genome. Furthermore, population division means that migration – or interpopulational gene flow – needs to be integrated into the model.

These two opposing positions, with their purely empirical backdrop (the size and structure of natural populations), revealed the contours of two very different philosophies of evolution. Both approaches are thoroughly 'Darwinian', but they lead to very different visions of natural selection.

The Fisherian view of selection is thoroughly 'genic'.[13] Fisher conceived of natural selection as acting not on organisms, nor on the genome, nor on genetic complexes, but on genetic 'atoms'. For Fisher, the distribution of what he called the 'gene ratio' provided the ultimate expression of all selective effects, because in sexual organisms the gene remains the only trace of the existence of an individual in a population. In asexual reproduction, the organism transmits all its genes to its offspring. This means selection acts on the organism as a totality, and thus on a given combination of genes, all of which are transmitted. In sexual reproduction, however, this reasoning does not hold. In each generation gene combinations are broken up during meiosis by the segregation of chromosomes and by crossing over. Individual genomes are thus continually dissociated. Given that, by definition, selection acts on the heritable, all that counts in the final analysis is the 'mean effect' of each allele, all interactions pooled.[14] According to Fisher, the unit of selection cannot be anything other than the gene, and, strictly speaking, it is the genetic structure of the population, taken as a cloud of particles, that evolves.

There is no place for 'levels' of selection in such a representation of evolution. There is only one kind of selection, the selection of genes. Fisherian natural selection is thus a mass selection, and takes into account only the additive effects of genes. Its main limiting factor is the availability of new variation (mutation). This is why the high point of Fisher's evolutionist thought was a 'fundamental theorem of natural selection' that expressed the speed of action of selection as a function of additive genetic variance. This theorem, which will be discussed below (see section 9.2.1), is a sophisticated way of saying that, in the final analysis, only mutation and the selection of genes are pertinent factors in the modification of species. This implies that only determinist factors regulate evolution, even though mutation and selection can only be analysed statistically.

The selection of genes has implications for the significance of variation in a population. If selection acts specifically on genes, only genes have a selective value. Thus variation is a (perpetually) transient phenomenon: it is simply the diffusion of 'good' alleles in the population. Polymorphism is thus merely the indication of a process of diffusion which normally results in the fixation of the 'good allele'. Whether this process is slow or fast is another question. Haldane, later supported by Muller, tended to think that selection rates are high, that fixation occurs rapidly, and that there are thus relatively few polymorphic loci in a population. Fisher thought that selection intensity was extremely weak, and that at any given moment there are therefore many polymorphic loci. However, in both cases the atomistic representation of evolutionary change is presented against a backdrop of a population that, *ideally*, is homozygous at all loci. Both sides agreed that if the population is polymorphic it was only because selection takes time. But 'in the great majority of loci the normal condition is one of genetic uniformity', as Fisher put it.[15]

To sum up, the 'genic' view of selection illustrates what was described in the preceding chapter as one of the key aspects of the modern concept of selection – a dynamic of gene replacement. This in fact flows from the view of mutation taken by the first geneticists. Under experimental conditions, most mutations are harmful; this is why the 'wild' allele is classically called '+' or 'normal'. Projected onto evolutionary time, this suggests that there is always an 'optimal type' and that mutation, while it is indeed the fuel that drives the evolutionary engine, is a 'necessary evil' in life's perpetual runaway search for optimality.

Sewall Wright's view of selection makes a striking contrast. The American geneticist called his conception 'organismic', in order to emphasise what Fisher wrote out – interaction. Wright's great contribution was to articulate the interaction between evolutionary factors within a population, and the interaction of genes within the individual organism. In other words, he linked the structure of the population to the structure of the genome.

Wright's model is fundamentally based on the interaction between factors of evolution. The diagrams reproduced in Figure 35 are taken from a 1932 lecture that summed up the key conclusions of Wright's canonical paper 'Evolution in Mendelian Populations' (1931).[16] In these diagrams, the frequency (x) of an allele in a locus in the population is on the abscissa. If $x = 0$, the allele has disappeared; if $x = 1$, it is fixed. In the absence of any perturbing factor, the frequency of the allele will remain constant; in other words, the genetic structure of the population is stable.

Wright's model introduces four such perturbing factors: mutation, selection, sampling error due to population size and migration. The effect of each of these factors can easily be represented in a diagram of gene frequencies – the direct mutation pressure (U) of a gene reduces its frequency, while reverse mutations (V) increase its frequency; a favourable selection (S) also increases the frequency of the gene, while migration (M) increases or decreases the frequency of the gene, depending on the structure of the incoming population. Finally, random variations are by definition erratic, but ultimately lead to the fixation or the disappearance of the allele (with an equal probability). They have a greater effect when the population size (N) is small.[17] In 1931 each of these factors was subject to a classic algebraic analysis of the kinetic description of fixation or elimination.[18]

What happens if the factors are combined? If there are only deterministic factors (mutation, selection, migration), and if their action had a

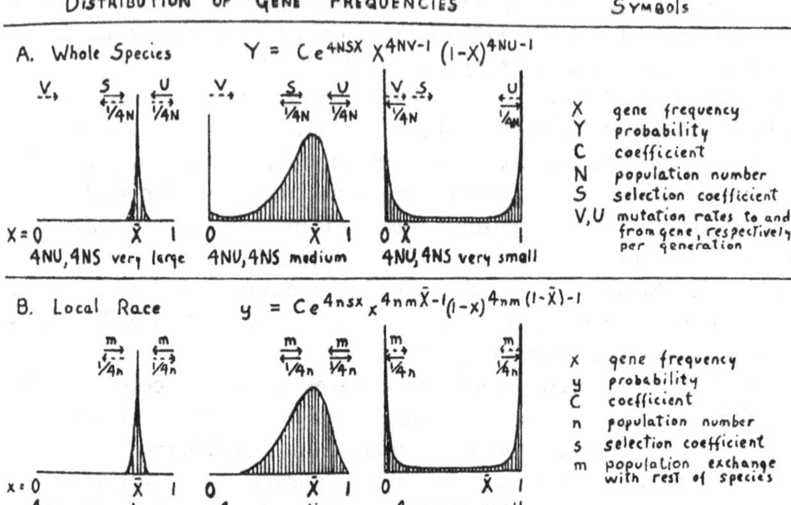

Figure 35 Random variability of a gene frequency under various specified conditions (S. Wright, 'The roles of mutation, inbreeding, crossbreeding and selection', 1932). The upper diagrams show the probability distribution of a gene frequency for a given locus as a function of various hypotheses of direct and reverse mutation pressure (V, U), direct selection pressure (S) and population size (N). The lower diagrams take into account the migration rate (m), which is formally analysed in a fashion similar to mutation pressure. In all cases, the key factor is the *relation* between the various determinist parameters (selection, mutation, migration) and the parameter that controls the stochastic effect (population size).

constant intensity, the population will tend towards equilibrium and thus a gene frequency that will not necessarily be equal to either 0 or 1. If there is random drift, all that can be defined is the probability that an allele will be fixed at a given value. This is exactly what Wright's diagrams show. Each gives the probable state of a population for a given locus, under various hypotheses. They can be interpreted as showing the state of a large number of populations subject to the same forces for the same locus, or (formally speaking this amounts to the same thing) to the state of a large number of loci subject to the same forces in a given population, or again, the mean state of an individual genome.

Using these schemata and their underlying algorithms, Wright gave a series of elegant definitions of the various possible evolutionary scenarios in a 'Mendelian population':

- If the population is very large (for example 10^{10} individuals), it evolves virtually deterministically. The amount of random variation is very small, and the fate of the system is determined by the equilibrium between mutation pressure and selection pressure. The rhythm of evolution is largely subordinated to the appearance of new variation by mutation; this rhythm is slow. This case is illustrated by the upper-left diagram in Figure 35.

- If the population is very small (upper-right diagram), stochastic factors will dominate, and the gene frequency will, in most cases, be fixed at either 0 or 1.
- The most interesting case is the intermediate one (middle diagram), where the deterministic equilibrium is modulated by a kind of freedom of exploration around an equilibrium point.

The meaning of these figures becomes clear only if one considers the implications for the whole genome. From this point of view, it is not genes that have a survival value, but groups of genes that are more or less well 'co-adapted'. When a species is divided into several small populations, each tends to become more or less homozygous; in other words, certain complexes of genes become stabilised. Many of these complexes are disadvantageous and the populations that carry them die out. But geographic division enables the species to explore a field of gene combinations. A variability of a different order of complexity than that associated with single genes thus becomes available for a second-order selection, which intervenes *between* populations. However, this interdemic selection is also expressed through migration rates – populations with the best-adapted gene complexes diffuse their genes into other populations. Thus, although the process is expressed in the canonical language of gene frequencies, it cannot be reduced to the selection of genes. This process is represented in the lower diagrams of Figure 35, which represent the combined effect of migration, of selection and of random frequency fluctuations in a local population. Formally speaking, the effects of migration are analogous to those of mutation. But there are nonetheless two different modes of selection – gene flow from one local population to another can be considered the measure of interpopulational selection.

According to Wright, the whole model illustrates the existence of a 'trial and error mechanism' that modulates the deterministic character of gene selection: '... there must be some trial and error mechanism on a grand scale by which the species may explore the region surrounding the small portion of the field which it occupies. To evolve, the species must not be under the strict control of natural selection.'[19]

The opposition between the two different levels of selection, and the notion of 'freedom' that it involves, is reminiscent of the famous quarrel that erupted in the behavioural sciences at the beginning of the century, at about the time Wright was a student. Jacques Loeb had explained the orienting movements of plants and lower organisms by what he called 'tropisms', that is, by reactions that were strictly determined by an anisotropy of the external medium (for example, gravity or light). Herbert S. Jennings opposed this deterministic concept by citing his observations that protists showed trial-and-error learning. Wright's philosophical writings reveal that he was familiar with this debate, that he viewed it as analogous to the dispute between himself and Fisher over the problem of the levels of selection, and that he had a hierarchical vision of reality in which levels of complexity are increasing degrees of 'liberty'.[20]

Wright described his conception of evolution as the 'shifting balance theory'. This expression sums up the idea that the deterministic equilibrium of a Mendelian population is subject to random variations. It also expresses the idea that evolution can take several different forms, each characterised by the weight of different factors of evolution. Given that the 'balance' of these factors is 'shifting', Wright's model implies that species shift between different evolutionary scenarios.

The most remarkable conceptual aspect of Wright's position was the link it forged between population structure and genome structure. To the extent that sexual species are divided into small populations, selection can operate on gene complexes as well as on genes. When there is an interaction between selection and drift, the physiological interaction of genes can be exposed to the action of selection because such populations are partially homozygous and their gene complexes are partially stabilised. Unlike Fisher's, this conception of evolution is clearly hierarchical. In a way, it resolves the veiled dialectical tension noted earlier in the first public formulation of the principle of selection. When Darwin and Wallace proposed their common hypothesis before the Linnean Society of London in 1858, they did not describe exactly the same thing. Darwinian selection implied the differential survival of inherited individual variations; Wallace's principle was more oriented to competition between populations. Wright's subtle articulation of the intra-populational selection of genes and selection between populations was a way of going beyond this debate. The nature of selection is a function of population structure. If the species has the structure of a large panmictic population, only the gene counts. The gene is the population's representation of the individual (or the gene is what the population retains of the individual). If the species is subdivided into partially isolated groups, selection can 'see' genomic complexity – the overall zygotic genotye can be seen as the species' representation of the elementary population (more or less inbred and homozygous).[21]

To summarise the two visions of natural selection developed by Fisher and Wright: they used the same methodological tools, and during the 1930s quickly agreed on the relevant algorithms. They studied the same theoretical object – the multidimensional field of gene frequencies – but developed different philosophical interpretations of this idealised object. For Fisher, the modification of the genetic structure of a population was the reality of evolution; for Wright, on the other hand, it was merely the phenomenal manifestation of that evolution. Although for Fisher, steeped in the model of statistical thermodynamics, the 'gene ratio' constituted the real state of the species, for Wright the modification of the genetic composition of populations was the expression rather than the ultimate reality of evolution.

9.2 Algorithms and images: the fundamental theorem of natural selection and adaptive landscapes

The genetic theory of selection rapidly divided into two competing interpretations of population genetics, clearly expressed in the theses, methodologies and beliefs of Fisher and Wright. Fisher's central thesis was that the evolutionary process is dominated by the mass selection of genetic atoms; Wright's central thesis was that the action of selection is modulated by random variations of various types and sizes, which have to be understood hierarchically. Fisher's method (in genetics as in statistics) isolated additive effects so that factors with a macroscopic determinism could emerge out of the chaos. Wright's method, on the contrary, emphasised all kinds of interactions and then tried to find algebraic, experimental and rhetorical ways of studying them. These theses and methodological approaches also corresponded to two visions of the meaning of science, which will be discussed at the end of this chapter.

These positions culminated in two remarkable theoretical constructions, known as the 'fundamental theorem of natural selection' and 'adaptive landscapes'. Their underlying ambition was to understand under what theoretical conditions selection can produce adaptation, and thus be the basis of a causal theory of the modification of species. These two theoretical constructions were based on a common question – is the dynamic of selection a maximisation of a magnitude, 'fitness'?[22]

In the 'fundamental theorem' and in the notion of 'adaptive surface', Fisher and Wright sought to condense their whole vision of the genetic theory of evolution. This makes an epistemological analysis particularly appropriate.

9.2.1 *Fisher's 'fundamental theorem of natural selection' (1930)*

In 1930, in a book that was presented as the first synthesis of what was soon to be called 'population genetics', Fisher proposed to give natural selection the form of a theorem. This theorem was extremely simple – a differential equation that expressed the fact that, at any given moment, the rate of increase of mean fitness in a population is strictly equal to the genetic variance of fitness in the population. Fisher called this theorem 'fundamental' and argued that it was destined to occupy a place in biology akin to that of the second principle of thermodynamics in the physical sciences.[23]

Since 1930, many commentators have criticised Fisher's extravagant pretension, the more obscure parts of the demonstration and the large number of restrictive conditions that affect the theorem's validity. A substantial number of 'refutations' have been proposed, together with a far smaller number of attempts at 'redemonstrating' the theorem.

Reading this literature is, it must be admitted, not easy. It can even be frankly discouraging. Nevertheless, study of Fisher's extremely abstract and badly worked-out demonstration[24] leads to two linked conclusions:

- Most of the 'refutations' and 'redemonstrations'[25] are based on initial hypotheses that are not Fisher's, and that follow deductive paths very different from his.
- Both kinds of study generally conclude that the theorem is valid only under a series of preconditions that are so restrictive that the theorem becomes pointless. Fisher, however, continually repeated that the whole point of his theorem was that it was independent of the restrictive conditions that people sought to impose upon it.

These points lead to the suspicion that, as Motoo Kimura and George Price[26] have pointed out, most of Fisher's readers have simply not understood the theorem. If this is the case, much of the blame must lie with Fisher's presentation. To make matters easier, an exegesis of the theorem will be provided here. First, the terms of the theorem will be explained, and then its specificities will be shown to imply a number of extremely instructive epistemological lessons that are just as important as discussions of the deduction of the theory.

9.2.1.1 *The exposition of the 'fundamental theorem of natural selection'*

> 'The rate of increase in fitness of any organism at any time is equal to its genetic variance in fitness at that time.'[27]

This is the verbal exposition that concludes Fisher's mathematical demonstration of 1930. Most discussions of the theorem in fact refer to this formulation, which contains two elements that are clarified by the immediate context:

- By 'fitness of any organism', Fisher in fact means 'the mean fitness of a population' (more exactly, the mean fitness of the species[28] envisaged as an idealised panmictic population),[29] which can thus be considered 'a well-defined statistical attribute of the population'.[30] In 1941 Fisher indeed substituted 'population' for 'organism'.[31] Despite Albert Jacquard's suggestions to the contrary,[32] this does not affect the meaning of the theorem as it was set out in 1930 – the populational context is quite explicit. 'Fitness' is not individual 'adaptation' (this would be meaningless), but a demographic parameter which describes the growth rate of the population. Fisher explicitly distinguishes between the qualitative notion of 'adaptation' and the statistical notion of 'fitness'.[33]

- By 'genetic variance in fitness', Fisher always meant additive genetic variance, that is, the variance that can be attributed to the additive effects of genes, as against the variation due to the interaction of alleles (dominance) and to the interaction of genes (epistasis). This precise use of the term 'genetic' is one of the keys to understanding the difficulties of the text. This will be dealt with in greater detail.

Now that I have made these preliminary points, an unusual aspect of the presentation of the theorem needs to be noted. Fisher gives only a verbal exposition, despite the fact that the demonstration is in algebra. Fisher never gave an algebraic expression of the theorem, neither in its original presentation nor in any of his subsequent reexaminations of the question.

What form might such an expression take? Using the same symbolism proposed by Fisher in chapter 2 of *The Genetical Theory of Natural Selection* (1930), where he demonstrates the theorem, the verbal exposition clearly implies a differential equation such as:

$$dM / dt = W$$

where M = the mean fitness of the population, and
W = the genetic variation in fitness.[34]

However, Fisher never wrote such a formula, and only ever expressed his theorem in terms of natural language. This omission, confirmed in various revised versions from 1941 to 1958, is in fact linked to the meaning of the theorem.

Before dealing with this point, the exact meaning of 'fitness' needs to be understood. Fisher introduced this term into population genetics when he elaborated his fundamental theorem, and in so doing considerably altered its meaning. In evolutionary theory, this term had its origin in a reversal of Spencer's expression 'survival of the fittest', which was transformed into 'fitness to survive' and was finally reduced to 'fitness'. This metamorphosis of the qualitative notion of 'fitness' (as a synonym of adaptation) into a measurable value is not easy to reconstruct.[35] Fisher appears to have based himself on two traditions – on the one hand that of Pearson, who was more or less determined to make fitness into a statistical value summarising differential viability and differential fecundity (see section 7.2.3), and on the other that of Lotka, who in 1914 had proposed that the demographic parameter 'r' or the 'growth rate of a species by head and by unit of time' (later called the 'innate rate' of growth of a population or the 'intrinsic rate') could be 'considered as a quantitative index of the fitness of an organism'.[36] This parameter simultaneously takes into account the probability of survival (life expectancy at a given age), the probability of reproduction at a given age and the overlap of generations (and thus of age distributions).

The parameter 'r' is also found in the demonstration of Fisher's fundamental theorem, but under another name. Fisher uses 'm' – 'Malthusian population growth parameter'. The parameter 'm' is the number that will solve the equation:

$$\int_0^\infty e^{-mx} l_x b_x dx = 1$$

where l_x is the probability of survival from birth to age x, and b_x is the rate of reproduction at age x.

In this equation, 'm' measures the 'present value' of a newborn; in the 1930 version, the mean value of this parameter, 'M', also appears. 'M' is the geometric rate of growth of the population, or the 'Malthusian parameter' itself. Fisher first presented this equation in 1927.[37] It is very similar to one previously proposed by Lotka[38] (the two men subsequently engaged in an unseemly squabble over priority).

This equation gives a precise operational definition to 'fitness'. Measured by the Malthusian parameter, fitness is a rate of population growth. It has no dimensions, because it is simply a rate of reproduction – a number of individuals per individual (offspring per parent).[39] Fitness is thus a 'well-defined statistical attribute of the population'.[40] Fisher often remarked that this parameter can take, and often does take, a negative value.[41] This implies that the fundamental theorem does not mean that dM/dt = W. Quite simply, given that W is a variance, this term can only ever be either positive or zero.

How did Fisher establish his theorem? The proof is entirely based on a model that he developed in 1916 to analyse the inheritance of quantitative characters under the Mendelian hypothesis.[42] This model, which still has canonical value in quantitative genetics, is based on two ingenious techniques. The first permits the analysis of the variance of a character into the additive effects of genes, dominance, epistasis and environmental effects. It was in this canonical study that Fisher first proposed the terms and the concepts 'variance' and 'covariance', as well as the analysis of variance. The second aspect of Fisher's model is a way of evaluating the 'mean effect' of a gene substitution on the mean value of a quantitative character. This classic technique, which is included in all quantitative genetics textbooks, will not be analysed in any detail.[43] We can simply say that it isolates the effect of an allelic substitution (for example, the substitution for 'A' by 'a') by neutralising the effects of interaction (dominance and epistasis). The key idea is that only genes, and not their interaction, are transmitted to offspring.

The demonstration of the fundamental theorem is based on the application of this analysis to fitness itself, treated as a quantitative character like weight or height. Fisher's explicit objective – too often misunderstood – was to isolate something in the evolution of a population that could be attributed to natural selection, and only to natural selection, under the Mendelian hypothesis. In other words, to find the level of analysis at which something is transmitted (the gene, not the genotype) that influences the probability of the production of offspring.

We can now return to the exposition of the fundamental theorem. Despite its rigorous demonstration, it contains a paradox. Taken at face value ($dM/dt = W$), it apparently means that, necessarily and under all conditions, the size of any population increases as soon as it contains a potential genetic variation that affects fitness. However, this clearly does not apply to natural populations. Fisher was well aware of this, because he accepted that the mean value of the Malthusian parameter (M) could not be much greater than 0, and that it could decrease, because of the 'constant deterioration of the environment' (the effect of the evolution of other species) and because of overpopulation.

It is highly significant that it is only at this point that Fisher provides an algebraic formula. However, this formula, which relates to a situation dealing with more parameters, is not that of the theorem. Nevertheless, it is essential for understanding the theorem. This formula is to be found several pages after the demonstration of the theorem, in a kind of appendix which in fact constitutes an explanation of the meaning of the theorem.[44] The formula given by Fisher is as follows:[45]

$$\frac{dM}{dt} + \frac{M}{C} = W - D$$

where M is 'the mean of the Malthusian parameter'[46] and measures the fitness of the population;

C is a constant (expressed in units of time), expressing 'the relation between fitness and population increase, and defined as the increase in the natural logarithm of the population, supposed stationary at each stage, produced by unit increase in the value of M'.[47] C measures the ease with which the population grows, when its fitness grows. The ratio M/C expresses the deleterious effects of overpopulation.

W is the additive genetic variation in fitness. But given the fundamental theorem established a few pages earlier, Fisher now describes W as 'the rate of actual increase in fitness determined by natural selection'.[48]

D is the 'rate of loss due to the deterioration of the environment'.[49]

The somewhat open-ended formula reproduced above provides the key to the fundamental theorem. It can be rewritten as follows:

$$\frac{dM}{dt} = W - \left(D + \frac{M}{C}\right)$$

It now becomes obvious why the fundamental theorem cannot be written as:

$$\frac{dM}{dt} = W$$

Fisher's explicit assertion is that the rate of increase of fitness at any given moment (dM/dt) is the result of a balance between two types of forces: on the one hand the additive genetic variance in fitness (W), on the other the totality of effects that can be attributed to the environment (the limiting effect of overpopulation, and the deterioration of the environment). Nothing suggests that the rate of increase must always be positive. Quite the opposite; the deterioration of the environment 'will tend, in the majority of species, constantly to lower the average value of m'.[50] In most cases, the increase in the Malthusian parameter will in fact have a value close to zero.[51]

This leads to an initial conclusion: the fundamental theorem (the exposition of which, it should be recalled, is purely verbal) does not deal with a rate of total increase of fitness and thus cannot be represented by dM/dt = W.

The theorem nevertheless deals with a value that is continually maximised, because it posits an equality between a rate of increase of something and a variance, which by definition is always positive (or zero). What is this mysterious value, if not total fitness? And, correlatively, why did Fisher not give a symbolic formulation of his theorem, given that the demonstration is algebraic and is symbolically expressed?

The answer to this question lies in the critical discussions that the fundamental theorem has provoked for over sixty years.[52]

Most attempts at redemonstrating the theorem have led to an

accumulation of restrictive conditions. Schematically, the theorem can be proven only if the following conditions exist:

- Panmixia
- Frequency-dependent genotypic selection coefficients
- No dominance
- No epistatic interaction
- No linkage disequilibrium (which means that the theorem most probably has no significance whatsoever for any character involving several loci)

Fisher admitted only the first of these conditions.[53] The second refers to methods of demonstration that he did not employ, because he did not reason on the basis of genotypes but on the basis of the mean effect of an allelic substitution. As for the other conditions, they correspond to the list of factors that Fisher thought his theorem was independent of. This discordance is even more curious given that Fisher maintained his theorem without any modifications, despite such criticisms. His correspondence provides some clues – it shows that he made no retractions, but instead became increasingly annoyed with his critics, who, as far as he was concerned, were attacking something different from the 'fundamental theorem'.[54]

The mysterious character of Fisher's theorem is in fact linked to his completely unorthodox use of the terms 'natural selection', 'environment' and 'genetic variance'. The late George Price made a remarkable contribution to unravelling this mess by calling attention to the (verbal) variants of the theorem present in Fisher's work,[55] two of which show a particularly revealing similarity. These two variants are those in which Fisher, instead of dealing with the 'rate of increase in fitness', writes either of 'the rate of increase in fitness due to all changes in gene ratio'[56] or of 'the rate of increase in the average value of the Malthusian parameter ascribable to natural selection'.[57] These two formulations can justifiably be criticised for their confused formulations – typical of Fisher's discussions of changes in fitness due to factors other than 'natural selection', in particular changes in the environment, which are always presented as leading to its deterioration.[58]

If these formulations mean anything at all, it can only be one thing. This interpretation can be expressed in a series of linked assertions:

1. By 'environment', Fisher meant not only the external environment, but also the totality of interactions within the genome – dominance, epistasis, linkage. The 'environment' is the totality of interactions of the genetic 'atoms' with something other than themselves, in the space-time of a panmictic population.

2. By 'natural selection', Fisher meant a process that can only be

properly understood and formalised by reference to a domain defined by the field of gene frequencies. Fisher did not deny that allelic interaction (dominance) and epistatic interaction had played a role in evolution. Indeed, he was the first to work out algorithms enabling the calculation of the populational effects of heterosis and dominance.[59] Of course, interaction between genes has a fundamental importance, like the interaction with the external environment, but both are on the side of the forces that orient the process of natural selection, not on the side of the process itself. Fisher argued that 'natural selection' should be considered with reference to the distribution of gene frequencies, and not to the field of gene combinations.[60] Failure to understand this subversion of the concept of natural selection in Fisher's work leads to a failure to understand the theorem itself.

3. The two preceding remarks allow the identification (at least nominal) of the enigmatic value which, in the theorem, is permanently maximised. The increase in the Malthusian parameter has to be separated out into a component that can be attributed to 'natural selection' and a component that can be attributed to 'environmental changes' (which includes everything that does not consist of the additive effects of genes). George Price[61] gave a symbolic representation of this interpretation by separating out the differential increase dM:

$$d\,M = d_{SN}\,M + d_{CE}\,M$$

where SN is natural selection and CE is changes in the environment.

This symbolism is not very orthodox, which may explain why Fisher did not use it, but instead restricted himself to less precise verbal explanations. However, the theorem concerns only the first part of the equation. This is the part that is always positive (or zero), and which equals the 'genetic variance in fitness', an expression which, in Fisher's work, always denotes the additive part of this variance. What he thought made the theorem 'fundamental' was precisely the fact that it identified the level at which something was necessarily maximised by natural selection.

9.2.1.2 The epistemological profile of the 'fundamental theorem'. The exposition of the 'fundamental theorem of natural selection' raises several points which lead to a more rigorous definition of Fisher's biological philosophy.

1. Fisher's theorem is a manifesto in favour of a purely 'genic' conception of selection. In a 1955 letter to Kempthorne, who at the time was writing an article on the fundamental theorem, Fisher declared: '... the only evolutionary effect, either in increased fitness or in anything else, that I can recognize as such, is constituted by the changes in gene ratio'.[62]

Everything else, that is, everything which is traditionally described as interactions within the genome (dominance interaction and epistasis), is reduced to 'the environment'. Apart from the dynamic alteration of the 'gene ratio', there is merely noise or 'passive' reaction – nothing that could strictly speaking be described as an 'evolutionary effect':

> . . . if by the extinction of certain insects a plant were rapidly to become generally self-fertilized and homozygous through lack of means to cross-pollination, I should, so long as the gene ratios remained unchanged, consider that the plant had not evolved but was reacting passively to its changed environment.[63]

For Fisher, the mating system and the system of allelic and epistatic interactions have an evolutionary significance only if they are expressed in the realm of gene frequencies.

Faced with such dogmatism, it is tempting to see Fisher as the ancestor of those atomistic and simplistic views of natural selection that have become notorious over the last two decades. It should, however, be pointed out that the epistemological consequences of Fisher's attitude went far beyond the naive symbolism of the 'selfish gene'. After all, the fundamental theorem is an explicit way of expressing the canonical position which was to be the very basis of population genetics: all observable processes of evolution (mutation, migration, selection, random variations . . .) can be described in the homogeneous language of the statistical distribution of genes. In this respect, whatever one might think of its biological consequences, the theorem is marked by an extreme rigour. Unlike Fisher, too many contemporary evolutionary biologists have simultaneously and uncritically admitted both that 'evolution is a change in the genetic composition of populations' and that it is not genes *per se* that are selected.

2. Fisher's fundamental theorem merely expresses a *speed*, or more exactly an acceleration. (In fact, the 'Malthusian parameter' M measures population growth rate, and the theorem itself deals with the 'rate of increase' of this value.) This is both the elegance and the limitation of the differential equation. Fisher's work was marked by a leitmotif typical of physicists at the beginning of the century, and popularised amongst biologists by Lotka. The fundamental equations of the evolution of a system, Lotka liked to say, 'assume the simplest, the most perspicuous form, when they are written relative to rates of change of the state of the system, rather than to this state itself'.[64] Applied to population genetics, this implies that it is more pertinent to deal with changes in the distribution of gene frequencies than with modifications in the population of genotypes. For Fisher, the fundamental theorem occupied a methodological place similar to the fundamental principle of momentum in physics

($\vec{F} = m\vec{\gamma}$). The theorem means that at the level of the 'gene ratio' it is possible to identify the specific effect of the force that is natural selection: it is the rate at which M grows, M being a speed.

Fisher found his inspiration in physics. As he wrote in 1953, 'the frequencies with which the different genotypes occur define the gene ratios characteristic of the population, so that it is often convenient to view a natural population as an aggregate of gene ratios. Such a change of viewpoint is similar to that familiar in the theory of gases, where the specification of the population velocities is often more useful than that of a population of particles.'[65] This remark was at the heart of Fisher's criticism of Wright – because he saw selection as operating on a field of gene combinations, Wright did not seem to realise that the number of possible Mendelian combinations is so great that, in practice, the genome of each individual is unique. Calculating any kind of kinetic description of selection for such a massive field of variability is virtually impossible.[66] The same criticism does not apply to the field of gene frequencies, which is continuous.

3. The fundamental theorem is perhaps more significant because of what it leaves out. In theorising a mass selection acting indefinitely and gradually on populations sufficiently large not to be affected by random drift and isolation, the 'fundamental' theorem is the key point of a scenario that has very little to say about the diversification of species, their origin and their extinction.

4. As the reader may already suspect, this major biological limitation paralleled Fisher's well-known eugenic positions: the eugenicists were less interested in the origin and extinction of species than in the 'improvement' of existing species, in particular human populations, which form one of the best examples of a large panmictic population.[67]

9.2.2 *Sewall Wright's adaptive landscapes (1932)*

The fact that the most theoretically minded population geneticists were fascinated by Fisher's 'fundamental theorem' can probably be explained in two ways. First, its algebraic form shows it to be a theorem of maximisation (very different from the qualitative discussions of 'optimisation' which characterise the work of so many evolutionary biologists).[68] Second, the maximised magnitude ('fitness') is a demographic parameter. It is difficult to imagine a quantitative paraphrase that would be more redolent of the principle which Darwin had declared 'leads to the improvement of each creature in relation to its organic and inorganic conditions of life'[69] (second edition of the *Origin of Species* – December 1859).

The explicit limits of Fisher's method correspond to his ambitions. By concentrating on additive genetic variance, Fisher put himself in an ideal position where the effects of genes can be described by functions with independent variables. Under such conditions, one can be sure that there will be a fitness function and that it will have a defined maximum value. Second, by dealing with very large panmictic populations, and examining them only in terms of their growth, Fisher avoided all the problems linked to the restriction of sample size and to population structure.

It has been seen (section 9.1) how, from 1930 onwards, Sewall Wright explored the linked implications of the opposite starting point. The 'shifting balance theory' is based on the idea that species are not virtually infinitely large panmictic populations in which genes circulate completely freely. Even under the hypothesis that a species is a large population with a continuous geographical distribution, total panmixia is a fiction – distance prevents the completely free circulation of genes and produces isolation effects.[70] Thus species cannot be represented as isotropic Mendelian clouds; they are always finite and structured populations. As a result, natural selection has to deal with stochastic forces of nature and, given that it acts hierarchically, functions between elementary populations known as 'demes'.

Wright thought it useful to express this model in terms of an image known as an 'adaptive landscape'. Apart from the name, this image was first presented in a 1932 lecture given to a non-mathematical audience.[71] Although the context suggests that it was a pedagogic device, its influence on modern evolutionary theory has been colossal. But for Wright it was not merely an image. Beyond the immediate success of his 1932 lecture, Wright continued to return to the question, and to find a formal interpretation for his 'landscape'. He thus gave shape to the idea of a 'fitness surface' and of a 'surface of mean selective values'.[72] The place of these two concepts in Wright's work was comparable to that of the 'fundamental theorem of natural selection' in Fisher's career. What Fisher expressed in a relatively simple theorem based upon a purely additive model, Wright attempted to study in a hyperspace in which all the parameters can interact. This complexity, however, prevents the development of any kind of algorithm. The image of an adaptive landscape thus takes on a heuristic value; despite not being mathematically respectable, in Wright's work it plays the role of a figurative auxiliary that indicates the direction in which the theory should develop. This explains why people often speak of Wright's 'adaptive landscapes',[73] the expression referring at least as much to the image as to the formal schemata that lie behind it.

There is an intimate conceptual and historical link between Wright's adaptive landscapes and Fisher's fundamental theorem. In both cases, the

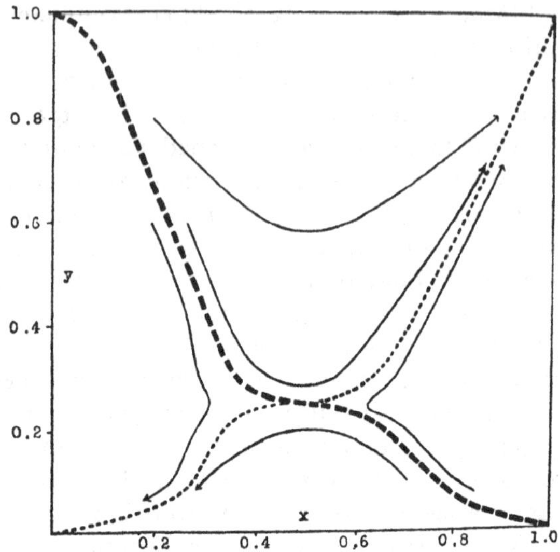

Figure 36 Evolution of a population in which selection operates on two epistatic diallel loci (the selection coefficients of each genotype for a given locus depend on the genotype of the other locus). The diagram represents a specific numerical example calculated and drawn by C. H. Waddington. Axes (x and y): frequencies of alleles a (against A) and b (against B) in the population. Solid lines: examples of trajectories. The dotted lines represent the frontiers of different classes of trajectories. There are an unstable equilibrium (at the point where the dotted lines cross) and two stable equilibria (upper right and lower left of the diagram). If the mean adaptive value of the population were represented in a third dimension, this would give a typical adaptive surface, with a col (0.5–0.2) and two summits (1–1 and 0–0).

key question is how fitness can be maximised. However, Wright's incredibly suggestive schema – perhaps the most elegant moment in the whole history of Darwinism (including Darwin) – was nevertheless merely an image: its ambiguities are due to fundamental methodological difficulties.

The history of adaptive landscapes begins some time before the 1932 lecture in which Wright first presented the idea.

The diagram presented in Figure 36 was devised by J. B. S. Haldane and Conrad Waddington in 1931. It is taken from one of Haldane's ten articles entitled 'A Mathematical Theory of Artificial and Natural Selection' (1924–34), that on 'metastable populations' (1931).[74] It contains the first model describing the evolution of a population subject to selection on two loci showing epistatic interaction. This simply means that the selective values of one locus depend on the state of the other locus. The two axes show the frequencies of the recessive alleles (a and b) of the two loci. The selection coefficients are defined by four possible phenotypes: [AB]: 1; [aB]: $1-k_1$; [Ab]: $1-k_2$; [ab]: $1+K$ (k_1 and k_2 are small and positive; K can be positive or negative, but its absolute value is less than k_1 or k_2). As a result, the selection coefficients of [AB] and [ab] are always greater than the other two. This system shows a deterministic evolution towards one of two equilibrium states: one of these corresponds to the fixation of the two recessive alleles a and b (top right), the other to the fixation of the dominant alleles A and B (bottom left). The actual evolution of the system will depend on its initial state. There is also an unstable equilibrium (see Figure 36).

The key point here is that Haldane was reasoning on the basis of a continuous field of gene frequencies – each point represents the state of a population. Indeed, Haldane suggests that the model should be extended to *m* genes. In this case, each population will be represented in an *m*-dimensional space and will be deployed in a hypercube with $2m$ summits. Each of these $2m$ summits represents one of the possible situations of homozygosity, and there are at most $2m - 1$ stable equilibrium positions. Of course, the situation can be complicated by assuming that loci have multiple alleles, with various modes of allelic interaction (semi-dominance and heterosis). But in these circumstances the equilibrium points will not necessarily correspond to summits within the hypercube – in other words, to the fixation of alleles in the population. This is intuitively the case for heterosis (heterozygote advantage). If the favoured genotype at a given locus is Aa, a polymorphic equilibrium is established.

From this point, it is easy to imagine the development of the idea of a surface of mean selective value. All that is necessary is to add to Haldane's diagram a third dimension which would measure the mean adaptive value (or fitness) of the population. The two equilibrium positions then correspond to the summits towards which the population can climb, and the unstable equilibrium corresponds to a col, or saddle.

However, this does not appear to have been Wright's route to the 'adaptive landscape'. William Provine, in his remarkable biography of Wright, has published a crucial letter of February 1931 in which Wright explained to Fisher his disagreement with the 'fundamental theorem of natural selection'. This letter also contains the first formulation of the idea of fitness surface:

> Think of the field of visible joint frequencies of all genes as spread out in a multi-dimensional space. Add another dimension measuring degree of fitness. The field would be very humpy in relation to the latter because of epistatic relations, groups of mutations which were deleterious individually producing a harmonious result in combination.[75]

The letter includes a drawing which schematically shows a line between peaks and valleys, symbolising the fitness maxima and minima. This representation enabled Wright to give a suggestive explanation of the reasons that led him to reject the fundamental theorem and its associated evolutionary scenario. If epistasis does indeed imply that there are not one but many equilibrium points (in other words, several possible fitness maxima), can a given species move from one adaptive summit to another, and if so, how? Fisher's theorem describes only how a species climbs a summit. Furthermore, the process described by the theorem is extremely slow, because it depends on the rate at which new mutations appear. But if the size and structure of the population are taken into account, a very different scenario appears. On the one hand, limiting the population size shows how a population can randomly deviate from the equilibrium point to which it had been carried by selection: stochastic fluctuations of gene frequencies can push the population from its present 'adaptive peak' towards another 'peak'. The repetition of this process in many isolated sub-populations implies that the species explores the adaptive peaks to which it has access. The best-adapted local races proliferate, their complexes of co-adapted alleles are diffused by migration, and an interpopulational selection is thus added to the effects of the selection of genes. In this scenario, the fundamental theorem represents merely one possibility – that in which the species, being very large and panmictic, remains fixed on an adaptive peak. The famous drawings shown in Figure 37 illustrate this idea. They need to be studied in the context of those in Figure 35, with which they were presented in the lecture given in 1932 to the Sixth International Congress of Genetics.

In his letter to Fisher of 3 February 1931, Wright gave an elegant summary of his disagreement with Fisher's theorem. First he declared that in his review of *The Genetical Theory of Natural Selection*[76] he had not understood that the theorem applied only under the condition of the constancy of the external milieu *and* of 'all internal factors'. This led him to reformulate the theorem as follows: 'The rate of increase of fitness of any population at any time is equal to its genetic variance in fitness at any time, except as affected by mutation, migration, change of environment and the effects of random sampling.'[77] The theorem having been reinterpreted, it was easy for Wright to define his own conception of the genetic evolution of species. Four factors intervene: 'First, an irregularly changing environment ... Second, novel mutations in an indefinitely extended series ... Third, limitation in the size of the population to such a figure that random variation in gene frequency becomes important ... Fourth, subdivision of an indefinitely large species into many small, not quite completely isolated groups.'[78] Wright made clear that the first two factors

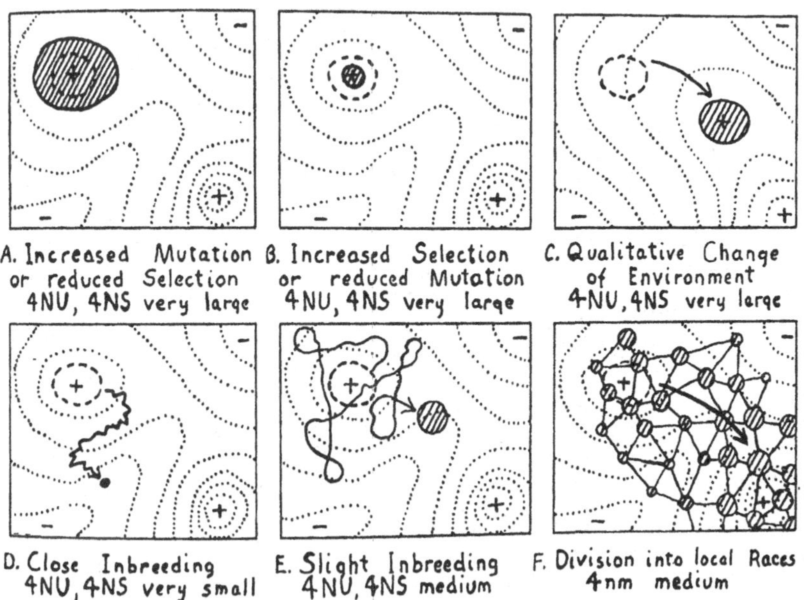

Figure 37 Field of gene combinations occupied by a population within the general field of possible combinations (original legend) (S. Wright, 'The roles of mutation, inbreeding, crossbreeding, and selection', *Proceedings of the 6th International Congress of Genetics*, **1**, 1932). N, U, m and S refer to population size, mutation rate, migration rate and the selection coefficient. The figure shows the interaction of these factors of evolution in a Mendelian species. Compare this representation with the diagrams reproduced in Figure 35.

modify the surface (the change of environment deforms it; mutation creates new valleys), while the other two provide a way of exploring the surface, stopping a species from getting trapped on a peak (random genetic drift enables a population to move from a peak; the geographical division of the species permits the simultaneous exploration of several peaks).

Up to this point, everything is quite straightforward. However, the text that accompanies the figures from Wright's 1932 lecture shows that Wright was discussing something different from what was raised in his letter to Fisher. According to both the text and the figure legends (see Figures 37 and 38), the populations are situated in a 'field of possible gene *combinations*' (added emphasis). The adaptive hypersurface is thus created by the total system of possible genotypes, taking into account all loci. If, for example, the genome contains a thousand diallel loci, with

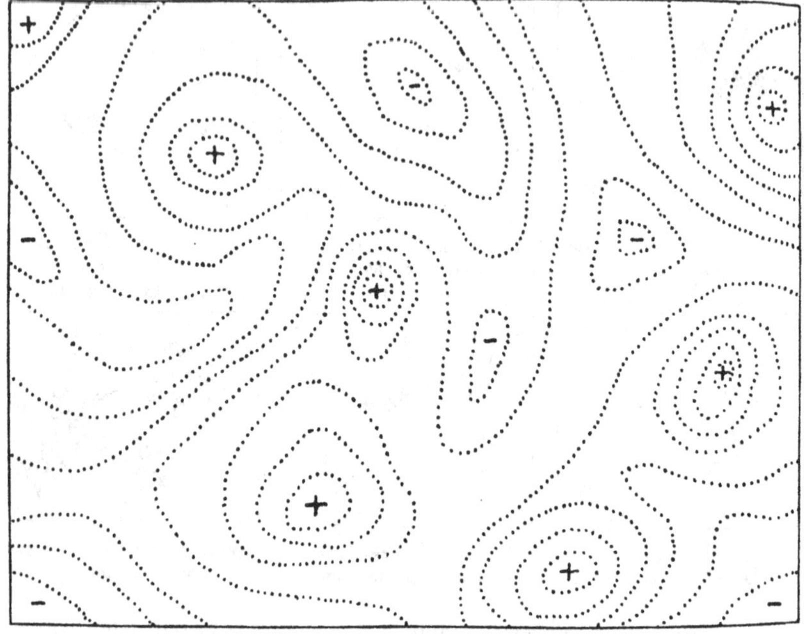

Figure 38 The first image of an adaptive surface (S. Wright, as Fig. 37). Original legend: 'Diagrammatic representation of the field of gene combinations in two dimensions instead of many thousands'. Dotted lines represent contours with respect to adaptiveness. Note that each point on the diagram represents a combination – the only parameter with a dimension is adaptive value.

three possible genotypes per locus, the total number of different possible genotypes, taking into account all loci, is 3^{1000}. Such a field does not have the same meaning as a field of gene frequencies. A field of gene frequencies enables a population to be situated at a precise point. A field of 'possible combinations', on the other hand, corresponds to the totality of possible individual genotypes. Given that, in practice, all individuals have different overall genotypes, the population is no longer represented by a point on a surface, but by a cloud of points grouped in a region of the space of possible combinations. Thus each individual total genotype has a global selective value, and the adaptive surface is such that each of its points represents the selective value of a particular genotype. As to the adaptive peaks, they correspond to clusters of similar genotypes, grouped around an optimal combination. The well-known drawings reproduced in Figure 37 show precisely this.

William Provine, who has provided an extremely clear analysis of this historic episode, makes a very interesting remark about this point. How can you have 'surface' produced by combinations of discrete entities? In a field of 'gene combinations', what would be the axes for which a certain value (fitness) would represent the degrees? Wright's procedure cannot produce a 'surface' in any meaningful sense of the term.[79]

In its initial version, therefore, the algebraic meaning of the notion of an adaptive surface was ambiguous. It was, however, a striking popularisation of the 'shifting balance theory of evolution', elaborated a year earlier in a far more rigorous (but also difficult) article.[80]

A few years later, in 1935 and then in 1937,[81] Wright discussed the idea of adaptive surfaces in terms of a multidimensional field of gene frequencies. The surface represented the variation of the mean fitness of the population plotted against the continuous field of gene frequencies. As to whether there was any point in representing the mean fitness of a population (which Wright called $\overline{W}$, the equivalent of Fisher's e^m)[82] as a function of all gene frequencies,[83] this led to some particularly bitter disputes. In fact, the problem involves such a large number of dependent variables that the very idea of trying to calculate them becomes somewhat improbable. This is probably the reason why Wright always described his surfaces of selective values in qualitative terms as 'harmonious combinations of genes'.[84] The result is a unique hybrid style, in which the mathematician's quiet confidence rubs shoulders with a poetic vision of natural economy.

The epistemological meaning of this style, and the reasons for its astonishing fate in contemporary evolutionary theory, are particularly important. Somewhere between algorithm and image, the adaptive landscape is without doubt one of the central *concepts* of contemporary Darwinism. To fully understand it requires a philosophical investigation in order, to paraphrase one of the 20th century's greatest philosophers, to 'make . . . clear' and 'give . . . sharp boundaries' to thoughts which 'are, as it were, cloudy and indistinct'.[85]

To be rigorous we have to take full account of the metaphorical aspect of Wright's central conception. If the 'landscape' and the 'surface' are images, what exactly are they the images of? There is an aspect of Wright's imagery that neither the current evolutionary literature nor contemporary historiography has previously discussed. In mathematical physics, relief is a classic image of the potential of an entropy-producing system (or, in today's terms, of a 'dissipative' system). In such a system all movement leads to a resting position, or 'equilibrium', and the 'potential function' is the mathematical function that describes this movement. This dynamic is one of the simplest that can be imagined – whatever the initial conditions, after a certain amount of time the system always finds

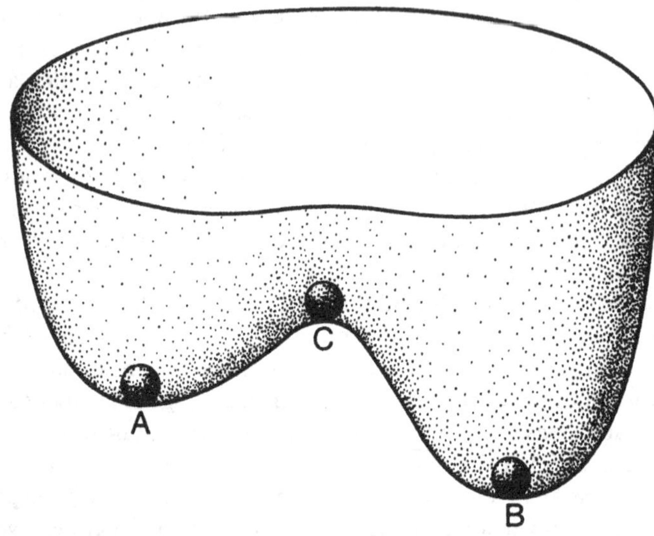

Figure 39 Concrete example of a dynamic system that can be described by a potential function – an asymmetrical bowl. Note that there are three equilibrium positions – two stable positions (A and B) and one unstable position (C) (I. Ekeland, *Le calcul et l'imprévu,* 1984).

itself close to an equilibrium. A trivial example would be the movement of a ball-bearing around a concave surface such as a bowl. Whatever the ball-bearing's initial movement parameters (position and speed), it will end up motionless at the bottom of the bowl, having lost all its kinetic energy through the friction produced by movement. There are, however, two kinds of equilibria in such systems: those that are stable and those that are unstable. In both cases, these are positions in which the system can remain indefinitely, as long as there is no disturbance. For example, the summit of the convex face of a bowl is an equilibrium point, as is the bottom of the concave face. But the first point is not reached by a trajectory, and thus has no dynamic significance. The bottom of the bowl, however, defines a stable equilibrium point. Entropy-producing systems can contain an infinite number of stable equilibrium points, each of which corresponds, in the physicists' jargon, to a 'potential' minimum. If, for example, the bowl is deformed such that it contains two dips of different depths, there will be two 'potential minima' – in other words, two stable equilibria (Figure 39). From this starting point, it is easy to understand how the potential of a dynamic system with many equilibrium points can be represented by a 'landscape' with as many 'valleys', 'peaks' and 'cols'. The contours of the landscape link points of the same potential energy, and the whole representation tends to present the dynamics of the system as a river valley. Finally, such dynamic systems can have an infinite number of dimensions.[86]

Up to a certain point, this physical image is curiously close to Sewall Wright's 'adaptive landscapes'. In the case of an adaptive surface, the system is a population that can be described by $n + 1$ dimensions. Allelic frequencies vary along n axes, with a supplementary axis for the mean selective value of the population ($\overline{W}$), which behaves as a potential function. Of course, in Wright's metaphor the polarity of the image is the opposite of that in the physicist's normal usage – instead of descending to a 'pit', species climb 'adaptive peaks'. From a formal point of view, this is of little importance, but it is clear that a climbing metaphor has a clear symbolic correspondence to the idea of progressive evolution. This system is deterministic, but when factors such as changing environment, mutation, migration and random drift are introduced, there are a series of perturbing forces that, at any moment, can either modify the surface, push the population from its equilibrium point or change its trajectory.

However, Wright never interpreted his adaptive landscapes in this way. Does this mean that Wright did not share this view, or that he was simply ignorant of the notion of a 'potential function'? The latter hypothesis does not fit what we know of Wright's education and scientific culture. Furthermore, Wright's image was a direct response to Fisher's 'fundamental theorem of natural selection' – a theoretical construction that was explicitly based on an analogy with the second principle of thermodynamics. However, even if Wright did not conceive of his 'adaptive landscapes' in terms of a physical model, at least one of his contemporaries understood them in this light, and criticised them because of this.

Fisher took it as read that the imagery of the adaptive landscape illustrated the analysis of the mean fitness of the population as a potential function, and it was precisely on this basis that he attacked it for being inconsistent.

This criticism was put forward in an important theoretical article of 1941, in which Fisher repeated in some detail the exposition of his 'fundamental theorem' and opposed it to Wright's conceptions. In this article, Fisher discusses Wright's famous formula of 1940,[87] which explains the variation of the frequency q of an allele as a function of the mean population fitness $\overline{W}$. This equation, which is today a basic element of all population genetics courses, sets out the relation between the change in the frequency of an allele in each generation (Δq), and the slope of the function $\overline{W}$. It can be formulated as follows:

$$\Delta p = \frac{pq}{2\overline{W}} \cdot \frac{d\overline{W}}{dp}$$

In this equation, $\overline{W}$ is classically interpreted as the 'fitness function' or the 'adaptive topography'.[88] This was Fisher's reaction in 1940:

Wright's conception embodied in equation (6) [Fisher alludes to the equation reproduced above] that selective intensities are derivable, like forces in a conservative system, from a simple potential function dependent on the gene ratios of the species as a whole, has led him to extensive but untenable speculations. For example, in the *New Systematics*, p. 170, we find: 'As already noted, $\overline{W}$ is a function of differences in only a few thousands, or even hundreds of loci, there are likely to be an enormous number of different harmonious combinations of characters. These would appear as peak values of $\overline{W}$, separated by valleys or saddles in a multidimensional surface.'

Prof. Wright here confuses the number of genotypes, e.g. 3^{1000}, which may be distinguished among individuals, with the continuous field of variation of gene frequencies. Even if a potential function, such as $\overline{W}$ is supposed to be, really existed, the larger number of genotypes supplies no reason for thinking that even one peak, maximal for variations of *all* gene ratios should occur in this field of variation.[89]

It should be noted in passing that the criticism of Wright made by the historian William Provine (that Wright confounded the space of genotypes with the space of gene frequencies) had already been raised by Fisher. More importantly, Fisher's criticism drew attention to a problematic aspect of 'fitness surfaces'. For Fisher, if Wright's 'fitness function' $\overline{W}$ was in fact a potential function (as he seems to have suggested), it had to be accepted that fitness surfaces represent the evolution of a conservative system. But such a conclusion was unacceptable both for Fisher and for Wright. As Fisher wrote in *The Genetical Theory of Natural Selection*, the increase of fitness due to genetic variance is comparable to the inverse of entropy in a physical system in which 'dissipative forces' act.[90] Thus, for Fisher as well as for Wright, populations evolving through mutation and selection had to be 'dissipative systems', or something similar to them. However, for Wright what was important was to incorporate into theoretical population genetics the idea that a species can explore indeterministically the surroundings of its 'adaptive peaks', which themselves could be modified by mutation and the changing environment. In other words, Wright underlined that both the adaptive topography and the occupation of 'peaks' by species were essentially unstable. This would appear to be in complete contradiction to everything Wright said about the metaphor. However, careful study of what Wright argued about the 'fitness function' reveals that such a function can be defined only in extremely idealised conditions – for example, by reasoning on the basis of a single locus or by admitting that selective values are constant. But this is merely a convenience – the mean fitness of a population is 'a function of all the gene frequencies', which obviously means that no fitness function can in fact be calculated. Or at least, this is the thesis Wright insistently repeated in *The Evolution and Genetics of Populations* (1969).[91] In fact, the adaptive landscape is a surface that is permanently subject to deformation. It is thus an image that illustrates an indefinite process of fluctuating equilibria.

The metaphor of adaptive peaks is thus clearly a qualitative metaphor. The peaks are not unequivocally situated in a parameter space, nor are they intended to be. Sometimes defined in terms of 'gene combinations', sometimes presented as a point in a hyperspace of gene frequencies, the 'adaptive peak' is an optimum more than a maximum. It is in fact at the intersection of two languages – the physiological vocabulary of harmonious combinations of genes and the statistical vocabulary of the distribution of Mendelian factors. In this particular case, Wright's genius perhaps lay not so much in his mathematical rigour as in the effective rhetoric and imagery that he used to take population genetics to its furthest conceptual limits. For Wright, the dynamic of gene frequencies was part and parcel of a biological philosophy that looked beyond the gene and the population towards the organism and the species. The organism is a physiological concept; the species is defined historically and geographically. Wright certainly remained in the methodological paradigm of population genetics, because it alone permits evolutionary predictions, but he was determined not to view either the gene or the population as the noumenal reality of evolution. Considered outside the totality of the organism, the gene is nothing, and the population is merely an ill-defined fraction of the species which has meaning only to the extent that it is separate from other fractions. To sum up, for Wright population genetics was always close to physiological genetics and to the theory of speciation.

9.3 The underlying philosophemes of the Fisher–Wright debate

There is no necessary and straightforward relationship between scientific ideas and the philosophical beliefs of the practising scientists involved. The philosophy of science is rarely particularly rich when it merely expresses the spontaneous philosophy of the scientist, precisely because such philosophy is generally unoriginal and borrowed. However, an epistemological–historical study such as this requires an examination of the philosophical context in which the producers of science explicitly situated their work, when they had such a context. Fisher and Wright both clearly and firmly indicated the philosophy of science in which they situated their respective conceptions of evolution.

It has already been pointed out how the fundamental theorem of natural selection could, with justification, be related to the model of the physical sciences that haunted Fisher. Fisher had a classical training in physics, which, for him, culminated in the statistical interpretation of thermodynamics and provided him with his own inimitable evolutionary style. The Fisherian vision of evolution was profoundly marked by Botzmann's statistical dynamics and kinetic theory of gases. Fisher argued that we cannot know the immediate state of each of the particles that interact in the cloud of genes that form a population, but we can know with absolute precision the statistical result of their interaction.[92] The fundamental theorem of natural selection defines the effect of selection, in the same way that the laws of gases or the principle of entropy can be deduced from the statistical analysis of a cloud of particles. It would, however, be an exaggeration to say that Fisher had a deterministic and Laplacian vision of evolution. As Jonathan Hodge pointed out on reading a draft version of this book, such an interpretation would not agree either with the kind of statistical mathematics to which Fisher devoted most of his career, or with his explicit philosophical convictions, which tended towards an 'indeterministic' and 'creative' view of causation.[93] For Fisher, this view was intimately linked to the idea of the fundamental importance of the principle of natural selection in biology:

> In an indeterministic theory intermediate events cannot be neglected or eliminated. In consequence scientific research is interested not merely in the manner in which precedent events determine or influence their consequences, but in locating in time and place the creative causation to which effects of especial importance are to be ascribed. In the present condition of evolutionary theory there can be little doubt that it is in the interaction of organism and environment – that the effective causes of evolutionary change must be located ... All that we mean by the designation creative is that certain happenings entail consequences, or entail systems of probability for various consequences, other than those that could have been foreseen from antecedent happenings.[94]

Nevertheless, Fisher always justified his indeterministic convictions by arguing that those events that were undetermined at the simplest level of analysis were expressed at a macroscopic level by mean effects that can be probabilistically predicted.[95] In the case of the fundamental theorem of natural selection, this had a very precise meaning. The theorem was a theorem of 'progress' which stated that despite random effects, natural selection is a cumulative factor that acts to maximise the mean fitness of species, in the same way that a casino's profit is maximised, even though it, too, is based on 'a succession of favorable chances'.[96] Philosophically speaking, the fundamental theorem states that in the contingent vision of natural history implied by the hypothesis of natural selection, there is nonetheless a law of progress. This is an important point in Fisher's comparison between the second law of thermodynamics and the fundamental theorem. Both laws are statistical; both describe the constant growth of a measurable quantity. They both occupy a 'supreme position' in their respective fields of physics and biology.[97] But one of the five differences outlined by Fisher – probably the most important from the point of view of a philosophy of nature – was as follows:

> Finally, entropy changes lead to a progressive disorganization of the physical world, at least from the human standpoint of the utilization of energy, while evolutionary changes are generally recognized as producing progressively higher organization in the organic world.[98]

Sewall Wright set out his personal philosophy of science and its link with his conception of evolution on several occasions.[99] On reading his articles, and noting the autobiographical secrets that can be perceived here and there, a strange similarity can be discerned between the 'theory of shifting balances' and quantum physics. Wright mentions that on learning of the new physics at the beginning of the century, he abandoned his initial convictions as to the Laplacian nature of reality. The new physics, he said, obliges us to consider the 'ultimate physical reality' as an 'oscillatory change' affecting 'singularities in space-time'.[100] Similarly, Wright thought (somewhat curiously) that Heisenberg's uncertainty principle was the expression of the 'emergence of genuine novelty at each instant, but novelty that is in statistical continuity with the past'.[101] Science thus cannot pretend to have a 'vision of the future'; it can only extrapolate past tendencies, knowing that in so doing it has not exhausted the rationality of the real: 'Existence is a real adventure, not a mere unfolding before our eyes of a "future" that really exists now and always has existed.'[102]

All these terms (oscillation, singularity, indeterminism, emergence of genuine novelty) are precisely those used by Wright to understand evolution. What was unique to Wright was his idea that the organism's evolution depends on a shifting balance between factors. In Wright's model, the key points are the unity of all individual genomes and the permanent eruption of stochastic events in population dynamics. Wright thought that the event, and not the element, was the fundamental category of any interpretation of nature. For Wright, kinetic laws (like Fisher's fundamental theorem) could at best describe only local tendencies.

In some of his 'philosophical' writings,[103] Wright explains how, in his youth, he had been profoundly marked by Bergson's *Creative Evolution*, and by the controversy between Jennings and Loeb over the behaviour of unicellular organisms. Against Loeb's 'tropisms', Jennings opposed the idea of a 'trial and error mechanism', implying a fundamental role for spontaneity in the simplest organisms. In 1932, Wright used Jennings' arguments to frame his opposition to Fisher. And it was the language of Bergson and of Lloyd Morgan, the language of 'creative evolution' and of 'emergence' that Wright – ambiguously – used to close a number of his lectures.[104] Wright believed that free will and choice existed at all levels of reality. This explains his 'dual-aspect or monistic panpsychism', which postulated that 'mind and matter are coextensive at all levels of reality'.[105] This led to a hierarchical representation of the real, in which non-determination, or freedom, varies in degrees. In this metaphysical schema, the status of statistics is the opposite to that given to it by Fisher – far from serving a deterministic conception, it is a window on the intelligibility of the singular.

The impact of Wright and Fisher's 'philosophical' digressions should not be overestimated. However, their philosophemes do shed light on the dialectic of algorithms and images that have marked 20th-century Darwinism. Fisher and Wright owed an enormous debt to the mathematical statistics of Karl Pearson, which they both developed and enriched. Both were partisans of the central epistemological thesis put forward by Pearson in *The Grammar of Science*, according to which scientific laws only ever have statistical value. But while for Fisher statistics was a key to defending a conception of evolution as a cumulative and progressive history deployed in a supposedly real space of 'gene ratios', Wright used statistics to winkle out a reality that was always singular, hierarchical and emergent. Both men were profoundly convinced of the need for the theory of natural selection to give way to an element of creative indetermination, but while Fisherian evolution located 'spontaneity or creative causality' in individual organisms,[106] Wright's version viewed it as a fundamental attribute of all levels of organisation.

In many sciences, this kind of opposition verges on the banal, and it would perhaps be better to describe these philosophemes as recurrent mythical elements. It is, however, useful to note that, in this too, neo-Darwinism is not so different from many other episodes in the history of science.

10 The empirical and the formal

In science as in philosophy, powerful ideas are not necessarily true; the fundamental point is that they push back the limits of thought and create new frontiers. Fisher and Wright's evolutionary ideas were of this sort – they were extreme ideas, with a mutual tension that defined the key sites on the map of contemporary neo-Darwinism. However, a science cannot be purely speculative: to close this study, an examination of the empirical content of the genetic theory of selection is thus required.

There are several ways of approaching this problem. The most obvious is to see if and how the formalisms of theoretical population genetics were applied. The genetics of experimental populations and the genetics of natural populations (in other words, ecological genetics) could provide a first response. On a larger scale, the vast galaxy of what is called the 'Modern Synthesis' could provide another. Nevertheless, despite the fundamental importance of the Modern Synthesis in the development of Darwinism, this *collection of proofs* does not exhaust the empirical implications of the genetic theory of selection.

A second aspect concerns one of the most problematic features of the founding models of this theory – their concentration on the effect of very small selection intensities. The selection coefficients that interested (and still interest) theoreticians are often below the threshold of experimental detection.[1] This has led to the frequently voiced suspicion that those effects that could be spectacularly demonstrated in nature (in the genetics of natural populations, and more widely in the disciplines involved in the Modern Synthesis) were not in fact an important part of the selection process.

A third way of looking at the relationship of the formal and the empirical in the genetic theory of selection implies paying particular attention to the theoretical challenge represented by empirical investigations. Since the 1950s, experimental studies have increasingly focused on measuring the amount of genetic variability in natural populations, and, from the 1960s onwards, have revealed hitherto-undreamed-of levels of genetic polymorphism. From this point onward, the key question was no longer to apply the models, but to rework them in the light of the immense genetic variability that was now known to exist.

The following sections deal with these three aspects of the relation of the formal and the empirical in the modern theory of selection.

10.1 From the formal to the empirical: the genetics of natural and experimental populations

The analyses developed in the preceding chapter may have given the impression that modern Darwinism is highly speculative. Are things any simpler from the point of view of direct empirical studies of natural selection? Did the spectacular studies of industrial melanism in moths give a firmer vision of the selection process?

The following section shows that this is not the case. The logic of investigation of modern Darwinian studies should not be confused with their pedagogic products, no matter how effective these may be. Ecological genetics (or the genetics of natural populations)[2] has found its inspiration in theoretical population genetics. Indeed, the empirical study of selection took off *because* there was a theory. The genetics of natural and experimental populations has confirmed what the mathematical speculations of the 1920s and 30s had indicated: the theory of natural selection is not and could not be trivially simple. This will be illustrated by a few key moments which, as in the preceding chapter, will provide an overview of the key features of Darwinism's empirical bases rather than an exhaustive chronological reconstitution of events.

First, a few words on industrial melanism in moths. Studies of this phenomenon probably did the most to convince people of the efficacy of selection in natural conditions. The evocative power of the example is easy to understand. It deals with extremely rapid evolution, observable on a human time-scale, involving an (apparently) simple Mendelian character (the melanic forms are virtually always dominant), and both the quantitative aspect and the ecological significance of the selection process have been exhaustively demonstrated. No other example has had such an effect in imposing the paradigm of natural selection as a process of the substitution and fixation of genes in a population. The real history of studies of industrial melanism in moths is thus particularly significant.

The proliferation of melanic forms in industrial regions constitutes one of the most rapid and spectacular evolutionary phenomena that have ever been observed. First noticed in England for the Peppered Moth (*Amphydasys* – or *Biston* – *betularia*), and then for a large number of other Lepidoptera from different families in many other countries (Germany, Eastern Europe, North America . . .), the phenomenon affects insects that habitually settle on tree-trunks and rocks, and are thus exposed to predators. In non-polluted areas the predominant form is light, providing the insects with camouflage. Museum collections show that dark forms began to appear in the middle of the 19th century in regions particularly polluted by soot. The first known melanic form of the Peppered Moth was collected in 1848, near Manchester in England. Fifty years later, in the same region, and in many other polluted areas, virtually all the moths were dark. The phenomenon is all the more spectacular because there is no gradual transition between the two morphs – dark moths replace light moths.

At the beginning of the century this discontinuity attracted the attention of geneticists. It was quickly shown that the character was inherited in a typically Mendelian fashion – the light forms are recessive, while the dark forms are dominant. This conclusion, first drawn for *Biston betularia*, was subsequently extended to a large number of other Lepidoptera showing the same phenomenon. In a review published in 1964, E. B. Ford noted that more than eighty species showing this kind of effect had been found in the British Isles alone.[3]

From the beginning of the century onward, the melanic transformation of Lepidoptera was given as an example of selection acting on discontinuous, Mendelian variation. In his book *Mimicry in Butterflies* (1915), the English geneticist Reginald Punnett gave the Peppered Moth as a striking example of such a process,[4] and contrasted it with the mechanism of gradual transformation proposed by Darwin.[5] To illustrate his point, Punnett drew up the table, discussed above, which, for the very first time, gave a kinetic description of selection under the Mendelian hypothesis (see Figure 33). However, Punnett admitted that nothing was known about what might be the advantage conferred by the melanic form. The following passage from his book refers to the replacement of *betularia* (the light morph) by the melanic form *doubledayaria*.[6]

> What advantage this new dark form has over the older one we do not know. Some advantage, however, it must have, otherwise it could hardly supplant *betularia* in the way it is doing. From our present standpoint two things are of interest in the case of the peppered moth – the rapidity with which the change in the nature of the population has taken place, and the fact that the two forms exhibit Mendelian heredity, *doubledayaria* being dominant and *betularia* recessive ... The development of Mendelian studies has given us a method, rough perhaps but the best yet found, of testing for the presence, and of measuring the intensity, of natural selection.[7]

Punnett's remarks were the starting point of a story that spanned more than half a century – the exhaustive demonstration of an example of natural selection. Punnett had not proved that the observed changes in moth populations were due to natural selection, even as defined in mutationist terms (that is, as a process of replacement of one form by another rather than of the gradual modification of the form itself). An alternative hypothesis was that the mutagenic action of atmospheric pollutants, such as sulphuric acid or metallic salts, might have produced the effect. In the 1920s, many studies were carried out to try and prove the hypothesis of an oriented mutation.[8] But, as J. B. S. Haldane pointed out in 1924, this would require an extremely high rate of mutation. In order for the mutagenesis explanation to be consistent with the observed rhythm of phenotypic change, 20 per cent of recessive alleles would have to mutate into dominant forms.[9] Such a rate would be several orders of magnitude greater than the strongest rates observed under conditions of artificial mutagenesis. It was thus more parsimonious to interpret the darkening of Peppered Moth populations as being due to intense selection. Using data collected between 1850 and 1900, Haldane calculated that a selection pressure of around 0.3 would be sufficient for the frequency of the dominant genotypes to increase from 1 per cent to 99 per cent in the space of forty-eight years.

The next step was to identify the kind of selection pressure that could explain the replacement of the light morphs by the dark ones. For many years there was a classic hypothesis to explain the darkening of Peppered Moth populations in the most urbanised regions of England. From 1890 onwards, it was commonly argued in the entomological literature that melanism could provide protection against predators. In polluted areas, lichen disappeared from tree-trunks and rocks, and the moths' substrate became dark. Under these conditions, the dark form would have an advantage. Conversely, the light, speckled form is favoured in non-polluted areas.[10]

This was a particularly seductive explanation, but it had to be proven. Curiously enough, the research carried out on this subject by the father of 'ecological genetics', E. B. Ford, went in a very different direction. In two studies published in 1937 and 1940,[11] Ford argued that the explanation of industrial melanism as being due to selection in favour of the dark form because of increased protection from predators was false. Ford started from the observation that under experimental conditions the melanic forms were more robust (they used the available food better, were more resistant to aggression from the milieu, etc.). He went on to accept that cryptic coloration was of the utmost importance *for the light forms* – that is, in non-polluted regions. Finally, he noted that in the rare species of moths where the melanic form was recessive, the dark morph did not become established in industrial regions. This seemed to indicate that, in itself, the dark form was not selected as protection against predators. From these facts, Ford developed a hypothesis of an equilibrium between two kinds of selective forces: on the one hand, physiological vigour (which was greater in the dominant form); on the other, camouflage against predators. In polluted areas the danger of predation decreased (because, for example, there were fewer birds), and the greater vigour of the dark moths enabled them to resist a more hostile environment. Ford subsequently repeated this explanation of industrial melanism,[12] and, it appears, he never completely abandoned it.[13]

It was only in 1955 that Kettlewell's patient and ingenious experimentation imposed the interpretation which is now so widely accepted that it seems obvious. Kettlewell decided to directly test the hypothesis of selective elimination by birds. By marking and recapturing moths, and by directly observing birds, Kettlewell was able to measure what proportion of dark and light forms of the Peppered Moth were eliminated by birds in rural and industrial regions. In both cases, the cryptic form had a strong advantage, birds eating two or three times more dark moths where the tree-trunks were covered in lichen, and more light moths in polluted areas where the tree-trunks were not covered.[14] Under these conditions, it was difficult to avoid the conclusion that the diffusion of melanic forms was due to selection on a gene controlling the colour of the moth. Kettlewell's study further confirmed the idea that selection intensity in nature can be very strong, and that natural selection can thus produce extremely rapid changes.

Kettlewell's work can arguably be considered one of the first exhaustive proofs of a phenomenon of gene selection in nature, nearly a century after the publication of the *Origin of Species*, and fifty-five years after the rediscovery of Mendel's laws. This story raises a number of points.

We should first, of course, note the rapidity of the observed process. The enormous selection coefficients found for industrial melanism, and in some other cases, must have surprised those who were accustomed to the Darwinian idea that evolution is an extremely slow and gradual process. In the 1930s it was common to argue that because of this slowness, direct observation would probably not be able to provide data to support the vision of evolution developed by population genetics. Thus, as late as 1940, Ford wrote with regard to Fisher's conceptions: 'The extreme slowness of evolutionary progress in nature makes it exceedingly difficult to study. It is not to be expected, then, that much direct observation will be available in support of the theory just outlined.'[15]

The 'genic' vision of selection that apparently flows from such studies should, however, be qualified. As both Kettlewell and Ford showed, it is an approximation to say that the evolution of melanism is the result of selection on a given gene. It rapidly became apparent that the melanic form of the Peppered Moth was itself evolving: there was no necessary relationship between the melanic form and the dominance of the selected gene. In other words, selection has been continually remodelling not a single gene but a 'gene complex'. Natural conditions thus reraised the problems encountered by Castle in his artificial-selection experiments on rats at the beginning of the century. This problem implies a return to Darwinian gradualism, and the geneticist has to deal with the difficult problems of quantitative inheritance and of selection on such characters. Peppered Moth melanism is straightforward only when it is explained in introductory textbooks or in popularising pot-boilers. Taken in its phenomenalistic totality, it reraises the problems that faced the first population geneticists. This is why Ford, irrespective of his homage to Kettlewell, remained faithful to his own interpretation of industrial melanism.

Furthermore, whatever the brilliance and suggestive power of studies of rapid selection, they all encounter the same fundamental objection. If evolution always, or mainly, consists of the rapid selective diffusion and fixation of favourable mutations, there should be very little genetic variability in natural populations. Most loci should be homozygous, with only a few showing a 'transient polymorphism'. From the 1930s and 40s onwards, there were serious reasons to believe that natural populations in fact showed high levels of genetic variability. The most important argument was the very existence of the inheritance of quantitative characters – under the hypothesis of Mendelian heredity, the inheritance of quantitative variation requires many polymorphic loci. Given that most quantitative characters are heritable, it could reasonably be concluded that there is a great deal of genetic variability. This idea is omnipresent in the work of Fisher and Wright, but, curiously enough, has rarely been pointed out. Another argument, more directly experimental, went in the same direction. It dealt with 'hidden variability', that is the totality of recessive genes (generally deleterious) that exist in a heterozygous state in the population. In 1927, the Russian geneticist Chetverikov drew attention to the high levels of hidden variability that could be found in natural populations of *Drosophila*.[16] Dobzhansky, a pupil of the Russian school, was well aware of this.

Whilst most polymorphisms are more or less stationary (on a human time-scale), there are three ways of explaining this. Either the genes are 'neutral' from the point of view of selection, and the polymorphism depends only on repeated mutations and sampling variation in each generation (random drift); or they are not neutral, and the polymorphism can be explained by some form of stabilising selection (e.g. heterosis; but this is not the only possible hypothesis); or the population is subject to low-intensity directional selection that cannot be detected experimentally ('transient polymorphism'). Modelling these effects and their interaction was a major challenge for Fisher, Haldane and Wright, the three founders of theoretical population genetics.[17] Hence their interest in low levels of selection: beyond a certain threshold, selection is no longer a force that is necessarily stronger than the pressure of mutation and of random drift. If most polymorphism was in a quasi-stationary equilibrium, there would be very small differences between a model of evolution dominated by the deterministic play of weak selective forces and a model in which the action of selection would be constantly perturbed and contradicted by the pressure of mutation and random drift. This brings us back to the heart of the theoretical controversy between Fisher and Wright, which became one of the main issues in empirical studies at the end of the 1940s. Two examples will be given here.

A joint article by Fisher and Ford, published in 1947, marked the first attempt to submit the famous quarrel to experimental testing.[18] Between 1939 and 1946 Fisher and Ford studied an isolated and relatively small population of a few thousand butterflies (*Panaxia dominula* L.) in the countryside near Oxford and observed changes in the distribution of a morph (*medionigra*). The results showed that the population size varied between 3000 and 6000 individuals, and the frequency of the *medionigra* gene fluctuated around 5.7 per cent. This situation was *a priori* favourable to Wright's model, because the population was small, and there was no observed secular change in gene frequency, but only fluctuations. The authors speculated that these fluctuations might be due to genetic drift, but then showed that they were in fact much greater if the gene was neutral and its survival was purely random:

> Our analysis, the first in which the relative parts played by random survival and selection in a wild population can be tested, does not support the view that chance fluctuations in gene-ratios, such as may occur in very small isolated populations, can be of any significance in evolution.[19]

In their conclusion ('The Significance of Changes in Gene-Ratio'), the authors argued that the observed fluctuations, and in fact all gene fluctuations, were due to selection of variable intensity and directionality:

> The conclusion that natural populations in general, like that to which this study is devoted, are affected by selective action varying from time to time in direction and intensity, and of sufficient magnitude to cause fluctuating variation in all gene-ratios, is in good accordance with other studies of observable frequencies in wild populations.[20]

Once again, this reveals the inspiration behind Fisher's fundamental theorem of natural selection – natural selection exerts a deterministic and total control over the fate of the population in the n dimensions of the field of gene frequencies.

However, at the same time as Fisher (aided by Ford) sought to corroborate his own model of evolution, Maxime Lamotte, a young French researcher, was trying to prove the experimental plausibility of Wright's model. Lamotte's doctoral thesis, on the 'genetic population structure of *Cepaea nemoralis* (L.)' (1950),[21] was the first attempt to test Wright's model in the field, twenty years after it had first been proposed.[22] It is intriguing that it had taken so long – Wright's work was hardly confidential. From the beginning of the 1920s, Wright was world-famous for his studies of the inheritance of pigmentation in mammals. As to his model of the evolution of Mendelian populations, it was extremely well known in Britain and the USA. This is shown, for example, by Dobzhansky's 1938 book *Genetics and the Origin of Species*, which to a certain extent was the founding text of the 'synthetic theory of evolution'. Wright's model, reworked in a qualitative language easily accessible to non-mathematicians, formed the central argument of Dobzhansky's book, showing that it was accepted as a paradigm, at least in the North American biological community, long before being confronted by any empirical test.

The reason why Wright's model had to wait so long before being tested undoubtedly lay in its methodological constraints. Wright's model emphasised species' population structure – it was vital for the model that the species should be divided into a large number of partially isolated populations, because stochastic effects have any meaning only in such a variation space. Fisher and Ford's data from their 1947 study were thus of little relevance to Wright's model – they were taken from a single isolated population; whereas to test the hypothesis of random variation of gene frequencies, several hundred such populations would have to be studied. Furthermore, estimating the size of a natural population is extremely difficult, because it requires data that are very hard to collect from natural populations. Measuring the exact geographical distribution of a population, its degree of isolation and the level of migration is often impossible. It is hard to imagine E. B. Ford or Theodosius Dobzhansky measuring the population size and the migration levels of a thousand populations of *Drosophila* or of butterflies. As to Sewall Wright himself, he specialised in mammals, and was more a laboratory biologist than a field-worker. It is understandable that it took virtually a generation for a new wave of scientists, reared on Wright's mathematical prose, to orient their experimental research to the organisms appropriate for putting Wright's algorithms to work. William Provine has made a telling analysis of the reasons why Dobzhansky, despite his close collaboration with Wright in the 1930s and

40s, failed to construct a methodologically satisfactory test of Wright's model.[23]

As indicated above, the first effective confirmation of Wright's model came from elsewhere. Maxime Lamotte was a student of Georges Teissier, who, together with Philippe L'Héritier, invented 'population cages' (for use with *Drosophila*) and the first experimental applications of the models of theoretical population genetics. Teissier and L'Héritier were trained in quantitative methods, like Lamotte, who had published, shortly before his thesis, a manual of 'quantitative biology'.[24] Lamotte was familiar with Wright's clearest outline of the preconditions, the implications and the methodology of his model. For many years, this article by Wright, published in Paris in 1939 (in English) by Editions Hermann, remained virtually unknown to the English-speaking world.[25]

Encouraged by Teissier to carry out an experimental study of Wright's model, Lamotte was not initially decided on a particular organism. He took the time to find an animal that could be studied in Occupied France, where scientific laboratories, understandably enough, were in some disarray. It should be noted in passing that Lamotte (like Teissier), was a secret member of the French Resistance. Like Dobzhansky in the USA or Fisher and Ford in England, whose first studies in the genetics of natural populations were published at around the same time, Lamotte did not allow the war to stop him from collecting data, even though, no doubt, his work was put into perspective by the terrible events that were shaking the planet. As has already been noted, the most spectacular developments in the Modern Synthesis took place during the Second World War. This was not (of course) because the war was conducive to work on population genetics, but because the key questions had become so ripe that they had to be answered *despite* the war.

Whatever the case, Maxime Lamotte chose to study a small land snail, *Cepaea nemoralis*, found in large numbers throughout Europe, with an ecology and mode of displacement which produce a large number of small, partially isolated colonies that can be relatively easily located and counted. *Cepaea nemoralis* shows a striking polymorphism in the coloured bands on its shell which was known to be determined by a Mendelian factor. However, given that previous studies had been affected by various biases produced by the snail's mating system, Lamotte began his work with a classic Mendelian study of the genetics of polymorphism. Once this preliminary work had been carried out, the key part of the study involved collecting a substantial number of individuals from a large number of local populations.

Lamotte applied Wright's interpretative framework to this material, using his equations to estimate the respective influences of mutation, migration, selection and random variation on the distribution of the 'no band' character. Various experimental tests and qualitative observations were also made.

Lamotte's main conclusion was that the polymorphism of the studied gene was primarily caused by random variations linked to small population size – the smaller the local population, the greater the random effects. In other words, Lamotte proved Wright right as against Fisher, and clearly took a position opposed to panselectionist views of evolution. However, the main historical interest of this work is not to be found in its conclusion, which both subsequent debate and the author himself were to qualify as overly hasty.

The key point resides in the methodological problems raised by this attempt to apply Wright's model. During the analysis, Lamotte introduced a number of factors that Wright had initially ignored because they made the algebraic interpretation too complicated, such as isolation through distance (which Lamotte called 'local inbreeding'), homogamy, the distinction between the measured population size and the 'effective population size' (that is, the population that really intervenes in the random fluctuations of gene frequencies), and finally heterosis (which Wright considered to be conceptually essential but which he did not include in his 1930s predictive models). Because certain formal aspects of the interpretation were somewhat difficult, Lamotte sought the aid of Gustave Malécot,[26] a mathematician who was well known for his contributions to the theoretical analysis of inbreeding. Together with Wright, Malécot was one of the main influences on one of the giants of population genetics in the 1970s and 80s, Motoo Kimura.

The main contribution of Lamotte's work on natural populations of *Cepaea nemoralis* was to show that Wright's model was not a vague general idea of evolution dressed up in a few equations, but a genuine instrument for data analysis. Lamotte's work showed that if it was difficult to apply the fundamental ideas of theoretical population genetics to empirical data, this was not because they were too theoretical, but rather because they were not theoretical enough. The fact that Malécot's skills were required to test the model fully is clear evidence of this.

The *Cepaea nemoralis* polymorphism rapidly became the subject of a major debate. Whilst Lamotte was establishing the role of genetic drift in the maintenance of the polymorphism, two English naturalists, A. J. Cain and P. M. Sheppard, tried to explain the same polymorphism by the subtle and differentiated action of natural selection.[27] Their work was based on the precise identification of the ecological forces that might alter the distribution of genes through selection. Much as Kettlewell had done for moth melanism, Cain and Sheppard showed that the appearance of the snail's shell played a key role in defence against predation. However, given that the polymorphism had apparently been maintained for several thousand years, and in a large number of localities, this explanation was insufficient. In fact, as the controversy developed, it appeared that neither genetic drift nor selection, nor the two factors together, were sufficient to explain the observed distributions. In fact, a balance of various forces (migration, mutation, drift and different selective forces) is responsible for the maintenance of the polymorphism. In particular, the ecological differentiation of local races, under the pressure of different selective forces, plays a very important role, together with long-term migratory factors. In the 1930s and 40s Wright's model was accepted, not as the truth, but as a methodological framework that could articulate the data and modulate the scenarios. Today, evolutionary biologists agree that the environment plays a major role in shaping populations. But because this action is fluctuating, and the environment changes indefinitely, the theory reintroduces the role of hazard at the higher level of ecological dynamics.[28]

There was a third route from the formal to the empirical: the classic experimental road, which involved studying the phenomenon under artificial conditions. From Galileo to Claude Bernard, this method has been shown to have a unique advantage. Factors isolated one by one can be evaluated by a kind of material abstraction. In the case of the Darwinian science of evolution, this implies the study of natural selection in artificial conditions. The terminological criticism that this procedure cannot grasp the 'natural' character of selection carries no weight whatsoever. Natural selection, if it is indeed a natural phenomenon, is no more or less 'natural' than the force, mass and acceleration of physicists. It should thus be possible to analyse it under artificial conditions. This is what Weldon had tried to do in 1898, in the most elegant biometrical study of selection. A comparable attitude can be found in L'Héritier and Teissier's studies of the 1930s.

It is utterly astonishing that these simple, elegant, precocious and theoretically crucial studies were met by general indifference and have left so few traces in the literature. It appears that L'Héritier and Teissier's work is mainly remembered for the 'technical' innovation of population cages, which were to be fully exploited by Dobzhansky and others fifteen years later. There are a number of conjunctural explanations of this strange amnesia that has afflicted both evolutionary biologists and historians. It is true that French biology, and French culture in general, have never completely rid themselves of a certain aversion to Darwinism.[29] But why then did the numerous 'notes' published in the *Comptes Rendus Hebdomadaires des Séances de l'Académie des Sciences* (widely available in academic libraries throughout the world) go unnoticed outside France? These articles were short and easily understandable; L'Héritier had a close relationship with American researchers; Teissier was a friend of Haldane's, and despite his membership of the Communist Party, received the support of the Rockefeller Foundation as the Director of the Centre National de la Recherche Scientifique (CNRS). The war was also undoubtedly a factor, coupled with the fact that in the postwar years both L'Héritier and Teissier, for different reasons, distanced themselves from their earlier experimental work on selection. All these factors are part of the story, but the key point lies in philosophical resistance – in the 1930s biologists were not inclined to believe that 'natural selection' could be studied in containers barely larger than a shoe box. They were even less inclined to read extremely simple conclusions, and, in a world increasingly modelled on North American scientific production, to take seriously articles that were generally no longer than a couple of pages. The spirit of this work, brief but exceptional, is reconstituted in the following paragraphs, I hope without chauvinist excesses.

In 1932, Philippe L'Héritier joined up with Georges Teissier at the Ecole Normale Supérieure to study *Drosophila* populations in artificial experimental conditions. The link with the theoretical work of the founders of population genetics could not be clearer. As L'Héritier later explained in an autobiographical lecture:

> In October 1931, having finished my studies at the Ecole Normale and my military service, I left . . . for the United States, funded by a grant from the Rockefeller Foundation. I went to learn genetics and above to all to find research subjects that I could continue on returning to France. I think I was the first young French scientist to follow this route . . . It was during my stay in the United States that I discovered population genetics, which I don't think anybody had spoken to me about before. I read R. A. Fisher's book, which had just been published, and also got to know Sewall Wright's work. I met and heard these men at one of the first International Congresses of Genetics, which took place at Cornell University in July 1932. I also met other founders of modern genetics and evolutionary biology, in particular Theodosius Dobzhansky and H. J. Muller.[30]

Returning to France in autumn 1932, L'Héritier, together with Georges Teissier, invented 'demometers', or 'population cages'. These are boxes with a moveable glass lid and a bottom with a number of circular holes, through which small cups containing fly food can be poked. Every day, a cup containing fresh food is introduced and the oldest cup is removed. The flies breed at such a rate that there is not enough food for all the eggs to develop into flies. Thus there is severe competition between larvae for food resources, and the adult population becomes stationary at around two to three thousand individuals. This arrangement provided 'one of the best laboratory representations of a process of natural selection, the selective factor here being very strong competition for food'.[31] If two genetically different samples are introduced at the beginning of the experiment (in the simplest case, two strains homozygous for different alleles of a given gene), the spontaneous evolution of the gene frequencies can be followed and a genuine process of natural selection can be observed in artificial conditions.

While these facts are well known, they are worth recalling in order to underline exactly when this work was carried out: 1932. L'Héritier and Teissier thus began their study of selection dynamics in experimental populations immediately after the publication of *The Genetic Theory of Selection* (Fisher, 1930) and 'Evolution in Mendelian Populations' (Wright, 1931), and of Haldane's book *The Causes of Evolution* (1932).

In general, their experiments involved two strains, one of which was wild-type for a given gene, the other mutant. The two strains would be introduced in equal proportions, and the evolution of the frequency of the mutant allele would be followed over a large number of generations (the experiments generally lasted two or three years, or around fifty generations). The first clear result was that while in some cases the mutant allele was eventually eliminated, in others it could be stabilised at quite a high level.[32] In such cases, the population shows a stationary polymorphism, a phenomenon that is well known in natural populations. Teissier also showed that this could occur without experimental intervention, following the *spontaneous* appearance of a favourable mutation in a population.[33] Several years later, Teissier discussed this observation in the following terms:

> One of the most interesting facts I ever observed I came upon by accident, although in reality it was the product of a great deal of patience – I had been vainly looking for this result for five years. In February 1942, a stationary population of *Drosophila* was found to contain a new gene affecting eye colour, which was rapidly identified as an allele of *sepia*. The frequency of this gene gradually increased and for around twenty generations was stable at around 0.225. This is a remarkable result, because it shows that a new gene can find a place within a population following its appearance by mutation, that it can spread despite strong larval competition, and that it can finally be stabilised at a relatively high frequency.[34]

After the arid tomes of the theoretical evolutionary literature of the 1930s, this passage is like a breath of fresh air.

The results from the population cages led Teissier to clearly pose the reasons for the existence of polymorphism. The classic explanation was that the selective value of the heterozygote was high (heterosis). However, this is simply a verbal sleight of hand, a way of restating the phenomenon that has to be explained. Teissier put forward three remarkable propositions which were not in themselves novel – they had all been expressed as possibilities in the founding documents of theoretical population genetics. Teissier's originality was to link them to clear experimental evidence.

1. In 1947, Teissier showed that the frequency of a gene can stabilise, depending on the genetic background of the organisms in which it is found. For example, the *ebony* gene, placed in a population homozygous for the *sepia* mutation, will stabilise at a frequency three times higher than in a population that is wild-type for the same mutation. Selective values of genotypes depend on the genotypic milieu.[35]

2. Teissier also showed (1947, 1954) that when a polymorphic population stabilised for a given gene is maintained for a long time, the equilibrium may break down and the gene frequency will show rapid oscillations. The form of these variations is different in each population, even when they have been rendered as similar as possible. These fluctuations cannot be explained by the environment (constant), or by random drift (the effect is too large), and are particularly problematic because they oscillate. The conclusion was that a separate allelic system had perturbed a certain number of associated polymorphic systems. This implies that the selective value of a genotype in a population depends on the frequency of other genotypes. Under these conditions, selective value is no longer an intrinsic character of the genotype.[36]

3. Teissier put forward a third selective schema that could explain polymorphism. The heterozygote is not necessarily superior because of some unknown physiological factor. Nor are genotypic selective values necessarily constant. A polymorphism will be maintained indefinitely, on condition that the heterozygote has a higher selective value than the most abundant homozygote. In such a model, the selective value of a genotype depends on the frequency of the alleles that constitute the genotype.[37] Today, this phenomenon is called 'frequency dependent selection'; there are a number of suggestive analogous cases in economics and ecology. Claudine Petit, a pupil of Teissier's, was the first to prove the existence of this phenomenon, as well as offering an outline of some of its possible physiological meanings.[38]

To sum up, Teissier's (and L'Héritier's) work on experimental populations of *Drosophila* clearly demonstrated (1) an example of innovative selection; (2) the existence of a balanced polymorphism at the level of an individual gene in controlled and constant conditions; and (3) the effect of genetic background on the selective value of a genotype, the fact that this value depends on the frequency of other genotypes, and the existence of frequency-dependent selection.[39]

It could be argued that this overestimates the impact and the originality of the school in which the present author learnt his evolutionary genetics. Teissier's short notes take up only a few dozen pages and deal with concepts and simple formal schemas that had previously been widely discussed in the theoretical literature. And, from an empirical point of view, all this work pales in comparison with the weighty monographs of the English and especially of the American schools. It could also be pointed out that Teissier and L'Héritier's work had very little impact outside France, and that their important discoveries were also discovered by others – somewhat later – on the basis of an impressive exploration of polymorphism and the dynamics of natural populations.

Although such remarks are historically justified, they miss the essential epistemological point.

It is true that what Teissier and L'Héritier discovered with their population cages others had previously constructed in theory, but this is hardly surprising. As shown in the preceding chapter, the models of population genetics contain virtually every possible evolutionary scenario under the Mendelian hypothesis. That is the splendour of this theoretical science, but it is also its limitation. To repeat a phrase of Punnett's, on judging Fisher's first theoretical article in 1916, theoretical population genetics resembles a science that 'deals with weightless elephants upon frictionless surfaces, where at the same time we are largely ignorant of the other properties of the said elephants and surfaces'.[40] The fundamental importance of the use of population cages was that by ingenious experimentation they permitted the isolation and proof of various effects.

From the point of view of empirical science, it is also true that the concepts developed by Teissier were rediscovered, argued out and developed by others (especially Dobzhansky and his pupils) with a weight of evidence that apparently reduces Teissier's pioneer work to the status of an intelligent suggestion. The response to this is that, in science, the most interesting experiments are those that point the way. The 'demometers', or 'population cages', were not merely a piece of technical apparatus, but were based on a far from trivial theoretical vision. They were the basis upon which, from 1946,[41] the Dobzhansky school were able to provide a necessary experimental rigour to support their excellent observations of natural populations *in situ*. It should also be pointed out that while Dobzhansky always accepted that selective value depended on genetic background, up until the 1960s he refused to accept that the selective value of a genotype could depend on either the gene frequencies at other loci or on its own gene frequencies.[42] These two hypotheses, relatively different in their vague organismic intuitions, are today viewed as commonplace but fertile working hypotheses of population biology.

In the final analysis, population cages were to evolutionary genetics what Galileo's inclined plane was to the study of falling bodies. In a population cage, natural selection is present in its totality, in the same way that the law of falling bodies is present in the movement of a rolling marble. In both examples the experimental setting removes all that is autonomous and impalpable in the 'natural' situation.

To close this discussion of 'empirical' evolutionary genetics, it is worth repeating that this was not intended to be a chronological or historical tableau, but rather a way of showing that through its rich and complex investigation of real populations, evolutionary genetics explored natural selection in three ways. By studying examples of rapid evolution such as industrial melanism, the causal chain of a given selective process could be reconstructed. As we have seen, this was a particularly onerous task, which, with Kettlewell's studies of *Biston betularia*, had the value of a case study, continuing Weldon's pioneer work from the 1890s. Strictly speaking, the first exhaustive direct proof of a case of natural selection was found only in the 1950s.

A second approach was to study all aspects of the formalisms of population genetics, and to subject them to experimental testing. In the 1940s, the empirical investigations of Dobzhansky and Wright, Fisher and Ford, and Lamotte, all went in this direction and underlined that there can never be too much theory. A similar requirement has recently become apparent in the interface between population ecology and population genetics.

Finally, the technique of population cages brought natural selection into the realm of the experimental method.

10.2 Selection coefficients: not only an empirical issue

There are a number of obvious links between the empirical bases and the formal constructions of evolutionary genetics. However, there are also more problematic areas. One of the most striking aspects of the early models is their concentration on extremely small selection coefficients. This is a problem here because it is hard to imagine how a selective advantage of 10^{-3} to 10^{-6} (and some articles deal with coefficients of 10^{-9}) could ever be shown to exist by direct observation. Fisher, Haldane and Wright were well aware of this difficulty, but nevertheless considered that natural selection posed the most important theoretical problems for the transformation of species at just such low levels. The present section puts forward some hypotheses concerning the reasons that pushed the first theoreticians of population genetics down this road. Some reflect real theoretical or experimental constraints; others have a more rhetorical value and were linked to the attempts by population geneticists in the 1930s to convince other biologists of the validity of Darwinism.

Before examining the ideas of Fisher, Haldane and Wright on this question, the concept of 'selection coefficient' needs to be clear. There is a small but important operational difference between the rate of selective mortality (or selective fertility) that was used by the biometricians and is still used today in quantitative genetics, and the 'selection coefficient' used in population genetics. For Weldon and Pearson, the rate of selective mortality (or fertility) was defined by reference to a continuous distribution curve and thus varied continuously as a function of the character. There is always a value for which selection intensity is zero, and other values that have increasing selection intensities. The overall percentage of the population that will die because of selection can be calculated, but this figure only defines the cost of selection for the population. The differential mortality (or fertility) rate is a function, not a number. The origin of this idea in Galton's work and its use by Weldon were discussed in Chapter 7.

The founding models of population genetics, however, deal with a field of discrete entities in which the selection coefficient is defined for genotypes. The coefficient thus represents the numerical relation of the survivors of one type to the survivors of another type from one generation to the next. The key figure measures the disadvantage of one form in relation to another. For example, the 'selective value' 1 is given to the advantaged form, and $1-k$ to the disadvantaged form, 'k' being the selection coefficient. There can be more than two forms, each with its respective coefficient: k_1, k_2 etc. The selection coefficient is, by definition, a precise value in a given generation for the whole of the population. In species with overlapping generations this coefficient is defined at time *t* and not for a given generation. Although this makes the mathematics a bit more complicated, it does not change much from the point of view of a kinetic description of selection.[43]

From a mathematical standpoint, the fact that the selection coefficient is a precise number has a very interesting consequence. It means inferences can be made about the speed of the selective process, using formal methods that are reminiscent of the differential equations of kinetic chemistry. For this reason, the terms 'kinetic description' and 'stoichiometry' have been used in the present study, despite the fact that they are not part of the normal vocabulary of population geneticists.[44] The reason why these terms are not used is simple: it would imply admitting that population genetics does not measure a genuine dynamic of selection.

In an insightful article published in 1926, S. S. Chetverikov clearly explained the implications of the concept of 'selection intensity' under the Mendelian hypothesis.[45] In this paper, the Russian geneticist argued that elaborating a quantitative understanding of selection intensity was *the* most important question for the genetic theory of selection. He pointed out, however, that 'the actual computation of selection intensity is at present [i.e. in 1926] an unapproachable problem'.[46] Nevertheless, he added, as long as a Mendelian perspective was adopted (that is, the existence of discrete characters), selection intensity could be characterised by a coefficient, which would enable *a priori* predictions to be made about the rhythm and the effectiveness of the selective process. Discussing Norton's famous table, published by Punnett in 1915 (see Figure 33), Chetverikov applied this approach to that well-known case of rapid evolution, the Peppered Moth. Like Haldane two years earlier, he noted that, in such a case, the selection coefficient was very high. The figure he arrived at was $k > 0.5$, which was even higher than that given by Haldane ($k > 0.332$).[47] A selection intensity of 0.1 or 0.2 means that 'the probability of survival in the struggle for existence of individuals not having a favourable trait is 10 per cent, 20 per cent, etc., lower than individuals having it'.[48]

But the key point lies less in the application of a predictive schema to this or that spectacular example than in its theoretical consequences. Norton's table, argued Chetverikov, showed that under the Mendelian hypothesis the selective process 'was an efficient one', even in the case of very weak coefficients (Norton's table dealt with values of 0.50, 0.25, 0.10 and 0.01).[49]

This table merits more careful inspection. First of all, attention is attracted to the fact that in both cases – when the selected trait is dominant as well as when it is recessive – the process of the *transformation of the species*, that is, of the *complete replacement* of a former unadapted form by the more adapted one, proceeds, practically speaking, to an *end*. This process of complete replacement of one form by another goes on even under the very weakest intensity of selection of 1 per cent, the only contrast being in *different rates of the process*. [Original emphasis][50]

The data in the table allowed one to reason in terms of speed, and in particular to deal with the thorny question of the initiation of the process, in a time-scale that would be geologically meaningful. Norton's table (like Haldane's 1924 versions) showed that the initial diffusion of a slightly advantaged gene would be very slow, especially in the case of a recessive gene:

Should the favourably selected trait be dominant, the process of replacement of the less adapted form by the more adapted one proceeds very rapidly from the very beginning. For instance, with an intensity of selection of 10 per cent, 305 generations are sufficient to pass from a condition in which 99.9 per cent lack the selected character to a condition in which, on the contrary, 90.7 per cent will possess the character ... On the contrary, should the newly appearing favourably selected form be recessive, the beginning of the process of transformation of the population proceeds extremely slowly. At this same intensity of selection of 10 per cent, after almost 18,000 generations, more than 90.7 per cent of the population will still consist of forms not displaying the favourable variation ... Naturally, with even weaker degrees of selection, this process must be protracted to hundreds of thousands of generations.[51]

The formal framework of Mendelism thus gave a quantitative meaning to the Darwinian notion of minimal advantage:

... under the conditions of Mendelian heredity, every, even the slightest, improvement of the organism has a definite chance of spreading throughout the whole mass of individuals comprising the freely crossing population (species). Here Darwinism, in so far as natural selection and the struggle for existence are its characteristic features, received a completely unexpected and powerful ally in Mendelism.[52]

At the same time, this rudimentary calculation could imply that natural selection was unlikely in the case of weak selective advantages. In reply, Chetverikov correctly pointed out that in nature the initial phases would probably not be as slow as the theory suggested, because of the existence of a vast pool of unobservable variation. Rather than arguing that evolution always proceeds by the sudden appearance and diffusion of a favourable mutation, Chetverikov considered it more likely that species generally have a large amount of hidden (recessive) variability that can acquire a higher selective value in a new 'genotypic milieu', and can thus be rapidly diffused.[53]

However, while these points amply illustrate the conceptual climate in which the first population geneticists began their discussions of selection coefficients, they do not directly answer the question posed earlier: why did the theoretical Darwinism of 1920–40 postulate the existence of infinitesimally small selection coefficients (even smaller than those discussed by Norton and Chetverikov) and argue that these were the key to understanding evolution? To put the same question in terms of an image: why did the case of the Peppered Moth, which has been so important and so effective in Darwinian teaching throughout the 20th century, have such little exemplary value (in fact, the opposite) for *theoretical* population genetics in the period 1920–40?

To find the answer, a systematic study of the founding texts dealing with the problem of selection intensity is required. And in this respect, the imaginations of Fisher, Haldane, Wright and Ford seem to have known no bounds. Some of their arguments can be considered tactical; others, however, were fundamental.

The first reason to concentrate on weak selection intensities can be explained by the tactical use of population genetics to convince biologists of the superiority of the Darwinian explanation of evolution over rival models. Fisher, Haldane, Ford and Huxley repeatedly stated that selection intensities as low as 1 per cent or even 0.1 per cent would be an evolutionary factor that would dominate by several orders of magnitude the strongest evolutionary pressures imagined by the mutationists, the Lamarckians and the partisans of orthogenesis. This argument, dealt with in section 9.1, will not be repeated here.

The mathematical texts of this period also contain a more subtle algebraic argument. Haldane's articles are particularly revealing in this respect, because he pushed the theoretical kinetic description of selection furthest. In a long series of papers on 'the theory of natural and artificial selection',[54] Haldane essentially dealt with two examples:[55] artificial selection, where the selection coefficient is 1 (the unwanted genotypes are eliminated) and slow natural selection, where $k < 10^{-3}$. This value of 10^{-3} is systematically repeated in the calculations throughout the series of articles, for a simple methodological reason: Haldane's approach was to construct recursive equations which he resolved by expressing them as a differential equation. This analytical procedure is based on an approximation that is justified only if the selection coefficient is small (less than $10^{-3)}$).[56] In the case of more rapid selection, major mathematical problems appear. This explains why Haldane dealt with this problem only much later.[57] Similar considerations can be found in the work of both Fisher and Wright – practically all the models presuppose small selection coefficients. The fact that in the first article that explained his recursive method Haldane used the Peppered Moth as an example is particularly surprising, given that the selection coefficient is extremely high (> 0.3).

Beyond these rhetorical and algebraic arguments, there were five fundamental reasons why Fisher, Haldane and Wright based their work on small or minuscule selection coefficients.

The first argument dealt with the speed of evolution. If selection is always rapid it will soon exhaust the available genetic variation. The result is that the rhythm of evolution depends on the frequency of favourable mutations. In such a scenario, species would have only a limited power of transformation.

Behind this frequent and somewhat finalist argument, there is another, more rigorous, position. It is difficult to imagine how high selection pressures could act on many genes simultaneously. In the pioneer papers this argument is relatively vague, but it became widespread following Muller's use of the term 'genetic load' and its formalisation by Haldane in the 1950s. Any selective process involves a certain cost for the population in terms of 'genetic deaths'. Beyond a certain threshold (which is not very high), intense selection simultaneously applied on several independent characters condemns the population to extinction.[58] It is thus reasonable to think that if natural selection acts on several loci at once, and if, in an extreme case, it controls all characters at all times as Darwin and Weismann thought, its action must be infinitesimally small on most of them at any given moment.

To this difficulty we have to add the problem of 'genetic complexes'. Wright, and also Fisher and Haldane, quickly recognised that the notion of a constant selective value was a fiction.[59] The selective value of a gene in fact always depends on the 'genotypic milieu', and also on its own frequency. This type of idea has often been attributed to Wright, but it was in fact common to all three founders of population genetics. A few lines by Fisher illustrate this point: 'Each successful gene which spreads through the species, must in some measure alter the selective advantage or disadvantage of many other genes. It will thus affect the rates at which these other genes are increasing or decreasing, and so the rate of change of its own selective advantage.'[60] Despite this carnival of interaction, a coherent doctrine nevertheless emerges; most species are, at any given moment, in a metastable equilibrium, and on average a gene taken at random has a very strong chance of being subject to merely minimal selection pressure.

Fisher put forward another argument, wholly speculative but extremely interesting, which was related to a question that was inevitably raised after a moment's thought about 'mutations': what does it mean for a mutation to be 'favourable'? Without 'favourable' mutations, Darwinian evolution is impossible. But experimental genetics shows us that most mutations are deleterious. In *The Genetical Theory of Natural Selection,*[61] Fisher proposed not so much an answer to this question as a method that could provide a useful starting point. Let us consider the *a priori* probability that any variation depending on *n* parameters will be favourable. There is an optimal value for each of these parameters in a given environment, which can be represented geometrically as a point in an *n*-dimensional space. Motoo Kimura has provided a schematic representation of Fisher's idea. Figure 40A shows this representation in a two-dimensional space; the optimum is given by 'O'. Any organism can be represented as being at a certain distance (OA) from the optimum, and all those organisms with the same viability are contained within a circle surrounding the optimal point. In *n*-dimensional space, this circle becomes a hypersphere. In this abstract situation the appearance of variation simply means that 'if A is shifted through a fixed distance, *r*, in any direction, its translation will improve the adaptation if it is carried to a point within this sphere, but will impair it if the new position is outside'.[62] Or, in other words, any variation

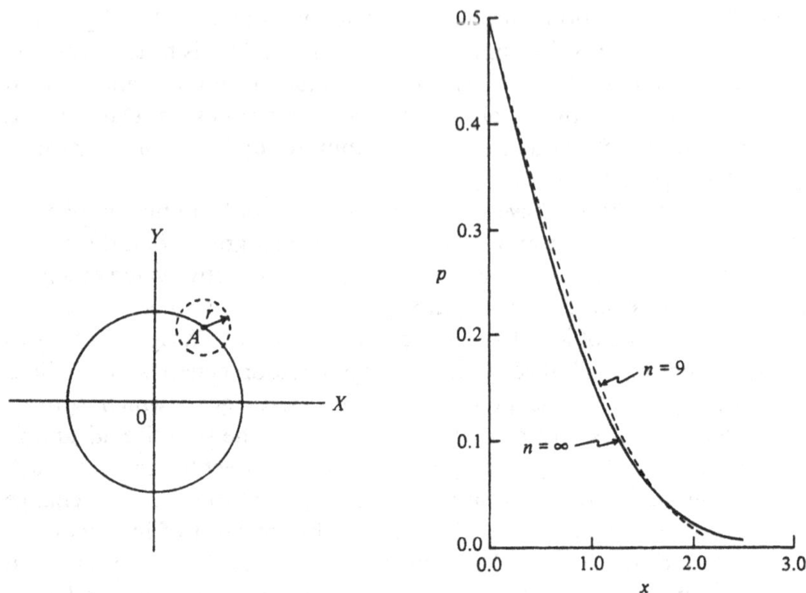

Figure 40 Illustration by Kimura of Fisher's (1930) model of the adaptive process. On the left, a model in two-dimensional space – the individual '*A*' is at a distance *OA* from the adaptive optimum. The circle represents all those individuals having the same adaptive value, and *r* measures the size of a variation (obtained by mutation or recombination). On the right, the probability *p* of a favourable variation as a function of the size *x* of the variation in an n-dimensional space ($x = r\sqrt{n}/d$), where d = the diameter of the circle (M. Kimura, *The Neutral Theory of Molecular Evolution*, 1983).

greater than the diameter of the hypersphere will be disadvantageous, because it will always take the organism outside this sphere of variation. For variation to be advantageous it must be both smaller in size than the diameter of the sphere and take place within the sphere. The probability that the point A approaches the maximum is in inverse proportion to the size of the deviation. This probability is in fact maximal (½) for an infinitesimally small *r*, however great the number of dimensions in the system. If the number of dimensions increases, the probability of improvement rapidly decreases with the size of the deviation (see Figure 40B). Fisher liked to illustrate this thought experiment with a concrete example. When you want to focus a microscope, the possibility of turning the knobs in the right direction is greater if the movements are small. A large movement will, in all probability, increase the distance from the correct adjustment. Furthermore, the greater the number of adjustments that have to be made, the greater the utility of proceeding by small movements. Applied to mutations, the same reasoning implies that the smaller the effect of a mutation (and therefore the greater the difficulty in detecting it), the greater its chance of being advantageous. If some mutations are advantageous, and thus selectionable, they will generally be very small, verging on neutrality.[63]

This leads us to the final reason why population geneticists dealt with weak rates of selection. In their famous controversy, Fisher and Wright argued about selection rates of the order of 10^{-5} to 10^{-7} and sometimes even lower. This kind of discussion might seem wholly unrealistic, because it is difficult to see what experimental conclusions it might lead to. But the determination with which Fisher and Wright concentrated on such levels had a fundamental theoretical significance. Very weak selection inevitably implies taking into account the role of various other evolutionary forces, in particular mutation, migration and random drift. Wright's view of a 'balance' among the 'pressures' of mutation, migration, selection and drift thus has meaning only if selection pressure is of the same order of magnitude as the pressure of mutation or drift. In large populations, drift is negligible, and the only pertinent evolutionary agent is the balance between selection and mutation. In this case, selection has a deterministic effect even if selection pressure is very weak (this is what Fisher argued). If, on the other hand, populations are small, weak selection pressures can be easily disturbed by random variations, and selection does not act in a strictly deterministic fashion. This is Wright's schema, which can equally be found in the later work of Motoo Kimura, for whom the question of weak selection rates was fundamental.[64] In this kind of discussion, it is important to understand that population geneticists are interested in the role not of a few rare characters that may rapidly alter a species, but of the immense mass of characters which, for long periods, are in a virtually stationary state. For such characters, the question of the interaction between the infinitesimal pressures of mutation, drift and selection is theoretically important.

The question of selection intensity thus constituted a 'speculative residue' – it was a question that had not been exhausted by the effective measurement of selection rates in nature, but which in fact involved the general theory of the genetic evolution of species. Theoretical population genetics, as its name suggests, seems to have always had the ambition of being a genuine theoretical discipline – a science of the conditions of the possibility of evolution under the Mendelian hypothesis. Fisher, Haldane and Wright knew perfectly well that 'selection coefficients' were methodological fictions. Haldane sought to devise algorithms that could be used by naturalists and experimental scientists for predictive purposes, while Fisher and Wright oriented their formal researches in a more transcendental (or more 'explanatory' and less 'predictive') direction. But all those who developed the genetic theory of evolution had in mind the same fundamental question: what happens in the initial phase of evolutionary change, when mutation and random drift can interact with selection?

In posing this question, the theory was perhaps less unrealistic than might at first appear. When considering the interaction between infinitesimally small selection, mutation and random drift due to small population size, at least two parameters – mutation rate and population size – have a rigorous and measurable empirical significance. This implies that in their ethereal models, the pioneers of population genetics required that selection be situated between two realities that could in principle be measured, even if, in practice, the natural mutation rate and especially the size of populations can generally be measured only indirectly, through the use of the idealised models of population genetics.

10.3 From the empirical to the formal: the challenge of molecular evolution

In the 1950s and 60s, the measurement of genetic polymorphism became a fundamental challenge for population geneticists. In 1966, the 'struggle to measure variation' (Lewontin)[65] was resolved: biochemical methods of analysis revealed an unexpected degree of molecular polymorphism. This led to the strongest and most precise question that experimental science had ever addressed to the genetic theory of evolution: Why is there so much genetic variation? In answer to this question, Motoo Kimura, without doubt one of the most inventive theoreticians since Fisher, Haldane and Wright, proposed what he called 'the neutral mutation–random drift theory of molecular evolution', or the 'neutral' theory of evolution. This final section deals with the remarkable conceptual implications of this theory, in particular its 'Darwinian' implications.

In 1950, five years after Hiroshima and two years after his resignation from the Soviet Academy of Sciences, the American geneticist Herman J. Muller published an article entitled 'Our Load of Mutation'.[66] In this fundamental paper the Nobel prize-winner (1946) put forward a method for estimating the mean number of deleterious genes present in a heterozygous state in each individual human. Muller estimated that this number ranged between eight and eighty, and that this was also a reasonable estimate of the total number of heterozygous loci per individual. This implied that genetic polymorphism essentially took the form of recessive genes which, in a homozygous state, constituted a 'load' for the population.

This paper is typical in several respects. Strictly speaking, its author was not a population geneticist, but the article opened a controversy that has preoccupied population genetics ever since. Muller was a laboratory scientist. A collaborator of Morgan's, in the early 1920s he provided an experimental basis for distinguishing between a mutation and a rare recombination. In both theory and practice, Muller was responsible for developing the genetic concept of mutation as a specific alteration of a Mendelian gene.[67] Using X-rays to induce mutations in massive experimental populations, Muller also mapped around five hundred genes in *Drosophila*. His experimental work made him the paradigmatic theoretician of the composition of the genome.

Muller's article threw down the gauntlet before the population geneticists: they had to find a way of measuring genetic variation in order to resolve the dispute. A new quarrel thus emerged: that of the real size of genetic polymorphism and its structural significance. Muller was genuinely convinced that his work resolved the problem at the same time as he posed it. The doctrine he put forward can be summarised in a few lines. Muller thought that the amount of genetic variation was low (he argued it would affect at most around one locus in a thousand, and probably fewer);[68] that it mainly consisted of recessive lethal genes; that loci were normally homozygous; that it was right to speak of a 'normal' or 'wild-type' state of a locus, as had always been the case in laboratory genetic studies. He finally argued that favourable mutations appeared rarely in species and are established rapidly, replacing the ordinary allele and becoming the 'normal' form of the gene. This representation of the genome implied that the optimal state of a species is that of genetic uniformity and – in passing – that eugenic programmes were justified.[69]

Muller's position was put forward after a decade in which the Modern Synthesis had appeared, completely suffused with the spirit of the 'new systematics'[70] – a populational systematics, Mendelian in its methods and driven by the conviction that natural populations contained a large amount of genetic polymorphism. For a whole generation of naturalists who had founded their methodology and their view of evolution on 'polymorphism', Muller's paper was viewed as an experimental refutation. It also constituted a serious threat to theoretical population genetics – if genetic variation was in fact relatively weak and limited to recessive deleterious genes, the subtle models of the founders of the discipline, with their infinitely small selection coefficients and their schemes of 'fluctuation' in a hyperspace of gene frequencies, could turn out to be nothing other than a series of interesting but ultimately pointless mathematical games.

Theodosius Dobzhansky speedily organised the counteroffensive. At the beginning of the 1950s, Dobzhansky's name was closely linked to a vision of evolution inspired by Wright's 'shifting balance' theory. In his remarkable studies of the population genetics of natural populations of *Drosophila*, Dobzhansky had taken on the mantle of champion of the notion of 'balanced' selection in which genetic diversity, far from being reduced, is maintained, for example by heterosis (heterozygote advantage) or by ecological variations within the species' habitat.

In 1955, Dobzhansky characterised the difference between two hypotheses with regard to the genetic structure of populations – the 'classical' hypothesis (Muller) and the 'balance' hypothesis (Dobzhansky).[71] In American population genetics this division rapidly became conventional (and impassioned). Richard Lewontin has vigorously summarised this legendary quarrel in *The Genetic Basis of Evolutionary Change*.[72] The 'classical' hypothesis admitted that individual genomes are homozygous at most of their loci, with only a few rare loci being heterozygous. These loci are not the same from one individual to another. Most heterozygous loci correspond to genes that would be deleterious or lethal in a homozygous state. A few others (extremely rare) correspond to favourable genes that tend to spread through the population. To sum up, according to the 'classical' conception, most genes in an individual can be considered to be in a 'normal' state.

The 'balance hypothesis', however, argued that individuals in a panmictic population are heterozygous for a large number of loci. Bruce Wallace put forward the most radical version of this attitude:

> Subject to the limitations imposed by chance elimination of alleles, by mating of close relatives, and by the finite number of alleles at a locus, we feel that the proportion of heterozygosis among gene loci of representative individuals of a population tends towards 100 percent.[73]

Dobzhansky's original formulation of the 'balance hypothesis' was more cautious (but not much):

> According to the balance hypothesis, the adaptive norm is an array of genotypes heterozygous for more or less numerous gene alleles, gene complexes, and chromosomal structures. Homozygotes for these genes and gene complexes occur in normal outbred populations only in a minority of individuals, and make these individuals more or less inferior to the norm in fitness.[74]

However, Dobzhansky thought that both situations occurred in nature.[75] The 'balance' hypothesis thus accepts that certain loci are homozygous, but places them under severe functional constraints. No allele can in fact be designated as 'normal' or 'wild-type' (the '+' allele of laboratory geneticists). Finally, for each locus, there are a large number of alleles present in any given population.

Richard Lewontin summed up the two hypotheses as follows.[76] According to the classical conception, the genetic description of the same chromosome fragment in two individuals chosen at random from a natural population would have the following appearance:

$$\frac{++++m+\cdots+++}{++++++\cdots+++} \qquad \frac{++++++\cdots+m+}{++++++\cdots+++}$$

while under the hypothesis of balanced polymorphism (in its most extreme form) it would appear thus:

$$\frac{A_3\ B_2\ C_2\ D\ E_5\cdots\ Z_2}{A_3\ B_2\ C_2\ D\ E_5\cdots\ Z_5} \qquad \frac{A_2\ B_4\ C_1\ D\ E_2\cdots\ Z_1}{A_3\ B_5\ C_2\ D\ E_3\cdots\ Z_1}$$

These two different representations of genetic structure are linked to two very different visions of selection. In the classical hypothesis, selection is essentially normalising and eliminatory, maintaining the frequency of deleterious recessive alleles at a very low level. When a favourable mutation appears, selection drives out the competing allele. From this point of view, it is reasonable to say that it is the gene as such that is the target of selection. The balance hypothesis, on the other hand, has to account for the maintenance of a high degree of heterozygosis. This hypothesis takes its name from Dobzhansky's view that alternative alleles are maintained by some form of 'balancing selection' or other stabilising selection. This maintains variation, and, over time, can even be argued to promote it, although it does not strictly speaking create it.[77]

According to the 'classical' conception, therefore, selection and variation are antithetical. The balanced-polymorphism school, on the other hand, saw selection and variation as being two aspects of a single dynamic: selection is diversifying, because it favours the appearance of novel, heterogeneous gene complexes. This theoretical tension had a number of interesting minor consequences. Muller, like Fisher and Haldane, was convinced of the importance of eugenics, although the involvement of the three men took place in very different national and political contexts. Before and after the Second World War, Muller was in favour of artificial insemination, he opposed American and Russian nuclear tests because of their mutagenic effects, and he argued that evolution was over as far as man was concerned. Dobzhansky, however, fought against eugenics throughout his life; he refused to denounce nuclear testing on the basis of purely biological arguments; he thought that evolution was an endless process, and, as is well known, he was an active member of the New York Society of the Friends of Teilhard de Chardin.[78]

The Muller–Dobzhansky quarrel was not merely verbal. Apart from its personal features (teacher and pupil) and its academic aspects (experimental biology versus field biology) the difference was open to experimental testing. It merely required finding answers to the following two questions (not so simple as it sounds):

- What proportion of loci are polymorphic in natural populations?
- What is the level of heterozygosity per individual?

According to Muller's hypotheses, both these numbers should be small (Muller estimated the level of heterozygosity in humans to be less than 10^{-3}, or fewer than 100 genes out of an estimated total of 100,000 in the human genome). According to the balanced-polymorphism hypothesis, these numbers should be greater than 10^{-1}.

From 1950 to 1966, a large number of different indirect techniques were used to measure genetic variation: listing morphological mutations, revealing the existence of hidden lethals by crossing, variability and fertility distributions, etc.[79] All these attempts encountered the objection that their results were not statistically significant. The main positive consequence of this period was limited to a realisation of the large number of theoretical and experimental biases that could affect population-genetic methods.

Over the same period, biochemistry and molecular biology developed to the extent that around 1960 it became apparent that it should be possible to measure the size of genetic polymorphism directly. In the 1960s, it was impossible to sequence a large number of samples of DNA, or even of proteins, but the fundamental dogma of molecular genetics (one gene = one protein) left open the possibility of developing relatively direct techniques for measuring gene variability.

In 1966 'the struggle to measure variation' came to an end, as two estimates of what has since become known as 'electrophoretic polymorphism' were published simultaneously. The English biochemist H. Harris published a study of a human population (from England), while Richard Lewontin and J. L. Hubby measured biochemical polymorphism in natural populations of *Drosophila*.[80] Hubby was a biochemist; Lewontin was a population geneticist, often presented as Dobzhansky's most brilliant pupil.

The principle of the two studies was as follows. Although it was not yet possible to directly compare nucleotide sequences in the genes of loci chosen at random, it was nonetheless possible to compare proteins, and thus the primary effects of genes. However, the techniques involved in protein sequencing were still too complex to be used for such a study. Hence the idea of using the physico-chemical properties of proteins that are linked to changes in their primary structure (that is, to their linear sequence of amino acids). The ingenious technique used by this trio of researchers was to detect variations in electric charge by subjecting samples to electrophoresis. Of the twenty amino acids that make up all proteins, only four are ionised. If one of these amino acids is replaced by a non-ionised amino acid at any point in the peptide chain, the protein's electric charge is altered. Electrophoresis takes advantage of this property by making the protein migrate in a gel under the action of an electric field. Depending on their degree of ionisation, proteins migrate more or less quickly on the electrophoresis plate. This technique was in fact the end-product of a highly complex physico-chemical theory,[81] but since the beginning of the 1960s it had become widely used and easy to use. In practical terms, it involved homogenising the tissues of an individual organism (generally whole fruit-flies), placing the resulting homogenate at one end of an electrophoresis gel and, by using various commercially available solvents, reactants and colorants, observing the bands produced following migration of a given enzyme (for example, an esterase) under the action of an electric field. A number of different forms of the enzyme could thus be visualised for the given population, heterozygous individuals being detected by the fact that they have two bands – each band representing a different protein. Matters are often complicated by the fact that many proteins are composed of sub-units, but this does not alter the principle involved. The kind of results obtained by this method are shown in Figure 41.

The bands produced by electrophoresis can be considered to be individual phenotypes, which biochemical theory explains by amino-acid substitutions. However, a classic Mendelian analysis is necessary in order to be certain that the observed differences really do correspond to genetic differences. In fact only around 30 per cent of amino-acid substitutions can be detected by this method – this proportion corresponds to that of the four ionised amino acids that are involved in the mean composition of proteins. The result is a measurable underestimate of the variation in the amino-acid sequence. Furthermore, even if this amino-acid sequence variation is known, it is an underestimate of the nucleotide variation in the DNA sequence, because of the existence of a number of 'synonymous' codons – different nucleotide sequences that code for the same amino acid. In other words, the protein polymorphism reveals only 70 per cent of the total DNA polymorphism. To sum up, electrophoretic polymorphism is an underestimate of protein polymorphism, which is itself an underestimate of DNA polymorphism.

This having been said, the results obtained by this method are particularly clear. Lewontin and Hubby's study was of a sample, as random as possible, of eighteen proteins in *Drosophila*. They showed that electrophoretic

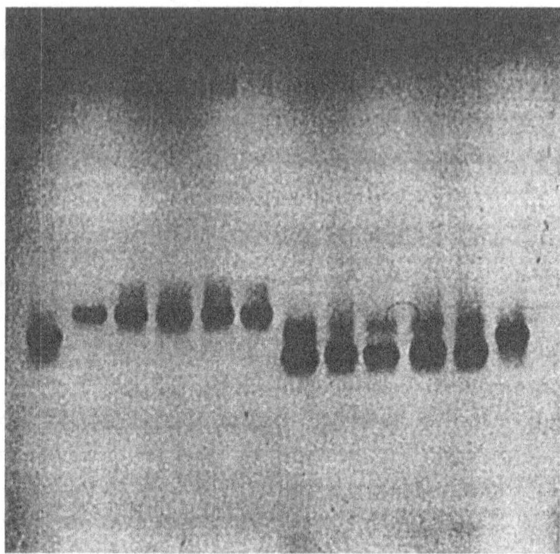

Figure 41 Migration bands produced by a gel electrophoresis of a *Drosophila* esterase. Each column corresponds to an individual fly. The bands show the point to which the esterase has migrated after a given amount of time. This particular experiment shows the existence of several forms of esterase in the same populations; some individuals are heterozygotes (as shown by the presence of two bands in the same column) (R. Lewontin, *The Genetic Basis of Evolutionary Change,* 1974).

polymorphism affected one-third of loci in this species, and that the rate of individual heterozygosity was around 12 per cent. In a very short space of time, similar studies were carried out on the whole of the animal kingdom, and showed that, on average, around 30 per cent of loci were polymorphic, and the mean level of heterozygosity was around 10 per cent.[82] Given that enzymatic polymorphism underestimates DNA polymorphism, it became quite clear that at the molecular level most loci are polymorphic. Direct studies of both protein and DNA sequences were later to confirm this interpretation.

At the end of the 1960s, the question of the amount of genetic variation was thus decided. The molecular approach had refuted the classical theory, and, strengthened by the support of the biochemists, Dobzhansky's school felt itself justified in repeating Dobzhansky's declaration of 1955, at the very outset of the controversy. In 1955, he had written:

An array of related genotypes consonant with the demands of the environment is the adaptive norm of a population. The 'norm' is, thus, neither a single genotype nor a single phenotype. It is not a transcendental constant standing above or beyond the multiform reality.[83]

In 1974, Lewontin, in some triumph, referred to the results of his work with Hubby in the following terms:

> The average heterozygosity per locus for natural populations of *D. pseudoobscura* is about 35 percent, and essentially every gene is polymorphic. At least on the face of it, the classical hypothesis of population structure is firmly and directly refuted, and the balance hypothesis is revealed as correct ... The first direct measurements of genetic variation in natural populations have proved, in Dobzhansky's words, that the norm 'is not a transcendental constant standing above or beyond the multiform reality' (1955).[84]

Does this mean that the theory of balanced polymorphism is true? Very much so. With the discovery of molecular polymorphism, the theory of balanced polymorphism was proven correct over the question of fact, beyond the wildest dreams of its advocates. But once the fact had been established, it had to be explained. It was at this point that, for the second time in its history, genetics provided Darwinism with a theoretical rebirth.

Understanding how and why the gene pool of a species should be polymorphic at a great number, if not the majority, of loci, is a particularly complex task. In a classic neo-Darwinian perspective, it would initially appear appropriate to look for an adaptive significance for this phenomenon. Traditionally, there are two main kinds of Darwinian explanations of polymorphism in population genetics, which Ford clearly summarised in various articles published in the 1940s. The first is that of 'transient polymorphism', where the locus is subject to a directional and secular selection pressure that has the effect of replacing one allele with another. The second possibility is that of saying that the polymorphism itself has an adaptive value and is thus the result of selection. However, given that the task is to explain the coexistence of two or more different forms (the alleles of a locus), it has to be accepted that this selection is 'balancing' and produces a 'balanced polymorphism'.[85]

Taken at face value, the idea of a transient polymorphism simultaneously affecting thousands or tens of thousands of genes is relatively unrealistic. It would imply that selection slowly diffuses favourable mutations that appear in the population. Population geneticists sometimes reason as though this were the case, but it must be realised that this is merely one technical possibility. From the 1930s onwards, empirical studies of polymorphism, in particular those of Dobzhansky and Ford, showed that in most cases gene frequencies vary. To explain this phenomenon it was necessary to postulate either random sampling effects or a shifting balance of opposing selective forces. This problem has already been evoked with regard to Fisher and Ford's article published in 1947.[86]

According to the hypothesis of 'balancing' selection, there are several classical models that can explain the maintenance of heterozygotes. The oldest model is that of heterosis, or hybrid vigour: if a genotype Aa has a selective value greater than that of AA and aa, both alleles will be maintained in the population. Although this phenomenon has been observed, and in a few rare cases explained (e.g. sickle-cell anaemia), it is doubtful that it could explain the vast amount of polymorphism that exists.[87] The persistence of variation can also be explained by the division of ecological space – one genotype may be found in the heart of the forest and another on the edges, but seasonal mating ensures that the polymorphism is maintained.[88] A third classical model is that of frequency-dependent selection, in which an allele is selected against if it is frequent, but favoured if it is rare.[89]

All these hypotheses have been modelled and have been shown to explain both experimental results and data from natural populations. In some cases, their physiological or ecological significance has been clearly analysed – there can be no doubt that balancing selection explains some cases of polymorphism. Nevertheless, the fundamental problem for the contemporary evolutionary biologist is to know whether such processes can explain the extraordinary variability that the biochemist can detect at the molecular level. In other words, is it reasonable to continue looking for an adaptive explanation of the structural variation that affects virtually all species? Is it necessary (is it even *possible*?) that variation is always controlled by natural selection?

In answer to this question, the Japanese geneticist Motoo Kimura (1924–94) put forward a fundamental problem. This discussion of Darwinism will close with a study of this radical proposition by a major theoretician which in its interrogative power is redolent of the ideas of the founders of evolutionary genetics.

In 1967 Kimura proposed an interpretation of molecular evolution known as the 'neutral theory', or, more precisely, the 'neutral mutation–random drift hypothesis of molecular evolution'.[90] This theory was termed 'non-Darwinian', first by J. L. King and T. H. Jukes, who had independently come to similar conclusions,[91] and then by Kimura himself, who sometimes argued that his conceptions illustrated 'the neutral theory of non-Darwinian evolution'.[92]

The precision of these formulations is important. Kimura never said that he was proposing a 'non-Darwinian theory of evolution', but only a 'theory of non-Darwinian evolution'. The nuance is fundamental. The 'neutral' theory of evolution is the theory of a level of organisation to which the Darwinian model does not apply. This theory, far from being part of an anti-Darwinian biological philosophy, is one of the most remarkable modern reinterpretations of the Darwinian theory of natural selection.

The starting point of Kimura's theoretical revision was J. B. S. Haldane's late and disconcerting discussion of 'the cost of natural selection'.[93] Because it is fundamental to the 'neutral theory', this idea needs to be explained. Haldane had put forward a quantitative estimate of the depressive effect of natural selection on the reproductive capacity ('fitness') of a species. The problem can easily be understood by considering the case of the Peppered Moth (*Biston betularia*). In 1956, Kettlewell had directly observed the rate of elimination of the phenotype most exposed to birds (the light moths if the environment is polluted, the dark moths in a rural non-polluted environment), showing that every day up to half the disadvantaged population was eliminated.[94] In his 1957 paper Haldane constructed a famous 'dilemma' on the basis of this estimate, known as 'Haldane's dilemma'. Let us suppose that at the same time as the Peppered Moth was subject to intense selection pressure for melanism, it was also subject to a comparable selection pressure on ten other independent characters, owing to an important change in the environment. In this case, it is no longer half of the original form that survive each day, but $(\frac{1}{2})^{10}$, or 1 out of 1024. It is very likely that such a species would rapidly become extinct. This fictitious example illustrates the idea that natural selection always initially reduces the reproductive capacity of the species, or, as demographers put it, its 'natural rate of growth'. The relative expansion of forms that are better adjusted to the environment no doubt leads in the long run to an increase in the reproductive capacity of the species. But for the selective process to accomplish anything, it should not be too strong.[95]

This raises a fundamental question for the genetic theory of selection according to which the elementary event is the substitution of one gene for another at a given locus in the course of evolution. This substitution has a 'cost'; in other words, the species has to pay for its improvement by a certain number of 'genetic deaths'. Or, put another way, the substitution of the normal allele by a mutation produces a reduction in the reproductive capacity (the natural rate of demographic growth), in the lapse of time in which it takes place. Commenting on Haldane's article on 'the cost of natural selection' in 1960, Kimura called this effect 'substitutional load', by analogy with Muller's 'mutational load'. The currently accepted term is 'genetic load'. This notion raised several fundamental questions for the theory of selection:

- What exactly is the cost of a gene substitution in a species?
- Is the cost affected by the intensity of selection?
- What are the consequences for the possible speed of evolution by selection?

In his formal treatment of the question published in 1957,[96] Haldane gave a very elegant response. He showed that whatever the intensity of selection, its cost was the same. Haldane estimated this cost to be around thirty times the population size, that is, in a panmictic population of around one million individuals, the substitution of one allele by another will result in about thirty million 'genetic deaths'. He also made a reasonable estimate of the rate of possible gene substitution needed for a species subject to selection to have a reasonable chance of not becoming extinct. Haldane concluded that in a given species there should not, on average, be more than one gene substitution per three hundred generations. Of course, the process may be more rapid (as in the case of industrial melanism), but a species that often took this road would seriously expose itself to the threat of extinction. An interesting interpretation of the value put forward by Haldane is that it provides a way of measuring how selection affects fitness, itself measured by the absolute rate of increase in the population.[97] If there is a gene substitution every three hundred generations, optimal fitness is reduced by around 10 per cent; if selection is a hundred times more rapid, or one substitution every three years, fitness is reduced to 0.0004 with respect to the optimal level (or reduced 20,000 times).[98]

Haldane's discussion of the 'cost' of selection enables us to understand the idea developed eleven years later, in 1968, by Motoo Kimura, in a paper entitled 'Evolutionary Rate at the Molecular Level'. In this famous article, barely three pages long, Kimura put forward the 'neutral hypothesis' and proposed an estimate of the rate of nucleotide substitutions in mammals.[99] The estimate, realised from a molecular standpoint, was based on the comparison of homologous polypeptide sequences from various groups of mammals. Using a few simple hypotheses, it was possible to calculate the number of amino-acid substitutions that had taken place since the divergence of the various taxonomic groups, and the number of nucleotide substitutions in the DNA sequence. By dividing the number of years which separate the mammals from their common ancestor, Kimura arrived at the rate of nucleotide substitution. Then, extrapolating to the whole of the mammalian genome (around 4×10^9 base pairs) he provided an estimate of the mean rate of nucleotide substitution in mammals – at least one nucleotide per genome every two years. This means that, at the molecular level, every two years a point mutation completes its diffusion throughout the species and becomes fixed.

Scaled up to the whole genome, such a rate might seem very small to the biochemist and insignificant to the naturalist. But it poses a fundamental problem for any biologist trained in selectionist argumentation. If the substitution of nucleotides is due to natural selection, it has a cost. This cost cannot be sustained. On the basis that one year equals one generation, the application of Haldane's model to Kimura's estimate suggests that the fitness of a species would be reduced to 3×10^{-7} of its optimal level.[100] Or, to put it another way, each parent would have to produce 3,270,000 offspring for the population size to remain the same!

Kimura's calculation leads to a remarkable theoretical conclusion: natural selection cannot explain the fixation of point mutations. Or, more precisely, natural selection is not a reasonable explanation of the vast majority of nucleotide substitutions. In other words, panselectionism has no meaning at the molecular level – most molecular evolution is 'non-Darwinian'.

Kimura's alternative explanation is well known: 'most mutations produced by nucleotide replacement are almost neutral in natural selection'.[101] This conclusion implied the rehabilitation of the idea, supported by Sewall Wright, that random drift plays an essential role in evolution:

> Finally, if my chief conclusion is correct, and if the neutral or nearly neutral mutation is being produced in each generation at a much higher rate than has been considered before, then we must recognize the great importance of random genetic drift due to finite population number [Kimura has a reference to Wright, 1931] in forming the genetic structure of biological populations.[102]

Kimura often repeated this conclusion in later articles. Probably the most dramatic formulation can be found in the last lines of a lecture given in France in 1985, and published in 1988:

> I conclude that the most prevalent evolutionary changes that have occurred at the molecular level (i.e. in the genetic material) since the origin of life on the earth are those that have been caused by random genetic drift rather than by positive Darwinian selection.[103]

Kimura has clearly indicated the impact of the neutral model of molecular evolution on the problem of genetic polymorphism. Against Lewontin, who argued that most molecular polymorphism should be able to be explained by a moderate heterozygote selective advantage, Kimura pointed out that this kind of selection also had its cost in terms of 'genetic deaths'. Would it not be more reasonable for molecular polymorphism to be nothing other than the transient manifestation of a process of random fixation of mutations? As Kimura put it:

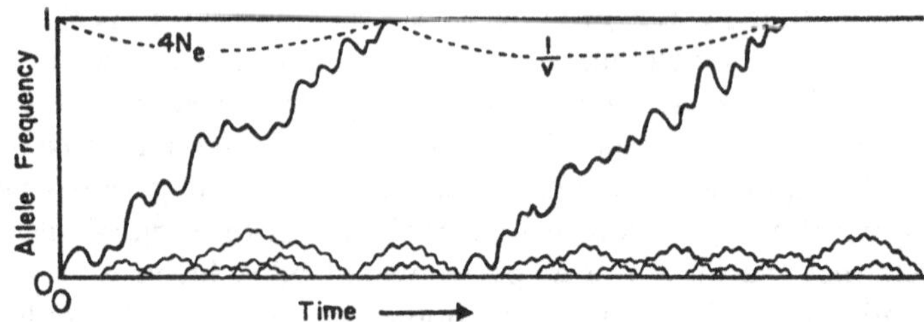

Figure 42 Behaviour of neutral mutant alleles in a population of limited size. The only factors that intervene are the mutation rate (v) and the effective population size (N_e). N_e determines the intensity of random drift. The mean fixation time (measured in generations) of a neutral allele equals four times the effective population size ($4N_e$). This theoretical result has been known since the 1920s. The mean interval between consecutive fixations (and thus the substitution rate for neutral alleles) is l/v. In terms of nucleotides, these rates are very high (M. Kimura, 'How genes evolve', 1976).

'polymorphism is a phase of molecular evolution'.[104] Figure 42 gives a simple presentation of this idea.

It is very important to understand Kimura's central assertion. He does not argue that all genetic polymorphism is a consequence of random drift, or that all fixation and all fluctuations are random: he simply states that most molecular variation is selectively neutral. In passing, it is worth noting a classic misunderstanding of the notion of 'selective neutrality'. Many biologists appear to have interpreted 'selective neutrality' as meaning that the mutated genes are genes without adaptive significance, or even without function. This misunderstanding reveals a singular ignorance of the Darwinian idea of adaptation. One of the pillars of the theory of natural selection is that adaptations are always relative. If natural selection can construct adaptations, it is precisely because there is never any absolute norm of adaptation. When the neutral theory states that elementary molecular variation is approximately neutral, this implies that the function of the protein is not altered: a substitution can be neutral in the sense that it does not affect the essential structural and functional qualities of the protein: 'What the neutral theory assumes is that the mutant forms of each gene participating in molecular evolution are selectively nearly equivalent, that is, they can do the job equally well in terms of survival and reproduction of the individual.'[105] In other words, evolution at the molecular level and evolution at the phenotypic level should not be represented in strictly identical terms, although they must obviously be connected:

We are confident that mutation is playing a much more direct and important role in controlling the rate and direction of evolution at the molecular level than we have been accustomed to think through decades of evolutionary studies at the phenotypic level, particularly under the flag of neo-Darwinism.[106]

The neutral theory has led to interminable debates between its supporters and those who have tried to rescue the universality of natural selection. The argument has yet to be fully resolved. Two kinds of critical arguments have been put forward. Some are based on the use of the notion of genetic load – there are many tacit postulates, biases and uncertainties involved in measuring this value. Indeed, it can be measured in a way that makes it compatible with the classic hypotheses of 'balancing' selection. Other arguments against neutralism are specifically based on polymorphism – there are different ways of explaining polymorphism without raising the spectre of 'genetic load'. For example, frequency-dependent selection maintains polymorphism with a genetic load which is zero at equilibrium. It appears that this kind of balancing selection is relatively frequent in nature.[107]

Of course, it is not the job of the historian of science to get involved in such an open debate, except to the extent that he or she is prepared to run a series of inevitable risks. But it is his or her task to notice major changes in concepts when these can be seen to occur.

On the whole, Kimura's hypothesis seems to be largely accepted today, in so far as it concerns only molecular evolution. Nevertheless, even if the neutral theory of molecular evolution should turn out to be false, it will have had the immense merit of having awoken evolutionary biologists from their dogmatic slumber. To close this convoluted study, which has gone from Darwin, through the crisis opened up by his hypothesis, and up to the 'genetic theory of selection', two (philosophical) reflections will be proposed. Both are suggested by Kimura's 'neutral theory'.

1. From the beginning of the century, neo-Darwinism has been dominated by the conviction that species evolve through the accumulation of beneficial gene mutations, however this might take place. When molecular biology revealed the material structure of genes, it seemed natural to extend to the molecular level the selectionist explanation that had proved itself at higher levels of organisation.

Biological macromolecules – DNA and proteins – are highly organised systems that will tolerate a large number of local perturbations. The message encoded in the DNA sequence will sometimes remain the same, even though the letters and even the words may change – these 'silent' mutations are due to the redundancy inherent in the genetic code (in other words, the existence of synonymous codons). At the protein level, tolerance is even greater. Away from catalytic, receptor or signalling sites, the sequence may alter quite radically without significantly altering form or function. This does not imply that any change whatsoever can take place. From a functional point of view, even a slightly defective protein will be eliminated; any slightly advantageous protein will spread throughout the population. Both processes have a cost: the 'mutational load', or 'substitutional load'. However, we need to distinguish fact from fantasy. The neutral mutations postulated by Kimura are not completely 'neutral'. They are in fact slightly unfavourable: Kimura argues on the basis of a selective disadvantage of 10^{-7}. These are mutations with an effect weak enough so that the pressures of mutation and drift can win out over negative selection.[108] Their evolution is thus 'non-Darwinian'.

But this theory of non-Darwinian evolution is not a non-Darwinian theory of evolution. Kimura never argued that the evolution of *organisms* takes place 'randomly'. It is difficult to imagine that this student of Sewall Wright, who was also the only theoretician to have thoroughly proved Fisher's theorem (as a good approximation)[109] was a non-Darwinian. Furthermore, the neutral theory of evolution had as its starting point Haldane's discussion of the cost of selection; besides Fisher's and Wright's speculative contributions around 1930, this was indeed a major step in the clarification of the most general conditions under which natural selection can succeed in transforming species on a massive timescale. Kimura's theory is the heir of the fundamental studies of the conditions necessary for evolution under the Mendelian hypothesis developed between 1920 and 1960 by the pioneers of theoretical population genetics.

Kimura's 'Darwinian' lesson, as I understand it, is as follows. There is no need for natural selection to control genes down to the last detail. The atomistic principle that proved so effective for both Mendelism and its neo-Darwinian corollary, need not be extended infinitely. Genes can show substantial changes in their elementary sequence without natural selection detecting anything at all. Selection sees only form and function, the interaction between the gene and the protein with other genes and proteins. This leads us to a biological philosophy which is very close to that of Darwin, give or take the scale of observation.

This biological philosophy is fully expressed in the explicit theory of natural selection that Kimura developed on a number of occasions.[110] Natural selection should essentially be understood at the phenotypic level. In other words, the empirical basis of the Darwinian principle is that of quantitative variation. When the environment is stable, the main effect of selection is to eliminate extremes and to concentrate the population around an intermediate optimum. When there is a change in the environment, species change rapidly by altering the mean values of many quantitative characters:

> Thus we are led to a general picture of evolution as follows. From time to time, the position of the optimum of a phenotypic character shifts as a result of change of environment, and the species tracks such a change rapidly by altering its mean. During this short period of change, extensive shift of gene frequencies is expected to occur at many loci, but this process itself will seldom cause gene substitutions. But most of the time stabilising selection predominates, under which neutral evolution or random fixation of mutant alleles occurs extensively, transforming all genes.[111]

To summarise, there is no need to extend Darwinian logic to the detail of genetic atoms (that is, to the nucleotides). Although natural selection does indeed build with 'bricks', and its effect could be described as the 'kinetics of brick substitution', at the end of the day it is relatively indifferent as to its materials. This is one of the possible lessons of the neutral theory of molecular evolution. At this point, it is useful to remember Darwin's deliberately anthropomorphic metaphor:

> Let an architect be compelled to build an edifice with uncut stones, fallen from a precipice. The shape of each fragment may be called accidental; yet the shape of each has been determined by the force of gravity, the nature of the rock, and the slope of the precipice, – events and circumstances, all of which depend on natural laws; but there is no relation between these laws and the purpose for which each fragment is used by the builder. In the same manner the variations of each creature are determined by fixed and immutable laws; but these bear no relation to the living structure which is slowly built up through the power of selection, whether this be natural or artificial selection.[112]

2. One of the most remarkable predictions of the neutral theory of molecular evolution is that for a large number of genes – or, more precisely, of nucleotide or peptide sequences – the evolution of DNA and proteins can function as a 'molecular clock'.[113] The principle of the idea is extremely simple: if most substitutions are selectively neutral, their rhythm is determined only by mutation pressure. Proteins (or DNA sequences) must thus behave like 'stochastic clocks', comparable to those in physics that measure the loss of radioactivity. Strictly speaking, such clocks show no periodic change, but instead a constant probability of change. Furthermore, each gene or protein is a separate, original clock, because their functional constraints are very different. For example, cytochrome changes only extremely slowly, while fibrinopeptides show a much higher substitution rate.

There has been much discussion of the 'precision' of the molecular clock, and of its meaning. And indeed it does not seem to be as precise as theory would like.[114] Nevertheless, it is increasingly used to construct phylogenies, which, taken as a whole, are remarkably congruent with those elaborated many years ago using macroscopic criteria.

This relatively recent orientation of the genetic theory of evolution contains another lesson. The non-Darwinian evolution of molecules provides geneticists with a window on the vast depths of geological time. But, at this scale, it is not the mechanism of evolution which is revealed, but rather the great tree of diversification or, if one prefers, the chronicle of the common descent of species.

Which leads us back to the beginning of this book. Nineteenth-century naturalists devoted their talents to an historical interpretation of classification, as is shown by the following passage from the *Origin of Species*:

> All the foregoing rules, and aids and difficulties in classification are explained, if I do not greatly deceive myself, on the view that the natural system is founded on descent with modification; that the characters which naturalists consider as showing true affinity between any two or more species, are those which have been inherited from a common parent, and, in so far, all true classification is genealogical.[115]

By posing 'descent' as the 'foundation' of classification, Darwin focused on the question of those individuals who, from generation to generation, transmit their characters to their offspring. The principle of 'descent with modification' was thus also a principle of 'inheritance with modification', and it is at this level that the causal hypothesis of natural selection has continuously sought its experimental meaning for nearly 140 years. However, after Darwin, morphological theory and the causal theory of evolution went their separate ways. The principle of natural selection became the foundation of a local theory of evolutionary mechanics. This tendency was developed to its fullest degree in its genetical elaboration. With molecular genetics and its utterly analytical interpretation of inheritance, it might have been expected that local intelligibility would reaffirm itself. But the opposite happened. The theory of biological atoms returns us to the fundamental question of the diversity of life. The principle of descent once again comes into its own and reminds us that the principle of selection, whatever the scale of its historic effects, only ever has a purely local significance.

There is no doubt that natural selection has drawn the tree of life step by step. But there is nothing in the theory of natural selection that enables us to reconstruct that tree *a priori*, and even less to understand the particular pattern it exhibits (this one rather than another one). We do understand, no doubt, why there is so much diversity, but not why there is one form of diversity rather than another. Indeed, it is quite possible that this is one of the limits of the theory. The classificatory tree is erratic – this was one of Darwin's key arguments in favour of the plausibility of the theory of modification by natural selection.[116] The theory of natural selection is the causal theory of a local process, but of universal occurrence; it is also a metatheory of classification. As a causal hypothesis, natural selection explains the path taken by a given life history, step by step; as the foundation of a general vision of the living world, it unifies this world in terms of process, and not of plan.

With these final considerations, we have clearly left the terrain of impartial historical analysis. But at the end of such a journey it is only right that the philosopher, empowered by the present, should choose to view the current situation from a particular peak in the conceptual landscape.

Conclusion

The most appropriate way of concluding this book is to return to its origins and its aim. As a philosopher trained in biology, I wanted to explore the rationality of natural selection. As often happens, epistemology led to history, and thence to a return to the key moments in the development of the subject under study.

In a way, Darwinism was born twice – it was only in the 1920s that the hypothesis of natural selection, as formulated by Darwin, took on even a semblance of validity. This leads to a paradox that can be expressed in two ways:

1) For the historian, there is a striking discord between Darwin's prestige in the half-century that followed the publication of the *Origin of Species* and the confusion that characterised research and discussions of selection in the same period. To use the terminology of the late Thomas Kuhn, the emergence of the Darwinian paradigm immediately led to a crisis over whether its central hypothesis was even possible. This crisis did not stop Darwin's work from taking on the status of an exemplar, nor did this status put a stop to the crisis. This is the first aspect of the paradox.

2. From an epistemological standpoint, the paradox is that one can *still* read the *Origin of Species*. Darwin did not have a clear operational understanding of a 'population' – a word that he did not even use in its modern sense.[1] As to his theory of inheritance, it was generally extremely confused, and when it was clear, it was a manifesto in favour of an extreme form of the inheritance of acquired characters. How is it that, despite these two limitations, Darwin's ideas are so familiar to the modern reader? Again to use terms inspired by Thomas Kuhn, this means that there is a 'paradox of commensurability' of Darwin's thinking with that of modern evolutionary biology.

This double paradox determined the structure of the current work. I wanted both to account fully for the crisis in spite of which (or because of) theoretical Darwinism developed, and to outline the form taken by the modern theory of selection once this crisis had been overcome.

At first I thought that the second of these tasks would be the more difficult. The modern theory of natural selection, as elaborated in population genetics, is sometimes extremely arcane in its mathematical abstraction (at least for a layman such as myself). However, this did not turn out to be the biggest problem. It was the initial crisis that raised the thorniest historical difficulties, necessitating a close study of heterogeneous and confused ideas and the identification of the crucial indices of what was to become a great theory. This in turn meant understanding the tangled confusion created by Darwin and the construction of a conjecture as to the chronology and the logic of the resolution of this confusion. To sum up, it meant understanding how the Darwinian schema triumphed at the price of a fundamental and profound weakening that lasted at least sixty years.

The spirit of the present work can be described quite simply: using an illustrious and enigmatic example, I have examined a paradox of commensurability. Today's Darwinism was indeed founded by Darwin, but the theoretical and experimental context of the Darwinian hypothesis had to be utterly changed for it to become even plausible. The key to the enigma is as follows: in the *principle* of natural selection we need to distinguish the *hypothesis* from the organising principle of a general *theory* of evolution. Darwin's genius lay in his visionary construction of a theory, in the understanding of the implications of a principle. What Darwin called the 'theory of natural selection' was in fact 'one long argument' that had the ambition of rethinking the structure and the unity of natural history. Darwin's limits can be found in the *hypothesis* of selection, in its preconditions. That is why, after Darwin, scientists had to examine the hypothesis and to go beyond indirect proof through explanatory power.

The current work has attempted to reconstitute the efforts made after 1859 to give a rational and realist consistency to the hypothesis of natural selection. It does not present the history of 'Darwinism' in general, or even that of the 'theory of natural selection'. Darwinism has been and remains a cultural configuration of such historical scale and richness that the historian-epistemologist is easily overwhelmed, although the genuinely 'cultural' and 'philosophical' aspects of Darwinism might, after all, turn out to be poor. As to the theory of natural selection in general, an exhaustive historical analysis would have meant taking into account the totality of disciplines and schools that defined themselves in terms of the Darwinian *paradigm*, that is, in terms of an exemplary model capable of organising a body of research. The present study is somewhat more modest, being limited to a few conjectures as to the rational history of Darwin's central hypothesis. This history cannot be understood retrospectively as an ordered development; its rationality, or more precisely that which is rational in its history, has rather consisted of a series of theoretical dilemmas, of experimental challenges and, in general, of conflicts the outcome of which did not depend only on the sociological contexts in which they were formulated.

This having been said, at this point it is appropriate to recapitulate (to repeat both an expression and a fundamental rhetorical requirement of Charles Darwin's). The present study contains three main threads. The first dealt with Darwin and his 'hypothesis', and the second with the profound crisis that marked the first six or seven decades of the selectionist explanation; the final thread examined the modern theory of the hypothesis – or the genetic theory of selection.

As far as Darwin is concerned, I have tried to reconstitute the conceptual constellation of the hypothesis of natural selection, by focusing both on the germs of self-destruction that the principle of selection contained within it and on what gave it its extraordinary power of anticipation.

The current relevance of Darwin's selectionist thinking can be rigorously described and revealed through three important themes:

1. Despite first impressions, Darwin's concept of natural selection was not that of the co-discoverer of the principle, Alfred Russel Wallace. For the young Wallace, selection (give or take the name) operated between species and varieties. He based his position on the fact that the Malthusian principle that limits population size is defined at the level of the community of species that inhabit a certain 'region' (in today's terms, on the scale of the 'ecological' community). For Darwin, selection is a process that takes place within a race (or 'variety') and consists of the 'accumulation' (or loss) of atoms of variation. This specifically Darwinian concept of selection manifests itself in two ways. First, on a demographic scale, those groups that are approximately stationary in size are local races and not communities. Second, and more important, the Darwinian concept of natural selection has as one of its explicit preconditions the inheritance of individual differences.

2. Darwin's work contains the outline of a general theory of selection, within which artificial selection, natural selection and sexual selection are particular modalities. This implies that natural selection is not a metaphor, unless it is understood that it is a *correct* metaphor, rooted in reality.[2] The unity of the Darwinian hypothesis of selection is based on the fact that selection – natural, artificial or sexual – is a process that acts on individual inherited variations that can be diffused and accumulate in 'races' and in species. What distinguishes the various modalities of selection is the nature of the forces that orient the result of the process. But this 'theory of forces' goes beyond the definition of the process – this very specific view of selection exposed the hypothesis to the threat of refutation. It

is precisely because natural selection can be seen as a particular modality of a general process of modification through the accumulation of inherited individual differences that it has an empirical impact, and is fundamentally different from Spencer's 'survival of the fittest'. Despite the fact that, here and there, Darwin used this formulation in the last two editions of the *Origin of Species*, he never accepted that the principle of natural selection was an undoubted *a priori* truth, as Spencer argued. Following his Newtonian representation of the scientific method, Darwin always considered natural selection a 'conjecture', highly probable and even 'nearly certain', but never a necessary truth.

3. Closely linked to the two preceding points, Darwin's thinking contains a cardinal and painstaking examination of what is today called the problem of the units of selection. Darwin's schema requires three and only three kinds of ordered entities – 'variations' (or 'individual differences'), 'individuals' and 'varieties' (or 'races'). Darwin's genius was that he tore the traditional vocabulary of variation from its semantic roots and obliged it to express something new: a 'variation' is not a 'variety'. Of course, the texts are not always so clear – quite often Darwin used 'variation' for 'variety' (but not vice versa). The effect of selection, be it natural or artificial, does not fundamentally consist of choosing between 'varieties' (or races), but of modifying varieties, by accumulating variations that are advantageous to the individuals that carry them. Variation thus represents the individual in the race – it is what remains beyond individual existence. Darwin's successors used the term 'population' to describe what Darwin had called 'varieties' or 'races'. If Darwin did not use the term 'population', it was undoubtedly because he knew only too well that Malthus used this word to mean 'the power of population', that is, a tendency to growth. But Darwin's model of selection precisely presupposed that the groups upon which selection operates were stationary; Wallace, who, on the contrary, argued that selection was a process of expansion and shrinkage of groups relative to each other, did indeed use the Malthusian term 'population' in his 1858 paper. After Darwin, of course, the term came to be used in today's statistical sense, and rapidly replaced 'race' and 'variety' in the vocabularies of both selectionist explanation and the science of heredity. In a sense, therefore, Ernst Mayr is right to portray Darwin as the father of 'population thinking', even though in another sense it is an anachronism. Whatever the case, Darwin's strictly 'individualist' conception of selection is another index of the resolutely empirical fashion in which he constructed his hypothesis and bravely exposed it to the threat of disproof and to the need for reformulation.

In each of these three themes, the reader will have noted the fundamental importance of the precondition of individual heredity, which is neither the type of a race or a species, nor the transmissible essence of an individual, but something intermediate, which Darwin more than anyone else made a key problem for biology. Darwin imposed the term and the concept of heredity in scientific discourse, to the extent that by the end of the 19th century they had become self-evident. Darwin turned a term that was more common in French than in English, and that had been generally restricted to medicine, into a necessary precondition of the theory of modification by selection.[3] There is an intrinsic link in Darwin's thinking between 'modification' and 'heredity', to the extent that he sometimes described his theory as a theory of 'heredity with modification'.[4] This expression is undoubtedly linked to an archaic representation of heredity in the language of 'descent', but it is more important to recognise this bias in Darwin's conception of individual heredity than to underline its weaknesses. In fact, it is because heredity was crucial to the Darwinian theory of selection (both artificial and natural) that most theoretical views of heredity that were put forward over the next fifty years were developed in the context of a critical examination of the Darwinian hypothesis of the modification of species by selection.

All these elements show how the Darwinian hypothesis of selection can still appear familiar. They also explain why no sooner had the selectionist explanatory model been formulated than it found itself exposed to an interminable principled crisis.

By linking the fate of selection with that of heredity, Darwin made the validity of his principle contingent on the experimental clarification of heredity. This was the key aspect of the internal crisis that affected theoretical Darwinism up until the 1920s. For more than half a century, the main objection to the hypothesis of selection acting on the continuous variation present in populations was that this hypothesis was simply not compatible with the concepts and the reality of heredity.

Furthermore, the lack of an operational concept of 'population' led Darwin to put forward two different representations of selection. Sometimes Darwin saw selection as a process of *diffusion* ('accumulation') of a character-atom in a population (a 'race'); sometimes, however, he argued that selection acts on a range of continuous variation affecting the whole of the population – in this case it takes the form of a *displacement* of the mean character of the population. But this distinction was never clear in Darwin's writings; this is amply confirmed by Jenkin's critique and Darwin's reaction. In fact, Darwin never used a statistical representation which alone could clarify the distinction between a dynamic of 'replacement' and a dynamic of 'displacement'. But Darwin's rhetorical exposition benefited from this confusion of the continuous and the discontinuous. This ambiguity is one of a number of reasons why Darwin's discussions of selection can seem extremely familiar to the modern reader; it is too easy to project onto Darwin's writings ways of thinking that are not in fact present in the original. One thing is certain – for Darwin's followers the opposition between the two possible interpretations of the hypothesis was to become fundamental.

The second thread of this book led to an analysis of the crisis of selectionist explanation that continued from the publication of the *Origin of Species* up to around 1920. Darwin's successors did not rest satisfied with natural selection as the unifying principle of a new general theory of the history of life. They needed the hypothesis to be empirically demonstrated, as a possibility and as a reality. In the post-Darwinian decades, this led to the development of three main strategies of argument.

The Darwinians' preferred approach can be called the strategy of indirect corroboration. Faithful to the method of Darwin and Wallace – Newtonian in its spirit – this strategy accumulated groups of facts that could only be explained by natural selection. For many years, mimicry was a paradigm of this kind of argument. The strategy of indirect corroboration was thus illustrated by showing how the very category 'mimicry' was constructed during Darwin's lifetime as the first empirical 'illustration' of the hypothesis of selection. This strategy was essentially marked by its heuristic fertility. Most Darwinian naturalists adopted the hypothesis of natural selection because it opened the road to a promising programme of research. Nevertheless, this justification of the hypothesis of natural selection could never be sufficient, because the principle of natural selection – unlike the 'principles' of the physical sciences – does not have the status of an axiom. On the contrary, it can be thought of as the product of a probable inference. This inference itself was made on the basis of a certain number of empirical premises involving competition, variation and the individual inheritance of characters. On its own, confirmation by consequences could not justify the hypothesis – it had to be seen whether it was compatible with the available experimental data on variation and heredity.

In response to this question, in the 1890s the English biometricians developed a strategy of direct proof. Aware of the absence of firmly established experimental facts as to the physiological mechanisms of heredity and of variation, they used statistical methods and had a phenomenalist conception of science. Their aim was to establish natural selection as a pure and simple fact (or in other words they sought to establish the existence of cases of natural selection), without any kind of physiological hypothesis. Although this phase in the history of Darwinism was doomed from the outset, it has nevertheless been analysed in some detail because it reveals an important epistemological facet of the selectionist explanation. The biometricians showed that statistics was an essential methodological weapon for the theory of selection. But they also showed that if seen solely from a statistical point of view, the hypothesis of selection was heuristically sterile.

A third strategy involved establishing whether the hypothesis of selection was compatible with the experimental data of the science of heredity. This attitude in fact undermined the Darwinian principle from within. In the name first of Galton's statistical theory of heredity, then of the Mutation Theory and finally of Mendelian genetics, post-Darwinian biologists argued against the idea that Darwin's fluctuating variation (called 'continuous variation' by William Bateson) could play any role whatsoever in the transformation of species. Considering that only discontinuous (or saltational) variation was important, they constructed a famous argument which stated that selection, both natural and artificial, was not able to form or shape, because it merely chose *a posteriori* between types that it had played no part in creating. This is why genetics, based on Mendel's methodological heritage and on the evolutionary tradition of the study of discontinuous variation, initially appeared to deliver the *coup de grâce* to the Darwinian concept of selection.

And yet it was in Mendelism that the selectionist explanation finally become operationally intelligible. Throughout this study it has been emphasised that Darwin had an ambiguous representation of the populational significance of selection, which he sometimes presented as a process of diffusion–accumulation of atoms of variation, sometimes in the language of a change in the mean character of a race. Mendelian genetics gave a precise meaning to these two facets of the hypothesis of selection. When genetics initially analysed isolated Mendelian factors, it imposed a vision of selection as a rate of replacement of a discrete entity. When, reconciled with biometrical methodology, it absorbed into its theoretical field the inheritance of quantitative characters, it relegitimised the same fluctuating variation that for Darwin had been the basis of the process of natural selection. Mendelism thus led to a twin theory of selection. Mendelised Darwinism partly deals with the formal kinetic description of the diffusion of genes in populations – in this respect it is a theory of *things* that are selected. When, however, it deals with quantitative characters, the genetic theory of selection focuses on the level at which the causal action of selection can be understood. Statistics enables us to understand why the two doctrines are compatible – because characters controlled by a large number of different genes show an approximately continuous spectrum of inherited variation. In many respects, however, the two theories developed relatively independently. It was only in the visionary speculations of theoretical population genetics that the meeting of these two viewpoints became possible.

These investigations of Darwin and the post-Darwinian crisis of the hypothesis of selection can also be seen from another standpoint. Historians tend to emphasise the weight of the ideologies of 'progress' in the general history of Darwinism. In terms of the hypothesis of natural selection, however, it appears that the Darwinian tradition was instead haunted by the reversibility of evolution. 'Reversion', 'forces of return', 'atavism', 'ancestral heredity', 'recessivity', 'anatomical regression', 'statistical regression' – these terms all express a conceptual configuration that threatened the theory of modification of species from all sides, both in Darwin's work and in that of his successors. Darwin's thinking was deliberately constructed in opposition to the doctrine of the 'return' of varieties to the 'type' of the species; it thus constitutes one of the most striking examples of a philosophy of natural history that mixes up the duality of progress and regression.[5] Darwinism and post-Darwinism were developed in the 19th century in a cultural space largely occupied by the philosophies and ideologies of decadence. It is thus not surprising that 'regression' should have repeatedly resurfaced in the key discussions of the early theoreticians of selection. This can be seen first in the work of Darwin and Wallace, and then in the work of those who – whether they were critics or apologists – gradually gave the concept of selection the form we know today (Jenkin, Galton, Weismann, Weldon, Bumpus and also the first Mendelians).

It was Mendelism that was to cut the Gordian knot of selection, by carrying out two fundamental conceptual displacements. By dissociating the operational notion of 'heredity' from those of descent and ancestrality, Mendelism forced the science of heredity away from the contemplation of genealogy and imposed a representation of heredity as a structure. Furthermore, the extrapolation of Mendelian calculations to the population level destroyed once and for all the representations of heredity as a force of return. The famous Hardy–Weinberg law, or the law of panmictic equilibrium, describes the 'straightforward' effect of heredity in a Mendelian population, abstracted from any force of change (mutation, migration, selection, sampling effect . . .). This law, despite the fact that it describes an idealised situation, provided the modern science of evolution with a schema comparable to the principle of inertia in physics. Heredity could no longer be viewed as a force of return, or as a biological analogy of *momentum* – it was stripped of any reference to its effects in terms of 'force' or 'energy'. Furthermore, as a theorem of inertia the Hardy–Weinberg equilibrium led to the development of a predictive kinetic description of evolutionary change, in which selection appeared as a force amongst others that could modify the distribution of gene frequencies in a population.

Finally, as far as the genetic theory of selection is concerned, the present study has outlined the key concepts and methodological points rather than present an exhaustive history. From this point of view, it was essential to clarify the epistemological status of the models of population genetics. The fundamental merit of population genetics was that it focused on the conditions governing the possibility of biological evolution under the Mendelian hypothesis. Theoretical population genetics is less a science about how evolution has taken place than a formal discipline that confronts the naturalist with the implications of various conceivable evolutionary scenarios, given the weight of this or that factor of evolution and of a certain number of limiting conditions.

In this great formal science, the first and most elegant example of a 'mathematised' biology, the principle of selection was refined and differentiated. And finally, as often happens in sciences that are enriched by the intervention of mathematics, evolutionary theory opened itself to serious experimental testing. The Darwinian principle avoided becoming a tautology, but at the price of an increased abstraction, which unexpectedly gave rise to a new world of testable deductions.

In reevaluating the famous debate between the two founders of theoretical population genetics, Fisher and Wright, an attempt was made to describe the conceptual space within which contemporary Darwinism spread its wings. Through their methodological disputes, and also through their tendency to extend these to the philosophical level, these two mathematicians defined the transcendental topics of the modern science of evolution. In particular, they revitalised the Darwinian vision of evolution by showing that, under the hypothesis that natural selection is the 'paramount power' in the evolution of species (in itself, this was not obvious *a priori*), it is possible to imagine extremely different evolutionary scenarios.

Fisher and Wright were both steeped in statistical methodology and shared an indeterministic vision of reality. This point has been clearly established by Jonathan Hodge, who has also underlined that both men were profoundly marked by their reading of Bergson.[6] However, these two Darwinian philosophers of evolution applied this common approach in very different ways. Fisher argued that the kinetic theory of matter was a model for the genetic theory of evolution – the details of the selection process are thus highly contingent, but its mass effects can be formalised and are rigorously predictable. Because he believed that, in the long term, species are very large panmictic populations and that the mixing of genes dissolves organic individuality, Fisher developed a vision of selection as a secular factor that indefinitely maximises that part of the Malthusian parameter that can be attributed to additive genetic effects. This led to a 'genic' representation of the target of natural selection, and to the conviction that at the macroscopic level selection is deterministic.

For Wright, the fundamental principle is that living populations are always of a limited size, and are, to varying degrees, fractionated and isolated. The very same mathematics used by Fisher led Wright to recognise elements of individuality and indeterminism at all levels of the selection process. For Wright, gene selection is tempered and modulated by stochastic effects which, repeated in many partially isolated populations, lead to an understanding of selection on a superior level, operating between populations and on genomic systems. This 'second order' selection is also expressed in the ordinary language of gene frequencies, because it is expressed by migratory flows of genes from the richest populations to the others. This explains why Wright argued that although the dynamic of gene frequencies is the universal idiom of the evolution of sexual species, strictly speaking it is not the ultimate reality, but merely an expression of that reality.

The current work does not include a detailed history of the genetics of natural and experimental populations, and deliberately so. We have not entered the maze of the Evolutionary Synthesis and its multiple, more or less happily federated, kingdoms. Given the ambition of this study, it seemed preferable to use a few key examples to focus on how contemporary Darwinism tried to articulate the empirical and the formal in its genetic reconstruction of the hypothesis of selection. For this reason the gap between the abstraction of the models and the relative poverty of experimental studies has been repeatedly underlined. Studies of melanism in moths or of polymorphism in the snail *Cepaea nemoralis* clearly show that simple phenomena (such as the replacement of one gene by another, or stochastic fluctuations of gene frequencies) were frequently found to be merely the expression of more profound and more complex phenomena that required not less modelling and less theory, but more of both.

This study therefore closed with a brief examination of the challenge posed by molecular evolution, and in particular the so-called neutral theory of Motoo Kimura. It goes without saying that, with these considerations, I breached the frontiers of mere historical narration. There was a good reason for this, however. Kimura's speculations on the 'neutral mutation–random drift theory of molecular evolution' throw a retrospective philosophical light on the whole history of theoretical Darwinism. One of the main postulates of what has been known for over a century as neo-Darwinism is that all organismic characters have ultimately been fashioned by natural selection. During the 20th century, this panselectionist postulate came to be expressed in the genetic language that is today so – indeed, too – familiar. Species, the argument goes, evolve through favourable genetic mutations such that genes determine, by complex and dimly understood routes, all inherited characters. When molecular biology revealed the material structure of genes, it seemed natural to extend selectionist rhetoric to the most elementary levels of organisation. Kimura, however, suggested that it was unreasonable to argue that nucleotide sequences and their polypeptide equivalents were strictly determined by natural selection down to the last detail. Any selection process is onerous for the species, and it is hard to see how billions of different units could be simultaneously and permanently subject to the continuous adjustment of selection. As Elliott Sober has elegantly pointed out, the principle of natural selection has the status of what the physicists call a 'consequence-law'.[7] There is no reason to think that it 'acts' at every point of biological space-time in the sense that the action of a fundamental force in physics can be thought of in terms of a 'field'. Natural selection does not have a 'field of action'; it is always a *resultant* force – closer to the function of molecular patterns than to the atomistic detail of the composition of those molecules. Or, in other words, while it may be correct to say that natural selection ultimately controls all the properties of living beings, it would be naive to think that it *determines* them at each and every moment and in the detail of all their constituent elements.

Kimura's work suggests what kind of biological philosophy is appropriate for dealing with the principle of selection. After Darwin, natural selection developed in an increasingly atomistic direction. Today, we undoubtedly have a far clearer representation of the kind of *things* (or material units) that selection controls the distribution of. But it is also obvious that the elementarist representation of selection is pointless unless it attempts some kind of causal interpretation. And from a causal point of view, it is not sufficient to provide a kinetic description of the selected entities; the functional properties that determine a differential advantage must also be identified. But, as Kimura repeatedly noted, this question fundamentally has meaning only at the level of quantitative and phenotypic characters. This implies that the question of the units of selection, essential from the point of view of effects, is strictly subordinated to the question of the morpho-functional pattern that explains the given advantage or disadvantage in a given context. That was one of Darwin's principal lessons, and in this sense the neutral theory of molecular evolution, which is not a neutral theory of evolution, is merely the latest rehabilitation of Darwin.

Notes

Introduction

1 See for example S. J. Gould, 'Is a new and general theory of evolution emerging?' (1980).
2 See J. Gayon, 'Critics and criticisms of the modern synthesis: the viewpoint of a philosopher' (1990).
3 P. J. Bowler, *The Eclipse of Darwinism* (1983), p. 5.
4 K. Popper (1935, 1972, 1976, 1978, 1984).
5 The title of the book by Eberhart Dennert – *At the Death-Bed of Darwinism* – is typical of the spirit of the time. The book was published in German in 1903 and translated into English the following year. P. J. Bowler (*The Eclipse of Darwinism*, 1983) has given a vivid description of this decline.
6 E. Nordenskiöld, *The History of Biology: A Survey*, part 3, chap. 16, pp. 562–564, 'Decline of Darwinism'.
7 *Ibid.*, p. 616.
8 The second edition of the *Origin of Species* contains clear statements suggesting the 'progression' of evolution: 'This principle of preservation, I have called, for the sake of brevity, Natural Selection; and it leads to the improvement of each creature in relation to its organic and inorganic conditions of life' (C. Darwin, *On the Origin of Species*, 2nd edn, 1860, chap. 4; quoted in M. Peckham, *The Origin of Species by Charles Darwin. A Variorum Text*, 1959, p. 271). In the fifth edition, Darwin added: 'and consequently, in most cases, to what must be regarded as an advance in organisation' (*ibid.*).
9 '. . . natural selection can act only by taking advantage of slight successive variations; she can never take a leap, but must advance by the shortest and slowest steps' (C. Darwin, *On the Origin of Species*, 1859, p. 194).
10 'Natural selection can act only by the preservation and accumulation of infinitesimally small inherited modifications, each profitable to the preserved beings' (*ibid.*, p. 95).
11 This thesis is set out in the Introduction to the first edition of the *Origin of Species* (1859, p. 6): 'I am convinced that Natural Selection has been the main but not exclusive means of modification.' All subsequent editions use this same formulation. In *The Variation of Animals and Plants under Domestication* (1868), Darwin speaks of 'selection' (without any adjective) as the 'paramount power' in the modification of domestic races and of species in the natural state (chaps. 21 and 28). Finally, in the final chapter of the last edition of the *Origin of Species* (1872) Darwin states that he had not changed his

opinion as to the respective roles of natural selection and of other factors involved in the modification of species, repeating the formulation used in the Introduction to the first edition (M. Peckham, *The Origin of Species by Charles Darwin. A Variorum Text*, 1959, p. 747). On Darwin's notion of the 'paramount power of selection', see J. Gayon, 'The paramount power of selection: From Darwin to Kauffman' (1997).

12 L. Cuénot, *La genèse des espèces animales*, 3rd edn (1932), p. 300. 'Floating variation' (*variation flottante*) is a synonym for 'fluctuating variation'.

13 The term 'modern synthesis' was coined by Julian Huxley in his book *Evolution: The Modern Synthesis* (1942).

14 *Ibid.*, chap. 1, §§ 22–28.

15 *Ibid.*, 3rd edn (1974), p. 22.

16 The term 'evolution' only appears in the conclusion to the sixth and final edition of the *Origin of Species* (1872). For many years Darwin did all he could to avoid his theory of 'the descent of species with modification' being labelled by a term that was associated with Spencer's philosophical system. (On this point, see the excellent analyses of E. Gilson, *D'Aristote à Darwin et retour*, 1971.) For 'population', see section 1.1.

17 For further discussion of this aspect of 'Darwinism', see *Darwinisme et société*, ed. P. Tort (1992).

18 Two collective works exist: *The Comparative Reception of Darwinism* (1974), ed. T. F. Glick, and *The Darwinian Heritage* (1985), ed. D. Kohn. There are a number of studies on various national traditions, in particular: Y. Conry, *L'introduction du darwinisme en France au XIXe siècle* (1974); W. M. Montgomery, *Evolution and Darwinism in German Biology* (1975); J. R. Moore, *The Post-Darwinian Controversies: A Study of the Protestant Struggle to Come to Terms with Darwin and America, 1870–1900* (1979); R. Moreno, *La polemica del darwinismo en México, siglo XIX* (1984); E. J. Pfeiffer, *The Reception of Darwinism in the United States, 1959–1880* (1957); D. P. Todes, *Darwin without Malthus: The Struggle for Existence in Russian Evolutionary Thought* (1989); A. Vucinich, *Science in Russian Culture, 1861–1917*, chap. 14 (1970). See also P. Tort (ed.), *Dictionnaire du darwinisme et de l'évolution*, (1996), vol. I, pp. 822–1108 (articles on Darwinism in various national contexts).

19 R. M. Burian has described this period in the history of evolutionism by suggesting that the 'Darwinism' of the end of the 19th century was a genuine 'paradigm' in the initial sense given to the word by Thomas Kuhn (an exemplary model of scientific research), but certainly not in the sense of a 'disciplinary matrix' with all that that implies for the methodological unification of a field of research (R. M. Burian, 'The influence of the evolutionary paradigm', 1989). For a more sceptical discussion of the usefulness of the epistemological category of 'paradigm' for the history of Darwinism, see J. C. Greene, 'The Kuhnian paradigm and the Darwinian revolution in natural history' (1971), and E. Mayr, 'The nature of the Darwinian revolution' (1971). For a further discussion by Kuhn of the two fundamental senses of the word 'paradigm', see the postface to the second English edition of *The Structure of Scientific Revolutions* (1970), 'Reflections on my critics' (1970), and 'Second thoughts on paradigms' (1974).

20 See note 3 above.

21 T. Kuhn, *The Structure of Scientific Revolutions* (1962), chaps. 1 and 2.
22 *Ibid.*, chap. 6.
23 Most studies of the history of anti-Darwinian theories have dealt with neo-Lamarckism. On the American origins of this school, see: R. W. Dexter, 'The development of A. S. Packard, Jr, as a naturalist and entomologist' (1957); E. J. Pfeiffer, 'The genesis of American neo-Lamarckism' (1965); P. J. Bowler, 'Edward Drinker Cope and the changing structure of evolution theory' (1977); S. J. Gould, 'The rise of neo-Lamarckism in America' (1980). French neo-Lamarckism was the subject of a special issue of the *Revue de Synthèse* (ed. J. Roger), under the title *Les néo-lamarckiens français* (1979). On psycho-Lamarckism, see B. Willey, *Darwin and Butler: Two Versions of Evolution* (1949). The excellent study by C. Zirkle on the ancient origins of the idea of the heredity of acquired characteristics and its history from antiquity to Darwin ('The early history of the ideas of the inheritance of acquired characters and pangenesis', 1946) should also be mentioned, as should L. I. Blacher's study of modern developments of the same idea, *The Problem of the Inheritance of Acquired Characters* (1982). Non-apologetic and non-polemical studies of the origins and developments of orthogenetic theories are rare. See D. Ospovat, 'Perfect adaptation and teleological explanation: Approaches to the problem of the history of life in the mid-nineteenth century' (1978); P. J. Bowler, 'Theodor Eimer and orthogenesis: Evolution by definitely directed variations' (1979). The question of a theist evolutionism also deserves attention. Although by the end of the 19th century this tradition had generally been reduced to a systematic criticism of Darwinian ideas, the outline of the doctrine can still be discerned. Part of the difficulty arises from the fact that historians have too often exaggerated the institutional separation of religion and science, in particular in 19th-century Britain, and even more so in the natural sciences. The reference work on this question is J. R. Moore, *The Post-Darwinian Controversies* (1979). The work and the school of Louis Agassiz give a good idea of the form taken by theist evolution in the second half of the 19th century. See E. Lurie, *A Life in Science* (1960); R. W. Dexter, 'The impact of evolutionary theories on the Salem group of Agassiz zoologists' (1979). There are a vast number of studies of pre-Darwinian natural theology. For a survey of these works, see C. Gillispie (1951) and G. Laurent (1987). P. J. Bowler's book *The Eclipse of Darwinism* (1983) is the best available study of anti-Darwinian evolutionism at the turn of the century.
24 Following Jean Piaget, I take this to be the touchstone of scientific knowledge: problems of a scientific nature are those which 'have been successfully isolated in such a way that their solution does not put everything into question' (J. Piaget, *Introduction à l'épistémologie génétique*, 2nd edn, vol. I, 1973, p. 14).
25 See for example S. Bachelard, 'Epistémologie et histoire des sciences' (1968).
26 I. Lakatos, 'Falsification and the methodology of scientific research programmes' (1970) and 'History of science and its rational reconstructions' (1971).
27 This point was clearly dealt with by L. Laudan in *Progress and Its Problems* (1977), chaps. 5–7.
28 The expression is that of S. Shapin (1982).

29 See the interesting remarks of E. Harrison in 'Whigs, prigs and historians of science' (1987).

30 D. M. S. Watson, 'A discussion on the present state of the theory of natural selection: Opening address' (1937), p. 45.

31 'As this volume is one long argument, it may be convenient to the reader to have the leading facts and inferences briefly recapitulated' (C. Darwin, *On the Origin of Species*, 1859, p. 459). The affirmation that 'the Origin of Species is merely one argument from the beginning to the end' is repeated in Darwin's *Autobiography* (ed. G. de Beer, 1983, pp. 84–85). Ernst Mayr thought the expression 'one long argument' sufficiently important for him to make it the title of a book (E. Mayr, *One Long Argument*, 1991).

32 This formulation is to be found in the last two editions of the *Origin of Species*. In the first edition, Darwin merely referred to a 'theory of descent with modification by natural selection' (M. Peckham, *The Origin of Species* by Charles Darwin. *A Variorum Text*, 1959, p. 719). The phrase is to be found at the beginning of the concluding chapter.

33 This appears clearly in the conclusion of the chapter of the *Origin of Species* that deals with 'natural selection' (chap. 4), where Darwin sets out the conditions under which natural selection could be possible. Apart from the existence of individual differences and of a struggle for existence, Darwin emphasises the 'powerful principle of heredity' (1859, p. 127).

34 See for example *On the Origin of Species*, 1859, chap. 1, p. 13: 'the laws governing heredity are virtually unknown'.

35 'On this principle of inheritance with modifications, we can understand how it is that sections of genera, whole genera, and even families are confined to the same area . . .' (ibid., chap. 11, pp. 350–351).

36 The expression is taken from C. Nowinski, 'Biologie, théories du développement et dialectique' (1967).

37 The expression '*vera causa*' was used by Darwin in chaps. 5 and 14 of the *Origin of Species* (1859) in the context of a critique of the theory of the independent creation of species. Community of descent and natural selection were '*verae causae*' in that they were not merely suppositions but causes which could be inferred to exist on the basis of independent empirical arguments. On the idea of the *vera causa* in the genesis of Darwin's arguments, and on the tradition that transmitted this idea to Darwin (Newton, Reid, Lyell), see D. Hull, *Darwin and His Critics* (1975); V. C. Kavaloski, *The* Vera Causa *Principle: A Historico-Philosophical Study of a Metatheoretical Concept from Newton through Darwin* (1974); M. Ruse, 'Darwin's debt to philosophy: An examination of the influence of the philosophical ideas of John F. W. Herschel and William Whewell on the development of Charles Darwin's theory of evolution' (1975); M. J. S. Hodge, 'Natural selection as a causal, empirical, and probabilistic theory' (1987) and 'Darwin's theory and Darwin's argument' (1989).

38 C. Darwin, *Autobiography* (ed. G. de Beer, 1983) p. 71.

39 *Ibid.*, pp. 82–84.

40 C. Darwin, *The Variation of Animals and Plants under Domestication* (2nd edn, 1875). Quoted in Darwin (1972), vol. VIII, pp. 350–351.

41 In his *Histoire générale des animaux* (1749), Buffon speculated about 'the

hidden means that nature employs for the reproduction of living beings' (Buffon, *Oeuvres philosophiques*, 1954, p. 243 A), and decided that questions of this type can only be answered by 'hypotheses' (*ibid.*, p. 243 A–B). It is in this context that he coined his famous phrase 'Gather the facts to get ideas' (*ibid.*, p. 238 B).

42 H. Bates, 'Contributions to an insect fauna of the Amazon Valley' (1862); A. R. Wallace, 'Mimicry, and other protective resemblances among animals' (1867).

43 'So that here we have an excellent illustration of the principle of natural selection' (C. Darwin, *On the Origin of Species*, 4th edn, 1866, chap. 13; quoted in M. Peckham, *The Origin of Species by Charles Darwin. A Variorum Text*, 1959, p. 668).

44 The three founders of population genetics each paid homage to Galton and especially to Pearson, despite the theoretical and ideological differences that separated them. Sewall Wright in the summary of his life's work entitled *Evolution and Population Genetics* (1968–78) wrote: 'the methods developed by the biometric school under Pearson's leadership constitute one of the pillars supporting modern population genetics' (vol. I, 1968, p. 373). It is also interesting to note that the first volume of Wright's four-volume magnum opus is called *Genetics and Biometric Foundations*. Haldane, in a mathematical appendix to the *Causes of Evolution* (1932), recalls the pioneering and fundamental role of Pearson in the development of the 'mathematical theory of natural selection' (pp. 171–172). Strangely enough, Fisher, who was far more closely linked with Pearson, both by his eugenicist beliefs and by the role he assumed in the realm of statistics, was generally much more reserved. Nevertheless, in his 1918 article which is generally taken to be the founding document of population genetics, Fisher began with a reference to Pearson (R. A. Fisher, 'The correlation between relatives on the supposition of Mendelian inheritance', 1918, pp. 400–401).

45 W. Bateson, *Materials for the Study of Variation, Treated with Special Regard to Discontinuity in the Origin of Species* (1894).

46 'In scientific investigations, it is permitted to invent any hypothesis, and if it explains various large and independent classes of facts it rises to the rank of a well-grounded theory' (C. Darwin, *On the Variation of Animals and Plants under Domestication*, 2nd edn [1875], quoted in C. Darwin (1972), vol. VII, p. 10). See below, sections 1.1 and 6.1.

1 Wallace and Darwin: a disagreement and its meaning

1 There are a large number of books and articles devoted to this question, in particular G. de Beer, Foreword to *Evolution by Natural Selection* (in C. Darwin and A. R. Wallace, 1958); G. Canguilhem, 'Les concepts de "lutte pour l'existence" et de "sélection naturelle" en 1858: Charles Darwin et Alfred Russel Wallace' (1959); B. G. Beddall, 'Wallace, Darwin and the theory of natural selection' (1968); C. Limoges, *La sélection naturelle: étude sur la première constitution d'un concept (1837–1859)* (1970); P. J. Bowler, 'Alfred Russel Wallace's concepts of variation' (1976); A. C. Brackman, *A Delicate*

Arrangement: The Strange Case of Charles Darwin and Alfred Russel Wallace (1980). A good synthetic presentation of Wallace can be found in G. Molina (1996b).

2 C. Darwin and A. R. Wallace, 'On the tendency of species to form varieties' (1859).

3 Cf. A. C. Brackman, *A Delicate Arrangement: The Strange Case of Charles Darwin and Alfred Russel Wallace* (1980).

4 For a chronology, see the excellent study by C. Limoges, *La sélection naturelle: étude sur la première constitution d'un concept (1837–1859)* (1970). Darwin's early notebooks have been published in two editions: G. de Beer (*Darwin's Notebooks on Transmutation of Species,* 1960) and P. Barrett et al. (*Charles Darwin's Notebooks, 1836–1844*, 1987).

5 Wallace, 'On the law which has regulated the introduction of new species' (1855).

6 C. Limoges, *La sélection naturelle*, p. 92.

7 R. C. Stauffer, *Charles Darwin's Natural Selection, Being the Second Part of His Big Species Book Written from 1856 to 1858* (1975).

8 See F. Darwin, *Life and Letters of Charles Darwin*, 1887, II, pp. 41 and 85. For further details, see R. C. Stauffer, *Charles Darwin's Natural Selection*... (1975), p. 5.

9 See the introduction to the *Variation* (1896, pp. 9–10).

10 C. Darwin, *On the Origin of Species,* 1859, Introduction, pp. 1–2.

11 See Introduction above, note 31.

12 See the letter from Wallace to Darwin dated 2 July 1866 and Darwin's reply dated 5 July 1866 (A. R. Wallace, *Alfred Russel Wallace. Letters and Reminiscences*, ed. J. Marchant, 1916, pp. 140–145). In this correspondence, Wallace urged Darwin to abandon the term natural selection for Spencer's term 'survival of the fittest'. Four years later, however, he published a book entitled *Contributions to the Theory of Natural Selection* (1870), thus accepting Darwin's term.

13 '... early in the summer of 1858 Mr Wallace, who was then in the Malay Archipelago, sent me an essay "On the tendency of varieties to depart indefinitely from the original type"; and this essay contained exactly the same theory as mine' (C. Darwin, *Autobiography* cited in G. de Beer, *Charles Darwin [and] Thomas Henry Huxley: Autobiographies* [1974], 1983, p. 72). There is a similar declaration in the Introduction to the *Origin of Species* (1859, pp. 1–2).

14 'There is a general principle in nature which will cause many varieties to survive the parent species, and to give rise to successive variations departing further and further from the original type' (A. R. Wallace, 'On the tendency of varieties to depart indefinitely from the original type' [1859], cited in P. H. Barrett, *The Collected Papers of Charles Darwin,* 1977, II, p. 11). 'This identical principle of selection [in nature and breeding]' (C. Darwin, 'Extract from an unpublished work on species' [1859], cited in *ibid.*, p. 7). The texts jointly presented to the Linnean Society in 1858 by Darwin and Wallace (but published in 1859) consisted of an 'extract' of the essay written by Darwin in 1844, an abstract of a letter from Darwin to Asa Gray dated 5 September

1857, and the draft sent by Wallace to Darwin in 1858 ('On the tendency . . .'). These texts are reproduced in the volume of Darwin's writings published by P. H. Barrett (1977).

15 This epistemological presentation of the structure of the theory of natural selection is given explicitly only in the Introduction to *Variation of Animals and Plants under Domestication.* However, it can be discerned in the arguments of both Darwin and Wallace in the texts of 1858. See quotation on pp. 31–32 above.

16 For Wallace, see 'Mimicry and other protective resemblances among animals' (1967, pp. 1–2). This is traditionally called the '*vera causa*' strategy of demonstration. See D. Hull, *Darwin and His Critics* (1973); V. C. Kavaloski, *The* Vera Causa *Principle: A Historico-Philosophical Study of a Metatheoretical Concept from Newton through Darwin* (1974); M. Ruse, 'Darwin's debt to philosophy: An examination of the influence of the philosophical ideas of John F. W. Herschel and William Whewell on the development of Charles Darwin's theory of evolution' (1975); J. Hodge, 'Natural selection as a causal, empirical, and probabilistic theory' (1987).

17 C. Darwin, 'Abstract . . .' [1859], in P. H. Barrett (1977), II, p. 9.

18 A. R. Wallace, 'On the tendency . . .' [1859], in P. H. Barrett (1977), II, p. 11. In the *Origin,* Darwin also finally opted for the term 'struggle for existence'.

19 C. Darwin, 'Extract . . .' [1859], in P. H. Barrett (1977), II, p. 5; A. R. Wallace, 'On the tendency . . .' [1859], *ibid.,* pp. 11–13.

20 C. Darwin, 'Extract . . .' [1859], *ibid.,* p. 5; A. R. Wallace, 'On the tendency . . .' [1859], *ibid.,* p. 11.

21 C. Darwin, 'Extract . . .' [1859], *ibid.,* p. 6; A. R. Wallace, 'On the tendency . . .' [1859], *ibid.,* p. 13.

22 C. Darwin, 'Extract . . .' [1859], *ibid.,* p. 6; A. R. Wallace, 'On the tendency . . .' [1859], *ibid.,* p. 12.

23 C. Darwin, 'Extract . . .' [1859], *ibid.,* p. 6; A. R. Wallace, 'On the tendency . . .' [1859], *ibid.,* p. 13.

24 C. Darwin, 'Extract . . .' [1859], *ibid.,* p. 6; A. R. Wallace, 'On the tendency . . .' [1859], *ibid.,* p. 14.

25 J. Hodge, 'Natural selection as a causal, empirical and probabilistic theory' (1987).

26 C. Darwin, 'Extract . . .' [1859], in P. H. Barrett (1977), II, p. 7; A. R. Wallace, 'On the tendency . . .' [1859], *ibid.,* pp. 15 and 18.

27 C. Darwin, *On the Origin of Species* (1859), pp. 115–116.

28 C. Darwin, 'Abstract . . .' [1859], in P. H. Barrett (1977), II, p. 10.

29 A. C. Brackman, *A Delicate Arrangement: The Strange Case of Charles Darwin and Alfred Russel Wallace* (1980).

30 A. R. Wallace, 'On the law which has regulated the introduction of new species' (1855).

31 A. R. Wallace, 'On the tendency . . .' [1859], in P. H. Barrett (1977), II, p. 15.

32 In fact, the term 'conversion of varieties into species' is only found later: 'how is it that varieties, which I have called incipient species, become ultimately converted into good and distinct species . . .?' (*On the Origin of Species,* 1859, p. 61). See also *Variation*: 'the conversion of varieties into species – that is, the augmentation of the slight differences characteristic of

varieties into the greater differences characteristic of species and genera' (*The Works of Charles Darwin* [1896], vol. VII, p. 5). However, from 1858 the idea can clearly be found in the work of both Darwin and Wallace (C. Darwin, 'Abstract...' [1859], in P. H. Barrett (1977), II, p. 10; A. R. Wallace, 'On the Tendency...' [1859], *ibid.*, p. 15.

33 P. J. Bowler, 'Alfred Russel Wallace's concepts of variation' (1976). Cf. also the same author's *Evolution: The History of an Idea* (1984), p. 174.

34 C. Darwin, 'Extract...' [1859], in P. H. Barrett (1977), II, p. 5.

35 *Ibid.*, p. 6.

36 See for example C. Darwin, *On the Origin of Species,* chap. 3 (1859, pp. 76ff).

37 *Ibid.*, p. 62.

38 *Ibid.*, p. 320.

39 'The struggle almost invariably will be most severe between the individuals of the same species, for they frequent the same districts, require the same food, and are exposed to the same dangers. In the case of varieties of the same species, the struggle will generally be almost equally severe, and we sometimes see the contest soon decided' (C. Darwin, *ibid.,* p. 75). On the basis of this passage it appears that the struggle between individuals is constant, whilst that between groups is not. The result of any given struggle between individuals is the disappearance of certain characters and the appearance of others. This struggle is constant because the pressure of the population size on individuals is constant. On the other hand, the struggle between varieties (or higher taxonomic groups) is episodic, and results in the extinction of one of the groups.

40 *Ibid.*, p. 76.

41 *Ibid.*, p. 78.

42 C. Darwin, 'Extract...' [1859], in P. H. Barrett (1977), II, p. 5. See also: *On the Origin of Species* (1859), pp. 63–76.

43 C. Darwin, *On the Origin of Species* (1859), p. 45.

44 *Ibid.*, p. 12.

45 'These individual differences are highly important for us, as they afford materials for natural selection to accumulate, in the same manner as man can accumulate in any given direction individual differences in his domesticated productions' (C. Darwin, *ibid.,* chap. 2, p. 45).

46 For more details of this ontological schema, see Chapter 2.

47 C. Darwin, 'Extract...' [1859], in P. H. Barrett (1977), II, pp. 6–7.

48 A. R. Wallace, 'On the tendency...' [1859], *ibid.*, pp. 11 and 13.

49 *Ibid.*, p. 11.

50 T. R. Malthus, *An Essay on the Principle of Population* (1798), chap 1.

51 A. R. Wallace, 'On the tendency...' [1859], in P. H. Barrett (1977), II, p. 15.

52 *Ibid.*, pp. 14 and 15.

53 *Ibid.*, p. 12.

54 *Ibid.*, p. 15.

55 *Ibid.*, p. 14.

56 *Ibid.*

57 *Ibid.*, p. 15.

58 It is worth reproducing the passage in which Wallace summarises his results: '... we have succeeded in establishing these two points – 1st, *that the animal*

population of a country is generally stationary, being kept down by a periodical deficiency of food, and other checks; and, 2nd, *that the comparative abundance or scarcity of the individuals of the several species is entirely due to their organization and resulting habits, which, rendering it more difficult to procure a regular supply of food and to provide for their personal safety in some cases than in others, can only be balanced by a difference in the population which have* [*sic*] *to exist in a given area*' (A. R. Wallace, *ibid.*, p. 14; original emphasis).

59 *Ibid.*, p. 11.

60 *Ibid.*, p. 14.

61 *Ibid.*

62 *Ibid.*, p. 15.

63 *Ibid.*, p. 10.

64 See M. Ruse, 'Charles Darwin and group selection' (1980), and E. Sober, *The Nature of Selection* (1984).

65 C. Darwin, *Variation* . . ., cited in *The Works of Charles Darwin* [1896], vol. VII, p. 9.

66 A. R. Wallace, 'On the tendency . . .' [1859], in P. H. Barrett (1977), II, pp. 10–11.

67 C. Lyell, *Principles of Geology* (1830–3), II, p. 23. Lyell admitted that 'the organisation of individuals is capable of being modified to a limited extent by the force of external causes', and that 'these modifications are, to a certain extent, transmissible to their offspring' (*ibid.*).

68 A. R. Wallace, 'On the tendency . . .' [1859], in P. H. Barrett (1977), II, p. 10.

69 *Ibid.*, p. 17.

70 *Ibid.*

71 Wallace did not use the term 'regulation', but he did make an explicit comparison with the control of speed in the steam engine (*ibid.*, p. 18).

72 A. R. Wallace, 'Mimicry and other protective resemblances among animals' (1867) and 'The problem of utility: Are specific characters always useful?' (1896).

73 A. R. Wallace, 'Mimicry and other protective resemblances among animals' (1867), pp. 2–3.

74 C. Darwin, *On the Origin of Species* (1859), p. 200.

75 C. Limoges, *La sélection naturelle: étude sur la première constitution d'un concept (1837–1859)* (1970), p. 144.

76 The term 'natural means of selection' appears for the first time in the draft of 1843. See: C. Darwin and A. R. Wallace, *Evolution by Natural Selection*, ed. F. Darwin and R. G. de Beer (1958), p. 46.

77 C. Limoges, *La sélection naturelle*, n. 242, pp. 165–185 and 88–101.

78 *Ibid.*, p. 148.

79 *Ibid.*

80 *Ibid.*, p. 149.

81 *Ibid.*, pp. 144–147.

82 The recent elegant study by B. G. Ogilvie should be consulted on this point. He also emphasises the very early importance of domestic variation in the development of Darwin's conceptions ('An unnatural history: Darwin's use of domesticated animals in early species theorizing, 1836–1838', 1991).

83 C. Darwin, *On the Origin of Species* (1859), p. 4.
84 *Ibid.*, p. 14.
85 A. R. Wallace, 'On the tendency...' [1859], in P. H. Barrett (1977), II, p. 17.
86 C. Darwin, *On the Origin of Species* (1859), p. 14.
87 'It seems to me not improbable, that if we could succeed in naturalising, or were to cultivate, during many generations, the several races, for instance, of the cabbage, in very poor soil ... that they would to a large extent, or even wholly, revert to the wild aboriginal stock' (C. Darwin, *ibid.*, p. 15).
88 'We may safely conclude that very many of the most strongly-marked domestic varieties could not possibly live in a wild state' (*ibid.*, p. 14).
89 *Ibid.*, p. 15.
90 See for example: *On the Origin of Species* (1859), p. 29: '... those naturalists who, knowing far less of the laws of inheritance than does the breeder...' See also: *Variation*... (1868), chap. 12.
91 C. Darwin, *On the Origin of Species* (1859), p. 4.
92 C. Darwin, *Journal of Researches*... (1839), p. 193. See also *Zoology of the Voyage of H.M.S. 'Beagle'*, part II: *Mammalia* (1839), p. 92.
93 C. Darwin, *Variation*..., cited in *The Works of Charles Darwin* [1896], vol. VIII, p. 411.
94 *Ibid.*, p. 6.
95 A. R. Wallace, 'On the tendency...' [1859], in P. H. Barrett (1977), II, p. 17.
96 C. Darwin, *On the Origin of Species* (1859), p. 4.
97 C. Darwin, *Variation*..., cited in *The Works of Charles Darwin* [1896], vol. VII, pp. 447–448.
98 *Ibid.*, p. 446.
99 *Ibid.*, p. 445.
100 *Systema naturae* opens with a set of 'Observations of the three kingdoms of Nature'. The fourth observation is as follows: 'Quum nullae dantur novae species (1); cum simile semper parit sui simile (2); cum unitas in omni specie ordinem ducit (3), necesse est, ut unitatem progeneratricem, Enti cuidam Omnipotenti et Omniscio attribuamus, Deo nempe, cujus opus Creatio audit' [Because there are no new species (1); because like breeds like (2); because unity keeps order in each species (3), we have to attribute the procreative unity to a certain all-powerful and omnipresent being, that is God, whose work is called Creation] (C. Linnaeus, *Systema naturae*, 1735).
101 C. Darwin, *Variation*..., cited in *The Works of Charles Darwin* [1896], vol. VII, p. 445.
102 C. Darwin, *On the Origin of Species,* 1st edn (1859), chaps. 6, 11, 13; 6th edn (1872), chaps. 6, 12, 14.
103 Darwin often used terms related to 'resemblance', in particular 'likeness', in order to draw a distinction between the phenomena of 'heredity' and what naturalists called 'type'. Thus he often wrote of the differing capacities of individuals to 'transmit their likeness' (*Variation*, 1896, vol. VII, pp. 58–59).
104 C. Darwin, *Variation*..., cited in *The Works of Charles Darwin* [1896], vol. VII, p. 460.
105 *Ibid.*
106 The term 'sport', often used after Darwin as a synonym for a sudden and

large variation, is in fact relatively rare in Darwin's writings. When he does use it, however, it is in the general sense of a 'sudden' and 'strongly marked' variation (see *ibid.*, pp. 223 and 397).

107 *Ibid.*, p. 449.

108 C. Darwin, *On the Origin of Species* (1859), pp. 12–13.

109 R. C. Stauffer, *Charles Darwin's Natural Selection*... (1975), pp. 480–481, n. 2.

110 C. Darwin, *Variation*..., cited in *The Works of Charles Darwin* [1896], vol. VII, p. 448.

111 *Ibid.*, p. 456.

112 *Variation*, chap. 13.

113 'On the doctrine of reversion, as given in this chapter, the germ becomes a far more marvellous object, for, besides the visible changes which it undergoes, we must believe that it is crowded with invisible characters, proper to both sexes, to both the right and left side of the body, and to a long line of male and female ancestors separated by hundreds or even thousands of generations from the present time: and these characters, like those written on paper with invisible ink, lie ready to be evolved whenever the organisation is disturbed by certain known or unknown conditions' (C. Darwin, *Variation* ..., cited in *The Works of Charles Darwin* [1896], vol. VIII, pp. 35–36).

114 *Ibid.*, pp. 25–31.

115 *Ibid.*, p. 368.

116 *Ibid.*

117 The conclusion to the chapters of the *Variation* on heredity shows clearly that Darwin did not consider heredity to be exclusively individual: 'The very same peculiarity, as the weeping habit of trees, silky feathers, &c, may be inherited either firmly or not at all by different members of the same group, and even by different individuals of the same species, though treated in the same manner. In this latter case we see that the power of transmission is a quality which is merely individual in its attachment. As with single characters, so it is with the several concurrent slight differences – which distinguish sub-varieties or races; for of these some can be propagated almost as truly as species, whilst others cannot be relied on' (C. Darwin, *ibid.*, pp. 57–58).

118 S. Berge has argued that this doctrine had its origin amongst British breeders, whilst its canonical formulation can be found in an 1848 article in the Berlin National College of Economy publication *Allgemeinen Gründsätzen* ('The historical development of animal breeding', 1961, pp. 113ff).

119 C. Darwin, *On the Origin of Species* (1859), p. 29.

120 *Ibid.*, p. 19.

121 *Ibid.*, p. 350.

122 C. Darwin, *Variation*..., cited in *The Works of Charles Darwin* [1896], vol. VIII, p. 37.

123 *Ibid.*

124 *Ibid.*, p. 38.

125 *Ibid.*

126 See the illuminating remarks by Y. Conry in *L'introduction du darwinisme en France au XIXème siècle* (1974), pp. 317–334.

127 C. Darwin, *On the Origin of Species* (1859), p. 15.

128 Letter from Wallace to Darwin, 2 July 1866, reproduced in J. Marchant, *Alfred Russel Wallace: Letters and Reminiscences* (1916), pp. 140–143.
129 *The Origin of Species by Charles Darwin. A Variorum Text*, ed. M. Peckham, 1959, p. 165.
130 C. Darwin, *Variation*..., cited in *The Works of Charles Darwin* [1896], vol. VIII, p. 176.
131 C. Darwin, *On the Origin of Species* (1859), p. 12.
132 *Ibid.*, p. 32.
133 C. Darwin, *Variation*..., cited in *The Works of Charles Darwin* [1896], vol. VIII, p. 236. See also the end of the general conclusion, *ibid.*, p. 426.
134 *Ibid.*, p. 177.
135 *Ibid.*
136 *Ibid.*, pp. 210–211.
137 *Ibid.*
138 C. Darwin, *On the Origin of Species* (1859), p. 224.
139 C. Darwin, *Variation* (1896), vol. VIII, pp. 209–219.
140 *Ibid.*, p. 209.
141 *Ibid.*, p. 211. Charollais bulls were finally successfully introduced into England in 1961.
142 *Ibid.*, p. 234.
143 *Ibid.*, p. 233.
144 *Ibid.*, pp. 221–222.
145 *Ibid.*, p. 211.
146 *Ibid.*, vol. VII, pp. 239–240 and 277.
147 *Ibid.*, vol. VIII, p. 219.
148 See S. Berge, 'The historical development of animal breeding' (1961), and M. Lerner and M. P. Donald, *Modern Developments in Animal Breeding* (1968), chaps. 5 and 7.
149 S. Berge, 'The historical development of animal breeding' (1961), p. 120.

2 The ontology of selection

1 H. Spencer, *The Principles of Biology*, vol. I (1864–7), §165.
2 See note 12 to Chapter 1. For a discussion of this question, see J. Gayon, 'Sélection naturelle ou survie des plus aptes? Eléments pour une histoire du concept de *fitness* dans la théorie évolutionniste' (1995).
3 In *First Principles* (1863), written when he had not yet coined the phrase 'survival of the fittest', Spencer introduced the principle of 'differentiation'. This marked an important point in his work on 'the great law of evolution'. This law consists of the 'abstract' postulate that throughout reality there is a necessary and progressive development from the indefinite to the definite or, as Spencer liked to put it, from the 'heterogeneous' to the 'homogeneous'. However, the genesis of complexity cannot be reduced to the appearance of a 'vague chaotic heterogeneity'. To avoid this, Spencer evoked a 'cause' that guarantees that the 'local differentiation' is also a 'local integration'. This cause is precisely 'segregation' which consists of a uniform force, acting on groups of heterogeneous units, and which has the effect of producing smaller, but more homogeneous, groups. For

example, the movement of water in a river, by moving in a certain direction and at a certain speed, provokes the deposit of sand at one place, small pebbles at another, and large stones at still another (*First Principles*, 1st edn, 1863, §123). This principle applies to everything that evolves and differentiates: atoms and celestial bodies (§124), geological changes (§125), organic development and the evolution of species (§126), 'mental evolution', or psychogenesis (§127), and 'social evolution' (§128). In the 1863 edition, this whole speculation was included in a chapter entitled 'Differentiation and Integration'. In the second and third editions (1867 and 1875), this chapter was renamed 'Segregation'. The law of 'differentiation' thus became the law of 'segregation'. But Spencer often also used the word 'selection' (and sometimes 'sorting') as a synonym. This linguistic shift is actually the main modification in the chapter, which remains virtually unaltered in all other respects. Spencer thus decided to rename one of his main concepts in a vocabulary that was close to that of Darwin. This decision came after Spencer had proposed to rename Darwin's 'natural selection' 'survival of the fittest' in *Principles of Biology* (1864; see above, note 1). In the 1867 and 1875 editions of *First Principles*, the paragraphs quoted above were renumbered as follows: §163 instead of §123, §164 instead of §124, etc.

4 C. Darwin, *On the Origin of Species* (1859), p. 81.

5 *Ibid.*, p. 194.

6 C. Darwin, *Variation* ... (2nd edn, 1875), quoted in C. Darwin (1972), vol. VII, p. 6.

7 C. Darwin, *On the Origin of Species* (1859), p. 320.

8 See for example *ibid.*, p. 15: 'When we look to the hereditary varieties or races of our domestic animals and plants ...' Darwin does not seem to have made the distinction – traditional since Kant – between hereditary varieties and races. Kant argued: 'Varieties are hereditary deviations within a genus ... If a variety maintains itself unchanged many generations when transplanted and, when crossed with other varieties, produces *intermediate offspring*, we call it a *race* ... Negroes and white people are different races, because they perpetuate their characters in all regions and produce unfailingly intermediate children and mongrels (mulattos). Blondes and brunettes are not different races, because when they intermarry, the children may all be blonde or all brunette' (I. Kant, 'On the different races of men, with the announcement of lectures on physical geography' [1775], reproduced in G. Rabel, *Kant* (1963), p. 98). This distinction was widespread amongst 19th-century naturalists (in large part owing to Blumenbach's work). It is difficult to tell why Darwin did not avail himself of this distinction. It is possible that the traditional variety/race distinction was more or less replaced by variation/variety in Darwin's writings.

9 C. Darwin, *On the Origin of Species* (1859), p. 84. A virtually identical formulation can be found in chap. 6: 'natural selection acts solely by and for the good of each' (*ibid.*, p. 201).

10 See for instance the interesting article by J. G. Lennox, 'Darwin *was* a teleologist' (1993).

11 C. Darwin, *On the Origin of Species* (1859), p. 170.

12 *Ibid.*, pp. 202–203.

13 C. Darwin, *The Descent of Man* (1871), I, pp. 161–162.

14 C. Darwin, *On the Origin of Species* (1859), chap. 8 (chap. 9 in later editions: 'Hybridisation').

15 The two aspects of the problem of the ontology of Darwinian natural selection distinguished here can be found in various forms in recent discussions by philosophers of biology. R. N. Brandon and R. M. Burian have proposed to distinguish between those problems related to the 'units of selection' (the entities that are selected, e.g. the gene) and those related to the 'levels of selection' (entities at the level of which selection acts, e.g. the organism). This distinction can be considered to be the same as that made by E. Sober between 'selection *of*' and 'selection *for*'. R. Lewontin, followed by R. Dawkins and D. Hull, pioneered these conventional distinctions, drawing attention to the two kinds of entities that can be found in the expression 'unit of selection': on the one hand those replicable entities which are modified in their populational distributions by selection ('replicators'), and on the other hand those entities that interact causally with the environment in the course of the selective process itself ('interactors' for Hull, 'vehicles' for Dawkins). Any process of natural selection requires the presence of these kinds of entities, which are not necessarily the same thing – and indeed, especially in sexual species, generally are not the same thing. This section has attempted to find traces of this debate – however ambiguous – in Darwin's writings, although, of course, our modern understanding of the problem is infinitely richer (R. C. Lewontin, 'The units of selection' (1970); R. Dawkins, 'Replication, selection and the extended phenotype' (1978); D. Hull, 'Individuality and selection' (1980) and 'The units of evolution' (1981); E. Sober, *The Nature of Selection* (1984); R. N. Brandon and R. M. Burian (eds.), *Genes, Organisms, Populations* (1984)).

16 See note 8 above.

17 C. Darwin, *On the Origin of Species* (1859), p. 45.

18 *Ibid.*, p. 44.

19 *Ibid.*, p. 12.

20 See Chapter 3.

21 C. Darwin, *On the Origin of Species* (1859), pp. 80–81.

22 *Ibid.*, p. 95.

23 *Ibid.*, 5th edn (1869), chap. 4; quoted in M. Peckham, *The Origin of Species by Charles Darwin. A Variorum Text* (1959), p. 166.

24 C. Darwin, *On the Origin of Species* (1859), p. 194.

25 'The term "variety" is almost equally difficult to define; but the community of descent is almost universally implied, though it can be rarely proved . . . Some authors use the term "variations" in a technical sense, as implying a modification directly due to the physical conditions of life; and "variations" in this sense are supposed not to be inherited: but who can say that the dwarfed condition of shells in the brackish waters of the Baltic, or dwarfed plants on Alpine summits, or the thicker fur of an animal from far northwards, would not in some cases be inherited for at least some few generations? and in this case I presume that the form would be called a variety' (*ibid.*, pp. 44–45). The Lamarckian tone of this quotation is obvious. But it also tells us something about the distinction – and relation – Darwin established between 'variation' and 'variety'.

26 *Ibid.*, p. 91.

27 H. Spencer, 'The inadequacy of natural selection' (1893), pp. 160–161.
28 This distinction has been particularly well analysed from the point of view of modern evolutionary theory by E. Sober, *The Nature of Selection* (1984).
29 C. Darwin, *On the Origin of Species* (1859), p. 83.
30 *Ibid.*, p. 84. See also p. 201.
31 *Ibid.*, p. 459.
32 'In social animals it [natural selection] will adapt the structure of each individual for the benefit of the community' (*ibid.*, p. 87. See also pp. 202–203).
33 *Ibid.* See also p. 200.
34 See for example E. Sober, *The Nature of Selection* (1984), pp. 189–193.
35 This link between demography and econometry has continued to strengthen since Darwin's time. See S. Kingsland, *Modelling Nature* (1985), and J. Gayon, 'Sélection naturelle ou survie des plus aptes? Eléments pour une histoire du concept de *fitness* dans la théorie évolutionniste' (1995).
36 C. Darwin, *On the Origin of Species* (1859), p. 242.
37 R. C. Stauffer (ed.), *Charles Darwin's Natural Selection... 1856–58* (1975), pp. 364–374; see also p. 510. The exact title of the chapter is: 'Chapter VIII: Difficulties of the Theory of Natural Selection in Relation to Passages from Form to Form'.
38 *Ibid.*, pp. 373–374.
39 C. Darwin, *On the Origin of Species* (1859), pp. 235–242.
40 *Ibid.*, p. 236.
41 R. C. Stauffer (ed.), *Charles Darwin's Natural Selection... 1856–58* (1975), pp. 365–366.
42 *Ibid.* This hypothesis, which for Darwin was pure speculation, has finally prevailed as the explanation of the allocation of the larvae of social insects to a given caste. Feeding 'royal jelly' will make a bee larva develop into a fertile female. Darwin was aware of this fact. (Cf. *ibid.*, p. 367).
43 *Ibid.*, p. 366.
44 *Ibid.*, p. 368.
45 'No amount of exercise, or habit, or volition, in the utterly sterile members of a community could possibly have affected the structure or instincts of the fertile members, which alone leave descendants. I am surprised that no one has advanced this demonstrative case of neuter insects, against the well-known doctrine of Lamarck' (C. Darwin, *On the Origin of Species*, 1859, p. 242).
46 *Ibid.*, p. 237.
47 *Ibid.*, 6th edn (1872), chap. 8; quoted in M. Peckham, *The Origin of Species by Charles Darwin. A Variorum Text* (1959), p. 417.
48 *Ibid.*
49 C. Darwin, *On the Origin of Species* (1859), p. 239.
50 R. C. Stauffer (ed.), *Charles Darwin's Natural Selection... 1856–58* (1975), pp. 373–374.
51 C. Darwin, *On the Origin of Species*, 5th edn (1869), chap. 9; quoted in M. Peckham, *The Origin of Species by Charles Darwin. A Variorum Text* (1959), pp. 470–471.
52 R. C. Stauffer (ed.), *Charles Darwin's Natural Selection... 1856–58* (1975), pp. 418–419.

53 C. Darwin, *On the Origin of Species,* 1859, p. 245.
54 *Ibid.*, 4th edn (1866), chap. 8; quoted in M. Peckham, *The Origin of Species by Charles Darwin. A Variorum Text* (1959), pp. 443–444.
55 *Ibid.*, p. 444.
56 C. Darwin, *On the Origin of Species,* 5th edn (1869), chap. 8; quoted in M. Peckham, *The Origin of Species by Charles Darwin. A Variorum Text* (1959), p. 445. This discussion is repeated in the same terms in *The Variation of Animals and Plants under Domestication.*
57 M. Ruse, 'Charles Darwin and group selection' (1980).
58 Letter from Wallace to Darwin, February 1868, reproduced in J. Marchant, *Alfred Russel Wallace: Letters and Reminiscences* (1916), p. 162.
59 See H. E. Gruber, *Darwin on Man* (1974); P. H. Barrett (ed.), *Metaphysics, Materialism and the Evolution of Mind: Early Writings of Charles Darwin* (1974); R. C. Stauffer (ed.), *Charles Darwin's Natural Selection... 1856–58* (1975).
60 C. Darwin, *The Descent of Man* (1871), p. 98.
61 *Ibid.*, pp. 98ff.
62 'It can hardly be disputed that the social feelings are instinctive or innate in the lower animals; and why should they not be so in man? Mr Bain (see for instance, 'The Emotions and the Will', 1865, p. 485) and others believe that the moral sense is acquired by each individual during his lifetime. On the general theory of evolution this is at least extremely improbable' (*ibid.*, p. 71, n. 5).
63 *Ibid.*, p. 82.
64 'The following proposition seems to me in a high degree probable – namely, that any animal whatever, endowed with well-marked social instincts, would inevitably acquire a moral sense or conscience, as soon as its intellectual powers had become as well developed, or nearly as well developed, as in man' (*ibid.*, pp. 71–72).
65 *Ibid.*, pp. 161–166.
66 *Ibid.*, pp. 391–394.
67 *Ibid.*, p. 166.
68 *Ibid.*, p. 82.
69 *Ibid.*, pp. 159–160.
70 *Ibid.*, p. 161.
71 *Ibid.*, chap. 3, § 'Sociability', pp. 74–84.
72 *Ibid.*, pp. 80–81.
73 *Ibid.*, p. 82.
74 A. R. Wallace, 'The origin of human races and the antiquity of man deduced from the theory of natural selection' (1864).
75 C. Darwin, *The Descent of Man* (1871), p. 163.
76 *Ibid.*, p. 166.
77 A. R. Wallace, 'The origin of human races...' (1864).
78 C. Darwin, *The Descent of Man* (1871), pp. 165–166.
79 *Ibid.*, p. 70.
80 C. Darwin, *Variation* (1896), vol. I, p. 6.
81 *Ibid.*, p. 7.
82 I. Lakatos, 'Falsification and the methodology of scientific research programmes' (1970); 'History of science and its rational reconstructions' (1971).

3 Jenkin's objections, Darwin's dilemma

1 Pronounced 'Fleming' (S. J. Gould, *Bully for Brontosaurus*, 1991, chap. 23, 'Fleeming Jenkin revisited', pp. 340–353).
2 F. Jenkin, 'The Origin of Species', *North British Review* (1867).
3 C. Darwin, *On the Origin of Species*, 5th edn (1869), chap. 4; quoted in M. Peckham, *The Origin of Species by Charles Darwin. A Variorum Text* (1959), p. 178.
4 Cited in F. Darwin and A. C. Seward, *More Letters of Charles Darwin* (1903), II, p. 379.
5 F. Darwin, *Life and Letters of Charles Darwin* (1887), III, p. 109.
6 L. Eiseley, *Darwin's Century: Evolution and the Man Who Discovered It* (1958).
7 P. J. Vorzimmer, *Charles Darwin: The Years of Controversy* (1970).
8 This manuscript, entitled 'Hypothesis of pangenesis', is to be found in Cambridge University Library, together with many other manuscripts by Darwin. On the content of this manuscript and the history of the idea of pangenesis in Darwin's thought, see R. C. Olby, 'Charles Darwin's manuscript of pangenesis' (1963), and *Origins of Mendelism* (2nd edn, 1985), pp. 84–85, 187–189.
9 E. Darwin, *Zoonomia* (1801). R. Olby, *Origins of Mendelism* (1985), p. 189, provides a fascinating excerpt from this work. Darwin's pangenesis seems to be already in existence!
10 P. J. Bowler, *Evolution: The History of an Idea* (1984), pp. 198–199.
11 E. Mayr, *The Growth of Biological Thought* (1982), pp. 512–513.
12 C. Lyell, *Principles of Geology*, vol. II (1832), p. 23.
13 See the interesting profile presented by S. J. Gould (*Bully for Brontosaurus*, 1991, chap. 23, 'Fleeming Jenkin revisited', pp. 340–353). Jenkin was a proponent of the laissez-faire school. Thus, in a sense, he was in a good position to assess Darwin's importation of Adam Smith's vision of economics into evolutionary biology.
14 W. Thomson, *On the Age of the Sun's Heat* (1862); 'On the secular cooling of the earth' (1864). A year after Jenkin's article, Lord Kelvin published an article in which he publicly questioned the validity of stratigraphic dating ('On geological time', 1868). This document, and the controversy it provoked, weighed heavily on transformism right up until the first decade of the 20th century.
15 A good summary of the whole controversy can be found in a lecture given by E. B. Poulton in 1896, 'A naturalist's contribution to the discussion upon the age of the earth'. The notes added by the author in 1906 when the text was revised prior to publication in a volume of essays (E. B. Poulton, *Essays on Evolution 1889–1907*, 1908) are particularly interesting. In these notes, Poulton explains how the realisation that there was radioactivity in the sun and in the earth led to Kelvin's estimations being multiplied a thousandfold, thus allowing the biologist a virtual free rein in estimating the moment that life appeared. Quoting the physicist John Perry ('we are now in a position to say that the physicist can make no calculation either as to the probable or possible age of life on the earth'), he adds: 'we [the biologists] are free to proceed, and to look for the conclusions warranted by our own evidence. In this matter, we are at one with the geologists' (pp. 15–16).

16 F. Jenkin, 'The Origin of Species' (1867), pp. 279–286.
17 *Ibid.*, pp. 279–280.
18 There appears to be only one passage in which Darwin states that variation, as such, can have no limit. This text (a concluding remark in *Variation*) postdates Jenkin's article by a year, and it is difficult not to consider it to be a response: 'It has been boldly maintained by some authors that the amount of variation to which our domestic productions are liable is strictly limited; but this is an assertion resting on little evidence. Whether or not the amount of change in any particular direction is limited, the tendency to general variability is, as far as we can judge, unlimited. Cattle, sheep, and pigs have varied under domestication from the remotest period, as shown by the researches of Rütimeyer and others; yet these animals have been improved to an unparalleled degree, within recent times, and this implies continued variability of structure' (*Variation,* 2nd edn [1875], quoted in Darwin (1972), vol. VIII, p. 411). In this quotation, there is a nuance between variation that is unlimited in direction, and continuously available variations. For Darwin, the second notion was the more important.
19 F. Jenkin, 'The Origin of Species' (1867), p. 282.
20 See section 1.3.3, around note 133.
21 See section 1.3.2.1, 'Heredity and individual variation'.
22 F. Jenkin, 'The Origin of Species' (1867), p. 285.
23 *Ibid.*, pp. 286–294.
24 *Ibid.*, p. 286.
25 This question explicitly appears only in the third part of the argument (*ibid.*, pp. 291ff). Prior to this point, Jenkin argues as though the 'intermediate' character of inherited characters were taken as given.
26 This situation is discussed in *ibid.*, pp. 291–293. In his own words, Jenkin writes of 'the mysterious faculty by which the divergence is transmitted unimpaired to countless descendants'. This example is opposed to that in which the character 'wanes' in proportion to the dilution of the blood in successive generations.
27 See for example C. Darwin, *Variation,* 2nd edn (1875), quoted in C. Darwin (1972), vol. VIII, pp. 40–47. Darwin used 'prepotency' or 'preponderance'. 'Dominance' was the term used by Mendel (1866).
28 F. Jenkin, 'The Origin of Species', p. 291.
29 *Ibid.*, p. 292.
30 *Ibid.*
31 *Ibid.*, p. 293.
32 Letter from Huxley to Darwin, 23 November 1859 (in *The Correspondence of Charles Darwin,* ed. F. Burkhardt, vol. VII, 1991). Strictly speaking, Darwin did not deny the existence of major, rare and abrupt inherited variations. He was well aware that breeders, and especially horticulturists, had often used such variants as the basis of selection (see *Origin of Species*, chap. 1), but he considered that this process was relatively unlikely to take place in nature (see *Origin of Species*, beginning of chap. 2). Furthermore, we know that Darwin had a strong philosophical aversion to discrete categories and entities. This is shown by his argumentation throughout his works: each time such a notion appears in his path, Darwin demonstrates the existence of an intermediate

form. The principle of continuity, inherited from 18th-century philosophy, constitutes an ontological preference that impregnates all of Darwin's work.

33 For a point of view opposed to that presented here, see S. Schweber, 'The origin of the *Origin* revisited' (1977).

34 F. Jenkin, 'The Origin of Species' (1867), pp. 287–288.

35 *Ibid.*, p. 286.

36 *Ibid.*, p. 291.

37 *Ibid.*, p. 289.

38 *Ibid.*, pp. 288–289.

39 *Ibid.*, pp. 290–291. 'As the numbers of the favoured variety diminish, so must its relative advantage increase, if the chance of its existence is to surpass the chance of its extinction, until hardly any conceivable advantage would enable the descendants of a single pair to exterminate the descendants of many thousands.'

40 *Ibid.*, p. 289.

41 *Ibid.*, p. 294.

42 In 1930 R. A. Fisher gave a thorough explanation of this criticism. Given characters showing continuous variation, and under the hypothesis of blending inheritance, the genetic variance of each character will be reduced by half at each generation. A continuous and superabundant source of new variation would have to exist for any modification to take place at the population level. Taken to its extreme, this would mean that each individual would have to be a mutant for a given character (R. A. Fisher, *The Genetical Theory of Natural Selection,* 1930 and 1958, chap. 1, pp. 4–5).

43 F. Jenkin, 'The Origin of Species' (1867), p. 291.

44 C. Darwin, *On the Origin of Species*, 5th edn (1869), chap. 4; quoted in M. Peckham, *The Origin of Species by Charles Darwin. A Variorum Text* (1959), p. 178.

45 C. Darwin, *Variation,* 2nd edn [1875], chap. 14, quoted in C. Darwin (1972), vol. VIII, pp. 40–47.

46 Letter from Darwin to Wallace, 22 January 1869, in J. Marchant, *Alfred Russel Wallace: Letters and Reminiscences* (1916), p. 191.

47 Letter from Wallace to Darwin, 30 January 1869, *ibid.*, p. 192.

48 Letter from Darwin to Wallace, 2 February 1869, *ibid.*

49 *Ibid.*

50 This argument is well summarised by Wallace in his letter of 30 January 1869 (*ibid.*): 'as all the more important structural modifications of animals and plants imply much co-ordination, it appears to me that the chances are millions to one against *individual variations* ever coinciding so as to render the required modification possible'.

51 C. Darwin, *On the Origin of Species*, 5th edn (1869), chap. 4; quoted in M. Peckham, *The Origin of Species by Charles Darwin. A Variorum Text* (1959), p. 178.

52 *Ibid.*

53 See R. Olby, *Origins of Mendelism* (2nd edn, 1985), pp. 84–86.

54 C. Darwin, *Variation,* 2nd edn [1875], chap. 27, quoted in Darwin (1972), vol. VIII, pp. 349–369.

55 *Ibid.*, p. 370.

56 See C. Zirkle, 'The early history of the idea of acquired characteristics and of pangenesis' (1946).

57 C. Darwin, *Variation*, 2nd edn [1875], chap. 27, quoted in C. Darwin (1972), vol. VIII, p. 369. In general, Darwin preferred to use 'unit', like many of his contemporaries. However, the reference to cell theory is explicit, as is shown by this passage, taken from the introduction to the section that presents the hypothesis of pangenesis.

58 C. Darwin, *On the Origin of Species*, 5th edn (1869), chap. 4; quoted in M. Peckham, *The Origin of Species by Charles Darwin. A Variorum Text* (1959), p. 178. Darwin here refers to Jenkin's example that was analysed above: the chances of survival of a variation are taken as one in 1,000,000, a selective value twice that of the mean, in a population where one offspring out of 10,000 survives to reproductive age.

59 *Ibid.*, p. 179.

60 C. Darwin, *On the Origin of Species*, 6th edn (1872), chap. 4; quoted in M. Peckham, *The Origin of Species by Charles Darwin. A Variorum Text* (1959), p. 179.

Part 2 Selection faced with the challenge of heredity: sixty years of principled crisis

1 F. Galton, 'Typical laws of heredity' (1877), p. 512.

2 W. B. Provine, *The Origin of Theoretical Population Genetics* (1971). On Galton, the English biometrical school and its dispute with Mendelism, the most important works are R. S. Cowan, 'Sir Francis Galton and the continuity of germ plasm: A biological idea with political roots' (1968), 'Francis Galton's contributions to genetics' (1972), 'Francis Galton's statistical ideas: The influence of eugenics' (1972) and 'Nature and nurture' (1977); P. Froggatt and N. C. Nevin, 'The "law of ancestral heredity" and the Mendelian–ancestrian controversy in England, 1899–1906' (1971); D. A. Mackenzie, *Statistics in Britain 1865–1930* (1981); B. J. Norton, *Theories of Evolution of the Biometric School* (1971, chap. 3), 'The biometric defense of Darwinism' (1973) and *Karl Pearson and the Galtonian Tradition: Studies in the Rise of Quantitative Social Biology* (1978); R. C. Olby, *Origins of Mendelism* (1966, 1985); R. G. Swinburne, 'Galton's law: Formulation and development' (1963). A number of studies also exist in French: J. Piquemal, 'Quelques distinctions à propos de l'informe et à propos de sa mathématisation' (1972); P. Sentis, *La naissance de la génétique au début du 20ème siècle* (1970); and the unpublished PhD dissertation by C. Lenay, 'Enquête sur le hasard dans les grandes théories biologiques de la deuxième moitié du dix-neuvième siècle' (1989).

3 C. Darwin, *Variation* ... (1896), pp. 426–427 and 236.

4 Galton and the concept of heredity

1 R. C. Olby, *Origins of Mendelism* (1966, 1985), pp. 63–64.

2 F. Galton, 'Hereditary talent and character' (1865).

3 F. Galton, 'A diagram of heredity' (1898).

4 See F. Galton, *Essays in Eugenics* (1909A). This volume contains articles published from 1901 to 1908.
5 R. G. Swinburne, 'Galton's law: Formulation and development' (1963).
6 Letter from Galton to Charles Darwin, 24 December 1869. Reproduced in C. P. Blacker, *Eugenics: Galton and After* (1952), p. 83.
7 F. Galton, *Hereditary Genius* (1869). This passage is virtually a word-for-word reproduction of a similar passage in 'Hereditary talent and character' (1865), p. 327.
8 The best source for these events remains Karl Pearson's biography of Galton: *The Life, Letters and Labours of Francis Galton* (1914–1930).
9 F. Galton, 'Hereditary talent and character' (1865), p. 321.
10 *Ibid.*, p. 322.
11 *Ibid.*
12 *Ibid.*, p. 319.
13 *Ibid.*
14 'Let us, then, give reins to our fancy and imagine a Utopia – or a Laputa if you will – in which a system of competitive examination for girls, as well as for youths, had been so developed as to embrace every important quality of mind and body, and where a considerable sum was yearly allotted to the endowment of such marriages as promised to yield children, who would grow into eminent servants of the state ... If a twentieth part of the cost and pains were spent in measures for the improvement of the human race that are spent in the improvement of the breed of horses and cattle, what a galaxy of genius might we not create! We might introduce prophets and high priests of civilisation into the world, as surely as we can propagate idiots by mating *crétins*. Men and women of the present day are, to those we might hope to bring into existence, what the pariah dogs of the streets of an Eastern town are to our own highly-bred varieties' (*ibid.*, pp. 165–166).
15 *Ibid.*, p. 326.
16 *Ibid.*, pp. 319–320.
17 F. Galton, *Hereditary Genius* (1869); *English Men of Science: Their Nature and Nurture* (1874). This book is a reply to Alphonse De Candolle's book *Histoire des sciences et des savants depuis deux siècles* (1872).
18 F. Galton, *Inquiries into Human Faculty* (1883).
19 On Galton and eugenics, see R. S. Cowan, 'Nature and nurture: The interplay of biology and politics in the work of Francis Galton' (1977) and *Sir Francis Galton and the Study of Heredity in the 19th Century* (1985).
20 A. Weismann, 'On heredity' [1st German edn 1883], published in English in *Essays upon Heredity* (1889), pp. 117–156.
21 F. Galton, 'Hereditary talent and character', p. 322, col. 2.
22 C. Darwin, *Variation* (1896), quoted in C. Darwin (1972), vol. VIII, p. 370.
23 *Ibid.*, pp. 370, 371, 373, 374, 379.
24 *Ibid.*, pp. 370, 371, 373, 379, 397.
25 'It appears probable that all external agencies, such as changed nutrition, increased use or disuse, &c., which induced any permanent modification in a structure, would at the same time or previously act on the cells, nuclei, germinative or formative manner, from which the structure in question were developed, and consequently would act on the gemmules or cast-off atoms'

(C. Darwin, *Variation*, 1st edn, vol. II, p. 382). This sentence appears only in the first edition.

26 F. Galton, *Hereditary Genius* (1869), p. 376.

27 'Mr Darwin maintains, in the theory of Pangenesis, that the gemmules of innumerable qualities, derived from ancestral sources, circulate in the blood, and propagate themselves, generation after generation, still in the state of gemmules, but fail in developing themselves into cells; because other antagonistic gemmules are prepotent and overmaster them, in the struggle for points of attachment, etc.' (F. Galton, *Hereditary Genius*, 1869, p. 367).

28 This correspondence is reproduced in K. Pearson, *The Life, Letters and Labours of Francis Galton*, vol. II (1924), pp. 156 et seq.

29 F. Galton, 'Experiments in pangenesis by breeding from rabbits of a pure variety, into whose circulation blood taken from varieties had previously been largely transfused' (1871), p. 404.

30 This letter is reproduced in K. Pearson, *The Life, Letters and Labours of Francis Galton*, vol. II (1924), pp. 163–164.

31 *Nature*, 4 May 1871. This letter is reproduced in K. Pearson, *The Life, Letters and Labours of Francis Galton*, vol. II (1924), pp. 164–165.

32 C. Darwin, *Variation* (1868), vol. II, p. 374. In the second edition (1875), the reference to circulation disappeared: the gemmules are simply 'dispersed'.

33 *Ibid.*, p. 379.

34 *Ibid.*

35 *Nature*, 4 May 1871, reproduced in K. Pearson, *The Life, Letters and Labours of Francis Galton*, vol. II (1924), p. 165.

36 This argument is included in 'A theory of heredity' (1875).

37 F. Galton, 'On blood-relationship' (1872).

38 *Ibid.*, p. 394.

39 *Ibid.*

40 'The observed facts of Reversion enable us to prove that the latent elements must be greatly more varied than those that are personal or patent' (*ibid.*, p. 395).

41 *Ibid.*, p. 402.

42 A. Weismann, 'On heredity' [1st German edn 1883], published in English in *Essays upon Heredity* (1889), pp. 71–105.

43 F. Galton, 'A theory of heredity' (1875).

44 Galton called this process 'class representation', by analogy with politics. The 'manifest person', in all its differentiated tissues, has the same relation to the patent elements as a parliament of deputies has to the nation it 'represents' ('On blood-relationship', 1872, p. 395).

45 'Before proceeding, I beg permission to use, in a special sense, the word "stirp", derived from the Latin *Stirpes*, a root, to express the sum-total of the germs, gemmules, or whatever they may be called, which are to be found, according to every theory of organic units, in the newly-fertilised ovum – that is, in the earliest pre-embryonic stage – from which it receives nothing further from its parents, not even from its mother, other than the mere nutriment' ('A theory of heredity', 1875, p. 81). This passage deserves close attention. First, Galton made a linguistic error: the correct Latin expression for 'a root' is *stirps*, not *stirpes* (plural). Second, Galton suggests that he is

using the word 'in a special sense', and indeed the *Oxford English Dictionary* (OED) credits Galton with originating a new meaning of the word in this article. However, it is also clear from the OED that Galton merely resuscitated a term that was in wide use until the 17th century, and which meant the stock of a family or a line of descent. This is probably the origin of the use of the term 'stirpiculture' to describe a programme of human reproduction by selective marriage, proposed by a certain Noyes in the late 1860s. *The Circular*, a newspaper produced by Noyes in Oneida (New York State, USA), referred to 'stirpiculture' in its issue dated 3 April 1865. Inspired by Galton's work, Noyes even tried to put this programme into action in 1869 (see D. Kevles, *In the Name of Eugenics: Genetics and the Uses of Human Heredity*, 1985, chap. 5). A similar word, 'stirps', had both a legal meaning as a branch of a family and in the 19th century, prior to Galton's coinage, was used in both zoology and botany to denote a family or race of animals or plants. Indeed, the Latin *stirps* was even employed in the 18th century by botanists such as Linnaeus to describe hereditary varieties. Its French equivalent – *stirpe* – was used at the beginning of the 19th century (for example by P. A. De Candolle). For Galton, the 'stirp' gives rise to the latent elements of the sex cells, and also to the development of the differentiated tissues. Jean-Marc Drouin is thanked for pointing out the use of *stirpe* in French prior to Galton.

46 F. Galton, 'On blood-relationship' (1872), p. 400.

47 *Ibid.*, pp. 400–401.

48 See S. Schweber, 'The origin of the Origin revisited' (1977).

49 F. Galton, *Natural Inheritance* (1889), p. 37.

50 F. Galton, *Memories of My Life* (1908), p. 304.

51 This section owes a heavy debt to the excellent analyses of S. Stigler, *The History of Statistics* (1986), pp. 203–220, and D. A. Mackenzie, *Statistics in Britain* (1981), pp. 51–72. See also H. Walker, *Studies in the History of Statistical Method* (1929), and K. Pearson, *The Life, Letters and Labours of Francis Galton*, vol. II (1924).

52 Its origin can apparently be found in the work of W. Lexis (1837–1914). In an 1877 representation of the frequency distribution of mortality as a function of age, Lexis distinguished 'normal deaths', which are distributed according to a Gaussian function, from 'infant deaths' and 'premature deaths', which are not. The 'normal' distribution can be analysed as the result of a large number of heterogeneous causes (see S. Stigler, *The History of Statistics*, 1986, pp. 222–225 and 229–233). For a detailed historical analysis of the terms 'normal distribution' and 'normal curve', see S. Stigler, 'Stigler's law of eponymy' (1980).

53 S. Stigler, *The History of Statistics* (1986), pp. 203–229.

54 F. Galton, *Natural Inheritance* (1889), p. 66.

55 *Ibid.*, p. 164. The image is based on a precise meteorological observation: on one side of a mountain, rising warm winds carry an invisible vapour that condenses at the summit. On the other side, particles are drawn to the base of the mountain and return to their invisible state when they meet the layers of warm air. A cloud is thus a system that is perpetually criss-crossed by a flow of particles.

56 F. Galton, 'Typical laws of heredity' (1877), p. 512, col. 2.
57 *Ibid.*
58 These experiments are reported in *ibid.* See below.
59 Galton was always careful to surround himself with professional mathematicians who could give an analytical interpretation of his statistical intuitions. The technical appendices which sometimes accompany his articles are particularly revealing. Written or inspired by mathematicians, they show how Galton's reflections naturally took their place in the development of statistics.
60 F. Galton, 'Statistics by intercomparison' (1875), p. 38. For the purely mathematical aspects of this episode, see the penetrating analysis in S. Stigler, *The History of Statistics* (1986).
61 F. Galton, 'Statistics by intercomparison' (1875), p. 40.
62 F. Galton, 'Typical laws of heredity' (1877), p. 493, col. 2. In this passage, Galton claims to have used the quincunx publicly in 1874 in a lecture to the Royal Institution. The original model, still on display in the Galton Laboratory at University College, London, was apparently built in 1873 (S. Stigler, *The History of Statistics*, 1986, pp. 276–277).
63 F. Galton, *Natural Inheritance* (1889), p. 63.
64 The term is used in 'Typical laws of heredity' (1877), p. 532, col. 1. 'Model' has a technical connotation, but it is clear that the apparatus gives a mechanical illustration of the general properties of the law of deviations (*ibid.*, p. 495, col. 1). Pierre Duhem has pointed out that the use of mechanical models was a characteristic of 19th-century English physics. Both Thomson and Maxwell frequently used them; Galton's approach is thus easily understood (P. Duhem, *La théorie physique. Son objet – sa structure*, 1906, chap. 4, 'Les théories abstraites et les modèles mécaniques').
65 K. Pearson, *The Life, Letters and Labours of Francis Galton* (1914–1930), vol. III B, pp. 465–466.
66 F. Galton, 'Typical laws of heredity' (1877), p. 513, col. 2.
67 *Ibid.* The numerical data from the sweet-pea experiments are contained in an appendix to 'Regression towards mediocrity in hereditary stature' (1885), pp. 258–260.
68 F. Galton, 'Typical laws of heredity' (1877), p. 513, col. 1.
69 *Ibid.*
70 *Ibid.*, col. 2.
71 *Ibid.*, pp. 532–533.
72 The term 'variance' comes from R. A. Fisher, 'The correlation of relatives on the supposition of Mendelian inheritance' (1918). Galton took the theorem of the additivity of the squares of two probable errors from G. Airy, *On the Algebraic and Numerical Theory of Errors of Observations and the Combination of Observations* (1861), §43.
73 The 'modulus' was one of the many measures of dispersion that were used at the end of the 19th century, before the widespread use of the concepts of variance and standard deviation. The 'modulus' is equal to 2.097 times the 'probable error' (that is the value such that half the 'errors' are greater than this value, and half are less).
74 The equation is explicitly given in 'Typical laws of heredity' (1877), p. 533.

Unlike the formula given in 'Regression towards mediocrity...' (1885), p. 256, the 1877 version is derived in a way that is mathematically clear.

75 See the list of studies in note 2 of the introduction to Part 2 above, in particular Swinburne (1963), Olby (1966, 1985), Froggatt and Nevin (1971), Provine (1971), Cowan (1972a and 1972b) and Norton (1971).

76 K. Pearson, 'Mathematical contributions to the theory of evolution. On the law of ancestral heredity' (1898). Galton did not use the term.

77 F. Galton, 'Regression towards mediocrity in hereditary stature' (1885). See also 'Inheritance and regression' (1885); 'Family likeness in stature' (1886); *Natural Inheritance* (1889).

78 F. Galton, 'Regression towards mediocrity...' (1885), pp. 247–249; 'Family likeness in stature' (1886), p. 53; *Natural Inheritance* (1889), p. 87.

79 F. Galton, 'Regression towards mediocrity...' (1885), p. 246.

80 'We can define the law of regression very briefly. It is that the height-deviate of the offspring is, on the average, two thirds of the height-deviate of its mid-parentage' (*ibid.*, p. 252). Galton qualified this statement as a 'law' because he believed that the $\frac{2}{3}$ ratio of 'regression' applied to any species. 'The average regression of the offspring to a constant fraction of their respective mid-parental deviations, which was first observed in the diameters of seeds, and then confirmed by observations on human stature, is now shown to be a perfectly reasonable law which might have been deductively foreseen' (*ibid.*, p. 253).

81 'It is more frequently the case that an exceptional man is the somewhat exceptional son of rather mediocre parents, than the average son of exceptional parents' (*ibid.*, p. 254).

82 F. Galton, 'Family likeness in stature' (1886), pp. 54–57.

83 F. Galton, 'Regression towards mediocrity' (1885), p. 247.

84 K. Pearson, 'Mathematical contributions to the theory of evolution. On the law of ancestral heredity' (1898).

85 F. Galton, 'The average contribution of each several ancestor to the total heritage of the offspring' (1897), p. 401.

86 F. Galton, 'Regression towards mediocrity in hereditary stature' (1885), pp. 260–261. The appendix is entitled 'Separate contribution of each ancestor to the heritage of the offspring'.

87 F. Galton, 'Family likeness in stature' (1886), pp. 61–62: 'Separate contribution of each ancestor'.

88 F. Galton, *Natural Inheritance* (1889), pp. 134–136: 'Separate contribution of each ancestor'.

89 F. Galton, 'Regression towards mediocrity in hereditary stature' (1885), p. 260.

90 *Ibid.*, pp. 252–253.

91 F. Galton, 'Regression towards mediocrity...' (1885).

92 *Ibid.*, p. 261.

93 Respectively, these regressions mean the probable height of children for parents of a given stature, and the probable stature of parents of children of a given height. The asymmetry of the two regressions ($\frac{2}{3}$ and $\frac{1}{3}$) comes from the fact that the mid-parent value is being used. This is a statistical invention that implicitly includes the variance of the two parents. If only one parent's

values were used, both regressions would be $\frac{1}{3}$. This artefact, of which Galton was not aware in 1885, comes from the fact that the distribution of the mid-parents is more contracted than that of the individual parents (male and/or female).

94 F. Galton, 'Regression towards mediocrity…' (1885), p. 261. For the sake of clarity 'a' has been replaced by 'D'.

95 *Ibid.*

96 This methodological error was pointed out by Pearson. See K. Pearson, *The Life, Letters and Labours of Francis Galton,* vol. III A (1930). For a critical review, see R. G. Swinburne, 'Galton's law: Formulation and development' (1963).

97 See the citation given at note 15 above.

98 B. J. Norton, *Theory of Evolution of the Biometric School* (1971); W. B. Provine, *The Origins of Theoretical Population Genetics* (1971). Provine is particularly convincing on this question.

99 F. Galton, 'The average contribution of each several ancestor to the total heritage of the offspring' (1897), p. 402.

100 *Ibid.*, p. 403.

101 *Ibid.*, p. 402.

102 K. Pearson, *The Life, Letters and Labours of Francis Galton,* vol. III A (1930), p. 60.

103 F. Galton, *Natural Inheritance* (1889), p. 136.

104 K. Pearson, 'Mathematical contributions to the theory of evolution. On the law of ancestral heredity' (1898); 'Mathematical contribution… On the law of reversion' (1900); 'The law of ancestral heredity' (1903); 'The theory of ancestral contributions in heredity' (1909); *The Life, Letters and Labours of Francis Galton,* vol. III A (1930). The definitive study of the question can be found in B. J. Norton, *Theories of Evolution of the Biometric School* (1971), chaps. 2 and 5.

105 K. Pearson, 'Mathematical contribution to the theory of evolution. On the law of ancestral heredity' (1898), pp. 411–412.

106 K. Pearson, 'The law of ancestral heredity' (1903), p. 226.

107 Galton's 1897 article does not give the formula, but merely the previous version (with D_1, D_2, D_3…). Nevertheless, it is indeed the former formula which is implied in the first verbal formulation in the article (see note 100).

108 F. Galton, 'A diagram of heredity' (1898).

109 *Ibid.*, p. 293, col. A.

110 *Horseman* (Chicago), 28 December 1897.

111 F. Galton, 'The average contribution…' (1897), p. 403.

112 F. Galton, *Hereditary Genius* (1869); 'On blood-relationship' (1872); 'A theory of heredity' (1879); 'The average contribution…' (1897).

113 F. Galton, 'The average contribution …' (1897), p. 401.

114 *Ibid.*

115 *Ibid.*, p. 403.

116 R. C. Olby, *Origins of Mendelism* (1985), p. 53.

117 F. Galton, 'Family likeness in stature' (1886), p. 63. In only one passage did Galton try to apply ancestral heredity to discontinuous hereditary variation – in the 1897 article where ancestral heredity was presented as being subject to

a 'general law'. However, in the case of 'exclusive' heredity (for example a pea being smooth or wrinkled) there is no real regression of the character towards the mean. The law of ancestral heredity is shown by maintenance of the variability from one generation to the next. It is remarkable that it was this kind of character that gave Galton his only seriously empirical confirmation of his law of ancestral heredity. In his 1897 article, the 'verification' of the law is carried out on the basis of data on the coat colour of basset hounds ('tricolour' or 'non-tricolour'). Galton measured the *number* of dogs in each category in successive generations. The *proportion* (of relative frequencies) appears in the formula of the law of ancestral heredity ('The average contribution...', 1897, pp. 403–410).

118 See for example the attitude of S. Wright in *Evolution and the Genetics of Populations*, vol. I, *Genetics and Biometric Foundation* (1968). The same could be said of Georges Teissier in France.

5 Post-Darwinian views of selection and regression

1 See the impressive collection of articles by P. A. Taguieff on this theme: *Les pensées modernes de la décadence* (1997).

2 To the extent that eugenics was explicitly developed on the basis that natural selection no longer leads to the improvement of man, it merits particular attention in this respect. See J. Gayon, 'Comment le problème de l'eugénisme se pose-t-il aujourd'hui?' (1992); 'Entre eugénisme et théorie mathématique de l'évolution' (1992); and 'L'eugénisme' (1997); and J. P. Thomas, *L'eugénisme* (1995).

3 P. J. Bowler, *The Eclipse of Darwinism* (1983). This is the reference work on this topic.

4 C. Darwin, *Variation*, end of chap. 21 and 28 (1972, vol. VIII, pp. 236 and 426). On this expression, see J. Gayon, 'The paramount power of selection: From Darwin to Kauffman' (1997).

5 A. Weismann, 'On heredity' [1st German edn 1883], in A. Weismann, *Essays upon Heredity and Kindred Biological Problems* (1889).

6 A. Weismann, 'The all-sufficiency of natural selection: A reply to Herbert Spencer' (1893).

7 G. Romanes, *An Examination of Weismannism* (1893). See also his *Life and Letters* (1896); *Darwin and After Darwin*, 3 vols. (1892–7).

8 A. Weismann, 'La régression dans la nature' (1886), in *Essais sur l'hérédité et la sélection naturelle* (1892), pp. 399–400. The German original reads: 'Man kann den Vorgang, der die Rückbildung eines über flüssigen Organs zu Stande bringt, vielleicht ganz passend mit dem griechischen Worte "Panmixie" oder "Allgemein-Kreuzung" bezeichnen, weil sein Wesen eben darin besteht, dass nicht nur diejenigen Individuen zur Fortpflanzung gelangen, welche das betreffende Organ in grösster Volkommenkeit besitzen, sondern celle, ganz unabhängig davon, ob dasselbe besser oder schechter bei ihnen beschaffern ist' (A. Weismann, 'Ueber den Rückschritt in der Natur', 1887, in *Aufsätze über Vererbung und verwandte biologische Fragen* (1892), pp. 547–548). This essay is not included in the 1889 English volume of Weismann's writings.

9 A. Weismann, 'On heredity' [1st German edn 1883], in A. Weismann, *Essays upon Heredity and Kindred Biological Problems* (1889), p. 90.
10 See for example W. Bateson, *Materials for the Study of Variation* (1894), p. 573; K. Pearson, 'Regression, heredity and panmixia' (1896), pp. 314–318; L. Cuénot, *La genèse des espèces animales* (1921), pp. 360–363.
11 C. Darwin, *On the Origin of Species* (1859), p. 152.
12 'There may be truly said to be a constant struggle going on between, on the one hand, the tendency to reversion to a less modified state, as well as an innate tendency to further variability of all kinds and, on the other hand, the power of steady selection to keep the breed true' (*ibid.*, pp. 152–153).
13 'If a considerable number of improved cattle, sheep or other animals of the same race, were allowed to breed freely together, with no selection, but with no change of their conditions of life, there can be no doubt that after a score or [*sic*] hundred they would be very far from excellent of their kinds' (*Variation,* 2nd edn [1875], quoted in Darwin (1972), vol. VIII, p. 225).
14 'Highly-bred animals when neglected soon degenerate; but we have no reason to believe that the effects of long-continued selection would, if the conditions of life remained the same, be soon and completely lost' (*ibid.*, p. 235).
15 C. Darwin, *On the Origin of Species* (1859), pp. 151–153. *Variation,* 2nd edn [1875], quoted in Darwin (1972), vol. VIII, pp. 225–226 .
16 C. Darwin, *On the Origin of Species* (1859), p. 134.
17 *Ibid.*, pp. 134–135.
18 *Ibid.*, pp. 135–136.
19 It is interesting to note the change in the title of the paragraph in the *Origin* (chap. 5). In the first edition, the title is simply 'Use and Disuse'. In the fourth edition (1866), this becomes: 'Effects of Use and Disuse, as Controlled by Natural Selection'.
20 In a note, Weismann attributes the idea of the 'power of conservation' of natural selection to Seidlitz, *Die Darwin'sche Theorie* (Leipzig, 1875) (A. Weismann, 'On heredity [1st German edn 1883], in A. Weismann, *Essays* . . ., 1889, p. 86 n. 1). This indicates that there was a debate on this question.
21 A. Weismann, 'On heredity' [1st German edn 1883], in A. Weismann, *Essays* . . . (1889), p. 83.
22 H. Spencer, 'The inadequacy of natural selection' (1893).
23 A. Weismann, 'The duration of life' (1st German edn 1881), in A. Weismann, *Essays*. . . (1889).
24 F. Galton, 'Discontinuity in evolution' (1894), p. 364.
25 F. Galton, 'Typical laws of heredity' (1877), p. 512, col. B.
26 *Ibid.*
27 *Ibid.*, p. 514.
28 *Ibid.*
29 'Natural selection is measured by the percentage of survival among individuals born with like characteristics' (*ibid.*, p. 514, col. A). Galton extends this vocabulary to 'productiveness' which 'is measured by the average number of children from all parents who have like characteristics, but it may physiologically be looked upon as the percentage of survival of a vast and unknown number of possible embryos, producible by such parents' (*ibid.*). The distinction between 'natural selection' *sensu stricto* and 'productiveness' corresponds

to what would today be termed the two elements of natural selection: viability and fertility.

30 *Ibid.*, p. 532, col. B.
31 See note 29 above.
32 F. Galton, 'Typical laws of heredity' (1877), p. 532, col. B.
33 'I have confined myself in this explanation to purely typical cases, but it is easy to understand how the actions of the processes would be modified in those that were not typical' (*ibid.*, p. 532, cols. A, B).
34 *Ibid.*, col. B.
35 *Ibid.*, pp. 532–533.
36 In fact, Galton deals first with the case of 'productivity', then with 'natural selection'. The symbols in the equations have been slightly simplified.
37 At the end of 'Typical laws of heredity', Galton makes a discreet (and equivocal) allusion to this problem: 'Reversion might not be directed towards the mean of the race, neither productiveness nor survival might be greatest in the medium classes, and none of their laws may be strictly of the typical character. However, in all cases the general principles would be the same. Again, the same actions that restrain variability would restrain the departure of average values beyond certain limits' (p. 532, col. B).
38 Galton returned to this theme on a number of occasions. The most significant passages are: *Natural Inheritance* (1889), chap. 7, pp. 119–124, and chap. 12, p. 192. In the second of these (this is in fact the conclusion to the whole book), Galton states: 'The investigation now concluded is based on the fact that the characteristics of any population which is in harmony with its environment, may remain statistically identical during successive generations. This is true for every characteristic whether it be affected to a great degree by natural selection, or only so slight as to be practically independent of it.'
39 F. Galton, 'Typical laws of heredity' (1877), p. 514, col. A.
40 Sexual selection is explicitly excluded in both the mathematical model and the mechanical model of the quincunx. *Ibid.*, p. 514, col. A; p. 532, col. B; p. 533, col. B.
41 F. Galton, *Hereditary Genius,* 2nd edn (1892), Preface.
42 F. Galton, 'Discontinuity in evolution' (1894), p. 362.
43 *Ibid.*
44 F. Galton, *Hereditary Genius,* 2nd edn (1892), Preface; *Finger Prints* (1892), p. 20; 'Discontinuity in evolution' (1894), pp. 366, 368.
45 F. Galton, 'Rate of racial change that accompanies different degrees of severity in selection' (1897), p. 605.
46 The term is Pearson's, in a direct allusion to Galton: 'Regression, heredity and panmixia' (1896), pp. 306–307. The focus of regression is understood as 'the mean of the general population from which selection has originally taken place'. For Pearson, this was clearly a pseudo-concept.
47 *Ibid.*, pp. 314 ff.
48 K. Pearson, 'Mathematical contributions to the theory of evolution. On the law of ancestral heredity' (1898).
49 This calculation can be found in *ibid.*, pp. 400–401.
50 *Ibid.*, p. 401.
51 *Ibid.*, p. 402. See also 'Regression, heredity and panmixia' (1896), pp. 306–318.

52 F. Galton, 'Discontinuity in evolution' (1894), p. 368.
53 *Ibid.*, p. 363.
54 *Ibid.*, p. 365. See also *Natural Inheritance* (1889), p. 30; *Finger Prints* (1892), p. 20.
55 F. Galton, 'Discontinuity in evolution' (1894), p. 369.
56 F. Galton, *Natural Inheritance* (1889), p. 198.
57 F. Galton, 'Discontinuity in evolution' (1894), pp. 367–368.
58 *Ibid.*, p. 367.
59 *Ibid.*, p. 366.
60 *Ibid.*, p. 368.
61 F. Galton, 'Typical laws of heredity' (1877), p. 514, col. A. This is repeated in the mathematical appendix, p. 532 B.
62 C. Darwin, *The Descent of Man and Sexual Selection* (1871), II, chap. 8, p. 256. See also *On the Origin of Species* (1859), pp. 87–90.
63 K. Pearson clarified the difference between sexual selection *sensu stricto* (preferential mating) and homogamy (assortative mating). See 'Regression, heredity and panmixia' (1896), pp. 257–258.
64 *Ibid.*, pp. 311–312.
65 F. Galton, 'Hereditary talent and character' (1865), p. 319, col. A.
66 F. Galton, 'The possible improvement of the human breed under the existing conditions of law and sentiment' (1901).
67 F. Galton, 'Segregation (of the feeble-minded)' (1909).
68 F. Galton, *Natural Inheritance* (1889), chap. 3, § 'Evolution not by minute steps only', pp. 32–34.
69 F. Galton, *Natural Inheritance* (1889), p. 22. Chap. 3 of this book is entitled 'Organic Stability'.
70 F. Galton, *Natural Inheritance* (1889), pp. 123–124.
71 G. Cuvier, *Recherches sur les ossements fossiles des quadrupèdes (Discours préliminaire)*; introduction reprinted several times under the title *Discours sur les révolutions du globe*. See §§ 'Les os fossiles des quadrupèdes sont difficiles à déterminer' and 'Principe de cette détermination' (Paris, 1881, pp. 62–72).
72 'Correlation of Growth. I mean by this expression that the whole organisation is so tied together during its growth and development, that when slight variations in any one part occur, and are accumulated through natural selection, other parts become modified' (C. Darwin, *On the Origin of Species*, 1859, p. 143). In the fifth edition (1869), 'Correlation of Growth' becomes 'Correlated Variation'.
73 C. Darwin, *On the Origin of Species* (1859), p. 143.
74 F. Galton, *Natural Inheritance* (1889), p. 123.
75 *Ibid.*, p. 192.
76 F. Galton, *Hereditary Genius*, 2nd edn (1892), pp. 421–422.
77 W. Bateson, *Materials for the Study of Variation, Treated with Special Regard to Discontinuity in the Origin of Species* (1894); H. de Vries, 'Ueber halbe Galton-Curven ab Ziechen diskontinuierlicher Variation' (1894).
78 F. Galton, *Natural Inheritance* (1889), chaps 2 and 3.
79 'There are probably no heritages that perfectly blend or that absolutely exclude one another' (*ibid.*, p. 12). As for 'sports', 'they are often transmitted to successive generations with curious persistence' (*ibid.*, p. 30). The two

distinctions (individual variations/sports and blending/exclusive inheritance) thus do not completely coincide.

80 From the Latin *transilio* (jump). The OED gives a usage in geology as early as 1811. The noun 'transilience' was in use as early as the 17th century. 'Divergent' alludes to Darwin (*Origin of Species*, chap. 4).

81 F. Galton, 'Discontinuity in evolution' (1894), p. 368.

82 F. Galton, *Finger Prints* (1892), p. 20.

83 '. . . a race does sometimes abruptly produce individuals who have a distinctly different typical centre' (F. Galton, 'Discontinuity in evolution', 1894, p. 368).

84 W. Bateson, *Materials for the Study of Evolution, Treated with Especial Regard to the Discontinuity in the Origin of Species* (1894).

85 H. de Vries, 'Ueber halbe Galton-Curven ab Zeichen diskontinuierlicher Variation' (1894). On the two meanings of 'discontinuous variation' in the 1890s, see H. de Vries, *The Mutation Theory* (1909), vol. I, p. 51 n. 3.

86 W. Bateson, *Materials for the Study of Evolution, Treated with Especial Regard to the Discontinuity in the Origin of Species* (1894), p. 573.

87 F. Galton, *Natural Inheritance* (1894), p. 192.

88 In his autobiography, *Memories of My Life* (1909), Galton recounts how he tried to measure the geographic distribution of female beauty: 'Whenever I have occasion to classify the persons I meet into three classes, "good, medium, bad", I use a needle mounted as a pricker, wherewith to prick holes, unseen, in a piece of paper, torn rudely into a cross with a long leg. I use its upper end for "good", the cross arm for "medium", the lower end for "bad". The prick holes keep distinct, and are easily read off at leisure. The object, place, and date are written on the paper. I used this plan for my beauty data, classifying the girls I passed in the street or elsewhere as attractive, indifferent or repellent. Of course this was a purely individual estimate, but it was consistent, judging from the conformity of different attempts in the same population. I found London to rank highest for beauty; Aberdeen lowest' (quoted in S. J. Gould, *The Flamingo's Smile*, 1985, p. 302).

6 The strategy of indirect corroboration: the case of mimicry

1 C. Darwin, *On the Origin of Species* (1859), p. 459.

2 See section 1.1 above.

3 C. Darwin, *Variation*, Introduction, quoted in C. Darwin (1972), vol. VII, p. 9.

4 C. Darwin, *On the Origin of Species* (1859), p. 82. This presentation owes a great deal to that of Julian Huxley, *Evolution: the Modern Synthesis* (1942), chap. 1, §1.

5 C. Darwin, *Variation*, Introduction, quoted in C. Darwin (1972), vol. VII, p. 9.

6 *Ibid.*

7 *Ibid.*

8 The last chapter of the *Origin* and the chapter on pangenesis in the *Variation* show this clearly. Both begin by listing the facts against the theory, generally only in order to turn them into arguments in its favour. This is a constant theme in Darwin's work.

9 A. R. Wallace, 'Mimicry and other protective resemblances among animals' (1867), pp. 1–2. Note that these sentences open Wallace's review of four books on 'mimicry', the fourth being (quite surprisingly; see below), the 4th edition of Darwin's *Origin of Species*.

10 See L. Laudan, *Progress and Its Problems* (1977).

11 For a recent, and controversial, discussion of these problems, see S. J. Gould and R. Lewontin, 'The spandrels of San Marco and the Panglossian paradigm' (1979). However, this kind of objection is nearly as old as Darwinism itself. See A. R. Wallace, 'The problem of utility: Are specific characters always or generally useful?' (1896).

12 'Phenotype' was introduced by Johannsen (1905). The word means 'apparent type' (*Erscheinungstypus*), as opposed to the 'true type' (from the point of view of heredity – 'genotype').

13 G. Teissier underlined this change of meaning in 'Transformisme d'aujourd'hui' (1962), p. 367.

14 C. Darwin, *On the Origin of Species* (1859), p. 236.

15 H. Bates, 'Contributions to an insect fauna of the Amazon valley. *Lepidoptera: Heliconidae*'. The article was read to the Linnean Society of London on 21 November 1861 and published the following year.

16 For more details, see E. B. Poulton, 'Natural selection the cause of mimetic resemblance and common warning colours', in *Essays on Evolution 1889–1907* (1908), §1, pp. 220–270. This paper, first read before the Linnean Society of London in 1898, contains a useful 'Historical Introduction'.

17 This is clear from Darwin's review of Bates' monograph, which shows that Darwin had become aware of a literature and a set of data that he was not previously aware of. See C. Darwin, 'A review of "Contributions to an insect fauna of the Amazon valley" by Henry Bates, Esq.' (1863).

18 A. Murray, 'On the disguises of nature: Being an inquiry into the laws which regulate external form and colour in plants and animals' (1860). (See E. B. Poulton, 'Natural selection the cause of mimetic resemblance and common warning colours', in *Essays on Evolution 1889–1907*, 1908.)

19 H. B. Tristram, 'On the ornithology of Northern Africa' (1859), pp. 429–430. For the most part, this monograph is a description of the bird species of North Africa.

20 *Ibid.*, pp. 431–432.

21 Later on in the same article, Tristram puts forward a concrete example: 'in course of time the longer-billed variety [of Lanko] would steadily predominate over the shorter, and in a few centuries they would be the sole existing ones, their shorter-billed fellows dying out until that race was extinct' (*ibid.*, p. 431).

22 H. W. Bates, 'Contributions to an insect fauna of the Amazon valley. *Lepidoptera: Heliconidae*' (1862), pp. 495–566. The paper was read in 1861.

23 C. Darwin, 'A review of "Contributions to an insect fauna of the Amazon valley" by Henry Bates, Esq.' (1863).

24 'I think it clear that the mutual resemblance in this and other cases cannot be due entirely to similarity of habits and the coincident adaptation of the two analogues to similar physical conditions' ('Contributions to an insect fauna of the Amazon valley. *Lepidoptera: Heliconidae*', 1862, p. 508).

25 *Ibid.*
26 *Ibid.*, p. 512.
27 *Ibid.*
28 *Ibid.*, p. 513.
29 C. Darwin, 'A review of "Contributions to an insect fauna of the Amazon valley" by Henry Bates, Esq.' (1863), in P. H. Barrett, *The Collected Papers of Charles Darwin* (1977), vol. II, p. 89.
30 C. Darwin, *On the Origin of Species*, 4th edn (1866), chap. 13; quoted in M. Peckham, *The Origin of Species by Charles Darwin. A Variorum Text* (1959), p. 668.
31 See note 28.
32 A. R. Wallace, 'Mimicry and protective resemblances among animals' (1867), pp. 36–39.
33 C. Darwin, *The Descent of Man and Sexual Selection* (1871), I, chap. 11.
34 C. Darwin, *On the Origin of Species* (1859), p. 88.
35 C. Darwin, *The Descent of Man and Sexual Selection* (1871), I, chap. 11, p. 421.
36 *Ibid.*, p. 296.
37 *Ibid.*
38 Letter from Darwin to Wallace, 15 April 1868. Quoted in F. Darwin and C. Seward, *More Letters of Charles Darwin*, vol. II (1903), pp. 73–74.
39 *Ibid.*, p. 74.
40 *Ibid.*, pp. 56–97.
41 *Ibid.*, p. 74.
42 C. Darwin, *The Descent of Man*, 2nd edn (1874), chap. 9, end of paragraph 'Imitation'.
43 A. R. Wallace, 'Mimicry and protective resemblances among animals' (1867).
44 *Ibid.*, p. 39.
45 See for example the remarks of E. B. Ford on non-mimetic polymorphism in butterflies (*Ecological Genetics,* 1964).
46 For a recent synthesis of the problem, see the remarkable article by J. R. G. Turner, 'Why male butterflies are non-mimetic: Natural selection, sexual selection, group selection, modification and sieving' (1978).
47 W. Bates, 'Contributions to an insect fauna of the Amazon valley. *Lepidoptera: Heliconidae*' (1862), p. 507.
48 F. Müller, '*Ituna* and *Thyridia*; a remarkable case of mimicry in butterflies' (1879). This text is the English translation by R. Meldola of an article that originally appeared in the German journal *Kosmos*.
49 *Ibid.*, pp. xx–xxiii.
50 *Ibid.*, pp. xxiii–xxv.
51 *Ibid.*, p. xxvii.
52 *Ibid.*
53 *Ibid.*, p. xxviii.
54 F. Müller was the author of *Für Darwin* (1864).
55 See R. Meldola, comments added to his translation of '*Ituna* and *Thyridia*...' (1879), p. xxix, and F. B. Poulton, *Essays on Evolution* (1908), pp. 213 and 223.
56 R. Meldola, 'The president's address' (1896).
57 See for example F. Le Dantec, 'Mimétisme et imitation' (1898). Le Dantec

sought to reestablish the literal significance of mimicry as an effort of imitation.

58 For more details, see W. C. Kimler, 'Mimicry: Views of naturalists and ecologists before the modern synthesis' (1983).

7 The search for direct proof: biometry

1 On the history of English biometry, see P. Froggatt and N. C. Nevin, 'The "law of ancestral heredity" in the Mendelian–ancestrian controversy' (1971); W. B. Provine, *The Origins of Theoretical Population Genetics* (1971); B. J. Norton, 'The biometric defense of Darwinism' (1973); J. Gayon, 'Biométrie' (1996).

2 C. Darwin, *On the Origin of Species* (1959), chap. 13, paragraph entitled 'Embryology' (in the 6th edn this is chap. 14). For a discussion of this aspect of Darwin's thought, see J. Gayon, 'Le concept de récapitulation à l'épreuve de la théorie darwinienne de l'évolution' (1991).

3 For a discussion of this important aspect of Weldon's biography, see W. B. Provine, *The Origins of Theoretical Population Genetics* (1971), p. 29.

4 K. Pearson, 'Walter Frank Raphael Weldon, 1860–1906' (1906).

5 C. Darwin, *On the Origin of Species* (1859), p. 143.

6 K. Pearson, 'Walter Frank Raphael Weldon, 1860–1906' (1906).

7 W. F. R. Weldon, 'Remarks on variation in animals and plants' (1895), p. 380.

8 F. Galton, 'Co-relations and their measurement, chiefly from anthropological data' (1888).

9 K. Pearson, *The Chances of Death and Other Studies in Evolution* (1897), p. 63.

10 W. F. R. Weldon, 'On certain correlated variations in *Carcinus moenas*' (1893), p. 329.

11 See W. F. R. Weldon, 'Report of the Committee . . .' (1895), p. 360.

12 For a detailed history of the controversy, see P. Froggatt and N. C. Nevin, 'The "law of ancestral heredity" and the Mendelian–ancestrian controversy in England, 1889–1906' (1971), and W. B. Provine, *The Origins of Theoretical Population Genetics* (1971), chaps. 2 and 3.

13 W. F. R. Weldon, 'The variations occurring in certain decapod Crustacea – I. *Crangon vulgaris*' (1890).

14 Weldon gives the following note: '*Natural Inheritance*, pp. 119–124'.

15 Weldon, 'The variations occurring . . .' (1890), p. 446.

16 *Ibid.*, p. 451.

17 W. F. R. Weldon, 'Certain correlated variations in *Crangon vulgaris*' (1892).

18 C. Darwin, *On the Origin of Species* (1859), p. 143.

19 K. Pearson contested Weldon's interpretation. His calculations suggested that the observed differences for *r* were all significant. See 'Heredity, regression and panmixia' (1896), p. 267.

20 W. F. R. Weldon, 'Certain correlated variations in *Crangon vulgaris*' (1892), p. 11.

21 W. F. R. Weldon, 'On certain correlated variations in *Carcinus moenas*' (1893).

22 *Ibid.*, p. 324.

23 K. Pearson, 'Contributions to the mathematical theory of evolution' (1894). This monograph appeared in *Philosophical Transactions of the Royal Society*. A year earlier, an abstract had appeared in *Proceedings of the Royal Society* as an appendix to Weldon's article on *Carcinus moenas*.

24 This is Pearson's term. Today we would say 'non-normal'.

25 See S. Stigler, *The History of Statistics* (1986), pp. 222–225 and 229–233.

26 The general problem of skewed distribution curves and homogeneous material is dealt with in a later text, with particular attention to infantile mortality: K. Pearson, 'Mathematical contributions to the theory of evolution – II. Skew variation in homogeneous material' (1895 and 1896). See also *The Chances of Death* (1897), chap. 1.

27 K. Pearson, 'Contributions to the mathematical theory of evolution (Abstract)' (1893), p. 332.

28 W. F. R. Weldon, 'On certain correlated variations in *Carcinus moenas*' (1893), p. 324.

29 K. Pearson, 'Contributions to the mathematical theory of evolution – II. Skew variation in homogeneous material' (1895).

30 Pearson recounts the episode in 'Walter Frank Raphael Weldon (1860–1906)' (1906).

31 W. F. R. Weldon, 'Report of the Committee . . . Part I. An attempt to measure the death-rate due to the selective destruction of *Carcinus moenas* with respect to a particular dimension' (1895).

32 K. Pearson, 'Walter Frank Raphael Weldon (1860–1906)' (1906), p. 25.

33 W. F. R. Weldon, 'Remarks on variation in animals and plants. To accompany the first report of the Committee for Conducting Statistical Inquiries into the Measurable Characteristics of Plants and Animals' (1895), p. 380.

34 *Ibid.*

35 *Ibid.*, pp. 360–367.

36 *Ibid.*, p. 367.

37 F. Galton, 'Typical laws of heredity' (1877). See section 5.2.1 above.

38 W. F. R. Weldon, 'Report of the Committee . . .' (1895), p. 371.

39 *Ibid.*, p. 379; see also 'Remarks on variation in animals and plants' (1895), p. 381.

40 W. F. R. Weldon, 'Report of the Committee . . .' (1895), p. 379).

41 A. R. Wallace, *Darwinism: An Exposition of the Theory of Natural Selection with Some of Its Applications* (1889); 'The problem of utility: Are specific characters always or generally useful?' (1896). This 'utilitarian' radicalisation of Darwinism had existed since the 1860s, as shown by this declaration by Wallace from 1867: 'Perhaps no principle has ever been announced so fertile in results as that which Mr Darwin so earnestly impresses upon us, and which is indeed a necessary deduction from the theory of Natural Selection, namely – that none of the definite facts of organic nature, no special organ, no characteristic form or marking, no peculiarities of instinct or of habit, no relations between species or between groups of species – can exist, but which must now be or once have been useful to the individuals or the races which possess them' ('Mimicry and other protective resemblances among animals', 1867, pp. 2–3).

42 W. F. R. Weldon, 'Remarks on variation in animals and plants' (1895), p. 381.
43 W. Bateson, *Materials for the Study of Variation, Treated with Especial Regard to Discontinuity in the Origin of Species* (1894).
44 E. Ray Lankester, letter to *Nature*, 16 July 1896 (vol. 54), p. 245; Weldon's reply can be found in *Nature*, 30 July 1896 (vol. 54), pp. 294–295.
45 At the end of his life Pearson explained frankly how he arrived at this idea, which has an important place in *The Grammar of Science* (1892). Like many other scientists, Pearson was inspired by Galton's *Natural Inheritance*. In Pearson's case, he realised that statistics constituted a method that would permit the application of mathematics to a wide range of problems, to the extent that causation was merely an extreme example of the category of correlation (K. Pearson, in *Speeches Delivered... in Honour of Professor Karl Pearson, 23 April 1934)*. For a useful discussion of this position, see M. Veuille, 'Corrélation, le concept pirate' (1987).
46 W. F. R. Weldon, 'Remarks on variation in animals and plants' (1895), p. 381.
47 W. F. R. Weldon, 'Report of the Committee... (1895), p. 371.
48 H. C. Bumpus, 'The elimination of the unfit as illustrated by the introduced sparrow, *Passer domesticus* (A fourth contribution to the study of variation)' (1898).
49 *Ibid.*, p. 209.
50 *Ibid.*
51 *Ibid.*, p. 217.
52 *Ibid.*, p. 214. The first sentence alludes to H. C. Bumpus, 'The variations and mutations of the introduced sparrow, *Passer domesticus*' (1896).
53 H. C. Bumpus, 'The elimination of the unfit as illustrated by the introduced sparrow, *Passer domesticus* (A fourth contribution to the study of variation)' (1898), p. 219.
54 *Ibid.*, pp. 218–219.
55 *Ibid.*
56 K. Pearson, 'Mathematical contributions to the theory of evolution. III. Regression, heredity and panmixia' (1896), p. 257.
57 W. F. R. Weldon, presidential address to the Zoological section of the British Association (1898).
58 W. F. R. Weldon, 'Report of the Committee...' (1895), p. 371.
59 W. F. R. Weldon, Presidential address... (1898), p. 900.
60 *Ibid.*, p. 901.
61 W. F. R. Weldon, 'A first study of natural selection in *Clausilia laminata* (Montagu)' (1901).
62 K. Pearson, 'Walter Frank Raphael Weldon 1860–1906' (1906).
63 See W. F. R. Weldon, 'On certain correlated variations in *Carcinus moenas*' (1893), p. 329; 'Report of the Committee...' (1895), p. 379; 'Remarks...' (1895), p. 381; 'Presidential address... (1898), p. 902.
64 The primary source on Pearson's biography and works is a book by his son, E. S. Pearson, *Karl Pearson: An Appreciation of Some Aspects of His Life and Work* (1936–1938). The spirit of Pearson's project has been well summarised by B. J. Norton in *Karl Pearson and the Galtonian Tradition: Studies in the Rise*

of *Quantitative Social Biology* (1978). See also D. A. MacKenzie, *Statistics in Britain 1865–1930* (1981), chaps. 4 and 7. An interesting judgement can be found in J. B. S. Haldane, 'Karl Pearson, 1857–1957' (1957).

65 On Pearson's contributions to the analysis of skewed curves, see S. Stigler, *The History of Statistics* (1986), pp. 329–338.

66 Weldon to Galton, 27 November 1892: 'either Naples is the meeting point of two distinct races of crabs, or a "sport" is in process of establishment. You have so often spoken of this kind of curve as certain to occur that I am glad to send you the first case which I have found' (cited in P. Froggatt and N. C. Nevin, 'The "law of ancestral heredity"...', 1971, p. 4).

67 K. Pearson, 'Contributions to the mathematical theory of evolution' (1894).

68 K. Pearson, 'Mathematical contributions to the theory of evolution. II. Skew variation in homogeneous material' (1895).

69 K. Pearson, 'Mathematical contributions to the theory of evolution. III. Regression, heredity and panmixia' (1896), pp. 268–318; 'Mathematical contributions to the theory of evolution. On the law of ancestral heredity' (1898).

70 K. Pearson, 'Mathematical contributions to the theory of evolution. III. Regression, heredity and panmixia' (1896), pp. 308 and 314.

71 *Ibid.*, p. 317.

72 K. Pearson, 'Further remarks on the law of ancestral heredity' (1911), p. 243.

73 K. Pearson, 'Mathematical contributions to the theory of evolution. XI. On the influence of natural selection on the variability and correlations of organs (Abstract)' (1902), p. 331. The theory is in fact set out in 'Mathematical contributions to the theory of evolution. IV. On the influence of random selection on variation and correlation' (1898). See also: 'On the criterion that a given system of deviations from the probable in the case of a correlated system of variables is such that it can be reasonably supposed to have arisen from random sampling' (1900).

74 'I have already shown in an earlier memoir of this series the effect of random selection, or what is better to term random sampling, on the characters of a population' (K. Pearson, 'Mathematical contributions to the theory of evolution. XI. On the influence of natural selection on the variability and correlations of organs (Abstract)', 1902, p. 331).

75 K. Pearson, 'Regression, heredity and panmixia' (1896), 'Definitions', pp. 254–261.

76 F. Galton, 'Typical laws of heredity' (1877).

77 C. Darwin, *On the Origin of Species*, 6th edn (1872), chap. 4; quoted in M. Peckham, *The Origin of Species by Charles Darwin. A Variorum Text* (1959), pp. 173–174.

78 K. Pearson, 'Regression, heredity and panmixia' (1896), pp. 257–258.

79 C. Darwin, *On the Origin of Species*, 3rd edn (1861); quoted in Peckham (1959), pp. 164–165.

80 F. Galton, 'Co-relations and their measurement, chiefly from anthropological data' (1888).

81 A. Bravais, 'Analyse mathématique sur les probabilités des erreurs de situation d'un point' (1846).

82 This possibility had in fact already been explored by F. Y. Edgeworth in 'The law of error and correlated averages' (1892) and 'A new method of treating correlated averages' (1893).

83 Pearson uses both expressions. See 'Regression, heredity and panmixia' (1896), §10a and 'Fundamental theorem in selection', pp. 254 and 298.

84 *Ibid.*, §10a, p. 298.

85 This explains Pearson's obscure phrase 'the *p* organs selected may be specially correlated together in selection, in a manner totally different from their "natural" correlation or correlation of birth'. This in fact means: in selection, 'correlation' refers to something *wholly* different from anatomical 'correlation'; that is, not correlation between parts of a given organism, but correlation between similar parts belonging to different generations. Both correlations can be statistically analysed, but they do not refer to the same level of biological phenomena. Correlation 'in selection' refers to hereditary or transgenerational correlation.

86 K. Pearson, 'Mathematical contributions to the theory of evolution. On the law of ancestral heredity' (1898).

87 See the remark by Pearson in 'The law of ancestral heredity' (1903), Appendix I, p. 228: 'I believe that my memoir of 1896 was the first in which the equations of multiple regression were worked out and applied to the problem of heredity. Such equations are, I presume, what Mr Yule refers to and says may be termed the law of ancestral heredity.'

88 K. Pearson, 'Regression, heredity and panmixia' (1896), §10d, pp. 306–318.

89 'The two hypotheses with which we have dealt give practically the two extremes; observation and experiment are perfectly able to determine between them, or to settle whether an intermediate theory is necessary which will give a progression, but a slower progression, to the focus of regression . . . What is wanted is a wide extension of the experimental and statistical work of Mr FRANCIS GALTON and Professor WELDON' (*ibid.*, p. 318).

90 K. Pearson, *The Grammar of Science*, 2nd edn (1900), chap. 7, §6, p. 205. Similar formulations can be found throughout the book: Preface, chap. 1, §§2, 4, 5, 11, 13, 14; chap. 2, §13; chap. 3, §§1n, 2, 10, 15; chap. 5, §§5, 7; chap. 6, §5; chap. 9, §§3, 11; chap. 10, §2.

91 *Ibid.*, chap. 9, §11, p. 356.

92 For a more detailed account of Pearson's philosophy of science, see J. Gayon, 'Karl Pearson ou: les enjeux du phénoménalisme dans les sciences de la vie autour de 1900' (1997).

93 K. Pearson, *The Chances of Death and Other Studies in Evolution* (1897), p. 65. See also 'Regression, heredity and panmixia' (1896), p. 259.

94 J. B. S. Haldane, *The Causes of Evolution* (1932), 'Outline of the mathematical theory of natural selection', p. 171.

95 K. Pearson, 'Mathematical contributions to the theory of evolution. On the law of ancestral heredity' (1898); *The Grammar of Science*, 2nd edn (1900), chap. 12, pp. 177–178.

96 This ambiguity appears clearly in an appendix to Pearson's 1903 article on the law of ancestral heredity. Replying to his colleague Yule, who criticised him for having introduced *post hoc* sexual selection, natural selection, reproductive selection and the effect of circumstances into his 'constants of

heredity', Pearson stated that, from his first studies of heredity and selection onward, he had continually dissociated these factors of evolution, precisely in order to measure their effect on the 'constants of heredity'! In other words, if Yule was wrong, it was because he did not realise that this was a deliberate procedure and not an afterthought (Cf. G. U. Yule, 'Mendel's laws and their probable relations to intra-racial heredity', 1902).

97 K. Pearson, 'Mathematical contributions to the theory of evolution. On the law of ancestral heredity' (1898), p. 412.

98 *Ibid.*, p. 411.

99 R. A. Fisher, *The Genetical Theory of Natural Selection* (1930), chap. 2.

100 C. Darwin, *On the Origin of Species* (1859), p. 194.

101 C. Darwin, *On the Origin of Species*, 5th edn (1869), chap. 6; quoted in M. Peckham, *The Origin of Species by Charles Darwin. A Variorum Text* (1959), p. 362.

102 C. Darwin, *On the Origin of Species* (1859), p. 62.

103 *Ibid.*, pp. 80–81.

104 C. Darwin, *On the Origin of Species*, 6th edn (1872), chap. 4; quoted in M. Peckham, *The Origin of Species by Charles Darwin. A Variorum Text* (1959), pp. 227–228.

105 F. Galton, *Hereditary Genius* (1869), p. 330.

106 F. Galton, 'On the probability of the extinction of families' (1874), p. 138.

107 M. Greg, 'On the failure of natural selection in the case of Man', *Fraser's Magazine,* September 1868. Darwin's reproduction of this declaration in chap. 5 of his *Descent of Man* made it famous.

108 K. Pearson, *The Chances of Death and Other Studies in Evolution* (1897), p. 83.

109 F. Galton, 'The possible improvement of the human breed under the existing conditions of law and sentiment' (1901).

110 On this point see D. A. Mackenzie, *Statistics in Britain 1865–1930* (1981); and J. Gayon, 'L'intelligence naturelle: mode de fabrication et mode d'emploi' (1988).

111 C. Darwin, *The Descent of Man* (1871), chap. 5.

112 F. Galton, 'Typical laws of heredity' (1877).

113 *Ibid.*, pp. 532–533.

114 *Ibid.*, p. 514.

115 K. Pearson, 'Regression, heredity and panmixia' (1896), pp. 257–259.

116 K. Pearson, A. Lee and L. Bramley-Moore, 'Mathematical contributions to the theory of evolution ... VI. Genetic (reproductive) selection' (1899), part II: 'On the inheritance of fertility in mankind'.

117 *Ibid.*, part III: 'On the inheritance of fecundity in thoroughbred racehorses'.

118 *Ibid.*, p. 314.

119 K. Pearson, *The Grammar of Science*, 2nd edn (1900), pp. 443–445.

120 'Reproductive selection supposes the fertility curve and correlation surfaces to embrace only homogeneous material, and it can accordingly never give rise to a new species; it is purely a source of progressive change in the same species' ('Note on reproductive selection', 1896, p. 301, note). 'Correlation surfaces' is another allusion to the law of ancestral heredity. For Pearson, all characters, including fertility, conformed to this law.

121 On this terminological point, see *ibid.*; K. Pearson, A. Lee and L. Bramley-

Moore, 'Mathematical contributions to the theory of evolution. VI. Genetic (reproductive) selection' (1899), p. 257, note.

122 K. Pearson, *The Chances of Death and Other Studies in Evolution* (1897).

123 *Ibid.*, p. 65. This affirmation is repeated in 'Contributions to the mathematical theory of evolution. VI. Genetic (reproductive) selection' (1899), p. 258.

124 K. Pearson, 'Regression, heredity and panmixia' (1896), p. 259.

125 *Ibid.* See also *The Chances of Death and Other Studies in Evolution* (1897), p. 65.

126 K. Pearson, 'Mathematical contributions to the theory of evolution. VI. Reproductive or genetic selection (Abstract)' (1898), p. 163.

127 K. Pearson, *The Grammar of Science*, 2nd edn (1900), p. 440.

128 On the early history of the term 'fitness' in evolutionary biology, see J. Gayon, 'Sélection naturelle ou survie des plus aptes? Eléments pour une histoire du concept de *fitness* dans la théorie évolutionniste' (1995).

129 K. Pearson, *The Chances of Death and Other Studies in Evolution* (1897), pp. 91–95; *The Grammar of Science*, 2nd edn (1900), p. 446.

130 K. Pearson, *The Grammar of Science*, 2nd edn (1900), pp. 446 and 448.

131 K. Pearson, 'Data for the problem of evolution in man. V. On the correlation between duration of life and the number of offspring' (1900), p. 171.

132 For the use of this term, see *The Grammar of Science*, 2nd edn (1900), p. 441.

133 K. Pearson, 'Mathematical contributions to the theory of evolution. VI. Genetic (reproductive) selection' (1899), pp. 314–315.

134 K. Pearson, *The Grammar of Science*, 2nd edn (1900), pp. 467–468.

135 See K. Pearson, A. Lee and L. Bramley-Moore, 'Mathematical contributions to the theory of evolution. VI. Genetic (reproductive) selection: Inheritance of fertility in man, and of fecundity in thoroughbred racehorses' (1898), p. 257, note.

136 K. Pearson, *The Chances of Death* (1897), p. 95.

137 See the interesting remarks by R. Olby, 'The dimensions of scientific controversy: The biometric–Mendelian debate' (1988).

8 Establishing the possibility of natural selection: the confrontation of Darwinism and Mendelism

1 H. de Vries, *Die Mutationstheorie* (1901–3).

2 F. Jenkin, 'The Origin of Species' (1867), p. 281.

3 H. de Vries, *Intracelluläre Pangenesis* (1889). Reference will be made to the 1910 American translation, *Intracellular Pangenesis*.

4 *Ibid.* (1910), p. 5.

5 *Ibid.*, p. 6.

6 *Ibid.*, p. 65.

7 *Ibid.*, p. 11.

8 *Ibid.*, p. 13. This statement contains a position which was to be characteristic of Mendelian genetics. The very term 'gene', which Johannsen proposed in 1909 to denote Mendelian factors, was nothing more than an abbreviation of de Vries' 'pangen'. There is thus a conceptual and terminological thread which, starting with Darwin's 'pangenesis' (1868), goes through its non-Lamarckian reinterpretation (de Vries, 1889), and

then through the appearance of 'genetics' as a 'physiology of descent' founded on Mendelian methodology (Bateson, letter to A. Sedgwick, 18 April 1905, reproduced in Carlson, 1966, p. 16), and finally ends up with the notion of a 'gene' as a Mendelian factor (Johannsen, 1909). The link between de Vries' early works and those of Mendel should not be exaggerated. In 1889 de Vries did not put forward Mendel's fundamental hypothesis: the existence of hereditary determinants in several different forms ('allelomorphs', or 'alleles' as they would later be called).

9 *Ibid.*, pp. 73–74.

10 *Ibid.*, p. 71.

11 *Ibid.*, p. 74.

12 *Ibid.*, p. 71.

13 H. de Vries, 'Sur la loi de disjonction des hybrides' (1900).

14 The article first appeared in German ('Ueber halbe Galton-Curven', 1894), and then in French (1895).

15 H. de Vries, *The Mutation Theory*, vol. I (English translation, 1909), part I, pp. 71–129.

16 'The modern theory of selection rests on two unproved hypotheses: 1. The advance brought about by selection may increase for an indefinite period. 2. The result of selection can become independent of selection. The experience of breeders stands in direct contradiction to both these hypotheses' (*ibid.*, pp. 85–86).

17 W. B. Provine, *The Origins of Theoretical Population Genetics* (1973), pp. 66–68.

18 H. de Vries, *The Mutation Theory*, vol. I (English translation, 1909), 'Introduction', p. 4.

19 *Ibid.*, pp. 211–212.

20 'Under the general term variation, then, are included two distinct phenomena: Mutability and fluctuation or ordinary variation' (*ibid.*, p. 5).

21 *Ibid.*, p. 4.

22 'The object of the present book is to show that species arise by saltations and that the individual saltations are occurrences which can be observed like any other physiological process' (*ibid.*, 'Preface to the first volume', p. viii).

23 For today's geneticists, it goes without saying that de Vries' 'mutations' were not what Mendelian genetics would call 'mutations'. The *Oenothera* mutations were eventually found to be a particular case of what H. J. Muller in 1918 called 'balanced lethals'. This phenomenon is very different from the punctual alteration of a Mendelian gene. Cytogenetic analysis was to show that these plants showed exceptional translocations which, with other factors, explain the stability of the filial generations.

24 H. de Vries, *The Mutation Theory*, vol. I (English translation, 1909), p. 257.

25 See the Introduction to the present work.

26 W. L. Johannsen, *Ueber Erblichkeit in Populationen in reinen Linien: Ein Beitrag zur Beleuchtung schwebender Selektionsfragen* (1903). This is the German translation of the Danish original, which appeared in *Acts of the Royal Academy of Sciences of Denmark*. Partial English translation can be found in H. Gall and E. Putschar, *Selected Readings in Biology for Natural Sciences* (1955).

27 W. Bateson and E. R. Saunders, 'Experimental studies in the physiology of heredity' (1902), pp. 126–130.
28 W. L. Johannsen, *Ueber Erblichkeit in Populationen in reinen Linien* (1903), p. 66.
29 *Ibid.*, pp. 57–58.
30 *Ibid.*, p. 58.
31 In 1905, Bateson proposed the term 'genetics' to denote the science of variation and heredity, or the 'physiology of descent', in an exchange of letters which discussed the title of a chair that he might take at Cambridge (the title adopted was in fact that of Chair of Protozoology). The following year, at the 'Third Conference on Hybridisation and Plant Breeding', Bateson proposed that the study of heredity and variation under the Mendelian hypothesis be called 'genetics'. The conference on 'hybridisation' was thus rebaptised 'Conference of Genetics'. On this legendary episode, see L. C. Dunn, *A Short History of Genetics* (1965), pp. 68–69.
32 F. Galton, 'On blood-relationship' (1872), p. 395. For more on the term 'stirp', see Chapter 4, note 45.
33 A. Weismann, 'De l'hérédité' (1883). Published in French in *Essais sur l'hérédité et la sélection naturelle* (1892), pp. 117–156.
34 See for example K. Pearson, 'Nature and nurture: The problem of the future' (1910). This contains an interminable series of correlation coefficients that are supposed to express the 'Strength of Nature' and the 'Strength of Nurture'.
35 W. L. Johannsen, *Elemente der exakten Erblichkeitslehre* (1909), pp. 162–163. Reproduced and translated in L. C. Dunn, *A Short History of Genetics* (1965), pp. 91–92.
36 W. L. Johannsen, 'Does hybridization increase fluctuating variability?' (1907).
37 G. Mendel, 'Experiments on plant hybrids' ([1865] 1966), pp. 40–41.
38 W. L. Johannsen, *Elemente der exakten Erblichkeitslehre* (1909), p. 165.
39 Although Johannsen publicly said that the word 'genotype' was 'entirely independent of any hypothesis', in a series of philosophically oriented papers that were not widely circulated he admitted that the notion of genotype was close to an Aristotelian form. See N. Roll-Hansen, 'The genotype theory of Wilhelm Johannsen and its relation to plant-breeding and the study of evolution' (1978).
40 W. L. Johannsen, *Ueber Erblichkeit in Populationen in reinen Linien* (1903), p. 64.
41 K. Pearson, 'Mathematical contributions to the theory of evolution. XII. On a generalised theory of alternative inheritance, with special reference to Mendel's laws' (1904), p. 60. The article is dated 11 September 1903.
42 *Ibid.*, pp. 84–85.
43 W. Johannsen, 'Does hybridization increase fluctuating variability?' (1907).
44 J. P. Lotsy, *Evolution by Means of Hybridization* (1916).
45 *Ibid.*, p. VII.
46 *Ibid.*, p. 118.
47 P. J. Bowler, *The Eclipse of Darwinism* (1983), chap. 8, p. 197.
48 W. Bateson, President's address to the British Association for the Advancement of Science (1914).

49 See W. Bateson, *Problems of Genetics* (1913), where the original phases of the story are told.

50 H. J. Muller, 'Genetic variability, twin hybrids and constant hybrids, in a case of balanced lethal factors' (1918). This article put an end to a controversy that had lasted a decade. The case of *Oenothera* was in fact extremely complex, and involved other phenomena apart from balanced lethals – ring translocations, polyploidy and even, in one case, a genuine 'mutation' in the current sense of the term.

51 Most of these studies were carried out in the 1920s. On the life and work of H. J. Muller, see E. A. Carlson, *Genes, Radiation and Society: The Life and Work of H. J. Muller* (1981). His key article on mutation was published in 1922: 'Variation due to change in the individual gene'. See also the eyewitness account of A. H. Sturtevant, *A History of Genetics* (1965), chap. 11, and the more sociological account of R. E. Kohler, *Lords of the Fly* (1994).

52 G. Mendel, 'Versuche über Pflanzen-Hybriden [Experiments on hybrid plants]' (1865), English translation in C. Stern and E. R. Sherwood (eds.), *The Origin of Genetics: A Mendel Source Book* (1966), p. 2. Unless otherwise indicated, the references to Mendel are to this English edition.

53 W. Bateson and E. R. Saunders, 'Experimental studies in the physiology of heredity' (1902), p. 12. This report was presented on 17 December 1901.

54 W. E. Castle, 'Mendel's law of heredity' (1903). The article is dated 20 December 1902.

55 See for example: L. C. Dunn, *A Short History of Genetics* (1965), chap. 1 §7; A. H. Sturtevant, *A History of Genetics* (1966), chap. 4; C. Stern and E. R. Sherwood (eds.), *The Origin of Genetics: A Mendel Source Book* (1966); R. Olby, *Origins of Mendelism* (1984), chap. 6.

56 H. de Vries, 'Das Spaltungsgesetz der Bastarde' (presented 30 March, published 25 April 1900 in *Berichte der Deutschen Botanischen Gesellschaft*).

57 H. de Vries, 'Sur la loi de disjonction des hybrides' (published 26 March in *Comptes rendus Hebdomadaires des Séances de l'Académie des Sciences*). English translation in C. Stern and E. R. Sherwood (eds.), *The Origin of Genetics: A Mendel Source Book* (1966).

58 C. Correns, 'G. Mendel's Regel über das Verhalten der Nachkommenschaft der Rassenbastarde' (published May 1900 in *Berichte der Deutschen Botanischen Gesellschaft*). English translation in C. Stern and E. R. Sherwood (eds.), *The Origin of Genetics: A Mendel Source Book* (1966).

59 E. von Tschermak, 'Ueberküntsliche Kreuzung bei *Pisum sativum*' (published July 1900 in *Berichte der Deutschen Botanischen Gesellschaft*).

60 C. Correns, English translation in C. Stern and E. R. Sherwood (eds.), *The Origin of Genetics: A Mendel Source Book* (1966), p. 119.

61 *Ibid.*, p. 120.

62 In his 1900 paper, Tschermak made no mention of Mendel's resolution of the 3 : 1 ratio into the more fundamental 1 : 2 : 1 ratio. He thus missed the most important point (R. Olby, *Origins of Mendelism*, 1966, pp. 120–124). This is why Curt Stern excluded Tschermak's articles from the anthology that he published on the origins of genetics (*The Origin of Genetics: A Mendel Source Book,* 1966, 'Foreword', pp. X–XII).

63 H. de Vries, 'Das Spaltungsgesetz...' (1900), English translation (1966), p. 110.

64 In a 'post script', Correns responded as follows to de Vries' German article: 'Mendel's Law of segregation cannot be applied universally' ('G. Mendel's Regel...', English translation, 1966, p. 132). Correns' article was explicit on this point: 'At present, however, this law is applicable only to a certain number of cases, i.e., those where one member of a pair of traits dominate, and probably only to hybrids between varieties. It seems possible that all pairs of traits of all hybrids should behave according to this law' (p. 131).

65 W. B. Provine cites an astonishing letter from de Vries to Bateson, dated 30 January 1901: 'I prayed you last time, please don't stop at Mendel ... it becomes more and more clear to me that Mendelism is an exception to the general rule of crossing. It is in no way THE rule' (*The Origins of Theoretical Population Genetics*, 1971, p. 68).

66 H. de Vries, 'Das Spaltungsgesetz der Bastarde' (1900), English translation (1966), p. 108; C. Correns, 'G. Mendel's Regel über das Verhalten der Nachkommenschaft der Rassenbastarde' (1900), English translation (1966), p. 131.

67 H. de Vries, 'Das Spaltungsgesetz der Bastarde' (1900), English translation (1966), p. 112 (original emphasis).

68 W. Bateson and E. R. Saunders, 'Experimental studies in the physiology of heredity' (1902), p. 12.

69 R. Olby, *Origins of Mendelism* (1985), pp. 241–247 and 250–254.

70 L. Cuénot, 'Les recherches expérimentales sur l'hérédité' (1902), p. LIX.

71 W. Bateson and E. R. Saunders, 'Experimental studies in the physiology of heredity' (1902), pp. 126–130.

72 H. de Vries, 'Das Spaltungsgesetz der Bastarde' (1900), English translation (1966), p. 111. The sentence quoted here opens the paragraph on 'the law of segregation of hybrids'.

73 W. Sutton, 'On the morphology of the chromosome group of *Brachystola magna*' (1902). For the details of this story, see the excellent L. C. Dunn, *A Short History of Genetics* (1965), chap. 11.

74 W. E. Castle, 'The laws of Galton and Mendel, and some laws governing race improvement by selection' (1903), p. 230.

75 H. de Vries, 'Das Spaltungsgesetz...' (1900), English translation (1966), p. 110.

76 *Ibid.*, p. 110; C. Correns, 'G. Mendel's Regel...', English translation (1966), p. 131.

77 F. Galton, 'Hereditary talent and character' (1865). Mendel's article was the result of a communication read to the Natural Sciences Society of Brünn (today Brno) on 8 February and 8 March 1865. It was published in the 1865 volume of the acts of the Society, which in fact appeared in 1866. This explains why bibliographies give both 1865 and 1866 as the year of publication.

78 Apart from the 1865 paper, see the letter to Carl Nägeli dated 27 September 1870, where the term *Anlage* is used in the same way as by Correns and de Vries (C. Stern and E. R. Sherwood (eds.), *The Origin of Genetics: A Mendel Source Book*, 1966, pp. 95–96).

79 Letter from Mendel to Nägeli, 31 December 1866, reproduced and translated in C. Stern and E. R. Sherwood (eds.), *The Origin of Genetics: A Mendel Source Book* (1966), p. 57.

80 For Mendel, dominance was an auxiliary hypothesis which was required to establish his law on the formation of hybrids. But he did not think it was universal, even for characters that conformed to the said law. In a letter to Nägeli dated 27 September 1870, Mendel admitted that the hybrid of two pure forms could have an appearance different from that of either parent (C. Stern and E. R. Sherwood (eds.), *The Origin of Genetics: A Mendel Source Book*, 1966, pp. 95–96).
81 G. Mendel, 'Experiments on plant hybrids' ([1865] 1966), p. 2.
82 *Ibid.*, pp. 3, 4, 5, 29 etc.
83 *Ibid.*, p. 16.
84 *Ibid.*, p. 22.
85 This appears clearly in the following formulation: 'The law of combination of differing traits according to which hybrid development proceeds . . .' A few lines later, Mendel writes of 'the law of development discovered for *Pisum*' (*ibid.*, p. 32).
86 J. Jindra, 'A possible derivation of the Mendelian series' (1971). See also R. Olby, *Origins of Mendelism* (1985), chaps. 5 and 6.
87 G. Mendel, 'Experiments on plant hybrids' ([1865] 1966), pp. 1–23.
88 Nägeli's remark is cited by Mendel in a letter dated 18 April 1867 (C. Stern and E. R. Sherwood (eds.), *The Origin of Genetics: A Mendel Source Book*, 1966, p. 63).
89 G. Mendel, 'Experiments on plant hybrids' ([1865] 1966), p. 63.
90 *Ibid.*, pp. 23–38.
91 *Ibid.*, pp. 30, 43. 'Union' is used on p. 42: '. . . a constant law based on the material composition and arrangement of the elements that attained viable union in the cell.'
92 *Ibid.*, p. 24.
93 *Ibid.*
94 *Ibid.*
95 *Ibid.*, pp. 24–29.
96 *Ibid.*, p. 30.
97 *Ibid.* Evoking the individuals produced by the associations A/A and a/a, Mendel clearly states: 'the products of [this] association must be constant, namely A and a'.
98 W. Bateson and E. R. Saunders, 'Experimental studies in the physiology of heredity' (1902), p. 12.
99 Letter from Mendel to Nägeli, 18 April 1867 (C. Stern and E. R. Sherwood (eds.), *The Origin of Genetics: A Mendel Source Book*, 1966, p. 62).
100 J. Heimans, 'Mendel's ideas on the nature of hereditary characters: The explanation of fragmentary records of Mendel's hybridizing experiments' (1971).
101 G. Mendel, 'Experiments on plant hybrids' ([1865] 1966), pp. 42–43.
102 R. Olby, 'Mendel no Mendelian?' (1979), reproduced in R. Olby, *Origins of Mendelism* (2nd edn, 1985).
103 *Ibid.*, pp. 251 and 258.
104 *Ibid.*, p. 2.
105 *Ibid.*, p. 44. The discussion of this question takes up the whole of the conclusion (pp. 39–48). On the tradition of 'hybridisers', the standard reference

work remains H. F. Roberts, *Plant Hybridization before Mendel* ([1929] 1965).

106 R. Olby, *Origins of Mendelism* (2nd edn, 1985), p. 44.

107 G. Mendel, 'Experiments on plant hybrids' ([1865] 1966), pp. 40–41.

108 *Ibid.*, pp. 37–38.

109 *Ibid.*, p. 43.

110 *Ibid.*, p. 47.

111 'With the views of Darwin which were at that time coming into prominence Mendel did not find himself in full agreement, and he embarked on his experiments with peas, which as we know he continued for eight years' (W. Bateson, *Mendel's Principles of Heredity*, 1909, p. 311). For a critique of this position, see R. A. Fisher, 'Has Mendel's work been rediscovered?' (1936).

112 Letter from Mendel to Nägeli, 27 September 1870 (C. Stern and E. R. Sherwood (eds.), *The Origin of Genetics: A Mendel Source Book*, 1966, p. 96).

113 On the material indices that suggest that Darwin did not directly or indirectly know of Mendel's work, see R. Olby, 'References to Mendel before 1900', in *Origins of Mendelism* (1985), pp. 219–234. Darwin's remarks on cultivated peas (*Pisum sativum*) in the *The Variation of Animals and Plants under Domestication* are particularly interesting in this respect. If Darwin had heard of Mendel's experiments, it is difficult to imagine that he would not have mentioned them in a passage like the following, which is present in both the 1868 and 1875 editions: 'Certain varieties [of peas] can be propagated truly, whilst others show a determinate tendency to vary; thus two peas differing in shape, one round and the other wrinkled, were found by Mr Masters within the same pod, but the plants raised from the wrinkled kind always evinced a strong tendency to produce round peas. Mr Masters also raised from a plant of another variety four distinct sub-varieties, which bore blue and round, white and round, blue and wrinkled, and white and wrinkled peas; although he sowed these four varieties separately during several successive years, each kind always reproduced all four kinds mixed together!' (2nd edn [1875], quoted in Darwin (1972), vol. VII, pp. 348–349).

114 C. Darwin, *Variation* (1868), chap. 22. M. J. S. Hodge has convincingly shown that the young Darwin first thought hybridization to be the major explanation of the modification of species (*Origins and Species*, 1991; *Darwin's Theory of Natural Selection*, in press).

115 Like Galton, Weismann had an 'ancestral' interpretation of heredity: 'Thus the nuclear substance must be the sole bearer of hereditary tendencies, and the facts ascertained by van Beneden in the case of *Ascaris* plainly show that the nuclear substance must not only contain the tendencies of growth of the parents, but also those of a very large number of ancestors. Each of the two nuclei which unite in fertilisation must contain the germ-nucleoplasm of both parents, and this latter nucleoplasm once contained and still contains the germ-nucleoplasm of the grandparents as well as that of previous generations ... While the germ-plasm of the father or mother constitutes half the nucleus of any fertilized ovum, that of the grandparent only forms a quarter, that of the tenth generation backwards only 1/1024 and so on. The latter can, nevertheless, exercise influence over the development of the offspring' (A. Weismann, 'The continuity of the germ-plasm as the foundation of a

theory of heredity' [1885], quoted in Weismann, *Essays upon Heredity*, 1889, p. 179).

116 K. Pearson, 'The law of ancestral heredity' (1903), p. 211.

117 W. F. R. Weldon, 'Mendel's laws of alternative inheritance in peas' (1901), p. 252.

118 See section 1.3.2.2, 'Heredity and modification'.

119 Weinberg had a slight precedence over Hardy. For the details of this story, see W. B. Provine, *The Origins of Theoretical Population Genetics* (1971), pp. 131ff.

120 G. Mendel, 'Experiments on plant hybrids' ([1865] 1966), p. 16.

121 *Ibid.*, p. 39.

122 The second and final 1869 paper is entirely devoted to *Hieracium*, as is a large part of the correspondence with Nägeli.

123 G. Mendel, 'On hieracium-hybrids obtained by artificial fertilization' ([1869] 1966), p. 51.

124 G. Mendel, 'Experiments on plant hybrids' ([1865] 1966), p. 16.

125 G. U. Yule, 'Mendel's laws and their probable relations to intra-racial heredity' (1902), p. 225 (original emphasis).

126 K. Pearson, 'Mathematical contributions to the theory of evolution. XII. On a generalised theory of alternative inheritance, with special reference to Mendel's laws' (1904), p. 60.

127 Reported in G. H. Hardy, 'Mendelian proportions in a mixed population' (1908), p. 49.

128 Hardy's demonstration differs from that to which we are accustomed. Hardy does not reason on the basis of the frequency of alleles, but only on the basis of genotype frequencies, starting from three genotypes (AA, Aa, aa), distributed according to the proportions $p:2q:r$. If mating is random, the next generation will have the structure $(p + q)^2:2(p + q)(q + r):(q + r)^2$, or $p_1:2q_1:r_1$. For this distribution to be the same as that of the preceding generation, it suffices for the frequencies of one genotype, AA for example, to be conserved. Thus: $p/(p + 2q + r) = p_1/(p_1 + 2q_1 + r_1)$. This expression itself can be written: $(p + q)^2/(p + q)^2 + 2(p + q)(q + r) + (q + r)^2$. By developing and simplifying, we find $q^2 = pr$. Replacing p_1, q_1, r_1 by their value in terms of p, q, r, it is clear that the second generation satisfies this condition: with $q_1 = (p + q)(q + r)$ and $p_1r_1 = (p + q)^2(q + r)^2$, we indeed have $q_1{}^2 = p_1r_1$. The population will thus conserve the same structure in the following generations. In fact, Hardy did not give the detail of the calculation. Although it is, as he says, 'easy to see' (*ibid.*), the biologists who read the article in *Science* might not have agreed.

129 *Ibid.*

130 W. Weinberg, 'Ueber den Nachweis der Vererbung beim Menschen' (1908). On the specificity of Weinberg's discovery, see C. Stern, 'Wilhelm Weinberg, 1862–1937' (1962), and W. B. Provine, *The Origins of Theoretical Population Genetics* (1971), pp. 134–136. From the outset, Weinberg generalised the law of equilibrium to *n* alleles. Long before Fisher, he also carried out a Mendelian analysis of quantitative characters.

131 W. E. Castle, 'The laws of heredity of Galton and Mendel, and some laws governing race improvement by selection' (1903), p. 237. (See section 8.2.2.3, 'The new shape of selection'.)

132 G. H. Hardy, 'Mendelian proportions in a mixed population' (1908), p. 50.

133 On this term, see the interesting discussion of A. Langaney, 'Panmixie, "Pangamie" et systèmes de croisement' (1969).

134 Strictly speaking, this term only appeared later, in Dobzhansky's article 'Mendelian populations and their evolution' (1950): 'Sexual reproduction does not erode and level off, but on the contrary, conserves hereditary variability. Every sexual species accordingly possesses a gene pool, in which each gene may be represented by a certain number of alleles, and each chromosome variant in the gene pool remains constant from generation to generation, unless mutation, selection, or genetic drift intervene to alter them' (pp. 575–576). This is the first time the term 'gene pool' was used. The link between this notion and the Hardy–Weinberg equilibrium is striking. For a detailed history of the origins and fate of this term, see M. B. Adams, 'From "gene fund" to "gene pool": On the evolution of the evolutionary language' (1979).

135 On this point, see the remarkable study of B. A. Kimmelman, 'Agronomie et théorie de Mendel: la dynamique institutionnelle et la génétique aux Etats-Unis, 1900–1925' (1990).

136 W. E. Castle, 'The laws of heredity of Galton and Mendel, and some laws governing race improvement by selection' (1903), p. 237.

137 W. E. Castle, 'The mutation theory of organic evolution from the standpoint of animal breeding' (1905).

138 W. F. R. Weldon, 'On the ambiguity of Mendel's categories' (1902).

139 P. J. Bowler, *The Eclipse of Darwinism* (1983).

140 T. H. Morgan, 'For Darwin' (1909), p. 375. This strange declaration is discussed by W. B. Provine in *The Origin of Theoretical Population Genetics* (1971), p. 120. For a more academic treatment of the same idea, see Morgan's *Experimental Zoology* (1907), chaps. 13–14.

141 On Morgan and evolution, see G. E. Allen, *Thomas Hunt Morgan: The Man and His Science* (1978). On Morgan's experimental model – *Drosophila* – see R. E. Kohler, *Lords of the Fly* (1994), part I.

142 The key texts are T. H. Morgan et al., *The Mechanism of Mendelian Heredity* (1915); H. J. Muller, 'Variation due to change in the individual gene' (1922).

143 R. C. Punnett, *Mimicry in Butterflies* (1915).

144 *Ibid.*, pp. 143–144.

145 The theoretical development takes up the whole of chap. 8 and appendix 1, pp. 93–103 and 154–156.

146 J. B. S. Haldane, 'A mathematical theory of natural and artificial selection. Part I' (1924). Haldane mentions Punnett's table as the only previous attempt to formalise the effect of selection in a Mendelian population.

147 R. C. Punnett, *Mimicry in Butterflies* (1915), p. 96.

148 *Ibid.*, p. 103.

149 T. H. Morgan, *A Critique of the Theory of Evolution* (1916).

150 C. Darwin, *On the Origin of Species* (1859), chap. 6, 'Summary of chapter', p. 206.

151 W. E. Castle, 'The mutation theory of organic evolution from the standpoint of animal breeding' (1905).

152 W. E. Castle, 'The heredity of albinism' (1903), pp. 699 and 617.

153 W. E. Castle, 'Yellow mice and gametic purity' (1906).
154 W. E. Castle, 'Recent discoveries in heredity and their bearing on animal breeding' (1905).
155 W. E. Castle and A. Forbes, 'Heredity in hair-length in guinea-pigs and its bearing on the theory of pure gametes' (1906), quoted in W. B. Provine, *Sewall Wright and Evolutionary Biology* (1986), p. 40.
156 See L. C. Dunn, 'William Ernst Castle' (1965); E. A. Carlson, *The Gene: A Critical History* (1966); and especially W. B. Provine, *The Origins of Theoretical Population Genetics* (1971), and *Sewall Wright and Evolutionary Biology* (1986), chaps. 2 and 3.
157 W. E. Castle and H. MacCurdy, 'Selection and cross-breeding in relation to the inheritance of coat-pigments and coat-patterns, in rats and guinea-pigs' (1907).
158 W. E. Castle and J. C. Phillips, 'Piebald rats and selection: An experimental test of the effectiveness of selection and of the theory of gametic purity in Mendelian crosses' (1914).
159 *Ibid.*
160 E. C. MacDowell, 'Piebald rats and multiple factors' (1916); R. Pearl, 'The selection problem' (1917); A. H. Sturtevant, 'An analysis of the effects of selection' (1918); A. Pictet, 'Sur l'existence de deux facteurs de panachure dissociables par croisement' (1925). On Wright's role, see W. B. Provine, *Sewall Wright and Evolutionary Biology* (1986), pp. 69–73. Castle finally accepted the hypothesis of multiple factors – 'Piebald rats and selection' (1919).
161 See for example R. A. Fisher, 'On some objections to mimicry theory: Statistical and genetic' (1927), pp. 274–275.
162 B. J. Norton and E. S. Pearson, 'A note on the background to, and refereeing of, R. A. Fisher's 1918 *On the Correlation between Relatives on the Supposition of Mendelian Inheritance*' (1976); B. J. Norton, 'Metaphysics and population genetics: Karl Pearson and the background to Fisher's multi-factorial theory of inheritance' (1975), and 'Fisher and the neo-Darwinian synthesis' (1978); W. B. Provine, *The Origins of Theoretical Population Genetics* (1971).
163 F. Galton, *Natural Inheritance* (1889), chap. 2.
164 G. U. Yule, 'Mendel's laws and their probable relations to intra-racial heredity' (1902), p. 235.
165 K. Pearson, 'Mathematical contributions to the theory of evolution. XII. On a generalised theory of alternative inheritance, with special reference to Mendel's laws' (1904).
166 G. U. Yule, 'On the theory of inheritance of quantitative compound characters on the basis of Mendel's laws – A preliminary note' (1907).
167 H. Nilsson-Ehle, 'Kreuzungsuntersuchungen an Hafer und Weizen' (1909–11).
168 T. Tammes, 'Das Verhalten fluktuierend variender Merkmale bei der Bastardierung' (1911).
169 R. C. Stauffer (ed.), *Charles Darwin's Natural Selection*... *1856–58* (1975), pp. 224–225.
170 'By an extension of prior usage in physical chemistry one may employ the term *stoichiometry* to denote that branch of the science which concerns itself

with the *material* transformation, with the relations between the *masses* of the components' (A. J. Lotka, *Elements of Mathematical Biology* [1924], 1956, p. 50). Lotka opposed this aspect to 'the *Energetic* or *Dynamics* of Evolution' (*ibid.*). Lotka's book gave J. B. S. Haldane's first paper on the mathematical theory of selection as an example of a 'kinetic' description of evolution, i.e. a theory describing changes in the distribution of various classes of elements (*ibid.*, pp. 122–127; Lotka referred to Haldane, 1924a).

Part 3 The genetic theory of selection

1 See section 5.1. For a useful discussion of the question, see J. R. Moore, *The Post-Darwinian Controversies* (1979), pp. 178–184.
2 See J. Gayon, 'What does Darwinism mean?' (1994); 'Neo-Darwinism?' (1995).
3 T. Dobzhansky, *Genetics and the Origin of Species* (1937), pp. 11–12.
4 *Ibid.*, p. 12.
5 See E. Mayr and W. B. Provine (eds.), *The Evolutionary Synthesis* (1980).
6 To use a term that was widespread in the first half of the 20th century. See for example E. B. Ford, *Mendelism and Evolution* (1931); J. Huxley, *Evolution: The Modern Synthesis* (1942).

9 The place of selection in theoretical population genetics

1 For biographies of each of the key personalities, see: J. Fisher Box, *R. A. Fisher: The Life of a Scientist* (1978); R. Clark, *The Life and Work of J. B. S. Haldane* (1968); W. B. Provine, *Sewall Wright and Evolutionary Biology* (1986).
2 R. A. Fisher, *The Genetical Theory of Natural Selection* (1930); S. Wright, 'The roles of mutation, inbreeding, crossbreeding and selection in evolution' (1932); J. B. S. Haldane, *The Causes of Evolution* (1932); E. B. Ford, *Mendelism and Evolution* (1931); J. Huxley, 'Natural selection and evolutionary progress' (1936); T. Dobzhansky, *Genetics and the Origin of Species* (1937).
3 J. B. S. Haldane, 'A mathematical theory of natural and artificial selection. Part I' (1924), p. 19.
4 R. A. Fisher, *The Genetical Theory of Natural Selection* (1930), chap. 4. See also 'Indeterminism and natural selection' (1936), pp. 113–114.
5 J. B. S. Haldane, *The Causes of Selection* (1932), passim (e.g. end of chap. 4); E. B. Ford, *Mendelism and Evolution* (1931); J. Huxley, *Evolution: The Modern Synthesis* (1942), chap. 2 §1.
6 W. B. Provine, 'The role of mathematical population geneticists in the evolutionary synthesis of the 1930s and 1940s' (1978).
7 See for example R. A. Fisher, 'Cuénot on preadaptation: A criticism' (1915), p. 60; *The Genetical Theory of Natural Selection* (1930), p. 39; 'Population genetics' (1953).
8 S. Wright, 'Genic and organismic selection' (1980).
9 A. L. and A. C. Hagedoorn, *The Relative Value of the Processes Causing Evolution* (1921), p. 294.

10 R. A. Fisher, 'On the dominance ratio' (1922), pp. 326–331.
11 R. A. Fisher, *The Genetical Theory of Natural Selection* (1930), pp. 38–39.
12 S. Wright, 'The effects of inbreeding and crossbreeding on guinea pigs. I. Decline of vigor. II. Differentiation among inbred families. III. Crosses between highly inbred families' (1922).
13 This description is that of M. Kimura. S. Wright also used this word. Fisher, however, always used 'genetic' to describe an effect on the gene-atom – that is to say, a specific gene, as opposed to an allelic or genetic interaction.
14 This is why Fisher paid particular attention to the problem of statistically defining 'the average effect of a gene substitution', in the contexts of both artificial and natural selection (R. A. Fisher, *The Genetical Theory of Natural Selection*, 1930, chap. 2; 'Average excess and average effect of a gene substitution', 1941).
15 R. A. Fisher, *The Genetical Theory of Natural Selection* (1930), chap. 6 (2nd edn, 1958, p. 137).
16 S. Wright, 'Evolution in Mendelian populations' (1931). The article is dated 20 January 1930. The diagrams are taken from 'The roles of mutation, inbreeding, crossbreeding and selection in evolution' (1932).
17 The notion of random genetic drift can easily be understood by reasoning from the point of view of the individual. Let us suppose that this individual leaves two children. For a given locus, while the probability that the individual does not transmit a given allele to one of the children is ½, the probability that this allele will not be transmitted to either child is ½ × ½ = ¼. By generalising to N individuals, it can be seen that this probability decreases as N increases. In the simplest case, the pressure of drift at each generation is ¼N.
18 See R. A. Fisher, 'On the dominance ratio' (1922); J. B. S. Haldane, 'A mathematical theory of natural and artificial selection', I (1924), II (1924), III (1925), V (1927).
19 S. Wright, 'The roles of mutation, inbreeding, crossbreeding and selection in evolution' (1932), p. 359.
20 See S. Wright, 'Biology and the philosophy of science' (1964), pp. 120–122.
21 This metaphorical representation is inspired by S. Wright, 'Classification of the factors of evolution' (1955).
22 For an examination of this too-little-discussed aspect of the debate between Fisher and Wright, see the interesting remarks of J. R. G. Turner, 'The basic theorems of natural selection: A naïve approach' (1969).
23 R. A. Fisher, *The Genetical Theory of Natural Selection* (1930), chap. 2 (2nd edn, 1958, p. 39).
24 Comparison of the 1930 and 1958 versions of Fisher's book yields an impressive list of typographical errors, omissions, mathematical mistakes and contradictions. Price, who made a close study of these points, came to the conclusion that they were generally accidental and that Fisher 'knew where he wanted to go'. We know that Fisher was extremely short-sighted and, in general, did not write but dictated to his wife while seated in his armchair. For a reconstitution of the demonstration of the theorem, see G. R. Price, 'Fisher's "Fundamental theorem" made clear' (1972), and M. Kimura, 'On the change of population fitness by natural selection' (1958). On Fisher's life, the reference work is that of his daughter, J. Fisher Box, *R. A. Fisher: The Life of a Scientist* (1973).

25 In chronological order: S. Wright (1930, 1931, 1955, 1968–78 (vol. II)), J. F. Crow (1955), C. C. Li (1955), O. Kempthorne (1957), M. Kimura (1958), P. A. P. Moran (1964), M. Nei (1964), A. W. F. Edwards (1967), J. R. G. Turner (1967, 1969, 1970), J. F. Crow and M. Kimura (1970), G. R. Price (1972), A. Jacquard (1974, 1977), M. Kojima and T. Kelleher (1981).

26 M. Kimura, 'On the change of population fitness by natural selection' (1958); G. R. Price, 'Fisher's "Fundamental Theorem" made clear' (1972).

27 R. A. Fisher, *The Genetical Theory of Natural Selection* (1930) (2nd edn, 1958, p. 37). Reference will be made to the 2nd edn of the book (Dover, 1958). The typography of this edition distinguishes the 1958 additions to the original 1930 text. *GTNS [1930]* refers to the 1930 text, cited in the Dover edition of 1958. *GTNS [1958]* refers to modifications made for the 2nd edn.

28 'The rate of increase of fitness of any species is equal to the genetic variance in fitness' (R. A. Fisher, *GTNS [1930]*, p. 50).

29 'Since the theorem is exact only for idealized populations, in which fluctuations in genetic composition have been excluded ...' (R. A. Fisher, *GTNS [1930]*, p. 38).

30 'The statement of the principle of Natural Selection in the form of a theorem determining the rate of progress of a species in fitness to survive (this term being used for a well-defined statistical attribute of the population) ...' (*GTNS [1930]*, p. 40).

31 'The rate of increase of average fitness of the population ascribable to a change in gene frequency ...' (R. A. Fisher, 'Average excess and average effect of a gene substitution', 1941, p. 57).

32 A. Jacquard, *Concepts en génétique des populations* (1977).

33 R. A. Fisher, *GTNS [1930]*, pp. 41–45.

34 Fisher's 'W' should not be confused with the 'W' conventionally used by Wright to signify selective value.

35 On this question, see J. Gayon, 'Sélection naturelle ou survie des plus aptes? Eléments pour une histoire du concept de *fitness* dans la théorie évolutionniste' (1995).

36 A. J. Lotka, 'An objective standard of value derived from the principle of evolution' (1914).

37 R. A. Fisher, 'The actuarial treatment of official birth records' (1927), p. 108.

38 A. J. Lotka, 'A natural population norm' (1913); 'The relation between birth rate and death rate in a normal population...' (1918); 'On the true rate of natural increase of a population' (with I. Dublin) (1925). The exhaustive algebraic treatment of the problem was first given in the 1925 article. A recapitulation can be found in *Elements of Physical Biology*, 1924 edn.

39 There is a simpler presentation of the meaning of 'm'. It is enough to reason on the basis of the population size in a given generation (N_t). In a continuous model like that of Fisher (continuous generations) we have:

$$N_t = N_{t-1} e^m$$

with: m = b – d (b = birth rate; d = death rate).
It can clearly be seen that:

$$m = \log_e N_t - \log_e N_{t-1}$$

This is a simplified situation (see J. F. Crow and M. Kimura, *Introduction to Population Genetics Theory,* 1970, pp. 18–19).

40 R. A. Fisher, *GTNS [1930]*, p. 40.

41 R. A. Fisher, 'The actuarial treatment of official birth records' (1927), pp. 103–108; *GTNS [1930]*, p. 26.

42 R. A. Fisher, 'The correlation between relatives on the supposition of Mendelian inheritance' (1918). This text was in fact written in 1916.

43 See for example D. S. Falconer, *Introduction to Quantitative Genetics* (1974), chaps. 7 and 8.

44 R. A. Fisher, *GTNS [1930]*, chap. 2 § 'Changes in population', pp. 45–49.

45 *Ibid.*, p. 46.

46 *Ibid.*

47 *Ibid.*

48 *Ibid.*

49 *Ibid.*

50 *Ibid.*, p. 45.

51 *Ibid.*, pp. 46 and 51.

52 See note 25.

53 R. A. Fisher, *GTNS [1930]*, p. 38.

54 See J. H. Bennett, *Natural Selection...* (1983). This book reproduces Fisher's most important letters. On this particular point, see the letters to Kempthorne and to Kimura.

55 G. R. Price, 'Fisher's "fundamental theorem" made clear' (1972).

56 'The rate of increase in fitness due to all change in gene ratio is exactly equal to the genetic variance of fitness W which the population exhibits' (R. A. Fisher, *GTNS [1930]*, p. 37).

57 R. A. Fisher, 'Average excess and average effect of a gene substitution' (1941), p. 57.

58 R. A. Fisher, *GTNS [1930]*, pp. 44–45.

59 R. A. Fisher, 'On the dominance ratio' (1922).

60 R. A. Fisher, 'Average excess and average effect of a gene substitution' (1941).

61 G. R. Price, 'Fisher's "fundamental theorem" made clear' (1972).

62 Letter from Fisher to Kempthorne, 18 February 1955 (in J. H. Bennett (ed.), *Natural Selection....*, 1983, p. 229).

63 *Ibid.*

64 A. J. Lotka, *Elements of Mathematical Biology* [1924], 1956 reprint, p. 42.

65 R. A. Fisher, 'Population genetics' (1953), p. 515.

66 R. A. Fisher, 'Average excess and average effect of a gene substitution' (1941), p. 58.

67 For a development of this idea, see B. J. Norton, 'Fisher's entrance in evolutionary science: The role of eugenics' (1983), p. 27.

68 On the theoretical, philosophical and anthropological implications of the theme of 'optimality', see the excellent collective work published by the philosopher John Dupré: *The Latest on the Best: Essays on Evolution and Optimality* (1987).

69 C. Darwin, *On the Origin of Species*, 2nd edn, 1860 [in fact December 1859], chap. 4; quoted in M. Peckham, *The Origin of Species by Charles Darwin. A*

Variorum Text (1959), p. 291. In the first edition (November 1859), the conclusion of chap.4 makes no allusion to improvement or to advance. In the third edition (1861), Darwin went further: 'This principle ... leads to the improvement of each creature in relation to its organic and inorganic conditions of life; and consequently, in most cases, to what must be regarded as an advance in organisation' (quoted in M. Peckham, *The Origin of Species by Charles Darwin. A Variorum Text*, 1959, p. 271).

70 S. Wright, 'Isolation by distance' (1943); 'Isolation by distance under diverse mating systems' (1946). The theory of isolation by distance, or – what amounts to the same thing – the theory of the effects of homogeneous and isotropic migration, was particularly developed by the French mathematician G. Malécot (*Les mathématiques de l'hérédité,* 1948, pp. 57–61).

71 S. Wright, 'The roles of mutation, inbreeding, crossbreeding, and selection in evolution', lecture given at 6th International Congress of Genetics (1932).

72 Strictly speaking, Wright did not use the expression 'landscape', which is in fact that of Waddington ('epigenetic landscape'). In 1932, Wright wrote of a 'rugged field of gene combinations' containing 'peaks' and 'pits'. Later he introduced the idea of a 'multidimensional surface of selection value for all possible genotypes' and of a 'surface of mean selective values for all possible populations' (S. Wright, 'Statistical genetics in relation to evolution', 1939). Sometimes he wrote of a 'surface of fitness function'.

73 J. R. G. Turner, 'The basic theorems of natural selection: A naïve approach' (1969).

74 J. B. S. Haldane, 'A mathematical theory of natural selection. Part VIII. Metastable populations' (1931).

75 Wright to Fisher, 3 February 1931, reproduced in W. B. Provine, *Sewall Wright and Evolutionary Biology* (1986), p. 272.

76 S. Wright, 'Review of *The Genetical Theory of Natural Selection*, by R. A. Fisher' (1930).

77 Wright to Fisher, 3 February 1931, reproduced in W. B. Provine, *Sewall Wright and Evolutionary Biology* (1986), p. 272.

78 *Ibid.*, p. 273.

79 *Ibid.*, p. 310.

80 S. Wright, 'Evolution in Mendelian populations' (1931). We know, from a summary of a text that has unfortunately been lost, that the schema was in place from 1929 (S. Wright, 'Evolution in a Mendelian population', 1929). The term 'shifting balance theory of evolution' appears to have become the conventional term only in 1970, although it was certainly used much earlier (S. Wright, 'Random drift and the shifting balance theory of evolution', 1970). The vocabulary permeated the whole of Wright's evolutionary thinking. For Wright, the genetic evolution of a species always depended on a certain 'balance' among various factors; the relation among factors (mutation rate, selection, inbreeding, migration) defines different evolutionary scenarios, and the modification of the relation leads to the shift from one evolutionary mode to another.

81 S. Wright, 'Evolution in populations in approximate equilibrium' (1935); 'The distribution of gene frequencies in populations' (1937).

82 S. Wright, *Evolution and the Genetics of Populations,* vol. II, *The Theory of Gene Frequencies* (1969), p. 31.
83 'Selective values depend on the interactions of the entire system of genes' (S. Wright, 'The distribution of gene frequencies in populations', 1937, p. 309).
84 The last article written by Wright, published in *American Naturalist* in the year of his death, is particularly interesting in this respect ('Surfaces of selective values revisited', 1988). It should be recalled that Wright was born in 1889 and that his first scientific publication appeared in 1912.
85 'Philosophy aims at the logical clarification of thoughts ... Without philosophy, thoughts are, as it were, cloudy and indistinct; its task is to make them clear and to give them sharp boundaries' (L. Wittgenstein, *Tractatus Logico-Philosophicus* [1918], 1961, §4.112).
86 This presentation is taken from I. Ekeland, *Le calcul et l'imprévu* (1984), pp. 99–104.
87 S. Wright, 'The statistical consequences of Mendelian heredity in relation to speciation', in *The New Systematics*, ed. J. Huxley (1941), p. 166.
88 See, for example, J. Roughgarden, *Theory of Population Genetics and Evolutionary Ecology: An Introduction* (1979), pp. 49–50.
89 R. A. Fisher, 'Average excess and average effect of a gene substitution' (1941), p. 58. Fisher's critique has recently been repeated and developed by A. W. F. Edwards: review of S. Wright, *Evolution and the Genetics of Populations,* vol. II, *The Theory of Gene Frequencies* (1971); *Foundations of Mathematical Genetics* (1977).
90 R. A. Fisher, *GTNS [1930]*, p. 39.
91 S. Wright, *Evolution and the Genetics of Populations,* vol. II, *The Theory of Gene Frequencies* (1969), pp. 120–122 and 140–142.
92 This philosophical bias appears already in Fisher's earliest writings. See, for example, 'Cuénot on preadaptation' (1914), pp. 60–61.
93 On Fisher's 'philosophy' and 'ideology', see the remarkable analyses presented by J. Hodge in 'Biology and philosophy (including ideology): A study of Fisher and Wright' (1990).
94 R. A. Fisher, 'Indeterminism and natural selection' (1934), p. 115.
95 *Ibid.*, p. 116.
96 R. A. Fisher, *The Genetical Theory of Natural Selection* [1930], 2nd edn (1958), p. 41. In chap. 2, Fisher repeats several times that the theorem applied to a 'rate of progress' (*GTNS [1930]*, pp. 39, 40, 50, 51), the fundamental theorem merely being an expression of a law of the conservation of individual advantages (*GTNS [1930]*, p. 51, penultimate paragraph of the conclusion).
97 *GTNS [1930]*, p. 39.
98 *Ibid.*, p. 40.
99 S. Wright, 'The emergence of novelty: A review of Lloyd Morgan's "emergent" theory of evolution' (1935); 'Biology and the philosophy of science' (1964); 'Panpsychism and science' (1975).
100 S. Wright, 'Biology and the philosophy of science' (1964), p. 115.
101 *Ibid.*
102 *Ibid.*
103 See in particular 'The emergence of novelty: A review of Lloyd Morgan's

"emergent" theory of evolution' (1935); 'Genes, the gene, and the hierarchy of biological sciences' (1959); 'Biology and the philosophy of science' (1961); 'Panpsychism and science' (1975).

104 See for example the end of 'Evolution in Mendelian populations' (1931), pp. 154–155.

105 S. Wright, 'Biology and the philosophy of science' (1964), pp. 116–117. Wright's avowed philosophical sources for his panpsychism are eclectic: Wundt, Fechner, Clifford, Pearson, Whitehead, Bergson, Hartshorne – at least some of whom he had undoubtedly read.

106 R. A. Fisher, 'Indeterminism and natural selection' (1934).

10 The empirical and the formal

1 This is one of the key points in R. Lewontin's book *The Genetic Bases of Evolutionary Change* (1974).

2 On the origin of this term, see E. B. Ford, *Ecological Genetics* (1964), chap. 2.

3 *Ibid.*, p. 248.

4 R. C. Punnett, *Mimicry in Butterflies* (1915), pp. 101–103.

5 *Ibid.*, pp. 1–7, 96, 152–153.

6 '*Doubledayaria*' probably corresponded to the variety known as *carbonaria*. Today, scientists simply speak of the melanic form of a single species, *Biston betularia*.

7 R. C. Punnett, *Mimicry in Butterflies* (1915), p. 102.

8 See for example J. W. H. Harrison and F. C. Garrett, 'The induction of melanism in the Lepidoptera and its subsequent inheritance' (1926).

9 J. B. S. Haldane, 'A mathematical theory of natural and artificial selection' (1924a), p. 26.

10 E. B. Poulton (*Essays on Evolution*, 1908, pp. 308–310) provides a sample of this literature.

11 E. B. Ford, 'Problems of heredity in the Lepidoptera' (1937); 'Genetic research in the Lepidoptera' (1940).

12 E. B. Ford, 'Polymorphism' (1945).

13 E. B. Ford, *Ecological Genetics*, 2nd edn (1971), chap. 14.

14 H. B. D. Kettlewell, 'Selection experiments on industrial melanism in the Lepidoptera' (1955); 'Further selection experiments on industrial melanism' (1956); 'The phenomenon of industrial melanism in the Lepidoptera' (1961).

15 E. B. Ford, *Mendelism and Evolution*, 3rd edn (1940), p. 76.

16 S. S. Chetverikov, 'Ueber die genetische Beschaffenheit wilder Populationen' (1927).

17 R. A. Fisher, 'On the dominance ratio'; *The Genetical Theory of Natural Selection* (1930), chap. 4; 'The distribution of gene ratios for rare mutations'. J. B. S. Haldane, 'A mathematical theory of artificial and natural selection. Part V. Selection and mutation' (1927); 'The part played by recurrent mutation in evolution' (1933); 'The equilibrium between mutation and random extinction' (1939). S. Wright, 'Evolution in Mendelian mutations' (1931). E. B. Ford (*Mendelism and Evolution*, 1931, chap. 4 §2, 'Mendelism and selection') well summarises the importance of these points (which he sees

essentially from Fisher's point of view) for the problem of genetic polymorphism.
18 R. A. Fisher and E. B. Ford, 'Spread of a gene in the moth *Panaxia dominula*' (1947).
19 *Ibid.*, p. 173.
20 *Ibid.*, p. 171.
21 M. Lamotte, 'Recherches sur la structure génétique des populations naturelles de *Cepaea nemoralis* (L.)'. Doctoral dissertation submitted in 1950, published in 1951.
22 Wright's canonical text ('Evolution in Mendelian populations') is dated 20 January 1930.
23 W. B. Provine, 'Origins of the genetics of natural populations series', in *Dobzhansky's Genetics of Natural Populations*, ed. R. C. Lewontin (1938–76); *Sewall Wright and Evolutionary Biology* (1986), chap. 10.
24 M. Lamotte, *Introduction à la biologie quantitative: présentation et interprétation statistiques des données numériques* (1948).
25 S. Wright, 'Statistical genetics in relation to evolution' (1939).
26 M. Lamotte, personal communication.
27 A. J. Cain and P. M. Sheppard, 'Selection in the polymorphic land snail *Cepaea nemoralis*' (1950); 'Natural selection in *Cepaea*' (1954).
28 M. Lamotte, 'Polymorphism of natural populations of *Cepaea nemoralis*' (1959); 'Les facteurs de la diversité du polymorphisme dans les populations de *Cepaea nemoralis*' (1966).
29 See Y. Conry, *L'introduction du darwinisme en France au XIXe siècle* (1974); G. Molina, 'Le darwinisme en France', in *Dictionnaire du darwinisme et de l'évolution*, ed. P. Tort (1996).
30 P. L'Héritier, 'Souvenirs d'un généticien' (1981), pp. 335–336.
31 G. Teissier, *Titres et travaux scientifiques* (1958), p. 64.
32 P. L'Héritier and G. Teissier, 'Une expérience de sélection naturelle: courbe d'élimination du gène "bar" dans une population de *Drosophila melanogaster*' (1934); 'Elimination des formes mutantes dans les populations de Drosophiles. Cas des Drosophiles "bar"' (1937); 'Elimination des formes mutantes dans les populations de Drosophiles. Cas des Drosophiles "ebony"' (1937).
33 G. Teissier, 'Apparition et fixation d'un gène mutant dans une population stationnaire de Drosophiles' (1943).
34 G. Teissier, *Titres et travaux scientifiques* (1958), p. 66.
35 G. Teissier, 'Variation de la fréquence du gène *ebony* dans une population stationnaire de Drosophiles' (1947).
36 G. Teissier, *ibid.*; 'Variation de la fréquence du gène *sepia* dans une population stationnaire de Drosophiles' (1947); 'Sélection naturelle et fluctuation génétique' (1954).
37 G. Teissier, 'Conditions d'équilibre d'un couple d'allèles et supériorité des hétérozygotes' (1954).
38 C. Petit, 'Le déterminisme génétique et psycho-physiologique de la compétition sexuelle chez *Drosophila melanogaster*' (1958).
39 For a detailed account of the origins of the French school of experimental population genetics, see J. Gayon and M. Veuille, 'The genetics of experi-

mental populations: L'Héritier and Teissier's population cages (1932–1954)' (to appear in R. Singh, C. Krimbas and J. Beatty (eds.), *Thinking about Evolution: Critical, Historical, Philosophical and Political Perspectives. Lewontin Festschrift*, Cambridge University Press).

40 R. C. Punnett, 'Report to the Royal Society of London on R. A. Fisher, "On the correlation between relatives on the supposition of Mendelian inheritance"' (1916). This report is reproduced in B. J. Norton and E. S. Pearson (1976), p. 155.

41 See S. Wright and T. Dobzhansky, 'Experimental reproduction of some of the changes caused by natural selection in certain populations of *Drosophila pseudoobscura*' (1946). Reproduced in T. Dobzhansky et al., *Genetics of Natural Populations* (1938–76), pp. 396–426. In this article, Wright and Dobzhansky report the spectacular results of their first study using population cages.

42 R. C. Lewontin, personal communication.

43 J. B. S. Haldane, 'A mathematical theory of natural and artificial selection. Part IV' (1927), p. 615.

44 The expression 'stoichiometry of evolution' can be found in Lotka, although in a more general context. Lotka opposed a 'stoichiometry' to an 'energetics' (or 'dynamic') of evolution (A. J. Lotka, *Elements of Mathematical Biology*, 1924, chap. 5, 1956 reprint, pp. 50–51).

45 S. S. Chetverikov, 'On certain aspects of the evolutionary process from the standpoint of modern genetics' (1926; English translation 1961).

46 *Ibid.*, (English trans.), p. 183.

47 *Ibid.*, pp. 181–182; J. B. S. Haldane, 'A mathematical theory of natural and artificial selection. Part I' (1924), p. 26.

48 S. S. Chetverikov, 'On certain aspects of the evolutionary process from the standpoint of modern genetics' (1926), English translation, 1961), p. 182.

49 *Ibid.*, p. 183.

50 *Ibid.*, p. 182.

51 *Ibid.*, p. 183.

52 *Ibid.*

53 *Ibid.*, pp. 186–191.

54 J. B. S. Haldane, 'A mathematical theory of natural and artificial selection. I–X' (1924–34).

55 See J. B. S. Haldane, 'Rapid selection' (1932).

56 The general principle of the method is set out in the first article in the series 'A mathematical theory of artificial and natural selection' (1924). Haldane reasons on the basis of the relation u_n between two forms *A* and *B*. Under certain conditions (e.g. complete dominance, panmixia etc.), he calculates the variation Δu_n of u from one generation to the next, under the effect of a selection intensity k (this coefficient implies that $(1 - k)B$s survive for every A. As long as k is small, Δu_n can be approximated by a differential equation du/dn, and the corresponding integral can be calculated. Haldane emphasises that the equation governing the diffusion of the favourable form 'is accurate enough for any practical problem when $|k|$ is small, and as long as k lies between ± 0.1' (Haldane, 1924a, p. 24). The same principle is used in the other articles in the series, except for the ninth paper, which deals with

rapid selection. In this case the problem is more complex mathematically, because the approximation of Δu_n by *dun/dn* is not possible ('A mathematical theory of artificial and natural selection. Part IX. Rapid selection', 1932).

57 Haldane, 1932c.

58 H. J. Muller, 'Our load of mutations' (1950); J. B. S. Haldane, 'The cost of natural selection' (1957).

59 S. Wright, 'Evolution in Mendelian populations' (1931), p. 156; R. A. Fisher, *The Genetical Theory of Natural Selection* (1930), 2nd edn (1958), pp. 102–103; J. B. S. Haldane, 'A mathematical theory of natural and artificial selection. Part VIII. Metastable populations' (1931).

60 R. A. Fisher, *The Genetical Theory of Natural Selection* (1930), pp. 102–103.

61 *Ibid.*, chap. 2 (2nd edn, 1958, pp. 41–44). This argument was taken up by J. B. S. Haldane (*The Causes of Evolution,* 1932, reprint 1966, pp. 175–179) and by M. Kimura (*The Neutral Theory of Molecular Evolution,* 1983, pp. 135–137).

62 R. A. Fisher, *The Genetical Theory of Natural Selection* (1930), chap. 2 (2nd edn, 1958, p. 47).

63 *Ibid.*, p. 44.

64 M. Kimura, 'Evolutionary rate at the molecular level' (1968).

65 R. C. Lewontin, *The Genetic Basis of Evolutionary Change* (1974), chap. 2, 'The struggle to measure variation', pp. 19–94.

66 H. J. Muller, 'Our load of mutation' (1950). On Muller's life, his scientific work, his political convictions and his thoughts on the effect of radiation on the human species, see E. A. Carlson, *Genes, Radiation and Society: The Life and Work of H. J. Muller* (1981).

67 H. J. Muller, 'Variation due to change in the individual gene' (1922).

68 The classic genetic estimate of the number of loci in the human genome varies between 30,000 and 100,000.

69 Muller's support for eugenics is well known (see H. J. Muller, *Out of the Night: A Biologist's View of the Future*, 1934). The way in which Muller's position related to American and Russian eugenics is analysed in D. Joravsky, *Soviet Marxism and Natural Science, 1917–1932* (1961); E. A. Carlson, *Genes, Radiation and Society: The Life and Work of H. J. Muller* (1981), chaps. 12 and 15–16; M. B. Adams, 'Eugenics in Russia, 1900–1940' (1990). See also D. Kevles, *In the Name of Eugenics* (1985), which shows how Muller was both one of the most virulent critics of eugenics (as a pseudo-science and an ideology of race or class) in the 1930s, and the most caricaturisable advocate of 'positive eugenics' (or the genetic improvement of the human race by judicious marriages) after the Second World War.

70 *The New Systematics* (1940), ed. J. Huxley. The term 'Modern Synthesis' is also Huxley's (1942). But the founding text of the Synthesis is classically accepted as Dobzhansky's *Genetics and the Origin of Species*. See *The Evolutionary Synthesis*, ed. E. Mayr and W. B. Provine (1980).

71 T. Dobzhansky, 'A review of some fundamental concepts and problems of population genetics' (1955).

72 R. C. Lewontin, *The Genetic Basis of Evolutionary Change* (1974), chaps. 2 and 5.

73 B. Wallace, 'The role of heterozygosity in *Drosophila* populations' (1958).

Cited in R. C. Lewontin, *The Genetic Basis of Evolutionary Change* (1974), p. 25.

74 T. Dobzhansky, 'A review of some fundamental concepts and problems of population genetics' (1955), p. 3.

75 *Ibid.*, p. 4.

76 R. C. Lewontin, *The Genetic Basis of Evolutionary Change* (1974), pp. 24–25.

77 *Ibid.*, pp. 26–27.

78 For many years Dobzhansky was the treasurer of this society. There is a large amount of correspondence archived under this title and conserved in the library of the American Philosophical Society at Philadelphia. An idea of Dobzhansky's interest in Teilhard can be gained by reading his book *The Biology of Ultimate Concern* (1969), the last chapter of which is entitled 'The Teilhardian synthesis' and contains a surprising mixture of the synthetic theory of evolution and Teilhard's mystical noogenesis.

79 These various attempts were well summarised by R. C. Lewontin in *The Genetic Basis of Evolutionary Change* (1974), chap. 2.

80 H. Harris, 'Enzyme polymorphism in man' (1966); J. L. Hubby and R. C. Lewontin, 'A molecular approach to the study of genic heterozygosity in natural populations. I. The number of alleles at different loci in *Drosophila pseudoobscura*' (1966); R. C. Lewontin and J. L. Hubby, 'A molecular approach to the study of genic heterozygosity in natural populations. II. Amount of variation and degree of heterozygosity in natural populations of *Drosophila pseudoobscura*' (1966).

81 For this aspect of the story, see C. Debru, *L'esprit des protéines: histoire et philosophie biochimiques* (1983), chaps. 2 and 4.

82 For a very complete recapitulation (with tables), see G. Pasteur, 'Génétique biochimique et populations ou: pourquoi sommes-nous multi-polymorphes?' (1972).

83 T. Dobzhansky, 'A review of some fundamental concepts and problems of population genetics' (1955).

84 R. C. Lewontin, *The Genetic Basis of Evolutionary Change* (1974), p. 113.

85 E. B. Ford, 'Polymorphism and taxonomy' (1940); 'Polymorphism' (1945).

86 See section 10.1 above.

87 It is surprising, for example, that R. C. Lewontin suggests that genetic polymorphism could be explained by a heterosis with an advantage of 10^{-2}, 10^{-3}, or perhaps less. This apparently represents a return to the relatively unrealistic discussions of the 1930s, when population geneticists discussed minute selection intensities.

88 H. Leven, 'Genetic equilibrium when more than one ecological niche is available' (1953).

89 S. Wright developed the first model of frequency-dependent selection ('Adaptation and selection', 1949, p. 375). The first experimental application was by C. Petit, 'Le déterminisme génétique et psycho-physiologique de la compétition sexuelle chez *Drosophila melanogaster*' (1958).

90 The hypothesis was first presented in M. Kimura, 'Evolutionary rate at the molecular level' (1968), but had previously been presented at the Congress of Genetics held in Fukuoka in November 1967 (cf. M. Kimura, *The Neutral Mutation Theory of Molecular Evolution*, 1983, p. 28). The expression 'neutral

mutation–random drift hypothesis of molecular evolution' was first used by Kimura in 1973 to describe retrospectively the ideas developed in 1968 ('Mutation and evolution at the molecular level', 1973, p. 19).

91 The identification of the 'neutral theory' with 'non-Darwinian evolution' was made by J. L. King and T. H. Jukes: 'Non-Darwinian evolution: Random fixation of selectively neutral mutations' (1969). The early history of the neutral theory was reconstituted by T. H. Jukes, 'Early development of the neutral theory' (1991). See also the article by N. Takahata in M. Kimura, *Population Genetics... Selected Papers* (1994), introductory essay for Part 18, pp. 559–563.

92 M. Kimura, *The Neutral Theory of Molecular Evolution* (1983), p. 29: 'the neutral theory of non-Darwinian evolution'.

93 J. B. S. Haldane, 'The cost of natural selection' (1957); 'More precise expressions for the cost of natural selection' (1961).

94 H. B. D. Kettlewell, 'A résumé of investigations on the evolution of melanism in the Lepidoptera' (1956).

95 J. B. S. Haldane, 'The cost of natural selection' (1957), p. 511.

96 *Ibid.*, pp. 520–524.

97 That is e^m, where $m = (b - d)$ ('b' for 'birth', 'd' for 'death'). The letter 'm' is an abbreviation for 'Malthusian parameter' and corresponds to Lotka's 'r' (the 'natural' rate of growth).

98 $e^{-30/n}$, or $e^{-10} = 0.00045$, if n (number of generations for one substitution) = 3. *Ibid.*, quoted in M. Kimura, 'Evolutionary rate at the molecular level' (1968).

99 M. Kimura, 'Evolutionary rate at the molecular level' (1968).

100 $e^{-30/n}$, with $n = 2$. Thus: $e^{-15} = 0.0000003$.

101 *Ibid.*, p. 624.

102 *Ibid.*, p. 626.

103 M. Kimura, 'Natural selection and neutral evolution, with special reference to evolution and variation at the molecular level' (1988), p. 281. This phrase is a stronger version of a statement in his 1983 book – 'a majority of nucleotide substitutions in the course of evolution must be the result of random fixation of selectively neutral or nearly neutral mutants rather than positive Darwinian selection' (*The Neutral Theory of Molecular Evolution*, 1983, p. 28).

104 M. Kimura, *The Neutral Theory of Molecular Evolution* (1983), p. 29.

105 *Ibid.*, p. 50.

106 M. Kimura and T. Ohta, 'Mutation and evolution at the molecular level' (1973), p. 33.

107 On all these objections, see R. C. Lewontin, *The Genetic Basis of Evolutionary Change* (1974).

108 M. Kimura and T. Ohta, 'On some principles governing molecular evolution' (1974), p. 2851. M. Kimura, *The Neutral Theory of Molecular Evolution* (1983) and 'The neutral theory of molecular evolution' (1983), p. 232.

109 M. Kimura, 'On the change of population fitness by natural selection' (1958).

110 *Ibid.*; M. Kimura and J. Crow, *An Introduction to Population Genetics Theory* (1970); M. Kimura, *The Neutral Theory of Molecular Evolution* (1983) and 'The neutral theory of molecular evolution' (1983).

111 M. Kimura, 'The neutral theory of molecular evolution' (1983), p. 233.

112 C. Darwin, *The Variation of Animals and Plants under Domestication*, 2nd edn [1875], quoted in Darwin (1972), vol. VIII, chap. 21, 'Summary on selection by man', p. 236.

113 The hypothesis of the 'molecular clock' was proposed by F. Zukerkandl and L. Pauling in 1965 ('Evolutionary divergence and convergence in proteins'). The hypothesis thus cannot be attributed to the neutralists, but the fate of neutralism is linked to that of the molecular clock. See M. Kimura, 'The rate of molecular evolution considered from the standpoint of population genetics' (1969); 'Molecular evolutionary clock and evolutionary theory' (1987).

114 On this point, see the cogent discussion by F. J. Ayala, *Biologie moléculaire et évolution* (1982), pp. 79–82. This book is based on five lectures given in 1979 at the Collège de France (Paris).

115 C. Darwin, *On the Origin of Species* (1859), p. 420.

116 *Ibid.*, chap. 4. The only diagram that appears in this chapter, and indeed the only diagram in the whole book, is intended to convince the reader that the genealogical ramification of living organisms is not a logical tree, but is nevertheless the effect of a causal principle.

Conclusion

1 'Population' is only rarely used in Darwin's writings. When he used it, it seems to have been most often (if not always) in the old sense of the word, that is the fact or the action of 'populating'. This was the way Malthus used the term in his *Essays on the Principle of Population* (1798). Darwin's Notebook 'E' shows the connection between Darwin's and Malthus' use of the words 'population' and 'depopulation'. In the following quotation, Darwin quotes Malthus: 'p. 529. "... since the world began, the causes of population & depopulation have been probably as constant as any of the laws of nature with which we are acquainted." – this applies to one species – I would apply it not only to population & depopulation, but extermination & production of new forms' (cited in P. H. Barrett et al., *Charles Darwin's Notebooks, 1836–1844*, 1987, E3, p. 397). On the reasons why Darwin did not use the word 'population', see above, section 1.1.

2 J. Gayon, 'Sélection naturelle et sélection artificielle: le principe darwinien est-il métaphorique?' (1994).

3 On this problem, see the remarkable work by C. López Beltrán, *Human Heredity 1750–1870: The Construction of a Domain* (1992); '"Les maladies héréditaires": 18th century disputes in France' (1995).

4 See above, Introduction, note 35.

5 J. Gayon, 'Sélection naturelle et régression dans la théorie évolutionniste darwinienne et post-darwinienne' (1997d).

6 J. Hodge, 'Biology and philosophy (including ideology): A study of Fisher and Wright' (1990).

7 E. Sober, *The Nature of Selection: Evolutionary Theory in Philosophical Focus* (1984).

Bibliography

Primary sources are indicated by †. When there is more than one author, the references are listed according to their first author. However, for some authors (for example Pearson) this procedure is inappropriate, because it hides the chronology of the principal author's work. In these cases the names of the first authors as they appear in the original reference appear after the principal author's name, in parentheses. E.g. Lee, Yule and Pearson (1900) appears as 'Pearson, K. (with Lee, A. and Yule, G. U.). 1900'.

Adams, M. B. 1979. From gene fund to gene pool: On the evolution of evolutionary language. *Studies in the History of Biology*, **3**, 241–285.

Adams, M. B. 1980. Sergei Chetverikov, the Kol'tsov Institute, and the evolutionary synthesis. In *The Evolutionary Synthesis*, ed. E. Mayr and W. B. Provine, pp. 242–278. Cambridge, MA: Harvard University Press.

†Airy, G. 1861. *On the Algebraic and Numeral Theory of Errors of Observations and the Combination of Observations*. London: Macmillan (2nd edn, 1875).

Allen, G. E. 1978. *Thomas Hunt Morgan: The Man and His Science.* Princeton: Princeton University Press.

Allen, G. E. 1980. The evolutionary synthesis: Morgan and natural selection revisited. In *The Evolutionary Synthesis*, ed. E. Mayr and W. B. Provine, pp. 356–382. Cambridge, MA: Harvard University Press.

Ayala, F. J. 1982. *Biologie moléculaire et évolution.* Paris: Masson.

Bachelard, S. 1968. Epistémologie et histoire des sciences. In *Actes du 12ème Congrès International d'Histoire des Sciences*. Paris: Blanchard.

Bajema, C. J. (ed). 1976. *Eugenics Then and Now.* Benchmark Papers in Genetics, vol. V. Stroudsburg, PA: Dowden, Hutchinson and Ross.

Balan, B. 1980. *L'ordre et le temps: l'anatomie comparée et l'histoire des vivants au XIX siècle*. Paris: Vrin.

Barrett, P. H. 1974. *Metaphysics, Materialism, and the Evolution of Mind: Early Writings of Charles Darwin.* Chicago: University of Chicago Press.

†Barrett, P. H. (ed.). 1977. *The Collected Papers of Charles Darwin.* 2 vols. Chicago: University of Chicago Press.

†Barrett, P. H. et al. (eds.). 1987. *Charles Darwin's Notebooks, 1836–1844*. Ithaca, NY: Cornell University Press.

†Bates, W. 1862a. Contributions to an insect fauna of the Amazon valley. *Lepidoptera: Heliconidae. Transactions of the Linnean Society of London*, **23**, 495–566.

†Bates, W. 1862b. Contributions to an insect fauna of the Amazon valley. – *Lepidoptera: – Heliconidae*. Abstract. *Journal of the Proceedings of the Linnean Society. Zoology*, **6**, 73–77.

†Bateson, W. 1894. *Materials for the Study of Variation, Treated with Especial Regard to Discontinuity in the Origin of Species*. London: Macmillan.

†Bateson, W. (with Saunders, E. R.). 1902a. Experimental studies in the physiology of heredity. In *Reports to the Evolution Committee of the Royal Society, Report 1 (17 Dec. 1901)*, pp. 1–160. London: Harrison and Sons.

†Bateson, W. 1902b. *A Defence of Mendel's Principles of Heredity*. Cambridge: Cambridge University Press.

†Bateson, W. 1903. *Problems of Genetics*. New Haven: Yale University Press.

†Bateson, W. 1909. *Mendel's Principles of Heredity*. Cambridge: Cambridge University Press.

†Bateson, W. 1914. President's address. *Report of the British Association for the Advancement of Science*, 3–38.

Beddall, B. G. 1968. Wallace, Darwin and the theory of natural selection. *Journal of the History of Biology*, **1**, 261–323.

Bennett, J. H. 1983. *Natural Selection, Heredity and Eugenics, Including Correspondence of R. A. Fisher with Leonard Darwin and Others*. Oxford: Clarendon Press.

Berge, S. 1961. The historical development of animal breeding. *Schriftenreihe des Max-Planck-Instituts für Tierzucht und Tierernahrung*, pp. 110–127.

†Bergson, H. 1907. *L'évolution créatrice*. Paris: Alcan (62nd edn, 1946). Paris: Presses Universitaires de France.

Blacher, L. I. 1982. *The Problem of the Inheritance of Acquired Characters: A History of a Priori and Empirical Methods Used to Find a Solution*. Washington: Smithsonian Institution Libraries. (1st Russian edn, 1971).

Blackher, C. P. 1952. *Eugenics: Galton and After*. London: Duckworth.

Blakley, G. R. 1967. Darwinian natural selection acting within populations. *Journal of Theoretical Biology*, **17**, 252–281.

Bowler, P. J. 1976. Alfred Russel Wallace's concepts of variation. *Journal of the History of Medicine*, **31**, 17–29.

Bowler, P. J. 1977a. Darwinism and the argument from design: Suggestions for a reevaluation. *Journal of the History of Biology*, **10**, 29–43.

Bowler, P. J. 1977b. Edward Drinker Cope and the changing structure of evolution theory. *Isis*, **68**, 249–265.

Bowler, P. J. 1979. Theodor Eimer and orthogenesis: Evolution by definitely directed variation. *Journal of the History of Medicine*, **34**, 40–73.

Bowler, P. J. 1983. *The Eclipse of Darwinism: Anti-Darwinian Evolution Theories in the Decades around 1900*. Baltimore: Johns Hopkins University Press.

Bowler, P. J. 1984. *Evolution: The History of an Idea*. Berkeley and Los Angeles: University of California Press.

Bowler, P. J. 1989. *The Mendelian Revolution: The Emergence of Hereditarian Concepts in Modern Science and Society*. Baltimore: Johns Hopkins University Press.

Box, J. Fisher. 1978. *R. A. Fisher: The Life of a Scientist*. New York: Wiley.

Brackman, A. C. 1980. *A Delicate Arrangement: The Strange Case of Charles Darwin and Alfred Russel Wallace*. New York: Times Books.

Brandon, R. N. and Burian, R. M. (eds.). 1984. *Genes, Organisms, Populations.* Cambridge, MA: MIT Press.

Brandon, R. N. 1990. *Adaptation and Environment.* Princeton: Princeton University Press.

Brandon, R. N. 1996. *Concepts and Methods in Evolutionary Biology*. Cambridge: Cambridge University Press.

†Bravais, A. 1846. Analyse mathématique sur les probabilités des erreurs de situation d'un point. *Mémoires présentés par divers savants à l'Académie Royale des Sciences de l'Institut de France,* **9**, 255–332.

†Buffon, G. L. Leclerc de. 1779. *Les époques de la nature.* Paris. Critical edition by J. Roger, Paris: Editions du Museum de Paris, 1962. (Reprinted 1988).

†Buffon, G. L. Leclerc de. 1954. *Oeuvres philosophiques,* ed. J. Piveteau. Paris: Presses Universitaires de France.

Buican, D. 1984. *Histoire de la génétique et de l'évolutionnisme en France.* Paris: Presses Universitaires de France.

†Bumpus, H. C. 1896. The variations and mutations of the introduced sparrow, *Passer domesticus. Woods Hole Museum Biological Laboratories: Biological Lectures,* 1–15.

†Bumpus, H. C. 1898. The elimination of the unfit as illustrated by the introduced sparrow, *Passer domesticus* (A fourth contribution to the study of variation). *Woods Hole Museum Biological Laboratories: Biological Lectures,* 209–226.

Burian, R. M. 1983. Adaptation. In *Dimensions of Darwinism,* ed. M. Grene. Cambridge: Cambridge University Press.

Burian, R. M. 1989. The influence of the evolutionary paradigm. In *Evolutionary Biology at the Crossroads,* ed. M. Hecht. New York: Queens College Press.

Burian, R. M., Gayon, J. and Zallen, D. 1988. The singular fate of genetics in the history of French biology. *Journal of the History of Biology,* **21**, 357–402.

Burkhardt, F. and Smith, S. (eds.). 1985–. *The Correspondence of Charles Darwin.* Cambridge: Cambridge University Press. (10 vols. published).

†Cain, A. J. and Sheppard, P. M. 1950. Selection in the polymorphic land snail *Cepaea nemoralis. Heredity,* **4**, 275–294.

†Cain, A. J. and Sheppard, P. M. 1954. Natural selection in *Cepaea. Genetics,* **39**, 89–116.

†Candolle, A. de. 1873. *Histoire des sciences et des savants depuis deux siècles, suivie d'autres études sur des sujets scientifiques, en particulier sur la sélection dans l'espèce humaine.* Geneva. 2nd edn, 1885. Geneva. Partial reprint under the title *Histoire des sciences et des savants depuis deux siècles, d'après l'opinion des principales académies ou sociétés scientifiques.* Paris: Fayard, 1987.

Canguilhem, G. 1959. Les concepts de 'lutte pour l'existence' et de 'sélection naturelle' en 1858: Charles Darwin et Alfred Russel Wallace. Paris: Conférences du Palais de la Découverte. Reproduced in G. Canguilhem, *Etudes d'histoire et de philosophie des sciences,* 3rd edn, pp. 99–110. Paris: Vrin, 1975.

Canguilhem, G. (ed.). 1972. *La mathématisation des doctrines informes.* Paris: Hermann.

Canguilhem, G. 1981. *Idéologie et rationalité dans les sciences de la vie.* 2nd edn. Paris: Vrin. English translation: *Ideology and Rationality in the History of the Life Sciences,* 1988. Cambridge, MA: MIT Press.

Canguilhem, G. 1994. *Selected Writings.* See Delaporte, F.

Canguilhem, G., Lapassade, G., Piquemal, J. and Ulmann, J. 1985. *Du développe-*

ment à l'évolution au XIXème siècle. Paris: Presses Universitaires de France.

Caplan, A. L. 1980. Philosophical issues concerning the modern synthetic theory of evolution. Unpublished doctoral dissertation, Columbia University, New York.

Carlson, E. A. 1966. *The Gene: A Critical History.* London: Saunders.

Carlson, E. A. 1981. *Genes, Radiation and Society: The Life and Work of H. J. Muller.* Ithaca, NY: Cornell University Press.

†Castle, W. E. 1903a. Mendel's law of heredity. *Proceedings of the American Academy of Arts and Sciences,* **38,** 535–548.

†Castle, W. E. 1903b. The laws of heredity of Galton and Mendel, and some laws governing race improvement by selection. *Proceedings of the American Academy of Arts and Sciences,* **39,** 223–242.

†Castle, W. E. (with Allen, G. M.). 1903c. The heredity of albinism. *Proceedings of the American Academy of Arts and Sciences,* **38,** 603–622.

†Castle, W. E. 1905a. Recent discoveries in heredity and their bearing on animal breeding. *Popular Science Monthly,* **66,** 193–208.

†Castle, W. E. 1905b. The mutation theory of organic evolution from the standpoint of animal breeding. *Science,* n.s., **21,** 521–525.

†Castle, W. E. 1906. Yellow mice and gametic purity. *Science,* n.s., **24,** 275–281.

†Castle, W. E. (with MacCurdy, H.). 1907. Selection and cross-breeding in relation to the inheritance of coat-pigments and coat-patterns in rats and guinea-pigs. *Carnegie Institution Publications,* **70.**

†Castle, W. E. 1914a. Variation and selection: A reply. *Zeitschrift für Induktive Abstammungs- und Vererbungslehre,* **12,** 255–264.

†Castle, W. E. (with Phillips, J. C.). 1914b. Piebald rats and selection: An experimental test of the effectiveness of selection and of the theory of gametic purity in Mendelian crosses. *Carnegie Institution Publications,* **195,** 1–56.

†Castle, W. E. 1915. Some experiments in mass selection. *American Naturalist,* **49,** 713–726.

†Castle, W. E. 1919. Piebald rats and selection. *American Naturalist,* **53,** 369–375.

†Chetverikov, S. S. 1926. On certain aspects of the evolutionary process from the standpoint of modern genetics. Original Russian publication: *Zhurnal Eksperimental'noi Biologii, A,* **2,** 3–54. English translation by M. Baker, *Proceedings of the American Philosophical Society,* **105,** 167–195 (1961).

†Chetverikov, S. S. 1927. Ueber die genetische Beschaffenheit wilder Populationen. *Proceedings of the 5th International Congress of Genetics,* pp. 1499–1500. Heidelberg: Springer-Verlag. English translation by R. A. Jameson, On the genetic constitution of wild populations, *Proceedings of the American Philosophical Society,* **105,** 263–264 (1961).

Clark, R. 1968. *The Life and Work of J. B. S. Haldane.* Oxford: Oxford University Press.

Cock, A. G. 1973. William Bateson, Mendelism and biometry. *Journal of the History of Biology,* **6,** 1–36.

Conry, Y. 1974. *L'introduction du darwinisme en France au XIXème siècle.* Paris: Vrin.

Conry, Y. 1981. Organisme et organisation: de Darwin à la génétique des populations. *Revue de Synthèse,* **102,** 291–330.

†Correns, C. 1900. G. Mendel's Regel über das Verhalten der Nachkommenschaft der Rassen Bastarde. *Berichte der Deutschen Botanischen*

Gesellschaft, **18**, 158–168. English translation: G. Mendel's law concerning the behavior of varietal hybrids. In *The Origin of Genetics: A Mendel Source Book*, ed. C. Stern and E. R. Sherwood, 1966, pp. 117–132. San Francisco: Freeman.

Cowan, R. S. 1968. Sir Francis Galton and the continuity of germplasm: A biological idea with political roots. *12ème Congrès International de l'Histoire des Sciences*, vol. VIII. Paris: Blanchard.

Cowan, R. S. 1972a. Francis Galton's contributions to genetics. *Journal of the History of Biology*, **5**, 389–412.

Cowan, R. S. 1972b. Francis Galton's statistical ideas: The influence of eugenics. *Isis*, **63**, 509–528.

Cowan, R. S. 1977. Nature and nurture: The interplay of biology and politics in the work of Francis Galton. *Studies in History of Biology*, **1**, 133–208.

Cowan, R. S. 1985. *Sir Francis Galton and the Study of Heredity in the 19th Century*. New York: Garland.

Coyne, J. A., Barton, N. H. and Turelli, M. 1997. Perspective: A critique of Sewall Wright's shifting balance theory of evolution. *Evolution*, **51**, 643–671.

†Crow, J. F. 1955. General theory of population genetics synthesis. *Cold Spring Harbor Symposia on Quantitative Biology*, **20**, 24–59.

†Crow, J. F. and Kimura, M. 1970. *An Introduction to Population Genetics Theory*. New York: Harper and Row.

†Cuénot, L. 1902. Les recherches expérimentales sur l'hérédité. *L'Année Biologique*, **7**, 58–77.

†Cuénot, L. 1911. *La genèse des espèces animales*. Paris: Alcan. 2nd edn, 1921. Paris: Alcan. 3rd edn, 1932. Paris: Alcan.

†Cuénot, L. 1914. Théorie de la préadaptation. *Scientia*, **16**, 60–73.

†Cuvier, G. 1812. *Recherches sur les ossements fossiles des quadrupèdes (Discours préliminaire)*. Paris: Deterville. Introduction reprinted in 1881 as *Discours sur les révolutions du globe*. Paris: Berche et Tralin.

Dagognet, F. 1970. *Le catalogue de la vie*. Paris: Presses Universitaires de France.

Dagognet, F. 1977. *Une épistémologie de l'espace concret: néo-géographie*. Paris: Vrin.

Dagognet, F. 1982. *Faces, surfaces, interfaces*. Paris: Vrin.

†Darwin, C. *Correspondence*. See Burkhardt, F.

†Darwin, C. 1838–9. Notebooks on transmutation of species. See de Beer, G. (1960) and Barrett, P. H. (1987).

†Darwin, C. 1839a. *Journal of Researches into the Geology and Natural History of the Various Countries Visited by H.M.S. Beagle, under the Command of Captain Fitz Roy from 1832 to 1836*. London: Colburn. 2nd rev. edn, 1845. London: Murray.

†Darwin, C. 1839b. *Zoology of the Voyage of H.M.S. Beagle. Part II: Mammalia*. London: Smith Elder.

†Darwin, C. 1859. *On the Origin of Species by Means of Natural Selection, or the Preservation of Favoured Races in the Struggle for Life*. London: Murray. (Facsimile: Cambridge, MA: Harvard University Press, 1964). 2nd edn, 1860 (in fact, 26 Dec. 1859). London: Murray. 3rd edn, 1861. London: Murray. 4th edn, 1866. London: Murray. 5th edn, 1869. London: Murray. 6th edn, 1872. London: Murray. Critical reference edition: M. Peckham (ed.), *A Variorum Text*, 1959. Philadelphia: University of Pennsylvania Press.

†Darwin, C. 1863. A review of 'Contributions to an insect fauna of the Amazon

valley', by Henry Bates, Esq. *Natural History Review: Quarterly Journal of Biological Science*, 219–224.

†Darwin, C. 1868. *The Variation of Animals and Plants under Domestication.* 2 vols. London: Murray. 2nd edn, 1875. London: Murray. Facsimiles: 1896, New York: Appleton; 1972, New York: AMS Press. Unless otherwise stated, quotations refer to the 1896 facsimile.

†Darwin, C. 1871. *The Descent of Man, and Selection in Relation to Sex.* 2 vols. London: Murray. (Facsimile: Princeton: Princeton University Press, 1981). 2nd edn, rev. 1 vol. London: Murray, 1874.

†Darwin, C. 1876. *The Effects of Cross and Self Fertilization in the Vegetable Kingdom.* London: Murray.

†Darwin, C. [Written 1876]. *Autobiography.* See de Beer (1983).

†Darwin, C. 1972. *The Works of Charles Darwin.* New York: AMS Press.

†Darwin, C. 1977. *Collected Papers.* See Barrett, P. H. (1977).

†Darwin, C. and Wallace, A. R. 1859. [Read 1 July 1858]. On the tendency of species to form varieties; and on the perpetuation of varieties and species by natural means of selection. *Journal of the Proceedings of the Linnean Society (Zoology)*, **3**, 45–62. Reproduced in *The Collected Papers of Charles Darwin*, ed. P. H. Barrett, 1977, vol. 2, pp. 3–19. Chicago: University of Chicago Press. See also Darwin and Wallace (1958a).

†Darwin, C. and Wallace, A. R. 1958a. *Evolution by Natural Selection. Darwin and Wallace. Darwin's Sketch of 1842, His Essay of 1844, and the Darwin–Wallace Papers of 1858.* Introduction by Sir Francis Darwin and foreword by Sir Gavin de Beer. Cambridge: Cambridge University Press.

†Darwin, F. (ed.). 1887. *Life and Letters of Charles Darwin.* 3 vols. London: Murray. (Reprinted 1969, New York: Johnson Reprint).

†Darwin, F. and Seward, A. C. 1903. *More Letters of Charles Darwin.* London: Murray.

†Dawkins, R. 1978. Replication, selection and the extended phenotype. *Zeitschrift für Tierpsychologie*, **47**, 61–76.

†de Beer, G. (R. Gavin). 1930. *Embryology and Evolution.* Oxford: Clarendon Press.

†de Beer, G. (R. Gavin). 1940. *Embryos and Ancestors.* Oxford: Clarendon Press. 2nd edn, 1951. Oxford: Clarendon Press. 3rd edn, 1958. Oxford: Clarendon Press.

†de Beer, G. (R. Gavin). 1960. Darwin's notebooks on transmutation of species. *Bulletin of the British Museum (Natural History), Historical Series*, **2** (2–5). Part I: First Notebook (July 1837–Feb. 1838); Part II: Second Notebook (Feb. 1838–July 1838); Part III: Third Notebook (July 1838–Oct. 1838); Part IV: Fourth Notebook (Oct. 1838–July 1839).

†de Beer, G. (R. Gavin). 1983. *Charles Darwin [and] Thomas Henry Huxley: Autobiographies* [1974]. Oxford: Oxford University Press.

†Debru, C. 1983. *L'esprit des protéines: histoire et philosophie biochimiques.* Paris: Hermann.

†Delage, Y. and Goldsmith, M. 1909. *Les théories de l'évolution.* Paris: Flammarion. English translation: *The Theories of Evolution.* New York: Huebsch, 1912.

Delaporte F. (ed.) 1994. *A Vital Rationalist: Selected Writings from Georges*

Canguilhem. Introduction by Paul Rabinow and critical bibliography by Camille Limoges. Translated by Arthur Goldhammer. New York: Zone Books.

Delsol, M., Sentis, P. and Flatin, J. 1989. Théorie synthétique et sélection. Analyse épistémologique. In *La vie*, ed. F. Tinland, pp. 19–53. Paris: Vrin.

†Dennert, E. 1904. *At the Death-Bed of Darwinism*. (Original German edition: 1903). English translation by E. V. O'Harra and J. H. Peschges. Burlington, IA: German Literary Board.

Depew, J. and Weber, B. H. (eds.). 1985. *Evolution at a Crossroads: The New Biology and the New Philosophy of Science*. Cambridge, MA: MIT Press.

Depew, D. J. and Weber, B. H. 1995. *Darwinism Evolving: Systems Dynamics and the Genealogy of Natural Selection*. Cambridge, MA: MIT Press.

†de Vries, H. 1889. *Intracelluläre Pangenesis*. Jena: Gustav Fischer. English translation: *Intracellular Pangenesis* (translated by C. S. Gager). Chicago: Open Court, 1910

†de Vries, H. 1894. Ueber halbe Galton-Curven ab Ziechen diskontinuierlicher Variation. *Berichte der Deutschen Botanischen Gesellschaft*, 7, 64ff. French version (1895): Les demi-courbes galtoniennes comme indice de variation discontinue. *Archives Néerlandaises des Sciences Exactes et Naturelles*, **28**, 442ff.

†de Vries, H. 1900a. Sur la loi de disjonction des hybrides. *Comptes Rendus Hebdomadaires des Séances de l'Académie des Sciences*, **130**, 845–847.

†de Vries, H. 1900b. Das Spaltungsgesetz der Bastarde (Vorläufige Mitteilung). *Berichte der Deutschen Botanischen Gesellschaft*, **18**, 83–90. English translation in *The Origin of Genetics: A Mendel Source Book*, ed. C. Stern and E. R. Sherwood, 1966, pp. 107–117. San Francisco: Freeman.

†de Vries, H. 1901–3. *Die Mutationstheorie: Versuche und Beobachtungen über die Entstehung der Arten im Pflanzenreich*. 2 vols. Leipzig: Veit. English translation: *The Mutation Theory: Experiments and Observations on the Origin of Species in the Vegetable Kingdom* (translated by J. B. Farmer and A. D. Darbishire). Chicago: Open Court (1909–10). Vol. I reprinted 1969, New York: Klaus Reprint.

Dexter, R. W. 1957. The development of A. S. Packard, Jr., as a naturalist and entomologist. *Bulletin of the Brooklyn Entomological Society*, **52**, 57–66, 101–112.

Dexter, R. W. 1979. The impact of evolutionary theories on the Salem group of Agassiz zoologists. *Essex Institute Historical Collections*, **115**, 144–171.

†Dobzhansky, T. 1937. *Genetics and the Origin of Species*. New York: Columbia University Press. 2nd edn, 1941. New York: Columbia University Press. 3rd edn, 1951. New York: Columbia University Press.

†Dobzhansky, T. et al. 1938–76. *The Genetics of Natural Populations, I–XLIII*, ed. R. C. Lewontin, J. A. Moore, W. B. Provine and B. Wallace. New York: Columbia University Press (a collection of 43 articles by Dobzhansky and 24 colleagues published in the journals *Genetics* and *Evolution*).

†Dobzhansky, T. 1950. Mendelian populations and their evolution. In *Genetics in the 20th Century: Essays on the Progress of Genetics during Its First 50 Years*, ed. L. C. Dunn. New York: Macmillan.

†Dobzhansky, T. 1955. A review of some fundamental concepts and problems of population genetics. *Cold Spring Harbor Symposia on Quantitative Biology*, **20**, 1–5.

†Dobzhansky, T. 1968. Adaptedness and fitness. In *Population Biology and Evolution*, ed. R. C. Lewontin, pp. 109–121. Syracuse, NY: Syracuse University Press.

†Dobzhansky, T. 1969. *The Biology of Ultimate Concern.* New York: New American Library.

†Dobzhansky, T. 1970. *Genetics of the Evolutionary Process.* New York: Columbia University Press.

†Dobzhansky, T. 1980. The birth of the genetic theory of evolution in the Soviet Union in the 1920s. In *The Evolutionary Synthesis*, ed. E. Mayr and W. B. Provine. Cambridge, MA: Harvard University Press.

†Dowdeswell, W. H., Fisher, R. A. and Ford, E. B. 1940. The quantitative study of populations in the Lepidoptera. *Annals of Eugenics*, **10**, 123–136.

Dronnamraju, K. R. 1968. *Haldane and Modern Biology.* Baltimore: Johns Hopkins University Press.

Duhem, P. 1906. *La théorie physique. Son objet – sa structure.* Paris: Chevalier et Rivière.

Dunn, L. C. 1965a. *A Short History of Genetics.* New York: McGraw-Hill.

Dunn, L. C. 1965b. William Ernst Castle. *Biographical Memoirs of the National Academy of Sciences*, **38**, 33–80.

Dupré, J. (ed.). 1987. *The Latest on the Best: Essays on Evolution and Optimality.* Cambridge, MA: MIT Press.

†Edgeworth, F. Y. 1892. The law of error and correlated averages. *Philosophical Magazine*, **34**, 192–204.

†Edgeworth, F. Y. 1893. A new method of treating correlated averages. *Philosophical Magazine*, **35**, 63–64.

†Edwards, A. W. F. 1967. Fundamental theorem of natural selection. *Nature*, **215**, 537–538.

†Edwards, A. W. F. 1971. Review of S. Wright, *Evolution and the Genetics of Populations*, vol. II, *The Theory of Gene Frequencies*. *Heredity*, **26**, 332–337.

†Edwards, A. W. F. 1977. *Foundations of Mathematical Genetics.* Cambridge: Cambridge University Press.

Eiseley, L. 1958. *Darwin's Century: Evolution and the Men Who Discovered It.* New York: Doubleday.

Ekeland, I. 1984. *Le calcul et l'imprévu: les figures du temps de Kepler à Thom.* Paris: Seuil.

†Eldredge, N. 1979. Alternative approaches to evolutionary theory. *Bulletin of the Carnegie Museum of Natural History*, **13**, 7–19.

†Falconer, D. S. 1960. *Introduction to Quantitative Genetics.* New York: Ronald Press.

Fancher, R. E. 1983a. Biographical origins of Francis Galton's psychology. *Isis*, **74**, 227–233.

Fancher, R. E. 1983b. Francis Galton's African ethnology and its role in the development of his ideas. *British Journal of the History of Science*, **16**, 67–79.

†Fisher, R. A. 1912–64. *Collected Papers*, ed. J. H. Bennett. 5 vols. Adelaide: University of Adelaide.

†Fisher, R. A. (with Stock, C. S.). 1915. Cuénot on preadaptation. *Eugenics Review*, 7, 46–61.

†Fisher, R. A. 1918. The correlation between relatives on the supposition of Mendelian inheritance. *Transactions of the Royal Society of Edinburgh*, **52**, 399–433.

†Fisher, R. A. 1922. On the dominance ratio. *Proceedings of the Royal Society of Edinburgh*, **42**, 321–341.

†Fisher, R. A. 1927a. On some objections to mimicry theory: statistical and genetic. *Transactions of the Entomological Society of London*, **75**, 269–278.

†Fisher, R. A. 1927b. The actuarial treatment of official birth records. *Eugenics Review*, **19**, 103–108.

†Fisher, R. A. 1928. The possible modification of the response of the wild type to recurrent mutations. *American Naturalist*, **2**, 115–126.

†Fisher, R. A. 1930a. The distribution of gene ratios for rare mutations. *Proceedings of the Royal Society of Edinburgh*, **50**, 205–220.

†Fisher, R. A. 1930b. *The Genetical Theory of Natural Selection*. Oxford: Clarendon Press. 2nd edn, 1958. New York: Dover.

†Fisher, R. A. 1934. Indeterminism and natural selection. *Philosophy of Science*, **1**, 99–117.

†Fisher, R. A. 1936. Has Mendel's work been rediscovered? *Annals of Science*, **1**, 115–137.

†Fisher, R. A. 1937. The relation between variability and abundance shown by the measurements of the eggs of British nesting birds. *Proceedings of the Royal Society of London*, ser. B, **112**, 1–26.

†Fisher, R. A. 1939. Selective forces in wild populations of *Paratettix texanus*. *Annals of Eugenics*, **9**, 109–122.

†Fisher, R. A. 1941. Average excess and average effect of a gene substitution. *Annals of Eugenics*, **11**, 53–63.

†Fisher, R. A. 1953. Population genetics. *Proceedings of the Royal Society of London*, ser. B, **141**, 510–523.

†Fisher, R. A. and Ford, E. B. 1947. The spread of a gene in natural conditions in a colony of the moth *Panaxia dominula* L. *Heredity*, **1**, 143–174.

†Ford, E. B. 1931. *Mendelism and Evolution*. London: Methuen.

†Ford, E. B. 1937. Problems of heredity in the Lepidoptera. *Biological Review*, **12**, 461–503.

†Ford, E. B. 1939. Genetic research in the Lepidoptera. *Annals of Eugenics*, **10**, 227–252 plus pl.

†Ford, E. B. 1940. Polymorphism and taxonomy. In *The New Systematics*, ed. J. Huxley, pp. 493–513. London: Oxford University Press.

†Ford, E. B. 1945. Polymorphism. *Biological Review*, **20**, 73–88.

†Ford, E. B. 1964. *Ecological Genetics*. London: Methuen.

†Ford, E. B. 1965. *Genetic Polymorphism*. London: Faber.

Foucault, M. 1966. *Les mots et les choses: une archéologie des sciences humaines*. Paris: Gallimard.

Foucault, M. 1969. *L'archéologie du savoir*. Paris: Gallimard.

Fox Keller, E. 1987. Reproduction and the central project of evolutionary theory. *Biology and Philosophy*, **2**, 383–396.

Fox Keller, E. and Lloyd, E. 1992. *Keywords in Evolutionary Biology*. Cambridge, MA: Harvard University Press.

Frogatt, P. and Nevin, N. C. 1971. The law of ancestral heredity and the Mendelian–Ancestrian controversy in England, 1899–1906. *Journal of Medical Genetics*, **8**, 1–36.

Gall, H. and Putschar, E. (eds.). 1955. *Selected Readings in Biology for Natural Sciences*, vol. III. Chicago: University of Chicago Press.

†Galton, F. 1865. Hereditary talent and character. *Macmillan's Magazine*, **1 (68)**, 157–166; **12 (71)**, 318–327. Facsimile in *Eugenics Then and Now*, Benchmark

Papers in Genetics, vol. V, ed. C. J. Bajema, 1976, pp. 14–32. Stroudsburg, PA: Dowden, Hutchinson and Ross.

†Galton, F. 1869. *Hereditary Genius: An Inquiry into Its Laws and Consequences.* London: Macmillan. 2nd edn, 1892. London: Macmillan.

†Galton, F. 1871. Experiments in pangenesis by breeding from rabbits of a pure variety, into whose circulation blood taken from other varieties had previously been largely transfused. *Proceedings of the Royal Society of London*, **19**, 394–410.

†Galton, F. 1872. On blood-relationship. *Proceedings of the Royal Society of London*, **20**, 394–402.

†Galton, F. 1874a. *English Men of Science: Their Nature and Nurture*. London: Macmillan.

†Galton, F. (with Watson, H. W.). 1874b. On the probability of extinction of families. *Journal of the Anthropological Institute of Great Britain and Ireland*, **4**, 138–144.

†Galton, F. 1875a. Statistics by intercomparison, with remarks on the law of error. *Philosophical Magazine*, 4th ser., **49**, 33–46.

†Galton, F. 1875b. A theory of heredity. *Contemporary Review*, **27**, 80–95.

†Galton, F. 1876. A theory of heredity. *Journal of the Anthropological Institute of Great Britain and Ireland*, **5**, 329–348.

†Galton, F. 1877. Typical laws of heredity. *Nature*, **15**, 492–495, 512–533.

†Galton, F. 1879a. The geometric mean in vital and social statistics. *Proceedings of the Royal Society of London*, **29**, 365–367.

†Galton, F. 1879b. The law of the geometric mean. *Proceedings of the Royal Society of London*, **29**, 367.

†Galton, F. 1883. *Inquiries into Human Faculty and Its Development*. London: Macmillan.

†Galton, F. 1884. *Record of Family Faculties*. London: Macmillan.

†Galton, F. 1885a. Inheritance and regression. *Transactions of the British Association* (Address to the Anthropological Section), **55**, 1206–1214.

†Galton, F. 1885b. Regression towards mediocrity in hereditary stature. *Journal of the Anthropological Institute of Great Britain and Ireland*, **15**, 246–263.

†Galton, F. 1886a. Family likeness in stature. *Proceedings of the Royal Society of London*, **40**, 42–73.

†Galton, F. 1886b. Family likeness in eye-colour. *Proceedings of the Royal Society of London*, **40**, 402–416.

†Galton, F. 1888. Co-relations and their measurement, chiefly from anthropological data. *Proceedings of the Royal Society of London*, **45**, 135–145.

†Galton, F. 1889. *Natural Inheritance.* London: Macmillan.

†Galton, F. 1892. *Finger Prints.* London: Macmillan.

†Galton, F. 1894a. Discontinuity in evolution. *Mind*, n.s., **3**, 362–372.

†Galton, F. 1894b. The part of religion in human evolution. *National Review* (August), 755–765.

†Galton, F. 1897a. Rate of racial change that accompanies different degrees of severity in selection. *Nature*, **55**, 605–606.

†Galton, F. 1897b. Hereditary colour in horses. *Nature*, **56**, 598–599.

†Galton, F. 1897c. The average contribution of each several ancestor to the total heritage of the offspring. *Proceedings of the Royal Society of London*, **61**, 401–413.

†Galton, F. 1898. A diagram of heredity. *Nature*, **57**, 293.

†Galton, F. 1901. The possible improvement of the human breed under the existing conditions of law and sentiment. Huxley Lecture of the Anthropological Institute. Reprinted in *Essays in Eugenics* (1909). Reproduced in *Nature*, **64**, 659–665 (1970). Facsimile in Bajema (1976), pp. 33–39.

†Galton, F. 1905. Eugenics: Its definition, scope and aims. *Sociological Papers* (London: Macmillan), pp. 45–50. Facsimile in Bajema (1976), pp. 40–45.

†Galton, F. 1908. *Memories of My Life*. London: Macmillan.

†Galton, F. 1909a. *Essays in Eugenics*. London: Eugenics Education Society.

†Galton, F. 1909b. Segregation (of the feeble-minded). In *The Problem of the Feeble-Minded*. London: King.

†Galton, F. 1910. The eugenic college of Kantsaywhere. Unpublished book. Extracts appear in K. Pearson (ed.), 1914–30, *The Life, Letters and Labours of Francis Galton*, vol. III A, pp. 411–425.

†Galton, F. 1924. *Life and Letters*. See K. Pearson (1914–30).

Gayon, J. 1987. Génétique des populations et évolutionnisme: esquisse d'une histoire générale (1976–1980). *Cahiers d'Histoire de la Philosophie des Sciences*, **22**, 86–110.

Gayon, J. 1988a. L'intelligence naturelle: mode de fabrication et mode d'emploi. *Milieux*, **31**, 60–67.

Gayon, J. 1988b. Le théorème fondamental de la sélection naturelle de R. A. Fisher: une approche historique. In *Biologie théorique. Solignac 1986*, ed. A. Kretzchmar, pp. 45–72. Paris: Editions du CNRS.

Gayon, J. (with Burian, R. M. and Zallen D.). 1988c. See Burian, Gayon and Zallen.

Gayon, J. 1989a. Epistémologie du concept de sélection. In *L'âge de la science*, vol. II, *Epistémologie*, pp. 201–207. Paris: Odile Jacob.

Gayon, J. 1990a. Critics and criticisms of the modern synthesis: the viewpoint of a philosopher. *Evolutionary Biology*, **24**, 1–49.

Gayon, J. 1990b. Le formel et l'empirique dans l'histoire de la génétique des populations. In *Histoire de la génétique: pratiques, techniques et théories*, ed. J. L. Fischer and W. H. Schneider, pp. 173–207. Paris: Editions Sciences en Situation.

Gayon, J. 1992a. Comment le problème de l'eugénisme se pose-t-il aujourd'hui?, in *La santé, la maladie*, ed. M.-A. Bernardis, pp. 290–295. Paris: Editions du Seuil et Cité des Sciences et de l'Industrie.

Gayon, J. 1992b. Entre eugénisme et théorie mathématique de l'évolution: la construction du concept de *fitness*. In *Darwinisme et société*, ed. P. Tort, pp. 489–499. Paris: Presses Universitaires de France.

Gayon, J. 1992c. L'individualité de l'espèce: une thèse transformiste? *Buffon 88. Actes du colloque international Paris–Dijon–Montbard*, ed. J. Gayon, pp. 475–489. Paris: Vrin.

Gayon, J. 1992d. L'espèce sans la forme. In *Les figures de la forme: philosophie biologique et esthétique*, ed. J. Gayon and J. J. Wunenburger, pp. 49–61. Paris: L'Harmattan.

Gayon, J. 1993. Le concept de récapitulation à l'épreuve de la théorie darwinienne de l'évolution. In *Histoire du concept de récapitulation: Ontogenèse et phylogenèse en biologie et sciences humaines*, ed. P. Mengal, pp. 79–92. Paris: Masson.

Gayon, J. 1994a. Sélection naturelle et sélection artificielle: le principe darwinien est-il métaphorique ? In *La maîtrise du milieu*, ed. P. Acot, pp. 133–147. Paris: Vrin.

Gayon, J. 1994b. What does 'Darwinism' mean? *Ludus Vitalis*, 2, 105–118.

Gayon, J. 1995a. Neo-Darwinism. In *Concepts, Theories, and Rationality in the Biological Sciences: The Second Pittsburgh–Konstanz Colloquium in the Philosophy of Science (Pittsburgh, October 1–4, 1993)*, ed. G. Wolters, J. G. Lennox and P. McLaughlin, pp. 1–25. Pittsburgh, PA: Universitätsverlag Konstanz and University of Pittsburgh Press.

Gayon, J. 1995b. Sélection naturelle ou survie des plus aptes? Eléments pour une histoire du concept de *fitness* dans la théorie évolutionniste. In *Nature, Histoire, Société: hommage à Jacques Roger*, ed. C. Blanckaert, J. L. Fischer and R. Rey, pp. 263–287. Paris: Klincksieck.

Gayon, J. 1996a. Biométrie. In *Dictionnaire du darwinisme et de l'évolution*, ed. P. Tort, vol. I, pp. 318–329. Paris: Presses Universitaires de France.

Gayon, J. 1996b. Entry for Fisher, Ronald Aylmer, 1890–1962. In *Dictionnaire du darwinisme et de l'évolution*, ed. P. Tort, vol. II, pp. 1671–1674. Paris: Presses Universitaires de France.

Gayon, J. 1996c. Entry for Weldon, Walter Frank Raphael, 1860–1906. In *Dictionnaire du darwinisme et de l'évolution*, ed. P. Tort, vol. III, pp. 4618–4620. Paris: Presses Universitaires de France.

Gayon, J. 1996d. Pearson: note sur les travaux évolutionnistes théoriques. In *Dictionnaire du darwinisme et de l'évolution*, ed. P. Tort, vol. III, pp. 3394–3398. Paris: Presses Universitaires de France.

Gayon, J. 1996e. Entry for Wright, Sewall, 1889–1988. In *Dictionnaire du darwinisme et de l'évolution*, ed. P. Tort, vol. III, pp. 4686–4698. Paris: Presses Universitaires de France.

Gayon, J. 1997a. L'eugénisme. In *Précis de génétique humaine*, ed. J. Feingold, M. Fellous and M. Solignac. Paris: Hermann (in press).

Gayon, J. 1997b. Karl Pearson ou: les enjeux du phénoménalisme dans les sciences de la vie autour de 1900. In *Conceptions de la science: hier, aujourd'hui et demain. Colloque international d'hommage à Marjorie Grene, Dijon, 17–19 mai 1995*, ed. R. M. Burian and J. Gayon. Brussels: Ousia (in press).

Gayon, J. 1997c. The paramount power of selection: From Darwin to Kauffman. In *Proceedings of the Xth International Congress of Logic, Methodology and Philosophy of Science – Florence, August 1995*, vol. II, *Structures and Norms in Science*, ed. M. L. Dalla Chiara, K. Doets, D. Mundici and J. van Benthem, pp. 265–282. Dordrecht: Kluwer.

Gayon, J. 1997d. Sélection naturelle et régression dans la théorie évolutionniste darwinienne et post-darwinienne. In *Les pensées modernes de la décadence*, ed. P. A. Taguieff. Paris: Berg International (in press).

Ghiselin, M. 1969. *The Triumph of the Darwinian Method.* Chicago: University of Chicago Press.

Gillispie, C. 1951. *Genesis and Geology: A Study in the Relations of Scientific Thought, Natural Theology, and Social Opinion in Great Britain, 1790–1850.* New York: Harper and Row. (Reprinted 1959).

Gilson, E. 1971. *D'Aristote à Darwin et retour.* Paris: Vrin.

Glick, T. F. 1970. Science and the revolution of 1868: Notes on the reception of Darwinism in Spain. In *La revolución de 1868: historia, pensamiento literatura*, ed. C. E. Lida and Im-Zavala, pp. 267–272. New York: Las Americas.

Glick, T. F. (ed.). 1988. *The Comparative Reception of Darwinism.* 2nd edn. Chicago: University of Chicago Press.

†Goldschmidt, R. 1940. *The Material Basis of Evolution.* New Haven: Yale University Press. (Reprinted 1982).

†Gould, S. J. 1977. *Ontogeny and Phylogeny.* Cambridge, MA: Harvard University Press.

†Gould, S. J. 1980. Is a new and general theory of evolution emerging? *Paleobiology*, **6,** 119–130.

Gould, S. J. 1982. Introduction. In T. Dobzhansky, *Genetics and the Origin of Species* (1937). New York: Columbia University Press. Reprint.

Gould, S. J. 1991. *Bully for Brontosaurus: reflections in Natural History.* New York: Norton.

†Gould, S. J. and Lewontin, R. C. 1979. The spandrels of San Marco and the Panglossian paradigm: A critique of the adaptationist programme. *Proceedings of the Royal Society of London*, ser. B, **205,** 581–585.

Greene, J. C. 1971. The Kuhnian paradigm and the Darwinian revolution in natural history. In *Perspectives in the History of Science and Technology*, ed. D. H. D. Roller. Norman, OK: University of Oklahoma Press.

Greg, W. 1868. On the failure of 'Natural Selection' in the case of man. *Fraser's Magazine*, **68** (September), 353–362.

Grene, M. (ed.). 1983. *Dimensions of Darwinism: Themes and Counterthemes in Twentieth-Century Evolutionary Theory.* Cambridge: Cambridge University Press.

Gruber, H. E. 1974. *Darwin on Man: A Psychological Study of Scientific Creativity.* Chicago: University of Chicago Press.

†Haeckel, E. 1866. *Generelle Morphologie der Organismen: allgemeine Grundzüge der organischen Formen-Wissenschaft, mechanisch begründet durch die Charles Darwin reformiste Descendenz-Theorie.* 2 vols. Berlin: Georg Reimer.

†Hagedoorn, A. L. and A. C. 1921. *The Relative Value of the Processes Causing Evolution.* The Hague: Martinus Nijhoff.

†Haldane, J. B. S. 1924a. A mathematical theory of natural and artificial selection. Part I. *Transactions of the Cambridge Philosophical Society*, **23,** 19–41.

†Haldane, J. B. S. 1924b. A mathematical theory of natural and artificial selection. Part II. The influence of partial self-fertilisation, inbreeding, assortative mating, and selective fertilisation on the composition of Mendelian populations, and on natural selection. *Proceedings of the Cambridge Philosophical Society, Biological Sciences*, **1,** 158–163.

†Haldane, J. B. S. 1926. A mathematical theory of natural and artificial selection. Part III. *Proceedings of the Cambridge Philosophical Society*, **23,** 363–372.

†Haldane, J. B. S. 1927a. A mathematical theory of natural and artificial selection. Part IV. *Proceedings of the Cambridge Philosophical Society*, **23,** 607–615.

†Haldane, J. B. S. 1927b. A mathematical theory of natural and artificial selection. Part V. Selection and mutation. *Proceedings of the Cambridge Philosophical Society*, **23,** 833–844.

†Haldane, J. B. S. 1929a. Natural selection. *Nature*, **124,** 444.

†Haldane, J. B. S. 1929b. The species problem in the light of genetics. *Nature*, **124,** 514–516.

†Haldane, J. B. S. 1930a. A note on Fisher's theory of the origin of dominance, and on a correlation between dominance and linkage. *American Naturalist*, **64,** 37–90.

†Haldane, J. B. S. 1930b. A mathematical theory of natural and artificial selec-

tion. Part VI. Isolation. *Proceedings of the Cambridge Philosophical Society*, **26**, 220–230.

†Haldane, J. B. S. 1930c. Natural selection intensity as a function of mortality rate. *Nature*, **125**, 583.

†Haldane, J. B. S. 1930d. A mathematical theory of natural and artificial selection. Part VII. Selection intensity as a function of mortality rate. *Proceedings of the Cambridge Philosophical Society*, **27**, 131–136.

†Haldane, J. B. S. (with Waddington, C. H.). 1931a. Inbreeding and linkage. *Genetics*, **16**, 357–374.

†Haldane, J. B. S. 1931b. A mathematical theory of natural and artificial selection. Part VIII. Metastable populations. *Proceedings of the Cambridge Philosophical Society*, **27**, 137–142.

†Haldane, J. B. S. 1932a. *The Causes of Evolution*. London: Longman. (Reprinted 1966. Ithaca, NY: Cornell University Press).

†Haldane, J. B. S. 1932b. A method for investigating recessive characters in Man. *Journal of Genetics*, **25**, 251–255.

†Haldane, J. B. S. 1932c. A mathematical theory of natural and artificial selection. Part IX. Rapid selection. *Proceedings of the Cambridge Philosophical Society*, **28**, 244–248.

†Haldane, J. B. S. 1933. The part played by recurrent mutation in evolution. *American Naturalist*, **67**, 5–19.

†Haldane, J. B. S. 1934. A mathematical theory of natural and artificial selection. Part X. Some theorems on artificial selection. *Genetics*, **19**, 412–429.

†Haldane, J. B. S. 1937. The effect of variation on fitness. *American Naturalist*, **71**, 338–349.

†Haldane, J. B. S. 1938. L'analyse génétique des populations naturelles. In *Réunion Internationale de Physique–Chimie–Biologie (Congrès du Palais de la Découverte, Paris: 1937)*, pp. 105–112. Paris: Hermann. (Lecture followed by a discussion between P. L'Héritier and G. Teissier).

†Haldane, J. B. S. 1939a. The equilibrium between mutation and random extinction. *Annals of Eugenics*, **9**, 400–405.

†Haldane, J. B. S. 1939b. The spread of harmful autosomal recessive genes in human populations. *Annals of Eugenics*, **9**, 232–237.

†Haldane, J. B. S. 1940. The conflict between selection and mutation of harmful recessive genes. *Annals of Eugenics*, **10**, 417–421.

†Haldane, J. B. S. 1942a. The selective elimination of silver foxes in Eastern Canada. *Journal of Genetics*, **44**, 296–304.

†Haldane, J. B. S. 1942b. Parental and fraternal correlations for fitness. *Annals of Eugenics*, **11**, 287–292.

†Haldane, J. B. S. 1942c. Selection against heterozygosis in Man. *Annals of Eugenics*, **11**, 333–340.

†Haldane, J. B. S. 1951. La méthode dans la génétique. *Compte rendu du 10ème Congrès International de Philosophie des Sciences (Paris: 1949)*, vol. VI, *Biologie*, pp. 41–45. Paris: Hermann.

†Haldane, J. B. S. 1954. The measurement of natural selection. *Caryologia*, supp. (*Atti del IX Congresso Internazionale di Genetica*), 480–487.

†Haldane, J. B. S. 1955. Natural selection. *Transactions of the Bose Research Institute of Calcutta*, **20**, 17–20.

†Haldane, J. B. S. 1956. The estimation of viabilities. *Journal of Genetics*, **54**, 294–296.

†Haldane, J. B. S. 1957a. The cost of natural selection. *Journal of Genetics*, **55**, 511–524.

†Haldane, J. B. S. 1957b. The elementary theory of population growth. *Journal of Madras University*, ser. B, **27**, 237–245.

†Haldane, J. B. S. 1957c. Karl Pearson 1857–1957. Being a centenary lecture by J. B. S. Haldane, delivered at University College London on 13 May 1957. *Biometry*, **44**, 303–313.

†Haldane, J. B. S. 1958a. The present position of Darwinism. *Journal of Scientific Indian Research*, **17**, 97–103.

†Haldane, J. B. S. 1958b. A defense of beanbag genetics. *Perspectives in Biology and Medicine*, **7**, 343–359.

†Haldane, J. B. S. 1958c. The theory of evolution, before and after Bateson. *Journal of Genetics*, **56**, 11–27.

†Haldane, J. B. S. 1959a. More precise expressions for the cost of natural selection. *Journal of Genetics*, **47**, 351–360.

†Haldane, J. B. S. 1959b. The theory of natural selection today. *Nature*, **183**, 710–713.

†Hardy, G. H. 1908. Mendelian proportions in a mixed population. *Science*, **28**, 49–50.

Hardy, G. H. 1985. *L'apologie d'un mathèmaticien.* Paris: Belin. (Original English edition: Cambridge University Press, 1940).

†Harris, H. 1966. Enzyme polymorphism in Man. *Proceedings of the Royal Society of London*, ser. B, **164**, 298–310.

Harrison, E. 1987. Whigs, prigs and historians of science. *Nature*, **329**, 213–214.

†Harrison, J. W. H. and Garrett, F. C. 1926. The induction of melanism in Lepidoptera and its subsequent inheritance. *Proceedings of the Royal Society of London*, ser. B, **99**, 241–263.

Heimans, J. 1971. Mendel's ideas on the nature of hereditary characters: The explanation of fragmentary records of Mendel's hybridizing experiments. *Folia Mendeliana*, **6**, 91–98.

Hodge, M. J. S. 1983. Darwin and the laws of the animate part of the terrestrial system (1835–1937): On the Lyellian origins of his zoonomical explanatory program. *Studies in History of Biology*, **6**, 1–106.

Hodge, M. J. S. 1986. Darwin as a life generation theorist. In *The Darwinian Heritage*, ed. D. Kohn. Princeton: Princeton University Press.

Hodge, M. J. S. 1987. Natural selection as a causal, empirical, and probabilistic theory. In *The Probabilistic Revolution*, ed. L. Krüger, G. Gigerenze and M. S. Morgan, vol. II, pp. 233–270. Cambridge, MA: MIT Press.

Hodge, M. J. S. 1989. Darwin's theory and Darwin's argument. In *What the Philosophy of Biology Is*, ed. M. Ruse, pp. 163–182. Dordrecht: Kluwer Academic.

Hodge, M. J. S. 1991. *Origins and Species: A Study of the Historical Sources of Darwinism and the Contexts of Some Other Accounts of Organic Diversity from Plato and Aristotle On*. New York: Garland.

Hodge M. J. S. 1992. Biology and philosophy (including ideology): A study of Fisher and Wright. In *The Founders of Evolutionary Genetics: A Centenary*

Reappraisal, ed. S. Sarkar. Boston Studies in the Philosophy of Science, vol. CXLII. Dordrecht: Kluwer.

Hodge, M. J. S. and Kohn, D. 1986. *The Immediate Origins of Natural Selection.* Princeton: Princeton University Press.

Hubby, J. L. and Lewontin, R. C. 1966. A molecular approach to the study of genic heterozygosity in natural populations. I. The number of alleles at different loci in *Drosophila pseudoobscura*. *Genetics*, **68,** 235–252. See also Lewontin and Hubby (1966).

Hull, D. 1973. *Darwin and His Critics.* Cambridge, MA: Harvard University Press.

Hull, D. 1974. *The Philosophy of Biological Science.* Englewood Cliffs, NJ: Prentice-Hall.

Hull, D. 1976. Are species really individuals? *Systematic Zoology*, **25**, 174–191.

Hull, D. 1978. A matter of individuality. *Philosophy of Science*, **45**, 335–360.

Hull, D. 1980. Individuality and selection. *Annual Review of Ecology and Systematics*, **11**, 311–332.

Hull, D. 1981. The units of evolution. In *Studies in the Concept of Evolution*, ed. U. J. Jensen and R. Harre, pp. 23–44. London: Harvester.

†Huxley, J. (ed.). 1940. *The New Systematics.* Oxford: Oxford University Press.

†Huxley, J. 1942. *Evolution: The Modern Synthesis.* London: Allen and Unwin. 2nd edn, 1963. London: Allen and Unwin. 3rd edn, 1974. London: Allen and Unwin.

†Huxley, J. 1945. La revanche du darwinisme. *Conférences du Palais de la Découverte*, pp. 5–19. Paris: Conférences du Palais de Découverte.

†Huxley, J. 1950. *La génétique soviétique et la science mondiale.* Paris: Stock.

†Huxley, L. 1900. *Life and Letters of Thomas Henry Huxley.* 2 vols. New York: Appleton.

Jacquard, A. 1970. *Structures génétiques des populations.* Paris: Masson.

Jacquard, A. 1974. *Structures génétiques des populations humaines.* Paris: Masson.

Jacquard, A. 1977. *Concepts en génétique des populations.* Paris: Masson.

†Jenkin, F. 1867. The Origin of Species. *North British Review*, **44**, 277–318.

†Jepsen, G. L., Mayr, E. and Simpson, G. G. (eds.). 1949. *Genetics, Paleontology and Evolution.* New York: Columbia University Press.

Jindra, J. 1971. A possible derivation of the Mendelian series. *Folia Mendeliana*, **6**.

†Johannsen, W. L. 1903. *Ueber Erblichkeit in Populationen und in reinen Linien: ein Beitrag zur Beleuchtung schwebender Selektionsfragen.* Jena: Gustav Fischer. Partial English translation: Concerning heredity in populations and in pure lines. In *Selected Readings in Biology for Natural Sciences*, ed. H. Gall and E. Putschar, vol. III, pp. 172–215. Chicago: University of Chicago Press, 1955.

†Johannsen, W. L. 1907. Does hybridization increase fluctuating variability? *Report of the Third International Congress of Genetics*, pp. 98–112. London: Spottiswoode.

†Johannsen, W. L. 1909. *Elemente der exakten Erblichkeitslehre.* Jena: Fischer.

Joravsky, D. 1961. *Soviet Marxism and Natural Science, 1917–1932.* London: Routledge and Kegan Paul.

Kavaloski, V. C. 1974. The *vera causa* principle: A historico-philosophical study of a metatheoretical concept from Newton through Darwin. Unpublished PhD dissertation, University of Chicago.

†Kellogg, V. L. 1908. *Darwinism Today: A Discussion of Present-Day Scientific*

Selection Theories, Together with a Brief Account of the Principal Other Proposed Auxiliary and Alternative Theories of Species-Formation. London: Bell.

Kelvin (Lord). See Thomson, W.

Kempthorne, O. 1957. *An Introduction to Genetic Statistics.* New York: Wiley.

†Kettlewell, H. B. D. 1955. Selection experiments on industrial melanism in the Lepidoptera. *Heredity,* **9,** 323–342.

†Kettlewell, H. B. D. 1956a. Further selection experiments on industrial melanism in the Lepidoptera. *Heredity,* **10,** 287–301.

†Kettlewell, H. B. D. 1956b. A résumé of investigations on the evolution of melanism in the Lepidoptera. *Proceedings of the Royal Society of London,* ser. B, **145,** 297–303.

†Kettlewell, H. B. D. 1961. The phenomenon of industrial melanism in the Lepidoptera. *Annual Review of Entomology,* **6,** 245–262.

Kevles, D. 1985. *In the Name of Eugenics: Genetics and the Uses of Human Heredity.* Berkeley and Los Angeles: University of California Press.

Kimler, W. C. 1983. Mimicry: Views of naturalists and ecologists before the modern synthesis. In *Dimensions of Darwinism,* ed. M. Grene, pp. 97–127. Cambridge: Cambridge University Press.

Kimmelman, B. A. 1990. Agronomie et théorie de Mendel. La dynamique institutionnelle et la génétique aux Etats-Unis, 1900–1925. In *Histoire de la génétique: pratiques, techniques et théories,* ed. J. L. Fischer and W. H. Schneider, pp. 17–42. Paris: Editions Sciences en Situation.

†Kimura, M. 1958. On the change of population fitness by natural selection. *Heredity,* **12,** 145–167.

†Kimura, M. 1960. Optimum rate and degree of dominance as determined by the principle of minimum genetic load. *Journal of Genetics,* **57,** 21–34.

†Kimura, M. 1962. On the probability of fixation of mutant genes in a population. *Genetics,* **47,** 713–719.

†Kimura, M. 1964a. Diffusion models in population genetics. *Journal of Applied Probability,* **1,** 177–232.

†Kimura, M. (with Crow, J. F.). 1964b. The number of alleles that can be maintained in a finite population. *Genetics,* **49,** 725–738.

†Kimura, M. 1968. Evolutionary rate at the molecular level. *Nature,* **217,** 624–626.

†Kimura, M. 1969. The rate of molecular evolution considered from the standpoint of population genetics. *Proceedings of the National Academy of Sciences of the USA,* **63,** 1181–1188.

†Kimura, M. and Ohta, T. 1973. Mutation and evolution at the molecular level. *Genetics,* **73,** (supp.) 19–35.

†Kimura, M. (with Ohta, T.). 1974. On some principles governing molecular evolution. *Proceedings of the National Academy of Sciences of the USA,* **71,** 2848–2852.

†Kimura, M. 1976. How genes evolve: a population geneticist's view. *Annales de Génétique,* **19,** 153–168.

†Kimura, M. 1983a. *The Neutral Theory of Molecular Evolution.* Cambridge: Cambridge University Press.

†Kimura, M. 1983b. The neutral theory of molecular evolution. In *Evolution of*

Genes and Proteins, ed. M. Nei and R. K. Koehn, pp. 208–233. Sunderland, MA: Sinauer.

†Kimura, M. 1987. Molecular evolutionary clock and evolutionary theory. *Journal of Molecular Evolution*, **27**, 24–33.

†Kimura, M. 1988. Natural selection and neutral evolution, with special reference to evolution and variation at the molecular level. In *L'évolution dans sa réalité et ses diverses modalités (Colloque international [novembre 1985] organisé par la fondation Singer-Polignac)*, ed. M. Marois, pp. 269–284. Paris: Masson.

†Kimura, M. 1994. *Population Genetics, Molecular Evolution, and the Neutral Theory: Selected Papers*. Edited and with introductory essays by N. Takahata, with a foreword by James F. Crow. Chicago: University of Chicago Press.

†King, J. L. and Jukes, T. H. 1969. Non-Darwinian evolution: Random fixation of selectively neutral mutations. *Science*, **164**, 788–798.

Kingsland, S. 1985. *Modeling Nature: Episodes in the History of Population Ecology*. Chicago: University of Chicago Press.

Kohler, R. E. 1994. *Lords of the Fly: Drosophila Genetics and the Experimental Life*. Chicago: University of Chicago Press.

Kohn, D. (ed.). 1985. *The Darwinian Heritage*. Princeton: Princeton University Press.

†Kojima, K. 1971. Is there a constant fitness for a given genotype? *Evolution*, **25**, 281–285.

†Kojima, K. and Kelleher, T. M. 1961. Changes of mean fitness in random mating populations when epistasis and linkage are present. *Genetics*, **46**, 527–540.

Kuhn, T. 1962. *The Structure of Scientific Revolutions*. Chicago: University of Chicago Press. (2nd edn, 1970).

Kuhn, T. 1970. Reflections on my critics. In *Criticisms and the Growth of Knowledge*, ed. I. Lakatos and A. Musgrave, pp. 231–278. Cambridge: Cambridge University Press.

Kuhn, T. 1974. Second thoughts on paradigms. In *The Structure of Scientific Theories*, ed. F. Suppe. Urbana: University of Illinois Press.

Lakatos, I. 1970. Falsification and the methodology of scientific research programmes. In *Criticisms and the Growth of Knowledge*, ed. I. Lakatos and A. Musgrave, pp. 91–196. Cambridge: Cambridge University Press.

Lakatos, I. 1971. History of science and its rational reconstructions. *Boston Studies in the Philosophy of Science*, **8**, 91–135.

†Lamarck, J. B. 1909. *Philosophie zoologique*. Paris: Dentu.

†Lamotte, M. 1948. *Introduction à la biologie quantitative. Présentation et interprétation statistiques des données numériques*. Paris: Masson.

†Lamotte, M. 1951. Recherches sur la structure génétique des populations naturelles de *Cepaea nemoralis* L. *Bulletin Biologique de France et de Belgique*, **suppl. 35**, 1–239.

†Lamotte, M. 1959. Polymorphism of natural populations of *Cepaea nemoralis*. *Cold Spring Harbor Symposia on Quantitative Biology*, **24**, 65–86.

†Lamotte, M. 1966. Les facteurs de la diversité du polymorphisme dans les populations de *Cepaea nemoralis*. *Lavori delle Società Malacologica Italiana*, **3**, 33–73.

†Lamotte, M. 1988. Phénomènes fortuits et évolution. In *L'évolution dans sa réalité et ses diverses modalités,* ed. M. Marois, pp. 241–268. Paris: Masson.

†Lamotte, M. and Coursol, J. 1974. Mutations, sélection diversifiante et fluctuations fortuites comme maintien du polymorphisme. *Mémoires de la Société Zoologique de France,* **37**, 443–471.

†Lang, A. 1906. *Ueber die mendelschen Gesetze, Arten und Varietätenbildung, Mutation und Variation, insbesondere bei unsern Hein- und Gartenschnecken.* Lucerne: Keller.

†Langaney, A. 1969. Panmixie, 'Pangamie' et systèmes de croisement. *Population,* **24**, 301–308.

Lankester, E. Ray. 1896. Letter, 16 July 1896. *Nature,* **54**, 245.

Laudan, L. 1977. *Progress and Its Problems.* Berkeley, and Los Angeles: University of California Press.

Laurent, G. 1987. *Paléontologie et évolution en France, 1800–1860: de Cuvier–Lamarck à Darwin.* Paris: Editions du Comité des Travaux Historiques et Scientifiques.

La Vergata, A. 1990. *L'equilibrio e la guerra della natura.* Naples: Morano.

†Le Dantec, F. 1898. Mimétisme et imitation. *Revue Philosophique,* **23**, 356–398.

†Le Dantec, F. 1899. *Lamarckiens et darwiniens.* Paris: Alcan.

†Le Dantec, F. 1904. *Les influences ancestrales.* Paris: Flammarion.

†Le Dantec, F. 1909. *La crise du transformisme.* Paris: Alcan.

Lenay, C. 1989. Enquête sur le hasard dans les grandes théories biologiques de la deuxième moitié du dix-neuvième siècle. Unpublished PhD dissertation, University of Paris I.

Lenay, C. 1995. Préhistoire de la génétique: Hugo de Vries et l'idée d'indépendance des caractères. In *Nature, histoire, société. Hommage à Jacques Roger,* ed. C. Blanckaert, J. L. Fischer and R. Rey, pp. 133–145. Paris: Klincksieck.

Lennox, J. G. 1993. Darwin *was* a teleologist. *Biology and Philosophy,* **8**, 409–421.

Lenoir, T. 1981. The Göttingen school and the development of transcendental naturphilosophie in the romantic era. *Studies in History of Biology,* **5**, 111–205.

†Lerner, I. M. 1954. *Genetic Homeostasis.* London: Oliver and Boyd.

†Levene, H. 1953. Genetic equilibrium when more than one ecological niche is available. *American Naturalist,* **87**, 131–133.

†Lewontin, R. C. and Hubby, J. L. 1966. A molecular approach to the study of genic heterozygosity in natural populations. II. Amount of variation and degree of heterozygosity in natural populations of *Drosophila pseudoobscura. Genetics,* **54**, 595–609.

†Lewontin, R. C. 1970. The units of selection. *Annual Review of Ecology and Systematics,* **1**, 1–18.

†Lewontin, R. C. 1974. *The Genetic Basis of Evolutionary Change.* New York: Columbia University Press.

Lewontin, R. C., Moore, J. A., Provine, W. and Wallace, B. (eds.). 1938–76. *Dobzhansky's Genetics of Natural Populations, I–XLIII.* New York: Columbia University Press.

†L'Héritier, P. 1934. *Génétique et évolution: analyse de quelques études mathématiques sur la sélection naturelle.* Paris: Hermann.

†L'Héritier, P. 1954. *Traité de génétique,* vol. II, *La génétique des populations.* Paris: Presses Universitaires de France.

†L'Héritier, P. 1981. Souvenirs d'un généticien. *Revue de Synthèse*, **102**, 331–350.

†L'Héritier, P. and Teissier, G. 1933. Etude d'une population de Drosophiles en équilibre. *Comptes Rendus Hebdomadaires des Séances de l'Académie des Sciences*, **197**, 1765–1766.

†L'Héritier, P. and Teissier, G. 1934a. Sur quelques facteurs du succès dans la concurrence larvaire chez *Drosophila melanogaster*. *Comptes Rendus de la Société de Biologie*, **116**, 306.

†L'Héritier, P. and Teissier, G. 1934b. Une expérience de sélection naturelle: courbe d'élimination du gène 'bar' dans une population de *Drosophila melanogaster*. *Comptes Rendus de la Société de Biologie*, **117**, 1049.

†L'Héritier, P. and Teissier, G. 1936. Contribution à l'étude de la concurrence larvaire chez les Drosophiles. *Comptes Rendus de la Société de Biologie*, **122**, 264–267.

†L'Héritier, P. and Teissier, G. 1937a. Elimination des formes mutantes dans les populations de Drosophiles. Cas des Drosophiles 'bar'. *Comptes Rendus de la Société de Biologie*, **124**, 880–882.

†L'Héritier, P. and Teissier, G. 1937b. Elimination des formes mutantes dans les populations Drosophiles. Cas des Drosophiles 'ebony'. *Comptes Rendus de la Société de Biologie*, **124**, 882–884.

†Li, C. C. 1955. *Population Genetics*. Chicago: University of Chicago Press.

Limoges, C. 1970. *La sélection naturelle: étude sur la première constitution d'un concept (1837–1859)*. Paris: Presses Universitaires de France.

†Linnaeus, C. 1735. *Systema naturae, sive regna tria naturae systematice proposita per classes, ordines, genera et species*. Leyden (Lugdunum Batavorum): Th. Hark. (This 1st edn is a sketch barely more than 10 pp. long.)

López Beltrán, C. 1992. Human heredity 1750–1870: The construction of a domain. Unpublished PhD dissertation, University of London.

López Beltrán, C. 1995. 'Les maladies héréditaires': 18th century disputes in France. *Revue d'Histoire des Sciences*, **48**, 307–350.

†Lotka, A. J. 1913, A natural population norm. *Journal of the Washington Academy of Sciences*, **3**, 241, 289.

†Lotka, A. J. 1914. An objective standard of value derived from the principle of evolution. *Journal of the Washington Academy of Sciences*, **4**, 409–418, 447–457.

†Lotka, A. J. 1918. The relation between birth rate and death rate in a normal population. *Quarterly Publications of the American Statistical Association*, **16**, 121.

†Lotka, A. J. and Dublin, I. 1925. On the true rate of natural increase of a population. *Journal of the American Statistical Association*, **20**, 305.

†Lotka, A. J. 1956. *Elements of Mathematical Biology: Formerly Published under the Title 'Elements of Physical Biology'*. New York: Dover. (Original edn 1924).

†Lotsy, J. P. 1916. *Evolution by Means of Hybridization*. The Hague: Martinus Nijhoff.

Lurie, E. 1960. *Louis Agassiz: A Life in Science*. Chicago: University of Chicago Press.

†Lyell, C. 1830–3. *Principles of Geology, Being an Attempt to Explain the Former Changes of the Earth's Surface, by Reference to Causes Now in Operation*. 3 vols. London: Murray. (Facsimile: Chicago: University of Chicago Press, 1990).

†MacDowell, E. C. 1916. Piebald rats and multiple factors. *American Naturalist*, **50**, 719–742.

†MacKenzie, D. A. 1981. *Statistics in Britain 1865–1930: The Social Construction of Scientific Knowledge*. Edinburgh: Edinburgh University Press.

Mahalanobis, P. C. 1938. Professor Ronald Aylmer Fisher. *Sankhya*, **4**, 265–272. Reprinted in *Biometrics*, **20** (1964), 238–250.

†Malécot, G. 1945. La diffusion des gènes dans une population mendélienne. *Comptes Rendus de l'Académie des Sciences de Paris*, **221**, 340; **222**, 841.

†Malécot, G. 1948. *Les mathématiques de l'hérédité*. Paris: Masson.

†Malthus, T. R. 1798. *An Essay on the Principle of Population, As It Affects the Future Improvement of Society*. London: Johnson.

Marchant, J. (ed.). 1916. *Alfred Russel Wallace: Letters and Reminiscences*. 2 vols. London: Cassell.

Mather, F. and Yates, K. 1963. Ronald Aylmer Fisher 1890–1962. *Biographical Memoirs of Fellows of the Royal Society of London*, **9**, 91–120. Reprinted in J. H. Bennett (ed.), *Collected Papers of R. A. Fisher*, pp. 23–52. Adelaide: University of Adelaide Press, 1971.

†Mather, K. and Jinks, J. L. 1977. *Introduction to Biometrical Genetics*. Ithaca, NY: Cornell University Press.

†Mayr, E. 1942. *Systematics and the Origin of Species*. New York: Columbia University Press. (Reprinted 1982).

†Mayr, E. 1947. Ecological factors of speciation. *Evolution*, **1**, 263–288.

†Mayr, E. 1954. Change of genetic environment and evolution. In *Evolution as a Process*, ed. J. Huxley, pp. 157–180. London: Allen and Unwin.

†Mayr, E. 1959. Where are we? *Cold Spring Harbor Symposia on Quantitative Biology*, **24**, 409–440.

Mayr, E. 1961. Cause and effect in biology. *Science*, **134**, 1501–1506.

Mayr, E. 1971. The nature of the Darwinian revolution. *Science*, **176**, 981–989. Reprinted in E. Mayr (1976), *Evolution and the Diversity of Life: Selected Essays*, pp. 277–296. Cambridge, MA: Harvard University Press.

Mayr, E. 1972. Lamarck revisited. *Journal of the History of Biology*, **5**, 55–94. Reprinted (with modifications) in E. Mayr (1976), *Evolution and the Diversity of Life: Selected Essays*, pp. 222–250. Cambridge, MA: Harvard University Press.

Mayr, E. 1976. *Evolution and the Diversity of Life: Selected Essays*. Cambridge, MA: Harvard University Press.

Mayr, E. 1980. Prologue: Some thoughts on the history of the evolutionary synthesis. In *The Evolutionary Synthesis*, ed. E. Mayr and W. B. Provine, pp. 1–48. Cambridge, MA: Harvard University Press.

Mayr, E. 1982. *The Growth of Biological Thought: Diversity, Evolution and Inheritance*. Cambridge, MA: Harvard University Press.

Mayr, E. 1985. How biology differs from the physical sciences. In *Evolution at a Crossroads*, ed. D. J. Depew and B. H. Weber. Cambridge, MA: MIT Press.

Mayr, E. 1988. *Toward a New Philosophy of Biology*. Cambridge, MA: Harvard University Press.

Mayr, E. 1991. *One Long Argument: Charles Darwin and the Genesis of Modern Evolutionary Thought*. Cambridge, MA: Harvard University Press.

Mayr, E. 1997. *This Is Biology: The Science of the Living World*. Cambridge, MA: Belknap Press of Harvard University Press.

Mayr, E. and Provine, W. B. (eds.). 1980. *The Evolutionary Synthesis: Perspectives on the Unification of Biology*. Cambridge, MA: Harvard University Press.

McKinney, H. L. 1972. *Wallace and Natural Selection.* London: Yale University Press.
Medawar, P. B. 1960. *The Future of Man.* New York: Basic Books.
†Meldola, R. 1888. Lamarckism versus Darwinism. *Nature*, **38**, 388–389.
†Meldola, R. 1896. The president's address. *Proceedings of the Entomological Society of London*, **1**, xii–xcii.
†Mendel, G. 1865. Versuche über Pflanzen-Hybriden. *Verhandlungen des Naturforschenden Vereines in Brünn*, **4**, 3–47. English translation: Experiments on plant hybrids. In *The Origin of Genetics: A Mendel Source Book*, ed. C. Stern and E. R. Sherwood (1966), pp. 1–48. San Francisco: Freeman.
†Mendel, G. 1866–73. Gregor Mendel's letters to Carl Nägeli. In *The Origin of Genetics: A Mendel Source Book*, ed. C. Stern and E. R. Sherwood (1966), pp. 56–102. San Francisco: Freeman. (An English translation of 10 letters from Mendel to Nägeli, published by C. Correns in 1905.)
†Mendel, G. 1869. Ueber einige künstlicher Befruchtung gewonnenen Hieracium Bastarde. *Verhandlungen des Naturforschenden Vereines in Brünn*, **8**, 26–31. English translation: On hieracium-hybrids obtained by artificial fertilization. In *The Origin of Genetics: A Mendel Source Book*, ed. C. Stern and E. R. Sherwood (1966), pp. 49–55. San Francisco: Freeman.
Milkman, R. (ed.). 1982. *Perspectives on Evolution.* Sunderland, MA: Sinauer.
Molina, G. 1996a. Le darwinisme en France. In *Dictionnaire du darwinisme et de l'évolution*, ed. P. Tort. Paris: Presses Universitaires de France.
Molina, G. 1996b. Wallace, Alfred Russel. In *Dictionnaire du darwinisme et de l'évolution*, ed. P. Tort, vol. III, pp. 4565–4586. Paris: Presses Universitaires de France.
Montgomery, W. M. 1975. Evolution and Darwinism in German biology, 1800–1883. Unpublished PhD dissertation, University of Texas – Austin.
Moore, J. R. 1979. *The Post-Darwinian Controversies: A Study of the Protestant Struggle to Come to Terms with Darwin in Great Britain and America, 1870–1900.* Cambridge: Cambridge University Press.
†Moorehead, P. and Kaplan, M. 1967. *Mathematical Challenges to the Neo-Darwinian Interpretation of Evolution.* Philadelphia: Wistar Institute.
†Moran, P. A. P. 1964. On the non-existence of adaptative topographies. *Annals of Human Genetics*, **27**, 383–393.
Moreno, R. 1984. *La polémica del darwinismo en México, siglo XIX.* Mexico City: Publicaciones de la Universidad Nacional Autónoma de México.
†Morgan, T. H. 1907. *Experimental Zoology.* New York: Macmillan.
†Morgan, T. H. 1909. For Darwin. *Popular Science Monthly*, **74**, 367–380.
†Morgan, T. H. 1916. *A Critique of the Theory of Evolution.* Princeton: Princeton University Press.
†Morgan, T. H., Sturtevant, A. H., Muller, H. J. and Bridges, C. B. 1915. *The Mechanism of Mendelian Heredity.* New York: Holt.
†Müller, F. 1864. *Für Darwin.* Leipzig.
†Müller, F. 1879. *Ituna* and *Thyridia*; a remarkable case of mimicry in butterflies. *Transactions of the Entomological Society of London*, pp. xx–xxix. (Originally published in German in *Kosmos*, May 1879, pp. 100–108; translated from German by R. Meldola).
†Muller, H. J. 1922. Variation due to change in the individual gene. *American Naturalist*, **56**, 32–50.

†Muller, H. J. 1950. Our load of mutations. *American Journal of Human Genetics*, **2**, 111–176.

†Murray, A. 1860. On the disguises of nature: Being an inquiry into the laws which regulate external form and colour in plants and animals. *Edinburgh New Philosophical Journal*, **68**, 66–90.

†Nei, M. 1964. Effects of linkage and epistasis on the equilibrium frequencies of lethal genes. II. Numerical solutions. *Japanese Journal of Genetics*, **39**, 7–25.

†Nei, M. 1975. *Molecular Population Genetics and Evolution*. Amsterdam: North-Holland.

†Nei, M. and Koehn, R. K. 1983. *Evolution of Genes and Proteins*. Sunderland, MA: Sinauer.

†Nilsson-Ehle, H. 1909–11. Kreuzungsuntersuchungen an Hafer und Weizen. *Lunds Universitets Årsskrift*, ser. 2, **5** (2).

Nordenskiöld, E. 1928. *The History of Biology: a Survey*. Translated from the Swedish by L. B. Eyre. New York: Tudor. (Original publication: *Biologins historia*, 3 vols., Stockholm: Björek and Börjesson, 1920–4).

Norton, B. J. 1971. Theory of evolution of the biometric school. Unpublished MPhil dissertation, University of London.

Norton, B. J. 1973. The biometric defense of Darwinism. *Journal of the History of Biology*, **6**, 283–316.

Norton, B. J. 1975. Metaphysics and population genetics: Karl Pearson and the background to Fisher's multi-factorial theory of inheritance. *Annals of Science*, **32**, 537–553.

Norton, B. J. 1978a. Karl Pearson and the Galtonian tradition: Studies in the rise of quantitative social biology. Unpublished PhD dissertation, University of London.

Norton, B. J. 1978b. Fisher and the neo-Darwinian synthesis. In *Human Implications of Scientific Advance: Proceedings of the XVth International Congress of the History of Science*, ed. E. G. Forbes, pp. 481–494. Edinburgh: Edinburgh University Press.

Norton, B. J. 1981. La situation intellectuelle au moment des débats de Fisher en génétique des populations. *Revue de Synthèse*, **102**, 231–250.

Norton, B. J. 1983. Fisher's entrance in evolutionary science: The role of eugenics. In *Dimensions of Darwinism*, ed. M. Grene. Cambridge: Cambridge University Press.

Norton, B. J. and Pearson, E. S. 1976. A note on the background to, and refereeing of, R. A. Fisher's 1918 paper 'On the correlation between relatives on the supposition of Mendelian inheritance'. *Notes and Records of the Royal Society of London*, **31**, 151–162.

Ogilvie, B. W. 1991. An unnatural history: Darwin's use of domesticated animals in early species theorizing, 1836–1938. Unpublished communication at the IVth International Society for the History, Philosophy and Sociology of Biology Conference, University of Chicago, 12 July 1991.

Olby, R. C. 1963. Charles Darwin's manuscript of pangenesis. *British Journal of the History of Science*, **1**, 251–263.

Olby, R. C. 1979. Mendel no Mendelian? *History of Science*, **17**, 53–72.

Olby, R. C. 1985. *Origins of Mendelism*. 2nd edn. Chicago: University of Chicago Press. (1st edn, 1966).

Olby, R. C. 1988. The dimensions of scientific controversy: The biometric–Mendelian debate. *British Journal of the History of Science*, **22**, 299–320.

†Olson, E. C. 1960. Morphology, paleontology and evolution. In *Evolution after Darwin*, vol. I, *The Evolution of Life*, ed. S. Tax, pp. 523–545. Chicago: University of Chicago Press.

Orel, V. 1984. *Mendel*. Oxford: Oxford University Press.

Ospovat, D. 1978. Perfect adaptation and teleological explanation: Approaches to the problem of the history of life in the mid-nineteenth century. *Studies in History of Biology*, **2**, 33–56.

Ospovat, D. 1981. *The Development of Darwin's Theory*. Cambridge: Cambridge University Press.

†Owen, R. 1849. *On the Nature of Limbs*. London: J. Van Voorst.

†Pasteur, G. 1972. Génétique biochimique et populations ou: pourquoi sommes-nous multi-polymorphes? In *Le polymorphisme dans le règne animal (hommage à G. Teissier)*, ed. M. Lamotte. *Mémoires de la Société Zoologique de France*, **37**, 473–533.

†Pearl, R. 1917. The selection problem. *American Naturalist*, **51**, 65–91.

†Pearson, E. S. 1936–8. Karl Pearson: An appreciation of some aspects of his life and work. *Biometrika*, **28**, 193–257; **29**, 161–248.

Pearson, E. S. 1974. Memories of the impact of Fisher's work in the 1920s. *International Statistical Review*, **42**, 5–8.

†Pearson, K. 1892. *The Grammar of Science*. London: Scott. 2nd edn, 1900. London: Black.

†Pearson, K. 1893. Contributions to the mathematical theory of evolution (Abstract). *Proceedings of the Royal Society of London*, **59**, 329–333.

†Pearson, K. 1894. Contributions to the mathematical theory of evolution. *Philosophical Transactions of the Royal Society of London*, ser. A, **185**, 71–110.

†Pearson, K. 1895a. Mathematical contributions to the theory of evolution. II. Skew variation in homogeneous material (Abstract). *Proceedings of the Royal Society of London*, **57**, 257–360.

†Pearson, K. 1895b. Contributions to the mathematical theory of evolution. II. Skew variation in homogeneous material. *Philosophical Transactions of the Royal Society of London*, ser. A, **186**, 343–414.

†Pearson, K. 1895c. Contributions to the mathematical theory of evolution. III. Regression, heredity and panmixia (Abstract). *Proceedings of the Royal Society of London*, **59**, 69–71.

†Pearson, K. 1896a. Mathematical contributions to the theory of evolution. III. Regression, heredity and panmixia. *Philosophical Transactions of the Royal Society of London*, ser. A, **187**, 253–318.

†Pearson, K. 1896b. Contributions to the mathematical theory of evolution. Note on reproductive selection. *Proceedings of the Royal Society of London*, **59**, 301–305.

†Pearson, K. 1897a. *The Chances of Death and Other Studies in Evolution*. London: Arnold.

†Pearson, K. 1897b. Mathematical contributions to the theory of evolution. On a form of spurious correlation which may arise when indices are used in the measurement of organs. *Proceedings of the Royal Society of London*, **60**, 489–498.

†Pearson, K. (and Lee A.). 1897c. Mathematical contributions to the theory of

evolution. On the relative variation and correlation in civilised and uncivilised races. *Proceedings of the Royal Society of London*, **61**, 343–357.

†Pearson, K. 1898a. Mathematical contributions to the theory of evolution. On the law of ancestral heredity. *Proceedings of the Royal Society of London*, **62**, 386–412.

†Pearson, K. 1898b. Contributions to the theory of evolution. IV. On the influence of random selection on variation and correlation. *Philosophical Transactions of the Royal Society of London*, ser. A, **191**, 234ff.

†Pearson, K. 1898c. Mathematical contributions to the theory of evolution. V. On the reconstruction of prehistoric races (Abstract). *Proceedings of the Royal Society of London*, **63**, 417–420.

†Pearson, K. (with Lee, A. and Bramley-Moore, L.). 1898d. Mathematical contributions to the theory of evolution. VI. Reproductive or genetic selection (Abstract). *Proceedings of the Royal Society of London*, **64**, 163–167.

†Pearson, K., Lee, A. and Bramley-Moore, L. 1898e. Mathematical contributions to the theory of evolution. VI. Reproductive or genetic selection. Part I. Theoretical. Part II. On the inheritance of fertility in man. Part III. On the inheritance of fecundity in thoroughbred race-horses (Abstract). *Proceedings of the Royal Society of London*, **64**, 163–167.

†Pearson, K. (with Beeton, M). 1899a. Data for the problem of evolution in man. I. A first study of the variability and correlation of the hound. *Proceedings of the Royal Society of London*, **65**, 126–134.

†Pearson, K. 1899b. Mathematical contributions to the theory of evolution. V. On the reconstruction of the stature of prehistoric races. *Philosophical Transactions of the Royal Society of London*, ser. A, **192**, 169–244.

†Pearson, K. (with Beeton, M.). 1899c. Data for the problem of evolution in man. II. A first study of the inheritance of longevity and the selective death-rate in man. *Proceedings of the Royal Society of London*, **65**, 290–303.

†Pearson, K. (with Lee, A. and Bramley-Moore, L.). 1899d. Mathematical contributions to the theory of evolution. VI. Genetic (reproductive) selection: Inheritance of fertility in man, and of fecundity in thoroughbred racehorses. *Philosophical Transactions of the Royal Society of London*, ser. A, **192**, 257–330.

†Pearson, K. 1900a. Data for the problem of evolution in man. III. On the magnitude of certain coefficients of correlation in man. *Proceedings of the Royal Society of London*, **66**, 22–32.

†Pearson, K. 1900b. On the criterion that a given system of deviations from the probable in the case of a correlated system of variables is such that it can be reasonably supposed to have arisen from random sampling. *Philosophical Magazine*, 5th ser., **50**, 157–175.

†Pearson, K. 1900c. Mathematical contributions to the theory of evolution. On the law of reversion. *Proceedings of the Royal Society of London*, **66**, 140–164.

†Pearson, K. 1900d. Note on the effect of fertility depending on homogamy. *Proceedings of the Royal Society of London*, **66**, 316–323.

†Pearson, K. 1900e. *The Grammar of Science*. 2nd edn. London: Black.

†Pearson, K. (with Beeton, G. and Yule, G. U.). 1900f. Data for the problem of evolution in man. V. On the correlation between duration of life and the number of offspring. *Proceedings of the Royal Society of London*, **67**, 159–171.

†Pearson, K. (with Lee, A.). 1900g. Mathematical contributions to the theory of

evolution. VII. On the application of certain formulae in the theory of correlation to the inheritance of characters not capable of quantitative measurement (Abstract). *Proceedings of the Royal Society of London*, **66**, 324–327.

†Pearson, K. (with Lee, A.). 1900h. Mathematical contributions to the theory of evolution. VIII. On the correlation of characters not quantitatively measurable. *Philosophical Transactions of the Royal Society of London*, ser. A, **195**, 1–47.

†Pearson, K. 1900i. Mathematical contributions to the theory of evolution. VIII. On the correlation of characters not quantitatively measurable (Abstract). *Proceedings of the Royal Society of London*, **66**, 241–244.

†Pearson, K. 1901. Mathematical contributions to the theory of evolution. IX. On the principle of homotyposis and its relation to heredity, to the variability of the individual, and to that of the race. Part I. Homotyposis in the vegetable kingdom. *Philosophical Transactions of the Royal Society of London*, ser. A, **197**, 285–379.

†Pearson, K. 1902. Mathematical contributions to the theory of evolution. XI. On the influence of natural selection on the variability and correlation of organs (Abstract). *Proceedings of the Royal Society of London*, **69**, 330–333.

†Pearson, K. 1903a. The law of ancestral heredity. *Biometrika*, **2**, 211–229.

†Pearson, K. 1903b. Mathematical contributions to the theory of evolution. On homotyposis in homologous but differentiated organs. *Proceedings of the Royal Society of London*, **71**, 288–313.

†Pearson, K. 1904a. Mathematical contributions to the theory of evolution. XII. On a generalised theory of alternative inheritance, with special reference to Mendel's laws. *Philosophical Transactions of the Royal Society of London*, ser. A, **203**, 53–86.

†Pearson, K. 1904b. On a criterion which may serve to test various theories of inheritance. *Proceedings of the Royal Society of London*, **73**, 262–280.

†Pearson, K. 1906. Walter Frank Raphael Weldon, 1860–1906. *Biometrika*, **4**, 1–52.

†Pearson, K. 1910. Nature and nurture: The problem of the future. In *Eugenics Laboratory Lecture Series*, vol. VI, pp. 27ff. London: Dulau.

†Pearson, K. 1911. Further remarks on the law of ancestral heredity. *Biometrika*, **8**, 239–243.

†Pearson, K. 1914–30. *The Life, Letters and Labours of Francis Galton.* 3 vols. Cambridge: Cambridge University Press.

†Pearson, K. 1924. Historical note on the origin of the normal curve of errors. *Biometrika*, **16**, 402–404.

†Pearson, K. 1934. *Speeches Delivered at a Dinner Held in University College, London: In Honour of Professor Karl Pearson, 23 April 1934.* Cambridge: Cambridge University Press.

Perelman, C. and Olbrechts-Tyteca, L. 1958. *Traité de l'argumentation, la nouvelle rhétorique.* Paris: Presses Universitaires de France.

†Petit, C. 1951. Le rôle de l'isolement sexuel dans l'évolution des populations de *Drosophila melanogaster. Bulletin Biologique de la France et de la Belgique*, **85**, 391–418.

†Petit, C. 1958. Le déterminisme génétique et psycho-physiologique de la compétition sexuelle chez *Drosophila melanogaster. Bulletin Biologique de la France et de la Belgique*, **92**, 248–329.

†Petit, C. and Zukerkandl, E. 1976. *Evolution. Génétique des populations. Evolution moléculaire*. Paris: Hermann.

†Petit, C. 1981. Problèmes actuels dans la génétique des populations. *Revue de Synthèse*, **102**, 351–379.

Pfeifer, E. J. 1957. The reception of Darwinism in the United States 1859–1880. Unpublished PhD dissertation, Brown Unversity.

Pfeifer, E. J. 1965. The genesis of American neo-Lamarckism. *Isis*, **56**, 156–167.

Piaget, J. 1977. *Introduction à l'épistémologie génétique*, vol. I. 2nd edn. Paris: Presses Universitaires de France.

†Pictet, A. 1925. Sur l'existence de deux facteurs de panuchure dissociables par croisement. *Comptes Rendus de la Société de Physique et d'Histoire Naturelle de Genève*, **42**, 20–24.

Piquemal, J. 1972. Quelques distinctions à propos de l'informe et à propos de sa mathématisation. In *La mathématisation des doctrines informes*, ed. G. Canguilhem, pp. 157–190. Paris: Hermann.

†Popper, K. 1935. *Logik der Forschung*. Vienna: Springer. English translation: *The Logic of Scientific Discovery*. London: Hutchinson, 1968.

†Popper, K. 1963. *Conjectures and Refutations: The Growth of Scientific Knowledge*. London: Routledge and Kegan Paul.

†Popper, K. 1972. *Objective Knowledge: An Evolutionary Approach*. Oxford: Oxford University Press.

†Popper, K. 1976. Darwinism as a metaphysical research programme. In *Unended Quest: An Intellectual Autobiography*, pp. 167–180. London: Fontana.

†Popper, K. 1978. Natural selection and the emergence of mind. *Dialectica*, **32**, 501–506.

†Popper, K. 1984. Evolutionary epistemology. In *Evolutionary Theory: Paths into the Future*, ed. J. W. Pollard. Chichester: Wiley.

Porter, T. M. 1986. *The Rise of Statistical Thinking 1820–1920*. Princeton: Princeton University Press.

†Poulton, E. B. 1897. Protective mimicry as evidence for the validity of the theory of natural selection. *Report of the British Association for the Advancement of Science*, **67**, 692–694.

†Poulton, E. B. 1898. Natural selection, the cause of mimetic resemblance and common warning colours. *Journal of the Linnean Society of London*, **26**, 558–612. Reproduced with modifications in E. B. Poulton (1908), *Essays on Evolution 1889–1907*, pp. 220–270. Oxford: Clarendon Press.

†Poulton, E. B. 1908. *Essays on Evolution 1889–1907*. Oxford: Clarendon Press.

†Price, G. R. 1972. Fisher's 'fundamental theorem' made clear. *Annals of Human Genetics*, **36**, 129–140.

Provine, W. B. 1971. *The Origins of Theoretical Population Genetics*. Chicago: University of Chicago Press.

Provine, W. B. 1978. The role of mathematical population geneticists in the evolutionary synthesis of the 1930s and 1940s. *Studies in History of Biology*, **2**, 167–192.

Provine, W. B. 1980. Epilogue. In *The Evolutionary Synthesis*, ed. E. Mayr and W. B. Provine, pp. 399–411. Cambridge, MA: Harvard University Press.

Provine, W. B. 1983. The development of Wright's theory of evolution: Systematics, adaptation and drift. In *Dimensions of Darwinism*, ed. M. Grene, pp. 43–70. Cambridge: Cambridge University Press.

Provine, W. B. 1986a. *Sewall Wright and Evolutionary Biology*. Chicago: University of Chicago Press.

Provine, W. B. (ed.). 1986b. Sewall Wright, *Evolution: Selected Papers*. Chicago: University of Chicago Press.

Provine, W. B. 1988. Sewall Wright and population structure in relation to evolution. *Evolution*, **42**, iii–iv.

†Punnett, R. C. 1915. *Mimicry in Butterflies*. Cambridge: Cambridge University Press.

†Quételet, A. 1835. *Sur l'homme et le développement de ses facultés, ou essai de physique sociale*. Paris: Bachelier.

†Quételet, A. 1846. *Lettres à Son Altesse Royale le Duc Régnant de Saxe-Cobourg et Gotha sur la théorie des probabilités, appliquée aux sciences morales et politiques*. Brussels: Hayez.

Rabel, G. 1963. *Kant*. Oxford: Clarendon Press.

Rheinberger, H.-J. and McLauglin, P. 1984. Darwin's experimental natural history. *Journal of the History of Biology*, **17**, 345–368.

Roberts, H. F. 1965. *Plant Hybridization before Mendel*. New York: Hafner. Facsimile of 1929 edn.

Roger, J. 1976. Darwin in France. *Annals of Science*, **33**, 481–484.

Roger, J. (ed.). 1979. *Les néo-lamarckiens français*. Special issue of *Revue de Synthèse*, 3rd ser., nos. 95–96.

Roger, J. (ed.). 1981. R. A. Fisher et l'histoire de la génétique des populations. (Colloque, Paris: 1980). *Revue de Synthèse*, **102**, 227–431.

Roger, J. 1983. Biologie du fonctionnement et biologie de l'évolution. In *L'explication dans les sciences de la vie*, ed. H. Barreau, pp. 135–158. Paris: Editions du CNRS.

Roger, J. 1989. L'eugénisme (1850–1950). In *L'ordre des caractères. Aspects de l'hérédité dans l'histoire des sciences de l'homme*, ed. C. Bénichou and J. L. Fisher, pp. 119–145. Paris: Vrin.

Roll-Hansen, N. 1978. The genotype theory of Wilhelm Johannsen and its relation to plant breeding and the study of evolution. *Centaurus*, **22**, 201–235.

†Romanes, G. J. 1889. Mr Wallace on Darwinism. *Contemporary Review*, **56**, 244–258.

†Romanes, G. J. 1892–7. *Darwin and after Darwin: An Exposition of the Darwinian Theory and a Discussion of Post-Darwinian Questions*. 3 vols. Vol. I, *The Darwinian Theory (1892)*. Vol. II, *Post-Darwinian Questions: Heredity and Utility* (1894). Vol. III, *Post-Darwinian Questions: Isolation and Physiological Selection* (1897). London: Longman.

†Romanes, G. J. 1893. *An Examination of Weismannism*. London: Longman.

†Romanes, G. J. 1896. *Life and Letters*. London: Longman.

†Roughgarden, J. 1979. *Theory of Population Genetics and Evolutionary Ecology: An Introduction*. New York: Macmillan.

Ruse, M. 1973. *Philosophy of Biology*. London: Hutchinson.

Ruse, M. 1975. Charles Darwin's theory of evolution: An analysis. *Journal of the History of Biology*, **8**, 219–241.

Ruse, M. 1975. Darwin's debt to philosophy: An examination of the influence of the philosophical ideas of John F. W. Herschel and William Whewell on the development of Charles Darwin's theory of evolution. *Studies in the History and Philosophy of Science*, **6**, 159–181.

Ruse, M. 1980. Charles Darwin and group selection. *Annals of Science*, **37**, 615–630.

Ruse, M. 1986. *Taking Darwin Seriously*. Oxford: Blackwell.

Ruse, M. 1989. *The Darwinian Paradigm*. London: Routledge.

Ruse, M. 1995. *Evolutionary Naturalism*. London: Routledge.

Ruse, M. 1996. *Monad to Man: The Concept of Progress in Evolutionary Biology*. Cambridge, MA: Harvard University Press.

†Schmalhausen, I. I. 1949. *Factors of Evolution*. Philadelphia: Blakiston.

Schweber, S. 1977. The origin of the *Origin* revisited. *Journal of the History of Biology*, **10**, 229–316.

Sentis, P. 1970. La naissance de la génétique au début du 20ème siècle. *Cahiers d'Etudes Biologiques*, **18–19**, 1–85.

Sentis, P. 1983. *Fortune ou richesse? La comptabilité du patrimoine*. Paris: Economica.

Shapin, S. 1982. History of science and its social reconstructions. *History of Science*, **20**, 157–211.

†Simpson, G. G. 1944. *Tempo and Mode in Evolution*. New York: Columbia University Press.

†Simpson, G. G. 1947. L'orthogenèse et la théorie synthétique de l'évolution. In *Paléontologie et transformisme: Colloque CNRS*. Paris: Albin-Michel.

Sober, E. 1984. *The Nature of Selection: Evolutionary Theory in Philosophical Focus*. Cambridge: MIT Press.

†Spencer, H. 1863. *First Principles*. London: William and Norgate. 2nd edn, 1867. London: William and Norgate. 3rd edn, 1875. London: William and Norgate.

†Spencer, H. 1864–7. *The Principles of Biology*. 2 vols. London: William and Norgate.

†Spencer, H. 1893. The inadequacy of natural selection. *Contemporary Review*, **63**, 153–164, 439–455, 743–760.

Stauffer, R. C. 1975. *Charles Darwin's Natural Selection, Being the Second Part of His Big Species Book Written from 1856 to 1858*. Cambridge: Cambridge University Press.

†Stebbins, G. L. 1950. *Variation and Evolution in Plants*. New York: Columbia University Press.

Stern, C. 1962. Wilhelm Weinberg, 1862–1937. *Genetics*, **47**, 1–5.

Stern, C. and Sherwood, E. R. 1966. *The Origin of Genetics: A Mendel Source Book*. San Franscico: Freeman.

Stigler, S. M. 1986. *The History of Statistics: The Measurement of Uncertainty before 1900*. Cambridge, MA: Harvard University Press.

†Sturtevant, A. H. 1918. *An Analysis of the Effects of Selection*. Washington, DC: Cambridge Institution Publications, no. 264.

†Sturtevant, A. H. 1965. *A History of Genetics*. New York: McGraw-Hill.

†Sutton, W. 1902. On the morphology of the chromosome group of *Brachystola magna*. *Biological Bulletin*, **4**, 24–39.

Swinburne, R. G. 1963. Galton's law: Formulation and development. *Annals of Science*, **21**, 15–31.

Taguieff, P.-A. (ed.). 1997. *Les pensées modernes de la décadence*. Paris: Berg.

†Tammes, T. 1911. Das Verhalten fluktuierend variender Merkmale bei der Bastardierung. *Recueil des Travaux Botaniques Néerlandais*, **8**.

†Teissier, G. 1942a. Vitalité et fécondité relative de diverses combinaisons génétiques comportant un gène léthal chez la Drosophile. *Comptes Rendus Hebdomadaires des Séances de l'Académie des Sciences*, **214**, 241–244.

†Teissier, G. 1942b. Persistance d'un gène léthal dans une population de Drosophiles. *Comptes Rendus Hebdomadaires des Séances de l'Académie des Sciences*, **214**, 327.

†Teissier, G. 1943. Apparition et fixation d'un gène mutant dans une population stationnaire de Drosophiles. *Comptes Rendus Hebdomadaires des Séances de l'Académie des Sciences*, **216**, 88–90.

†Teissier, G. 1944. Equilibre des gènes léthaux dans les populations stationnaires panmictiques. *Revue Scientifique*, **82**, 145–159.

†Teissier, G. 1947a. Variations de la fréquence du gène *sepia* dans une population stationnaire de Drosophiles. *Comptes Rendus Hebdomadaires des Séances de l'Académie des Sciences*, **224**, 676–677.

†Teissier, G. 1947b. Variations de la fréquence du gène *ebony* dans une population stationnaire de Drosophiles. *Comptes Rendus Hebdomadaires des Séances de l'Académie des Sciences*, **224**, 1788–1789.

†Teissier, G. 1954a. Conditions d'équilibre d'un couple d'allèles et supériorité des hétérozygotes. *Comptes Rendus Hebdomadaires des Séances de l'Académie des Sciences*, **238**, 621–623.

†Teissier, G. 1954b. Sélection naturelle et fluctuation génétique. *Comptes Rendus Hebdomadaires des Séances de l'Académie des Sciences*, **238**, 1929–1931.

†Teissier, G. 1958a. Mécanisme de l'évolution. *La Pensée*, **2**, 3–19; **3**, 15–31.

†Teissier, G. 1958b. *Titres et travaux scientifiques.* Paris: Prieur et Robin.

†Teissier, G. 1961. *Transformisme d'aujourd'hui.* Roscoff: Editions de la Station Biologique de Roscoff.

†Teissier, G. 1962a. Transformisme d'aujourd'hui. *L'Année Biologique*, 4th ser., **1**, 359–375.

Teissier, G. 1962b. *Supplément aux Titres et travaux scientifiques.* Paris: Robin et Mareuge.

Teissier, G. 1967. *Titres et travaux scientifiques. Supplément.* Paris: Robin et Mareuge. (This is a supplement to the previous supplement.)

Thomas, J. P. 1995. *L'eugénisme.* Paris: Presses Universitaires de France.

†Thomson, W. 1862. *On the Age of the Sun's Heat.* London: Macmillan.

†Thomson, W. 1864. On the secular cooling of the earth. *Transactions of the Royal Society of Edinburgh*, **25**, 157–169.

†Thomson, W. 1868. On geological time. *Transactions of the Geological Society of Glasgow*, **3**,

Todes, D. 1989. *Darwin without Malthus: The Struggle for Existence in Russian Evolutionary Thought.* Oxford: Oxford University Press.

Tort, P. (ed.). 1992. *Darwinisme et société.* Paris: Presses Universitaires de France.

Tort, P. (ed.). 1996. *Dictionnaire du darwinisme et de l'évolution.* Paris: Presses Universitaires de France.

†Tristram, H. B. 1859. On the ornithology of Northern Africa (Sahara). *Ibis*, **1**, 153–162, 277–301, 429–435.

†Tschermak, E. von. 1900. Ueber künstliche Kreuzung bei *Pisum sativum. Berichte Deutschen Botanischen Gesellschaft*, **18**, 232–239. English translation:

Concerning artificial crossing in *Pisum sativum*. *Genetics,* **35** (1950), supp. to no. 5, 'The Birth of Genetics', 2nd part, pp. 42–47.

†Turner, J. R. G. 1967. On supergenes. 1. The evolution of supergenes. *American Naturalist,* **101**, 195–219.

†Turner, J. R. G. 1969. The basic theorems of natural selection: A naïve approach. *Heredity,* **24**, 76–84.

†Turner, J. R. G. 1970. Changes in mean fitness under natural selection. In *Mathematical Topics in Population Genetics,* ed. M. Kojima. Heidelberg: Springer.

†Turner, J. R. G. 1978. Why male butterflies are non-mimetic: Natural selection, sexual selection, group selection, modification and sieving. *Biological Journal of the Linnean Society,* **10**, 385–432.

Turner, J. R. G. 1983. 'The hypothesis that explains mimetic resemblances explains evolution': The gradualist–saltationist schism. In *Dimensions of Darwinism: Themes and Counterthemes in Twentieth-Century Evolutionary Theory,* ed. M. Grene , pp. 129–169. Cambridge: Cambridge University Press.

Van der Waerden, B. L. 1967. *Statistique mathématique.* Paris: Dunod.

Veuille, M. 1986. *La sociobiologie.* Paris: Presses Universitaires de France.

Veuille, M. 1987a. Corrélation, le concept pirate. In *D'une science l'autre: des concepts nomades,* ed. I. Stengers, pp. 35–67. Paris: Seuil.

Veuille, M. 1987b. Sélection naturelle: passerelles conceptuelles entre l'ordre immobile et l'incertain. In *D'une science l'autre: des concepts nomades,* ed. I. Stengers, pp. 198–218. Paris: Seuil.

Vorzimmer, P. J. 1970. *Charles Darwin. The Years of Controversy: The Origin of Species and Its Critics.* Philadelphia: Temple University Press.

Vucinich, A. 1970. *Science in Russian Culture, 1861–1917.* Stanford, CA: Stanford University Press.

Walker, H. 1929. *Studies in the History of Statistical Method.* Baltimore: Williams and Wilkins.

†Wallace, A. R. 1855. On the law which has regulated the introduction of new species. *Annals and Magazine of Natural History,* **26**, 184–196.

†Wallace, A. R. 1859. On the tendency of varieties to depart indefinitely from the original type. *Journal of the Proceedings of the Linnean Society (Zoology),* **3**. Reprinted in *The Collected Papers of Charles Darwin,* vol. II, (1977), ed. P. H. Barrett, pp. 10–19. Chicago: University of Chicago Press.

†Wallace, A. R. 1864. The origin of human races and the antiquity of man deduced from natural selection. *Journal of the Anthropologial Society,* **2**, 170–186.

†Wallace, A. R. 1866. On the phenomena of variation and geographical distribution, as illustrated by the Papilionidae of the Malayan region. *Transactions of the Linnean Society,* **25**, 1–72.

†Wallace, A. R. 1867. Mimicry and other protective resemblances among animals. *Westminster Review,* n.s., **32**, 1–43.

†Wallace, A. R. 1870. *Contributions to the Theory of Natural Selection.* London: Macmillan.

†Wallace, A. R. 1889. *Darwinism: An Exposition of the Theory of Natural Selection, with Some of Its Applications.* London: Macmillan.

†Wallace, A. R. 1891. *Natural Selection and Tropical Nature*. London: Macmillan.

†Wallace, A. R. 1896. The problem of utility: Are specific characters always or generally useful? *Journal of the Linnean Society of London*, **25**, 481–496.

†Wallace, A. R. 1916. *Alfred Russel Wallace: Letters and Reminiscences*, ed. J. Marchant. New York: Harper Bros.

†Wallace, B. 1958. The role of heterozygosity in Drosophila populations. *Proceedings of the 10th International Congress of Genetics*, **1**, 408–419.

†Watson, D. M. S. 1937. A discussion of the present state of the theory of natural selection: Opening address. *Proceedings of the Royal Society of London*, ser. B, **121**, 43–45.

†Weinberg, N. 1908. Ueber den Nachweis der Vererbung beim Menschen. *Jahresschriften des Vereins für vaterländische Naturkunde in Württemburg*, **64**, 368–382. Translated in *Papers in Human Genetics*, ed. S. H. Boyer, pp. 4–15. Englewood Cliffs, NJ: Prentice-Hall, 1963.

†Weismann, A. 1883. *Ueber die Vererbung.* Jena: Fischer.

†Weismann, A. 1889. *Essays upon Heredity and Kindred Biological Problems*, ed. E. Poulton, S. Schönlans and A. E. Shipley. Oxford: Clarendon Press. Reprinted Oceanside, NY: Dabor Science, 1977.

†Weismann, A. 1892. *Aufsätze über Vererbung und verwandte biologische Fragen.* Jena: Fischer.

†Weismann, A. 1892. *Essais sur l'hérédité et la sélection naturelle*. Paris: C. Reinwald.

†Weismann, A. 1893. The all-sufficiency of natural selection: A reply to Herbert Spencer. *Contemporary Review*, **64**, 309–338, 596–610.

†Weldon, W. F. R. 1890. The variations occurring in certain decapod Crustacea. I. *Crangon vulgaris*. *Proceedings of the Royal Society of London*, **47**, 445–453.

†Weldon, W. F. R. 1892. Certain correlated variations in *Crangon vulgaris*. *Proceedings of the Royal Society of London*, **51**, 2–21.

†Weldon, W. F. R. 1893. On certain correlated variations in *Carcinus moenas*. *Proceedings of the Royal Society of London*, **54**, 318–329.

†Weldon, W. F. R. 1895a. Report of the Committee, consisting of Mr Galton (Chairman), Mr F. Darwin, Professor Macalister, Professor Meldola, Professor Poulton, and Professor Weldon, for Conducting Statistical Inquiries into the Measurable Characteristics of Plants and Animals. Part I. An attempt to measure the death-rate due to selective destruction of *Carcinus moenas* with respect to a particular dimension. *Proceedings of the Royal Society of London*, **57**, 360–379.

†Weldon, W. F. R. 1895b. Remarks on variation and plants. To accompany the first report into the measurable characteristics of plants and animals. *Proceedings of the Royal Society of London*, **57**, 379–382.

†Weldon, W. F. R. 1896. Letter, 30 July 1896. *Nature*, **54**, 294–295.

†Weldon, W. F. R. 1898. Presidential address to the Zoological section of the British Association. *Report of the British Association for the Advancement of Science*, pp. 887–902.

†Weldon, W. F. R. 1901a. A first study of natural selection in *Clausilia laminata* (Montagu). *Biometrika*, **1**, 109–124.

†Weldon, W. F. R. 1901b. Mendel's laws of alternative inheritance in peas. *Biometrika*, **1**, 227–254.

†Weldon, W. F. R. 1902. On the ambiguity of Mendel's categories. *Biometrika*, **2**, 44–55.

Willey, B. 1940. *Darwin and Butler: Two Versions of Evolution*. London: Chatto and Windus.

Wimsatt, W. C. 1980. Reductionist research strategies and their biases in the units of selection controversy. In *Scientific Discoveries: Case Studies*, ed. T. Nickles, pp. 213–259. Dordrecht: Reidel.

Wittgenstein, L. 1918. *Tractatus Logico-Philosophicus*. English translation by D. S. Tears and B. S. McGuinness. London: Routledge and Kegan Paul. (Reprinted 1961).

†Wright, S. 1922. The effects of inbreeding and crossbreeding on guinea pigs. I. Decline of vigor. II. Differentiation among inbred families. III. Crosses between highly inbred families. *U.S. Department of Agriculture Bulletin*, **1090**, 1–63; **1121**, 1–49.

†Wright, S. 1929. Evolution in a Mendelian population. *Anatomical Record*, **44**, 287.

†Wright, S. 1930. Review of *The Genetical Theory of Natural Selection*, by R. A. Fisher. *Journal of Heredity*, **21**, 349–356.

†Wright, S. 1931. Evolution in Mendelian populations. *Genetics*, **16**, 97–159.

†Wright, S. 1932. The roles of mutation, inbreeding, crossbreeding, and selection in evolution. *Proceedings of the 6th International Congress of Genetics*, **1**, 352–366.

†Wright, S. 1935a. Evolution in populations in approximate equilibrium. *Journal of Genetics*, **30**, 257–266.

†Wright, S. 1935b. The emergence of novelty: A review of Lloyd Morgan's 'emergent' theory of evolution. *Journal of Heredity*, **26**, 369–373.

†Wright, S. 1937. The distribution of gene frequencies in populations. *Proceedings of the National Academy of Sciences of the USA*, **23**, 307–320.

†Wright, S. 1939. *Statistical Genetics in Relation to Evolution*. Paris: Hermann.

†Wright, S. 1940. The statistical consequences of Mendelian heredity. In *The New Systematics*, ed. J. Huxley. Oxford: Clarendon Press.

†Wright, S. 1943. Isolation by distance. *Genetics*, **28**, 114–138.

†Wright, S. 1946. Evolution by distance under diverse systems of mating. *Genetics*, **31**, 39–59.

†Wright, S. 1948. Genetics of populations. *Encyclopaedia Britannica*. 14th edn rev. Vol. VIII, pp. 915–929.

†Wright, S. 1955. Classification of the factors of evolution. *Cold Spring Harbor Symposia on Quantitative Biology*, **20**, 16–24.

†Wright, S. 1959. Genetics, the gene and the hierarchy of biological sciences. *Proceedings of the 10th International Congress of Genetics*, **1**, 475–489.

†Wright, S. 1964. Biology and the philosophy of science. In *The Hartshorne Festschrift: Process and Divinity*, ed. W. L. Reese and E. Freeman, pp. 101–125. La Salle, IL: Open Court.

†Wright, S. 1968–78. *Evolution and the Genetics of Populations*. 4 vols. Vol. I, *Genetics and Biometric Foundations* (1968), Vol. II, *The Theory of Gene Frequencies* (1969), Vol. III, *Experimental Results and Evolutionary Deductions* (1977), Vol. IV, *Variability within and among Natural Populations* (1978). Chicago: University of Chicago Press.

†Wright, S. 1970. Random drift and the shifting balance theory of evolution. In *Mathematical Topics in Population Genetics*, ed. K. Kojima. Berlin: Springer.

†Wright, S. 1975. Panpsychism and science. In *Mind in Nature*, ed. J. E. Cobb and D. R. Griffin. Washington, DC: University Press of America.

†Wright, S. 1980. Genic and organismic selection. *Evolution*, **34**, 825–843.

†Wright, S. 1986. *Evolution: Selected Papers*, ed. W. B. Provine. Chicago: University of Chicago Press.

†Wright, S. 1988. Surfaces of selective value revisited. *American Naturalist*, **131**, 115–123.

†Yule, G. U. 1897. On the significance of Bravais' formulae for regression, &c., in the case of skew correlation. *Proceedings of the Royal Society of London*, **60**, 477–489.

†Yule, G. U. 1902. Mendel's laws and their probable relations to intra-racial heredity. *New Phytologist*, **1**, 193–220, 222–238.

†Yule, G. U. 1907. On the theory of inheritance of quantitative compound characters on the basis of Mendel's laws – A preliminary note. *Report of the 3rd International Congress on Genetics*, 140–142. London: Spottiswoode.

Zimmermann, F. 1975. Actualité de la pensée de Haldane. *Population*, **30**, 285–302.

Zirkle, C. 1946. The early history of the idea of the inheritance of acquired characters and of pangenesis. *Transactions of the American Philosophical Society*, n.s., **35**, 95–141.

Zirkle, C. 1951. The knowledge of heredity before 1900. In *Genetics in the 20th Century*, ed. L. C. Dunn, pp. 35–37. New York: Macmillan.

Zukerkandl, E. and Pauling, L. L. 1965. Evolutionary divergence and convergence in proteins. In *Evolving Genes and Proteins*, ed. V. Bryson and H. J. Vogel, pp. 97–116. New York: Academic Press.

Index

References to more important passages are italicised. Entries for authors are exhaustive and not thematic. The one exception is Darwin, for whom the entries are thematic and not exhaustive.